Fusion Methodologies in Crisis Management

Galina Rogova • Peter Scott
Editors

Fusion Methodologies in Crisis Management

Higher Level Fusion and Decision Making

Springer

Editors
Galina Rogova
State University of New York at Buffalo
Buffalo, NY, USA

Peter Scott
State University of New York at Buffalo
Buffalo, NY, USA

ISBN 978-3-319-22526-5 ISBN 978-3-319-22527-2 (eBook)
DOI 10.1007/978-3-319-22527-2

Library of Congress Control Number: 2015956615

Springer Cham Heidelberg New York Dordrecht London

Printed on acid-free paper

Springer International Publishing AG Switzerland is part of Springer Science+Business Media (www.springer.com)

Acknowledgements

The world is not getting any safer. The appalling costs in lives and property of natural and man-made disasters pose a great challenge to the technical community: find ways to use modern communications, sensor and digital technology to protect vulnerable populations from the ravages of these terrible events. The coeditors wish to express their gratitude to the authors whose work made it possible to produce this book. Jointly, they took a step in that direction.

We would also like to thank the Springer staff, particularly Mary James, Charles Glaser and Rebecca Hytowitz, for their support and patience throughout this process.

In Memoriam Dr. David Hall

We are very saddened by the sudden passing of Dr. David Hall. He served as Professor of the College of Information Sciences and Technology (IST) at the Pennsylvania State University and founding Director of the Center for Network Centric Cognition and Information Fusion. In 2010 he was appointed Dean of the College, a position he held until 2014. Dr. Hall had over 30 years of experience in multisource information fusion, software system development, and research and development. He published numerous papers and several archival books on information fusion. He also contributed his abundant energy as the Associate Director and Senior Scientist for the Penn State Applied Research Laboratory, Director of IR & D at HRB Systems, and Research Manager at the Computer Sciences Corporation. Dr. Hall consulted for the USAF TENCAP Program, the Joint IED Defeat Organization (JIEDDO), the National Security Agency, NASA, and acted as a Technical Advisor to the Defense Department's Joint Directors of Laboratories (JDL) Data Fusion Group. His honors include the DoD Joe Mignogna National Data Fusion Award and being named an IEEE Fellow for his important contributions to information fusion. David's passing is a tragic loss for the information fusion community. His lasting contributions in the field of information fusion and related disciplines, as well as his warm and generous spirit, will be fondly remembered by all who knew him.

Contents

Contributors

João Paulo A. Almeida Federal University of Espírito Santo (UFES), Vitória, Brazil

Youssif Al-Nashif Old Dominion University, Norfolk, VA, USA

Dorothea Aniola SAIL LABS Technology, Vienna, Austria

Gerhard Backfried SAIL LABS Technology, Vienna, Austria

Youakim Badr CNRS, LIRIS, UMR5205, INSA-Lyon, Université de Lyon, Lyon, France

Werner Bailer JOANNEUM RESEARCH, Graz, Austria

Marcello Basili DEPS and SEM, University of Siena, Siena, Italy

Erik Blasch Information Directorate, Air Force Research Laboratory, Rome, NY, USA

Laura Bonelli Theory & Technology group (T3), Istituto di Scienze e Tecnologie della Cognizione, Consiglio Nazionale delle Ricerche, Rome, Italy

Department of Psychology, University of Rome "La Sapienza,", Rome, Italy

Guoray Cai College of Information Sciences and Technology, The Pennsylvania State University, University Park, PA, USA

Alain Chateauneuf IPAG Business School and PSE-CES, University of Paris-I, Paris, France

Patrícia Dockhorn Costa Federal University of Espírito Santo (UFES), Vitória, Brazil

Paulo C.G. da Costa Department of Systems Engineering and Operations Research and C4I Center, George Mason University, Fairfax, VA, USA

Gerald Czech Austrian Red Cross, Vienna, Austria

Thomas Delavallade Thales, Gennevilliers, France

John Dowell University College, London, UK

Silvia Felletti Theory & Technology group (T3), Istituto di Scienze e Tecnologie della Cognizione, Consiglio Nazionale delle Ricerche, Rome, Italy

Department of Psychology, University of Rome "La Sapienza,", Rome, Italy

Simon Fossier Thales, Palaiseau, France

Phan H. Giang George Mason University, Fairfax, VA, USA

Markus Glanzer Austrian Red Cross, Vienna, Austria

Johannes Göllner National Defence Academy, Ministry of Defence and Sports, Vienna, Austria

Jake Graham College of Information Sciences and Technology, The Pennsylvania State University, University Park, PA, USA

David Hall College of Information Sciences and Technology, The Pennsylvania State University, University Park, PA, USA

Salim Hariri Center for Cloud and Autonomic Computing, The University of Arizona, Tucson, AZ, USA

John R. Josephson Computer Science and Engineering Department, The Ohio State University, Columbus, OH, USA

Anne-Laure Jousselme NATO Science and Technology Organization, CMRE, La Spezia, Italy

Alexey Kashevnik St. Petersburg Institute for Informatics and Automation of the Russian Academy of Sciences, St. Petersburg, Russia

Gert Kienast JOANNEUM RESEARCH, Graz, Austria

Dale A. Lambert Defence Science and Technology Group, Adelaide, SA, Australia

Kathryn B. Laskey Department of Systems Engineering and Operations Research and C4I Center, George Mason University, Fairfax, VA, USA

Claire Laudy Thales, Palaiseau, France

Tatiana Levashova St. Petersburg Institute for Informatics and Automation of the Russian Academy of Sciences, St. Petersburg, Russia

Gaëlle Lortal Thales, Palaiseau, France

Klaus Mak National Defence Academy, Ministry of Defence and Sports, Vienna, Austria

Henrique C. Marques Aeronautics Institute of Technology – ITA, São José dos Campos, Brazil

Patrick Maupin C2I Section, Defence R&D Canada, Québec, QC, Canada

Christian Meurers National Defence Academy, Ministry of Defence and Sports, Vienna, Austria

Alexis Morris Faculty of Computer Science, University of New Brunswick, Fredericton, NB, Canada

Fabio Paglieri Goal-Oriented Agents Lab (GOAL), Istituto di Scienze e Tecnologie della Cognizione, Consiglio Nazionale delle Ricerche, Rome, Italy

Andreas Peer National Defence Academy, Ministry of Defence and Sports, Vienna, Austria

Luís Ferreira Pires University of Twente, Enschede, The Netherlands

Isaac S.A Pereira Federal University of Espírito Santo (UFES), Vitória, Brazil

Birgit Pröll Institute for Application Oriented Knowledge Processing, Johannes Kepler University, Linz, Austria

Gerald Quirchmayr Multimedia Information Systems, Faculty of Computer Science, University of Vienna, Vienna, Austria

Werner Retschitzegger Department of Cooperative Information Systems, Johannes Kepler University, Linz, Austria

Galina L. Rogova State University of New York at Buffalo, Buffalo, NY, USA

William Ross Faculty of Computer Science, University of New Brunswick, Fredericton, NB, Canada

Andrea Salfinger Department of Cooperative Information Systems, Johannes Kepler University, Linz, Austria

Daniel R. Schlegel State University of New York at Buffalo, Buffalo, NY, USA

Christian Schmidt SAIL LABS Technology, Vienna, Austria

Wieland Schwinger Department of Cooperative Information Systems, Johannes Kepler University, Linz, Austria

Peter D. Scott State University of New York at Buffalo, Buffalo, NY, USA

Stuart C. Shapiro State University of New York at Buffalo, Buffalo, NY, USA

Nikolay Shilov St. Petersburg Institute for Informatics and Automation of the Russian Academy of Sciences, St. Petersburg, Russia

Marten van Sinderen University of Twente, Enschede, The Netherlands

Alexander Smirnov St. Petersburg Institute for Informatics and Automation of the Russian Academy of Sciences, St. Petersburg, Russia

ITMO University, St. Petersburg, Russia

Lauro Snidaro Department of Mathematics and Computer Science, University of Udine, Udine, Italy

Alan N. Steinberg Independent Consultant, Lake Ridge, VA, USA

Georg Thallinger JOANNEUM RESEARCH, Graz, Austria

Kerry Trentelman Defence Science and Technology Group, Adelaide, SA, Australia

Mihaela Ulieru School of Information Technology, Carleton University, Ottawa, ON, Canada

Nicolas A. Valcik Director of Institutional Research at West Virginia University, Public Administration, College of Arts and Sciences, West Virginia University, Morgantown, WV, USA

Mathukumalli Vidyasagar University of Texas at Dallas, Richardson, TX, USA

Indian Institute of Technology, Hyderabad, India

Ingrid Visentini Department of Mathematics and Computer Science, University of Udine, Udine, Italy

Chapter 1
Introduction

Galina L. Rogova and Peter D. Scott

Abstract Given the growth in variety and scope of computer-based tools developed this century to understand and manage crises, it may be somewhat perplexing that our recent track record in crisis mitigation and disaster management is decidedly mixed. This is partly explained by the implementation gap accompanying any complex evolving technology, but it goes beyond that. In particular, the promise of higher level fusion technology has not yet been widely realized outside the military and security domains. The principal goals of this volume are to explore this gap, identify strengths and weaknesses of existing theory and technology, and suggest the most promising avenues for future research and development. In this introductory chapter, the set of issues binding high level fusion and crisis management will be identified, and some recent disaster case studies discussed. These topics will be further developed in the chapters to follow.

1.1 Emergency Management

While the twenty-first century has seen steady, and in some areas dramatic, growth in the tools available to understand and manage crises, the track record on the ground in responding to natural or man-made disasters has not been particularly encouraging. On 9/11/2001 terrorists successfully avoided all security precautions, boarded three aircraft, took control, and flew them into major public buildings with the loss of 2996 lives. A fourth aircraft was diverted from its target, with the loss of all aboard, only by heroic actions of its passengers. In August 2005, Katrina bore down onto New Orleans as a Category 5 hurricane. Though the storm had been accurately tracked and scaled for more than a week, no evacuation order was issued until just hours before landfall, too late to be successful. No buses had been pre-positioned to move hurricane victims from the sports stadium where they had been marshalled, and where some died. With simple precautions, lives could have been

G.L. Rogova (✉) • P.D. Scott
State University of New York at Buffalo, Buffalo, NY, USA
e-mail: rogova@buffalo.edu; peter@buffalo.edu

G. Rogova, P. Scott (eds.), *Fusion Methodologies in Crisis Management*,
DOI 10.1007/978-3-319-22527-2_1

saved. The Deepwater Horizon oil rig in the Gulf of Mexico caught fire on April 20, 2010. It sank, and its oil well sustained a catastrophic blowout 2 days later. The well had been equipped with a blowout protector, which failed to operate, as did a supposedly fail-safe backup blowout device. Subsequently, in spite of continuous video evidence, responsible officials grossly underestimated the rate of oil release. Before it could be capped, this well discharged enough oil to do billions of dollars of damage to the marine environment and the nearby Gulf Coast. As the record of these incidents indicates, both human and technical systems are implicated in recent failures of emergency management systems to mitigate damage.

The record of response to twenty-first century disasters is not all disappointing, however. Notably, the rapid, well-coordinated and courageous response by first responders to the Fukushima Daichi nuclear power plant meltdown following the Tonoku tsunami on March 11, 2011, significantly limited the loss of life and ecological damage this incident caused. Other recent successes have been noted as well (Dell'Acqua et al. 2011; Teimouri et al. 2008; Yates and Paquette 2011).

Given the considerable technical advances in sensors, communications, networks and information devices, as well as conceptual advances in the modeling of risk and of crisis and emergency management systems themselves that have been achieved during this century, it is perhaps perplexing that the record of response to actual disaster incidents is so mixed. This is a matter whose causes are worth considering carefully when prioritizing research directions for the coming decades. A principal goal of the present volume is to contribute to that conversation by discussing the current state of affairs of high-level fusion in emergency management, and its strengths and limitations, and identifying the most promising avenues for future research and development.

Emergency management may be partitioned into four principal phases: preparedness, detection, response, and recovery. These phases are not strictly sequential: response may overlap detection, and recovery efforts often begin during the response phase. Here we briefly introduce some ideas relevant to these phases in current "All-Hazards" emergency management, the preferred framework for responsible organizations such as the Federal Emergency Management Agency (FEMA) for studying the broad scope of natural and man-made emergencies as a unified problem domain.

Preparedness begins with formulation of incident response plans, and includes training of emergency responders, decision makers, and other stakeholders in the procedures to be following as these plans are implemented in the event of a crisis. The adage that players act as they have been trained is as true for emergency response as it is for team sports or for the military. Recent progress in fine-grained and responsive simulation tools has been effective in enhancing training exercises and identifying weaknesses and limitations at different levels of granularity (d'Agostino 2002). Simulation also enhances risk recognition and facilitates prioritization planning, important elements of preparedness. In addition, for adequate preparedness response organizations need to be designated, and their roles both in strategic preparedness, such as pre-positioning of major resources, and procedures for tactical readiness, such as logistics of deployment of police and fire personnel, established. Evacuation plans need to be vetted well in advance, and in an all-hazards framework need to be agile. The goal of this phase of crisis management is mitigation: reducing risk, and containing the likely consequences of an incident.

Discovery is the phase in which a crisis is declared, identified, and the decisions made to initiate the first response plan and begin to deploy the available resources. Sometimes the incident is self-evident, such as an industrial explosion or hazardous material train derailment. For others, early discovery requires careful inference on the available data. The 9/11/2001 terrorist incident became discoverable when the first aircraft went silent and radically changed course, and earlier warning to likely target areas and target buildings could have been implemented at that time. Effective discovery requires an ongoing process of hypothesis generation and testing, with data derived from human reports, sensor data, situation assessment and threat assessment tools. While there has been much recent work in this area, there is a gap between feasibility demonstrations and field deployment. Discovery is still largely dependent on human observation and cognition, and is a fertile area for additional research.

Special consideration needs be given to the vexing problem of preparation for extreme events. Talib and collaborators (Makridakis and Taleb 2009a) have emphasized the central role played by "Black Swan" events, those of very low probability but high consequence, in determining the course of events. Recent work in this area has focused on designing flexibility and the capacity for improvisation into the response plan and training (Mendonca et al. 2001; Makridakis and Taleb 2009b).

The response phase has as its core goal saving of human lives. In this phase, search and rescue operations are the primary tools. Effective deployment of search and rescue resources requires effective management: an organization whose authority is acknowledged by all stakeholders, access to adequate data sources, tools to develop agile situation and threat assessments, and effective resource management capability. FEMA has established the Incident Command System, part of the National Incident Management System (Annelli 2006), towards this end for incidents under its jurisdiction. Similar systems have been created in other countries.

Another critical response-phase goal is the discovery and elimination of ongoing secondary threats, for instance gas line ruptures following an earthquake (Rogova et al. 2005). This process is logically related to the initial incident discovery phase. In the secondary threat case the likelihoods and potential consequences used to evaluate secondary hypotheses are conditioned upon the specific primary incident which has been discovered. As mentioned in the context of preparation, extreme event response requires building into the response plans a capacity for flexibility and improvisation. In Black Swan events, plans are likely to fail due to unpredicted infrastructure damage or unavailability of other expected resources. Successful response will likely require actions not considered in the pre-incident planning and training phase.

Recovery is the final phase, whose goal is the delivery of all necessary medical services, remediation of physical damage and return to normalcy. Both the time scale and resource demands of this phase of emergency management typically far exceed those of the other phases. Katrina's recovery costs to public and private sources exceeded $125 billion. The BP Corporation has spent $28 billion thus far in fines and remediation costs due to the Deepwater Horizon spill, with additional recovery costs to Gulf Coast residents and jurisdictions in the tens of billions of dollars. It is clear that paying mitigation costs forward in the pre-incident preparedness phase can return a high level of benefit to society. Specification of desirable mitigation has been advanced this century by "big data" initiatives, exposing relevant societal trends and determining hidden vulnerabilities. Finally, in many circumstances, private charities and NGOs such as the Red Cross, Red Crescent, and United Nations Relief Agency can be called on to share this burden.

1.2 Information Fusion[1]

Major ecological disasters such as earthquakes, tsunamis, and floods often involve a large population spread over a wide geographical swath. As the emergency unfolds, messages and reports of many kinds begin to propagate through a wide variety of channels: some from dedicated calibrated sensor networks, some by radio from uniformed responders with high reliability, and others by cell phone from the victims themselves, perhaps containing redundant or contradictory information. The decision makers of the incident response organizations must form an effective situation assessment, explicitly or implicitly mapping this heterogeneous collage of data into a coherent operational picture, one from which timely and effective decisions can be made. As the emergency unfolds and new data streams in, the picture must be rapidly updated to remain of value. This is the decision support role played by information fusion: combining multiple data streams into actionable knowledge in order to maintain a timely situation assessment, one from which a common operating picture can be made available to decision makers, responders, and other stakeholders.

Among the many models, both normative and descriptive, which have been advanced to characterize the data fusion process, one of the earliest and still most influential is the Joint Directors of Laboratories (JDL) functional model (Llinas et al. 2004). In its initial form, the JDL model contained operations aggregating entities at four increasing of abstraction: source preprocessing, object refinement, situation refinement, and impact or threat refinement. It also included an overarching function termed process refinement, a meta-process monitoring the underlying information fusion process. The model has since evolved to add at least one higher level, that of mission management (Schreiber-Ehle and Koch 2012) and model development continues to this date (Blasch et al. 2013). A useful property of the JDL model

[1] *Information fusion* usually refers to fusion operations applied to a set of entities including those beyond the object level (that is, JDL model levels 2 and above). The related terms of *data fusion* and *sensor fusion* usually refer to operations limited to the source and object levels (level 0 and level 1 of the JDL model).

structure is that it suggests a common sequence of operations on model entities at each level of abstraction. In one formulation, these are categorized as data alignment followed by data association and state estimation. When applied at the lowest level, the outputs are refined estimates of source data. When applied at the situation refinement level, the outputs are the evolving situation assessments and common operating picture. The JDL model is a functional rather than a process model since no specific order of processing is prescribed. In general, the data to be input at a given level includes tokens and parameters derived from a common bus connected to all other fusion levels, databases and human-computer interfaces. For instance, object refinement may exploit information derived from higher level evaluation of threats, as well as lower level sensor data estimates.

In contrast, Boyd's OODA Loop (Von Lubitz et al. 2008) is an information fusion process model not based on levels of data aggregation but directly on a cycle of productive behaviors engaged by a single agent accomplishing a complex task: Observe, Orient, Decide, Act. Many copies of the OODA Loop may be embedded at different time scales within a given scenario. Depending on where in the domain the loop is being applied, the goal could be understanding the nature of a single data source given the data and its context, such as a cell phone report from a first responder, or the overall current situation assessment of the entire domain. Originally designed as a model for use in military combat, this model has been extended to a broader class of information fusion applications as the Plan Do Check Act (PDCA) Loop, where it has found applications in design of complex industrial plants, and management science (Sobek and Smalley 2011).

In an effort to root information fusion more deeply in behavioral psychology than was apparent in other models available at that time, such as the JDL functional model and OODA Loop process models, Mica Endsley (1995) developed a descriptive model of human situation awareness which proved highly effective in guiding design of interfaces such as air traffic control displays and aircraft cockpit instrumentation layouts which must support time-bound critical decision making. In this model, the agent perceives elements of the current situation, processes and comprehends them to form an assessment, and then projects the future status based on that assessment and possible current actions. An action is decided and taken upon that basis and the results of that action create a new state of the environment to be perceived, closing the loop. Rather than prescribe normative behaviors, Endsley's model describes human behavioral capabilities and cognitive constraints on performance which should guide the design of decision support system to empower the agent to maximum performance and to minimize the risk of human error.

These models initially appeared in military applications but in the context of traditional warfare. Information models designed for dealing with natural and especially man-made disasters such as terrorist attacks have to account for a new trend in information fusion: human-centric fusion (Hall and Jordan 2010; Zarnowski et al. 2010). This is a result of the increased importance of human-based sensors ("every person is a sensor") and of Web-based information. There are two major differences between traditional military data fusion and data fusion in other disasters, especially man-made disasters. Ubiquitous cell phones, which record and transmit information, drones, and social networks, create massive amount of information of unknown trustworthiness and pedigree. There is a flood of data and information (GIS, pictures, video) with the quality of this information often unknown and methods of processing this information only emerging. Traditional information fusion designed principally for military application certainly used intelligence information but the source characteristics were generally much better understood. Another difference is the class of objects of interest in the information fusion: in human-centric applications they are mostly human not inanimate objects, which required designing new methods of processing social networks and association and fusion of WEB-based and sensor-based information.

Nowhere is the conceptual framework and are the goals of information fusion more clearly applicable than in the management of disasters and emergencies. Yet in spite of their use in a wide variety of other applications, technical information fusion systems have failed thus far to find broad acceptance in the emergency response community. This introduction began with a reminder that the response to recent disasters has been disappointing. Those outcomes might have been improved if effective emergency management systems based on information fusion had been in place. The failure to issue a timely evacuation order as Katrina approached may have been ameliorated if an effective threat assessment was being produced. A prompt, accurate situation assessment as the first aircraft went silent and radically changed course could have guided an earlier, and presumably more proactive, response on 9/11/2001.

Given the clear need, why is it that this tool is not in wider use in emergency management? The answer lies in the special and severe challenges faced by information fusion when applied in the emergency management domain. A primary challenge is the inter-jurisdictional and interorganizational distribution of emergency management resources. Within a given jurisdiction such as a city or county, police, fire-fighters, ambulance companies, emergency medical response teams, hospitals, National Guard units, and civic government officials all have important closely connected roles to play in the response phase. Coordination of their activities is hampered by a general failure of interoperability of communication equipment and protocols, and lack of a coordinated command structure bound effectively to all these organizations. The problem of coordination within a jurisdiction is exacerbated by the typical geographical extension of a major emergency to include multiple jurisdictions. Failure of interoperability of radios across such boundaries is often a legacy of history: they were purchased from different vendors by different offices, at different times, with different specifications. Another critical problem raised by the distributed nature of jurisdictions and organizations is information stovepiping. Public safety teams are trained to carefully guard the information they develop and store, limiting access through passwords, certifications, unpublished protocols and other firewalls. It may be difficult to bypass these stovepipe mechanisms in the emergency environment, during which the sharing of information across organizational and geographical boundaries is always essential to effective fusion and response. A third critical challenge due to this multiplicity of responding organization is the frequent lack of a common disaster ontology. This may become apparent during an "all-hands" training exercise, or not until orders and directives become misunderstood during an incident response.

Other special challenges faced by fusion systems in the emergency environment can be noted. Pre-incident plans are based on existing infrastructure, and in a disaster elements of the anticipated infrastructure might be severely disrupted. Circumstances change rapidly in the chaotic initial stage of an emergency incident, and thus the time-scale of required updates on situation and threat assessments is high. The data flowing into the input command center during an emergency tends to span the maximum range of data types, channels and information quality. The computational burden on a system merging, for instance, information gleaned from Twitter messages with those from police radios is much higher than one fusing data from similar seismographic sensors.

Research and development of systems meeting these challenges continues. Software suites such as CalMesh (Killeen et al. 2006), SLOSH (Jelesnianski et al. 1992), and incidentOS (Intorelli et al. 2009) address many of these issues directly. The ontology description language OWL and its descendants, such as the probabilistic variant PR-OWL, have proved quite useful in framing a common ontology for responders (Costa and Laskey 2006). While the challenges are severe and its record to date is checkered, information fusion is clearly an effective framework for emergency management decision support and with continuing investigation and testing, will no doubt become commonplace for civil emergencies, as it is already in the military domain, in the years to come.

1.3 Crisis Environment

Each stage of each crisis has specific characteristics, which define requirements for fusion architecture and processes. At the same time there is much similarity between them especially when underlying or unreported crisis is evolving during the post-declared-crisis phase. One of the most important characteristics of the declared crisis is a large amount of highly dynamic heterogeneous information of varying significance and reliability, often of low fidelity, uncertain, incomplete, contradictory and redundant. Although information characterizing the underlying evolving crisis environment is dynamic information of often poor quality this information is in many cases hidden and arrives as a lower rate. This information comes from various distributed sensors: physical sensors, human observers, traditional and social media, and intelligence reports (mostly for the primary crisis).

In the past, the main source of information obtained from human observers was limited to 911 calls, open source information, intelligence reports, and information from dedicated human sensors (e.g., police, ambulance drivers). Significantly expanded communication capabilities and social networks created a new paradigm, opportunistic human-centric sensing, allowing for virtually anyone to be a sensor. Human-centric sensing becomes an invaluable source of information while creating additional problems for processing it due to the avalanche of information often unknown quality and luck of existing adequate methods for association and fusion of such information with information coming from traditional physical sensors, so-called hard-soft fusion.

Another important characteristic of the crisis environment is the existence of multiple distributed decision makers and crisis responders who may differ by nations, agencies and application areas, decision, and mission types (tactical and strategic). Tactical mission is related to decisions in direct response to a declared crisis while strategic missions require higher level estimation of the crisis status, analysis, and discovery, and therefore deals with both declared and secondary crises. Multiple distributed crisis management actors have to collaborate through shared situation awareness, which requires reliable processes for sharing information needs, mental models, ontologies, and knowledge of current and predicted crisis evolution, which support their ability to interpret information in the same way and make accurate assessment of each other's decisions and actions. Designing such processes requires taking into account imperfection of mental models, cognitive biases and variable expertise of decision makers and observers.

In addition, all crisis situations are unique and therefore prior knowledge is always incomplete, ill defined, and even unsuitable for the evolving situations, which can be unpredictable and even unimaginable. Decision makers have to deal with resource and time constraints, delay in data transmission, and high cost of consequences such as lost lives, infrastructure damage, as well as adverse long-time effects (economic, environmental, etc.) as the result of poor or delayed decisions.

An important crisis environment factor is the degree to which resources required by the proactive response plans are actually available and infrastructure is intact. In some crises, such as an isolated railway derailment, the infrastructure is undamaged and the emergency plans can be executed as written. In others, like flood or earthquake, communications and transportation and perhaps even medical infrastructure are largely gone. In this case agility and flexibility needs to have been built into the planning and training.

These specific crisis domain characteristics call for a multi-agent distributed dynamic higher level fusion processes, which have to be scalable, adaptive to resource and time constraints, new and uncertain environments, and have to be reactive to uncertain inputs. These processes also have to accommodate heterogeneous information (both symbolic and numeric), allow for complex distributed system modeling, efficient information sharing, and incorporating qualitative experts' opinions and subjective judgements of variable quality. They also need to provide for effective communication and secure information exchange within and between systems, decision makers, and crisis responders. In the disaster environment, the Higher Level Fusion (HLF) processes exploit continually associated and fused information on single entities and the results of domain-specific simulations and models to produce a consistent estimate of the current and predicted state of the environment, which is then presented to users.

1.4 Higher Level Fusion

It is widely agreed that the great majority of successful data fusion methods to date have focused on low level (Level 0 and Level 1) fusion related to processing information about a single object of interest. Lower level fusion and answers the questions such as:

- Is there anything of interest here (object/background)?
- What are object characteristics? What is it (object class)?
- What is its space-time behavior (tracking)?

While effective fusion at the attribute and object levels producing object identification and characterization offers real performance gains in many applications, it does not provide for user situation awareness essential for effective decision making and actions. Situation awareness requires contextual understanding and interpretation of the events and behaviors of interest, which can be achieved by utilizing higher level fusion processes (situation assessment and impact prediction).

The purpose of higher level fusion (situation assessment and impact or threat assessment) is to provide multiple decision makers with essential knowledge by constructing a coherent integrated picture of the current situation along with prediction of the impact of the current situation (estimated future threat) to answer the following questions:

- What is going on?
- Is anything unexpected or suspicious going on? Where?
- What are the possible explanations?
- What can be expected in the future?

There are multiple definitions of situation used in the literature. For example, in Barwise and Perry (1980) situation is defined as "objects having properties and standing in relations to one another" while in the Merriam-Webster dictionary it is defined as "a relative position or combination of circumstances at a certain moment." In Steinberg and Bowman (2004) situation is described as a "structured part of reality that is discriminated by an agent" while "the agent should not be a part of the definition: a situation exists without being noticed or cared about." We can summarize these definitions by calling situation as a set of objects with their relations considered in relation to specific user goals, objectives, and functions defined within a specific context. Threat or impact is defined as an indication of something impending (Steinberg and Bowman 2004), i.e., possible crisis situation. It is important to understand the difference between threat and impact for the modeling purposes. Impact assumes plausible situations resulting from the dynamics of current situation (situation projection). Threat usually assumes human participation and represents an integrated whole of three components: intent, capability, and opportunity (imminent threat) or only two of these components (potential threat).

An important notion of HLF is the notion of event, which may be defined as a rapid change in situation over a defined spatial domain. Events may be threatening, such as the abandonment of a backpack in a crowded airport terminal, or mission supportive, such as the delivery of patients to a hospital. Events represent transition between states defining situations and are manifested by inconsistent/unusual attributes and behavior of situational items. Ontology of events is usually used during the preparedness phase for training and planning purposes. During discovery and response phases transient information is used for instantiation of the event characteristics and relations between events. Event discovery is important for the discovery phase of crisis management since it can lead to knowledge discovery about underlying causes and serves as a trigger for abductive reasoning about underlying/non-declared situations/threat. During the response phase event ontology can be used for prediction of obstacles to crisis response operations and mitigation of consequences.

The problem of situation, threat, and impact assessment is a *dynamic* process of *identifying* and *predicting* a subset of *plausible* states of the environment of *interest* for decision makers along with a measure of *plausibility* assigned to each state. It is recognized that decision makers do not have complete knowledge about all relevant states of the environment and do not exclude the existence of an unknown hypothesis (the open world assumption). Figure 1.1 shows an information flow in the HLF processing.

Thus situation, impact, or threat assessment involves dynamic generation of a set of hypotheses about the states of the environment and assessment of their plausibility via reasoning about situational items, their aggregates at different levels of granularity, relationships between them, and their behavior within a specific context.

The assessment processes comprise a combination of:

- Dynamic deductive reasoning about situational items and their characteristics for assessing current situations and their impact.
- Abductive reasoning leading to recognition of possible causes or explanations of the identified events and estimated states of the world.
- Recognition and reasoning about threat components and their relationships.

Situational hypotheses may include hypotheses about relations and properties, and behavior of situational items and events; hypotheses about possible hidden crisis, secondary incidents for declared disasters (undeclared disasters); and hypotheses about context and its characteristics (contextual variables). In general they can be selected according to context expectations, obtained by data mining, or a human in the loop.

The HLF processing is also complicated by variable and often unknown quality (uncertain, conflicting, contradictory, irrelevant, etc.) of transient data and information and their behavior; variable reliability of the data and information sources as well as erroneous or poor-quality context characteristics. Thus the reasoning methods used for evaluating plausibility of situational hypotheses have to be able to take this into account. Because of this and usually unknown or incomplete probability distributions of situational models, their characteristics, and behavior the

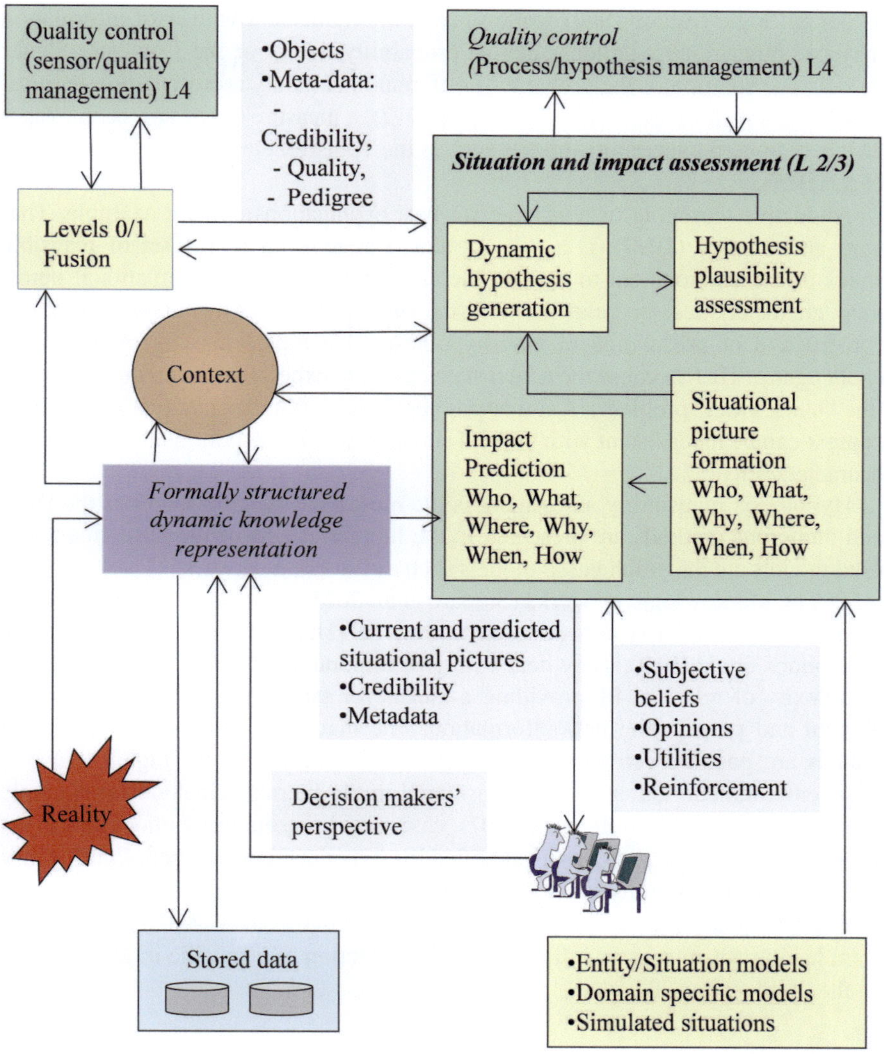

Fig. 1.1 Information flow of the HLF processing

reasoning models in most cases have to use belief theories such as the Possibility (Dubois and Prade 1988), Dempster-Shafer (Shafer 1976), and Transferable Belief (Smets 2000) Models instead.

Growing importance and volume of unstructured linguistic information coming from crowdsourcing, which has to be associated and combined with structures traditional physical sensor and fusion process outputs further complicate the problem of HLF, and especially the problem dealing with uncertainty in reasoning models (so-called soft-hard fusion problem). Uncertainty associated with soft information is usually modeled by fuzzy and possibility theories or non-monotonic logic such as default logic (Reiter 1980) while uncertainty associated with hard information may use other representations such as probability theory or the Dempster-Shafer model. The solution of the problem of different uncertainty representation requires designing transformation methods (Yager 2012), a measure based approach (Yager 2015), or unified uncertainty models such as the Transferable Belief Model (Rogova et al. 2015).

Abductive reasoning or reasoning for best explanations, see for example, Thagard and Shelley (1997), is a process aimed at revision of the set of plausible states of the environment to explain inconsistency of estimated situational items, their characteristics, or behavior with the ones provided by context. It requires construction or postulation of new hypotheses that represents one of the major challenges of HLF because the actual state can be unexpected or even unimaginable (the "black swan" problem). Another problem here is that expectations provided by context can be inconsistent with the real situation due to the poor quality of context characterization.

Hypothesis plausibility assessment is the most researched problem of the HLF, and numerous methods are presented in the literature. One of the most frequently used models are the graph based models such as Bayesian networks (Pearl and Russel 2011), Markov logic networks (Snidaro et al. 2015), Valuation tree (Shenoy and Shafer 1988), Similarity networks (Heckerman 1991), and Inexact Graph matching (Sambhoos et al. 2010). They deal with two important problems: uncertainty and complexity of relations by providing a means for structuring knowledge about a domain and propagating new information. The main problems with graph-based models are possible computation complexities, especially in the highly dynamic crisis environment. Other models considered in the literature include for example situation calculus (Mattioli et al. 2007), case-based reasoning (Aamodt and Plaza 1994; Bostwick et al. 2009), and quantified logic models such as a belief/probability basis argumentation introducing uncertainty into logic (Haenni et al. 2001; Rogova et al. 2006) or story-based belief argumentation (Rogova et al. 2015) with each of them having advantages and drawbacks. The selection of a specific model depends on the problem considered.

1.5 Process Refinement

Process refinement is a meta-process that monitors all other fusion processes to assess and improve their real-time performance and effectiveness. While standalone process refinement by sensor management is a well-researched problem, for HLF in general and especially for fusion in a very complex and highly dynamic crisis environment, it represents a major challenge.

One of the ways of measuring effectiveness of the HLF processes is to use domain specific models. For example, effectiveness of the assessment of casualty situation after a major natural disaster can be measured by the performance of routing algorithm utilizing the result of this assessment.

To measure the performance, process refinement requires methods of evaluation of and controlling the quality of the incoming information and results of all the processes in relation to the goals and objective of the users while users can be either human or other processes (Rogova 2009). This in turn requires nomination of quality attributes specific to the process under consideration and method of their evaluation. For example, relevance of incoming information can be measured by information gain. Another method is to utilize a domain and time variant threshold on quality characteristics and their combination. At the same time, growing role of information coming from the web creates a real problem for information quality evaluation due to unknown or not easily identifiable trustworthiness of the information source and possible malicious intent as well as the pedigree problem.

In general, quality control methods include but are not limited by identification of additional information to improve the fusion processing result along with means to collect this information; delaying the output of the process to achieve a better quality of the results by utilizing additional information, removing information of poor quality from consideration, or modifying fusion processes by modifying their parameters of using different models; considering methods based on utility theory. All these methods however face the problem of computational complexity, may require significant time, and require development of method of dealing with the problem of identifying quality of information obtained from crowdsourcing.

1.6 Domain Knowledge Representation

One of the major challenges of designing the HLF processes is a problem of providing a consistent and comprehensive representation of the domain under consideration. A combination of cognitive work analysis (CWA) and formal ontological analysis of a specific crisis domain is designed to overcome this problem and provide sufficient information about decision maker's goals, functions, information needs, types of objects, relations between them, and processes to support the domain-specific generation of situational hypotheses and high-level reasoning about these situational hypotheses (Bisantz et al. 2004). Users' goals, functions, and information needs are identified by the means of CWA (Rasmussen et al. 1994). CWA is a systems-based approach to the analysis, design and evaluation of an crisis management environment. The result of CWA provides answers to the following questions

- What are the decision makers expecting from a situational picture?
- What information is required for making decisions?
- What active alternative hypotheses about the environment can be expected?

The role of a formally structured ontology of the environment under consideration is to provide a comprehensively large and metaphysically accurate and consistent model of situations, through which specific tasks such as situation assessment, knowledge discovery, or the like, can be more effectively performed, since the information necessary for these decision-making aids is contained within the ontology's structure (Little and Rogova 2005). The formal ontology framework is necessary to provide a formal structure for ontological analysis of the specific environment and to assure a certain level of reusability of the designed domain-specific ontology in a different application domain. Domain-specific ontology also requires instantiation to represent dynamic operational context.

Any formal ontology of situations comprises two types of items: spatial (situational items) and temporal (processes), together with the relations between and among them. Spatial items, elements of the embedded *snap ontology*, and relations between them are defined by a set of spatial and mereological attributes. The values of these attributes define the state of these items and a corresponding state of the environment. Temporal items, i.e., processes, are elements of the related *span ontology*, which describe the temporal behavior of the situational items and dynamics of attributes and relations. Important characteristic of processes are events representing transition between states of the environment defining situations. Situation building blocks can be described by relations that fall into one of two basic categories: *inter-class relations* and *intra-class relations*. Intra-relations (i.e., internal relations) are spatial, temporal, or functional relations that exist within a given set of ontologically similar items while *inter-class relations* (e.g., external relations) exist *between* various items. Context and user dependent relations between these building blocks define derived situations at different levels of granularity and reflect the granularity of essential elements of information. A more detailed description of a formal ontology of crisis situation is presented in Little and Rogova (2008). An important element of crisis knowledge representation is ontology of methods processing crisis information, and expected critical situations (characteristics, planned actions, required resources). A challenge of designing a general ontology for representation of complex crisis domain is defining an appropriate level of details to assure scalability.

One of the major problems of the traditional ontology formalisms is the lack of consistent support for dynamic uncertainty instantiation since they represent only concepts and relations to describe the domain. Although different probabilistic (Laskey et al. 2008), fuzzy (Dey and Abulaish 2008), rough set (Kana and Akinkunmi 2014), and evidential (Bellenger and Gatepaille 2011) approaches exist, the challenge of incorporating uncertainty into ontology and a standard way of ontology instantiation to represent uncertain, imprecise, and vague dynamic operational context still exists.

1.7 Context in Fusion and Crisis Management

Context plays an important role in designing information fusion processes and crisis management. Context awareness offers decision makers important information about current situations and situation dynamics in relation to their goals, functions, and information needs to enable them to appropriately adapt their decision and actions. Context awareness offers important information for reasoning about situational items.

Context supports interrelated fusion and crisis management processes by (Steinberg and Rogova 2008; Rogova 2009):

- Establishing and representing an initial overall context ("declared crises" or "no crisis")
- Optimizing knowledge acquisition by constraining domain ontology and observations
- Influencing the reasoning about objects and situations of interest by providing fusion processes with constraints on situational items and relations between them, beliefs, and a set of hypotheses about the possible state of the environment
- Providing expectations for possible situational items, statistics, and rules, which ultimately can be used for detection and discovery of evolving crises by supporting inconsistency detection
- Optimizing decision and actions
- Facilitating effective communications between crisis management actors since it allows narrowing the possible meanings of the messages

Context exploitation in the uncertain crisis environment is a challenging problem. Designing a context aware fusion based crisis management system requires clear understanding of what context is and the relation between context on the one hand, and data, information, knowledge, and actions on the other. It is also necessary to understand (a) how to represent context in a formal way to support contextual reasoning in fusion processing; (b) how to deal with context dynamics, context recognition and discovery, and contextual reasoning under uncertainty inherent in data fusion problems; and (c) quality of context under consideration and its influence on quality of fusion processes and actions of crisis management actors.

1.8 High-Level Fusion Architecture

The initial class of information fusion systems were those limited to level 0 (data refinement) and level 1 (object refinement) processes, usually referred to as "data fusion" and "sensor fusion" systems. The primary goal of such systems is the identification and characterization of objects relevant to system goals. This analysis, most commonly applied in military settings, typically included identifying and tracking moving objects in a multi-target environment using radar and other imaging sensors.

Higher level fusion (HLF) includes operations at these levels, and importantly extends to the creation and maintenance of comprehensive situation assessment (level 2) and threat or impact assessment (level 3). The goal of HLF is creation and maintenance of situation and threat awareness, a common operating picture, and their timely dissemination to relevant decision-makers and emergency responders.

While consensus in design strategies for technical architectures of HLF systems lags well behind those for lower level data and sensor fusion systems (Blasch et al. 2012), there is general agreement on the basic network topologies and nodal architecture taxonomy.

Centralized HLF architectures focus computational resources on a single fusion node which has access to all contextual databases and receives all data inputs to the fusion system. The surrounding network can be star-configured or include direct communication channels between the other nodes, but fusion products are exclusively created in, and flow from, the central fusion node. This architecture allows for simple, fast fusion algorithms and permits fusion products to be simultaneously disseminated to neighboring nodes. The robustness of this architecture is limited however, since the system is vulnerable to compromise of the connectivity of the central node, or damage to the fusion node itself. But for spatially compact systems, such as a small industrial plant, centralized architectures are an attractive choice.

Decentralized architectures have local and global HLF fusion capabilities redundantly instantiated in multiple local nodes. Each node produces local HLF products from its locally available databases and local sensor suite, and shares this information with the network. The nodes then combine their local products with those from the other nodes to produce global HLF products (situation assessment, threat assessment). Differences in the global products can be resolved by a process of negotiation among the nodes, yielding a common operating picture. Decentralized architectures require more complex algorithms than centralized, but the lack of a unique fusion node makes them more robust against damage to the network or any given node. These architectures may be favored in the case in which core fusion assets may be considered vulnerable during the incident to which the system is responding.

Variants of these two basic architectural categories include distributed and hybrid architectures. Distributed architectures combine local fusion with an ad hoc distribution of global fusion tasks. While local nodes do not redundantly compute global HLF products, they may contribute unequally to their computation, with partial product computation allocated by the availability of local nodal resources. Hybrid architectures combine elements of the other forms, such as a star-configured centralized architecture in which some of the peripheral subnetworks are configured as decentralized systems.

Nodal architectures, the technical configurations of HLF processes and resources within the fusion nodes, can perhaps best be characterized by their logical design rather than topologies. Of the symbol-processing schemes, one of the earliest, and still one of the most popular, is the rule-based expert system (for instance, the KnoFuss system (Nikolov et al. 2008)). Semantic schemes find and exploit the semantic correlations within widely heterogeneous data inputs such as social media messages and calibrated sensor networks. Of the number-processing methods, parameter-updating systems employing probabilistic uncertainty modeling is most frequently employed. Other uncertainty models, such as evidence theory and rough sets, have also been used. Logical fusion node designs employing both symbol and number processing, such as agent-based systems, offer the advantage of more naturally modeling the blend between human and machine operations which characterizes HLF.

One reason for the limited current deployment of HLF systems is the broad question of how cultural and social issues can effectively be modeled and implemented within the scheme. For relatively machine-autonomous systems, such as fusion for fast emergency response of a high-speed engine, these need not be considered. But for the larger set of natural and man-made disasters characterized by human-computer interactions informing the decisions to be taken, cultural and social issues are often critical in supporting an effective response. Decision-makers not acculturated to system-proposed lines of action are less likely to take them, particularly in times of stress, and responders and at-risk populations less likely to follow directions which clash with their social norms. Even more broadly, closely focusing on the psychological and sociological nature of the human decision-maker, "human-based" higher level fusion architectures are the subject of current active research.

1.9 Human-Computer Interaction

If the machine element within a man–machine information fusion system is to be effective in its role of decision support, the information it offers to the decision-maker must be guided by an adequate descriptive model of human decision making. It is well known that people do not behave as predicted by simple behavioral models such as one based on maximization of expected utility, for instance Ellsberg (2001). Normative theories of decision making may have satisfying formal optimality properties, but fail to predict basic human decision-making behavior. Information exchange formats, content and rates based on the assumption they do behave in such an explicitly optimal fashion may in fact lead to degraded decision-making behaviors rather than improved.

A rational actor is a decision maker whose behavior exhibits consistent preferences which can be rationally justified. Behavioral decision theorists agree than the human decision maker a neither an optimal nor a rational actor. For instance, there is the critical problem of information overload. For an optimal or rational actor, increasing the amount of accurate information can never decrease the utility of decision making. Yet for human decision-makers, there is a maximum amount of information which can be profitably exploited over any given time interval. To expose the actor to information which exceeds this produces cognitive overload, which often leads to erratic, inconsistent decisions. The human-computer interaction (HCI) system designer must be mindful of this limit. Modern behavioral decision theories such as Kahneman's and Tversky's Prospect Theory describe some of the observed inconsistencies (Kahneman and Tversky 1979).

The particular environment the decision maker must work in during an emergency amplifies these human discrepancies. The high stakes and short time limits involved increase stress and the heterogeneous mix of messages and data strain the ability of the decision maker to populate familiar mental data patterns. The provenance and credibility of these data sources are often inadequately characterized or simply unknown, confounding the effort to properly set their weights.

These special difficulties of the crisis environment constrain the selection of HCI environments. Humans should be presented with the right information, not the most complete information. In this context, "right information" is the selection and formatting of HCI exchanges that empowers use of the best attributes of human decision making: judgment based on deep implicit knowledge and experience, agile improvisation, and induction under epistemic uncertainty. Good HCI encourages these positive attributes, while suppressing cognitive overload and other errors which unnecessarily limit the quality of human decision making.

HCI design studies have focused on graphical and video data formats as ways to reduce cognitive overload and generally enhance performance (Gertman 2009). A single map can contain many more bits of information than pages of textual data, and human visual pattern-making capabilities are well suited to this data format. A single glance at a well-constructed map might convey as much useful information as scanning of many pages of alphanumeric data. Visual metaphors such as iconic representations of complex objects and events contribute to the capacity of graphics to condense information in user-friendly ways. Virtual reality HCI permits deep exploitation of a visual display environment by immersing the decision-maker in it, a strategy to accelerate reactive decision making in gaming and training scenarios.

Other active areas of HCI research include the balancing of decision task complexity among decision makers, and, as mentioned in the context of HLF, the proper understanding of the role of cultural and societal issues in designing the interface of human with automated support systems.

1.10 Decision-Making Models

While the study of decision-making has a long and rich history, two simple questions endure at its core:

1. How *do* people make decisions?
2. How *should* people make decisions?

Focus on the first question results in *descriptive* models of human decision-making, models which describe and categorize observed human behaviors in gathering and processing information, making decisions and taking action on those decisions. Chief among researchers studying descriptive models are behavioral psychologists, behavioral economists and management scientists. Answers to the second question, how decisions should be taken, result in *normative* or *prescriptive* models. Here the goal is to specify preferencing relations by which the value of lotteries and decisions can be compared, axiomatize constraints on their selection by decision makers, and deduce the resulting properties of the preferred decisions. Contributions in this domain have come from a wide variety of investigators: mathematicians, philosophers, and computer scientists appear prominently.

The divergence of accurate descriptive models of human decision-making from the generally-accepted normative models was clearly demonstrated by the seminal work of Kahneman and Twersky (1979). Using abundant experimental data, they showed that biases and heuristics which drove human decision making contradicted the assumption that people behaved as rational agents. Key among these deviations from rationality are loss aversion, optimism bias, the availability, simulation and representativeness heuristics, and the conjunction and base rate fallacies. Each of these common human behaviors results in violation of strict rational agency in decision making. Their descriptive model, Prospect Theory (PT), predicts these behaviors and explained the data much better than the assumption of rational agency.

While a better predictor of human decision-making behavior than prevailing theories, PT failed to satisfy the "stochastic dominance" axiom that a shift of probability to a better outcome should increase the prospect. In addition, the observed behavior of diminishing sensitivity to increasing prospects was not well modeled. These defects were addressed with the introduction of Cumulative Prospect Theory (CPT) (Tversky and Kahneman 1992). In the mathematical framework of Choquet capacities and the Choquet expected utility, this model satisfied those requirements and resolved the Allais and Ellsberg paradoxes vexing the conventional normative theories. Further development of this approach continues, including clarification of CPT under risk, that is, with known probabilities, as compared to CPT under uncertain probabilities (Chateauneuf and Wakker 1999). Basili and Zappia (2009) grounded the Pecautionary Principle within CPT, the notion of optimism for windfall gains and pessimism for catastrophic losses which commonly characterizes human choice.

Other theories with elements of descriptive modeling have also found use in the design of decision-support systems. Naturalistic Theory, with recognition-primed decisions, emphasizes the role of experience and intuition in critical time-constrained decision making. Argumentation Theory supports a model of dialogic reasoning based on induction and weight of evidence often describing shared decision making.

The Von Neumann-Morganstern "rational actor" model long dominated the field of normative modeling (Von Neumann 1953). It proposed that any rational actor implicity has an interval-scaled utility function, and will always act to maximize the expected value of this utility. Rationality is defined in terms of four straight-forward axioms of the actor's preference function over lotteries $<$ ($\leq$):

1. Completeness: for all lotteries X and Y, either $X<Y$, $Y<X$, or $X \sim Y$
2. Transitivity: $X \leq Y$ and $Y \leq Z$ implies $X \leq Z$
3. Continuity: $X \leq Y \leq Z$ implies there is a $0 \leq p \leq 1$ such that $pX + (1-p)Z = Y$
4. Independence: $X < Y$ and $0 < p \leq 1$ imply $pX + (1-p)Z < pY + (1-p)Z$ for all Z

The power of this definition of rational agency is that while the individual axioms are relatively unobjectionable, any actor failing to satisfy them can be shown to behave in ways that guarantee his long-term loss. This model, as later extended to subjective probability, is still prevalent in most decision-support systems. Maximization of expected utility, with elicitation of subjective probability where necessary, is still the dominant approach.

Other normative models, such as multi-criteria decision-making, game-theoretic models, and axiomatized bounded rationality models, are the subject of vigorous development. But since the work of Kahneman and Tversky, the clear direction for research on decision-making models has been to incorporate elements of how humans actually prepare and make decisions with a prescriptive theory on how decisions should be made. In the joint human-computer environment of decision support systems, both matter. Sometimes to our benefit, and often to our peril, in our decision making we are not simply rational actors in the axiomatic sense. The design of human-computer systems to support the "right" decisions while acknowledging accurate descriptive models of human behavior is surely an important direction for future work.

1.11 Case Studies

Over the last two decades, the international community has suffered a wide variety of major natural and man-made disasters requiring coordinated emergency response efforts. In many of these incidents, technical decision support systems incorporating some level of information fusion capability have been employed. Evaluation of the effectiveness and limitations of the overall crisis response in general, and of the information fusion systems they employed in particular, have become available in the literature. Together they document the evolving state of this critical ongoing technical enterprise. We conclude this Introduction by noting a representative set of emergency response case studies, presented in chronological order of the incidents upon which they report.

In Ais Gill, Cumbria, UK, in 1995 a railroad accident with fatalities invoked the large-scale coordinated response of local municipal, fire, police, and medical services. A train was derailed due to a landslide, and a second train collided with it a few minutes later. An interagency communication network employing handheld and vehicular radios and was quickly established, and both formal and informal message-passing maintained close contact among the response organizations. A limiting factor in the response was the absence of a uniform "shared mental model," as reported by Smith and Dowell (2000). This was seen in the inefficient deployment of ambulances and other resources, leading to delays in medical assistance for some victims. More analysis of this incident can be found in the present volume.

With focus on clarifying decision support requirements for extreme events, Mendonca (Mendonca 2007) reviewed the response to the 9/11/2001 terrorist attack. As mentioned earlier, had more robust air traffic control situation awareness been operational, one which supported air piracy hypotheses, valuable time to evacuate and respond before any building was struck would have been gained. The significance of building-in space for unplanned contingencies by information fusion design supporting responder innovation is also emphasized by Mendonca in the aftermath of this incident.

The Indian Ocean Tsunami of 2004 led to the loss of over 200,000 lives and many billions of dollars of damage. Triggered by a historically powerful earthquake off the coast of Indonesia, the first catastrophic wave hit near Aceh, Indonesia, in 15 min, with damaging waves striking other coastlines up to 7 h later. An existing seismographic network to monitor such events in the Pacific Ocean recorded the earthquake and messaged response offices in the affected regions within a few minutes, but there was not an information dissemination system in place to transmit the warnings to vulnerable populations in time. As a consequence, the Indian Ocean Tsunami Warning System with similar capabilities to the Pacific System was stood up shortly thereafter, and proved effective at rapidly disseminating evacuation orders during a regional earthquake in 2010.

In the Summer of 2002 a great "hundred-year" flood swept the plains of Central and Eastern Europe. Kuhlike's case study (2013) of this response emphasized the rapid confounding of detailed flood response plans in the face of near-universal surprise at the scale of the disaster and the collapse of local communication infrastructures needed to support evacuation and rescue operations. And as is sometimes overlooked, conflicts of interest, norms and values among stakeholders underscore the importance of thoughtful constraints on "resilience," and an undisputed chain of incident response command.

Haiti sustained a catastrophic earthquake affecting densely populated areas in that island nation in 2010, suffering over 100,000 deaths. In the immediate aftermath, civil authority was largely absent in the most affected areas. as the government and security forces ceased organized operations. The hierarchical, centralized model of emergency management deployed by the government and by non-governmental organizations was of very limited effectiveness in the initial response period, since many of the "stovepipes" along which vital information normally flows were damaged or destroyed and the infrastructure to deploy resources badly damaged. The case study (Yates and Paquette 2011) points out that in the Haitian disaster, social media such as wikis and collaborative workspaces proved invaluable in re-establishing communications and disseminating critical information to vulnerable populations. But however robust, the open nature of such channels limits their reliability without support of social media tools specifically designed for collaborative communication in emergency situations. In addition, as Zook et al. (2010) discussed, during the Haiti disaster crowdsourced geolocation services such as Usahidi and GeoCommons effectively supplemented conventional mapping services in establishing detailed up-to-date geographical situation awareness and disseminating that critical information widely.

References

Aamodt A, Plaza E (1994) Case-based reasoning: foundational issues, methodological variations, and system approaches. AI Commun 7(1):39–59

Annelli JF (2006) The national incident management system: a multi-agency approach to emergency response in the United States of America. Rev Sci Tech 25(1):223

Barwise J, Perry J (1980) The situation underground. In: Barwise J, Sag I (eds) Stanford working papers in semantics, vol 1, Sec D. Stanford Cognitive Science Group, Stanford, CA, pp 1–55

Bellenger A, Gatepaille S (2011) Uncertainty in ontologies: Dempster-Shafer theory for data fusion applications. CoRR abs/1106.3876

Bisantz A, Rogova G, Little E (2004) On the integration of cognitive work analysis within information fusion development methodology. In: Proc of the human factors and ergonomics Society Annual Meeting, New Orleans, LA

Basili M, Zappia C (2009) Shackle and modern decision theory. Metroeconomica, 60(2), 245–282

Blasch EP, Lambert DA, Valin P, Kokar MM, Llinas J, Das S, Shahbazian E (2012) High level information fusion (HLIF): survey of models issues and grand challenges. IEEE Aero Elect Syst Mag 27(9):4–20

Blasch E, Steinberg A, Das S, Llinas J, Chong C, Kessler O, White F (2013) Revisiting the JDL model for information exploitation. In: 16th International Conference on Information fusion, FUSION 2013. IEEE, Washington, DC, pp 129–136

Bostwick D, Goldstein J, Stephenson T, Stromsten S, Tierno J, Torrelli M, White J (2009) PARSEC, an application of probabilistic case based reasoning to maritime surveillance. In: Proc of the IEEE conference on Technologies for Homeland Security, Waltham, MA

Chateauneuf A, Wakker P (1999) An axiomatization of cumulative prospect theory for decision under risk. Journal of Risk and Uncertainty, 18(2), 137–145

Costa PC, Laskey KB (2006) PR-OWL: a framework for probabilistic ontologies. Front Artif Intell Appl 150:237

Dell'Acqua F, Bignami C, Chini M, Lisini G, Polli DA, Stramondo S (2011) Earthquake damages rapid mapping by satellite remote sensing data: L'Aquila april 6th 2009 event. JSTARS 4(4):935–943

Dey L, Abulaish M (2008) Fuzzy ontologies for handling uncertainties and inconsistencies in domain knowledge description. In: Proc of the IEEE International Conference on Fuzzy systems, FUZZ-IEEE 2008, IEEE World Congress on Computational Intelligence. IEEE, Washington, DC

Dubois D, Prade H (1988) Possibility theory: an approach to computerized processing of uncertainty. Plenum, New York, NY

d'Agostino F, Farinelli A, Grisetti G, Iocchi L, Nardi D (2002) Monitoring and information fusion for search and rescue operations in large-scale disasters. Proceedings of the International Conference on Information Fusion, Fusion 2002, Annapolis MD USA, 672–679

Ellsberg D (2001) Risk ambiguity and decision. Taylor & Francis, London

Endsley MR (1995) Toward a theory of situation awareness in dynamic systems. Hum Fact 37(1):32–64

Gertman D (2009) Human factors and data fusion as part of control systems resilience. Human Systems Interactions Conference 2009, Catania, Italy, 642–647

Haenni R, Kohlas J, Lehmann N (2001) Probabilistic argumentation systems. In: Kohlas J, Moral S (eds) Handbook of defeasible reasoning and uncertainty management systems, vol 5. Kluwer, Dordrecht

Hall DM, Jordan JM (2010) Human-centric information fusion. Artech House, Norwood, MA

Heckerman DE (1991) Probabilistic similarity networks. MIT Press, Cambridge, MA

Intorelli A, Braig D, Moquin R (2009) Real-time data fusion and visualization in support of emergency response operations. In: IEEE Conference on Technologies for Homeland Security 2009 HST'09. IEEE, Washington, DC, pp 417–424

Jelesnianski CP, Chen J, Shaffer WA (1992) SLOSH: Sea lake and overland surges from hurricanes. National Weather Service, US Department of Commerce National Oceanic and Atmospheric Administration, Silver Spring, MD

Kahneman D, Tversky A (1979) Prospect theory: an analysis of decision under risk. Econometrica 47(2):263–291

Kana A, Akinkunmi B (2014) An algebra of ontologies approximation under uncertainty. Int J Intell Sci 4:54–64. doi:10.4236/ijis.2014.42007

Killeen JP, Chan TC, Buono C, Griswold WG, Lenert LA (2006) A wireless first responder handheld device for rapid triage patient assessment and documentation during mass casualty incidents. In: AMIA annual symposium proceedings American Medical Informatics Association, Washington, DC, p 429

Kuhlicke C (2013) Resilience: a capacity and a myth: findings from an in-depth case study in disaster management research. Nat Hazards 67(1):61–76

Laskey KB, Costa PC, Janssen T (2008) Probabilistic ontologies for knowledge fusion. In: Proc of the FUSION'2008-11th Conference on Multisource information fusion, Cologne

Little E, Rogova G (2005) Ontology meta-model for building a situational picture of catastrophic events. In: Proceedings of the Eighth international conference on Information Fusion. International Society of Information Fusion, Philadelphia, PA

Little E, Rogova G (2008) Designing ontologies for higher level fusion. Int J Inf Fus 10(11):70–82

Llinas J, Bowman C, Rogova G, Steinberg A, Waltz E, White F (2004) Revisiting the JDL data fusion model II. Space and Naval Warfare System Command, San Diego, CA

Makridakis S, Taleb N (2009a) Living in a world of low levels of predictability. Int J Forecast 25(4):840–844

Makridakis S, Taleb N (2009b) Decision making and planning under low levels of predictability. Int J Forecast 25(4):716–733

Mattioli J, Museux N, Hemaissia M, Laudy C (2007) Proceedings of International conference on information fusion, Quebec, Canada. IEEE, Washington, DC

Mendonça D (2007) Decision support for improvisation in response to extreme events: Learning from the response to the 2001 World Trade Center attack. Decision Support Systems, 43(3):952–967

Mendonca D, Beroggi GE, Wallace WA (2001) Decision support for improvisation during emergency response operations. Int J Emerg Manag 1(1):30–38

Nikolov A, Uren V, Motta E, De Roeck A (2008) Integration of semantically annotated data by the KnoFuss architecture. In: Gangemi A, Jérôme E (eds) Knowledge engineering: practice and patterns. Springer, Berlin, pp 265–274

Pearl J, Russel H (2011) Bayesian networks. Department of statistics, UCLA, Los Angeles, CA

Rasmussen A, Pejtersen J, Goodstein J (1994) Cognitive systems engineering. Wiley, New York, NY

Reiter R (1980) A logic for default reasoning. Artif Intell 13:81–132

Rogova GL (2009) Context-awareness in crisis management. In: Proceedings of the Military Communications Conference. IEEE, Washington, DC

Rogova GL, Scott PD, Lollett C (2005) Higher level fusion for post-disaster casualty mitigation operations. In: 8th International Conference on Information, Fusion 2005, vol 2. IEEE, Washington, DC, p 8

Rogova G, Scott P, Lollett C, Mudiyanur R (2006) Reasoning about situations in the early post-disaster response environment. In: Proc. of the FUSION'2006-9th Conference on Multisource information fusion, Florence, Italy

Rogova G, Llinas J, Yager R (2015) Argumentation and fusion of soft-hard Information for threat prediction. In: Proceedings of the NATO IST-SET-126 symposium on information fusion (hard and soft) for ISR Norfolk, VA, May 2015

Sambhoos K, Nagi R, Sudit M, Stotz A (2010) Enhancements to high level data fusion using graph matching and state space search. Inf Fusion 11(4):351

Schreiber-Ehle S, Koch W (2012) The JDL model of data fusion applied to cyber-defence—a review paper. In: 2012 Workshop on Sensor data fusion: trends solutions applications (SDF). IEEE, Washington, DC, pp 116–119

Shafer G (1976) A mathematical theory of evidence. Princeton University Press, Princeton, NJ

Shenoy PP, Shafer G (1988) An axiomatic framework for Bayesian and belief-function propagation. In: Proc of the AAAI Workshop on Uncertainty in AI, Minneapolis, MN, pp 307–314

Smets P (2000) Data fusion in the transferable belief model. In: Proceedings of the FUSION'2000-Third conference on Multisource-multisensor Information Fusion, Paris, June 2002, pp 21–33
Smith W, Dowell J (2000) A case study of co-ordinative decision-making in disaster management. Ergonomics 43(8):1153–1166
Snidaro L, Visentini I, Bryan K (2015) Fusing uncertain knowledge and evidence for maritime situational awareness via Markov logic networks. Inf Fusion 21:159–172. doi:10.1016/j.inffus.2013.03.004
Sobek DK II, Smalley A (2011) Understanding A3 thinking: a critical component of Toyota's PDCA management system. CRC Press, Chicago, IL
Steinberg AN, Bowman CL (2004) Rethinking the JDL data fusion model. In: Proceedings of the MSS National Symposium on Sensor and Data Fusion, vol. 1, Columbia, SC, June 2004.
Steinberg A, Rogova G (2008) Situation and context in data fusion and natural language understanding. In: Proceedings of the Eleventh International Conference on Information Fusion. Cologne, Germany
Teimouri M, Delavar MR, Kolyaie S, Chavoshi SH, Moghaddam HK (2008) A SDSS-based earthquake damage assessment for emergency response: case study in Bam. In: ISPRS 21st Silk road for information from imagery. Beijing
Thagard P, Shelley CP (1997) Abductive reasoning: logic, visual thinking, and coherence. In: Dalla Chiara M-L et al (eds) Logic and scientific methods. Kluwer, Dordrecht, pp 413–427
Tversky A, Kahneman D (1992) Advances in prospect theory: Cumulative representation of uncertainty. Journal of Risk and uncertainty 5(4), 297–323
Von Lubitz DK, Beakley JE, Patricelli F (2008) 'All hazards approach' to disaster management: the role of information and knowledge management Boyd's OODA Loop and network-centricity. Disasters 32(4):561–585
Von Neumann J (1953), Morgenstern 0 (1944) Theory of Games and Economic Behavior. Princeton NJ, Princeton University Press
Yager R (2012) Conditional approach to possibility-probability fusion. IEEE Trans Fuzzy Syst 20(1):46–55
Yager R (2015) A measure based approach for hard-soft fusion topic: uncertainty in fusion process. In: Proceedings of the NATO IST-SET-126 symposium on information fusion (hard and soft) for ISR, Norfolk, VA, May 2015
Yates D, Paquette S (2011) Emergency knowledge management and social media technologies: a case study of the 2010 Haitian earthquake. Int J Inf Manag 31(1):6–13
Zarnowski A, Levin E, Skubiev S (2010) Human centric approach to inhomogeneous geospatial data fusion and actualization. Archiwum Fotogrametrii Kartografii i Teledetekcji 21:523
Zook M, Graham M, Shelton T, Gorman S (2010) Volunteered geographic information and crowdsourcing disaster relief: a case study of the Haitian earthquake. World Med Health Policy 2(2):7–33

Part I
Knowledge Representation and Extraction

Chapter 2
Natural Language Understanding for Information Fusion

Stuart C. Shapiro and Daniel R. Schlegel

Abstract Tractor is a system for understanding English messages within the context of hard and soft information fusion for situation assessment. Tractor processes a message through text processors using standard natural language processing techniques, and represents the result in a formal knowledge representation language. The result is a hybrid syntactic-semantic knowledge base that is mostly syntactic. Tractor then adds relevant ontological and geographic information. Finally, it applies hand-crafted syntax-semantics mapping rules to convert the syntactic information into semantic information, although the final result is still a hybrid syntactic-semantic knowledge base. This chapter presents the various stages of Tractor's natural language understanding process, with particular emphasis on discussions of the representation used and of the syntax-semantics mapping rules.

2.1 Introduction

Tractor is a system for message understanding within the context of a multi-investigator, multi-university effort on "Hard and Soft Information Fusion" (Gross et al. 2012). Information obtained from physical sensors such as RADAR, SONAR, and LIDAR are considered hard information. Information from humans expressed in natural language is considered soft information. Tractor (Prentice et al. 2010) is a computational system that understands isolated English intelligence messages in the counter-insurgency domain for later fusion with each other and with hard information, all to aid intelligence analysts to perform situation assessment. In this context, "understanding" means creating a knowledge base (KB), expressed in a formal knowledge representation (KR) language, that captures the information in an English message.

This chapter is a slightly edited version of Shapiro and Schlegel (2013).

S.C. Shapiro (✉) • D.R. Schlegel
State University of New York at Buffalo, Buffalo, NY, USA
e-mail: shapiro@buffalo.edu; drschleg@buffalo.edu

G. Rogova, P. Scott (eds.), *Fusion Methodologies in Crisis Management*,
DOI 10.1007/978-3-319-22527-2_2

Tractor takes as input a single English message. The ultimate goal is for Tractor to output a KB representing the semantic information in that message. Later systems of the larger project combine these KBs with each other and with hard information. Combining KBs from different messages and different hard sources is done via a process of data association (Gross et al. 2012; Poore et al. 2009) that operates by comparing the attributes of and relations among the entities and events described in each KB. It is therefore important for Tractor to express these attributes and relations as completely and accurately as possible.

Many systems that are used for the same purpose as Tractor use information extraction techniques. For example, on its web site, Orbis Technologies, Inc. says, "Orbis Technologies, Inc. is a leader in providing cloud computing-based semantic text analytics, using MapReduce, to support *entity extraction,* relationship identification, and semantic search,"[1] and information extraction is defined as "the process of identifying within text instances of *specified* classes of entities and of predications involving these entities" (Grishman 2011, emphasis added). Rather than merely trying to identify certain pre-specified classes of entities and events (people, places, organizations, etc.) in a top-down fashion, by looking for them in the text, we want to faithfully identify and describe all the entities and events mentioned in each message in a bottom-up fashion, converting to a semantic representation whatever occurs there.

Our approach is to use largely off-the-shelf natural language processing software for text processing, to be discussed briefly in Sect. 2.3. The output of text processing is a hybrid syntactic-semantic representation, with semantic classification of entities contributed by named-entity recognizers. We translate the output of the text processing to the KR language we use. The KR language is introduced in Sect. 2.2, and the translator in Sect. 2.4. This KB is enhanced with relevant ontological and geographical information, briefly discussed in Sect. 2.5. Finally, hand-crafted syntax-semantics mapping rules are used to convert the mostly syntactic KB into a mostly semantic KB. This is still a hybrid syntactic-semantic representation, because the mapping rules do not yet convert all the syntactic information. The specific representation constructs we use are introduced in Sects. 2.6–2.8. The syntax-semantics mapping rules are discussed in Sect. 2.9, and some summary information drawn from a semantic KB are shown in Sect. 2.10. Although even the remaining syntactic information in the final KB is useful for data association, our intention is to add mapping rules so that, over time, the KBs that are produced are less syntactic and more semantic. The results of testing and evaluating the system are presented and discussed in Sect. 2.11.

This chapter constitutes an update and current status report on Tractor, which has been introduced and discussed in a previous set of papers (Prentice et al. 2010; Gross et al. 2012; Gómez-Romero et al. 2010; Kandefer and Shapiro 2011; Prentice and Shapiro 2011). An overview of the entire Hard and Soft Information Fusion project,

[1] http://orbistechnologies.com/solutions/cloud-based-text-analytics/ emphasis added.

and the architecture of the process is given in Gross et al. (2012). An introduction to Tractor and its initial architecture is given in Prentice et al. (2010). An introduction to the context-based information retrieval (CBIR) sub-process of Tractor, its proposed use of spreading activation, and how spreading activation algorithms might be evaluated is given in Kandefer and Shapiro (2011). A general overview of the role of contextual information in information fusion architectures is given in Gómez-Romero et al. (2010). Tractor's use of propositional graphs for representing syntactic and semantic information is introduced in Prentice and Shapiro (2011). That paper ends with the comment, "The graphs used in this paper have been hand-built using the mappings detailed in section IV. Automating this process to produce propositional graphs such as these is the major implementation focus of future work" (Prentice and Shapiro 2011, p. 527). That work has now largely been done. This chapter reports on the results of that work.

Tractor and the larger information fusion system of which it is a part have been developed by experimenting with several datasets, particularly the Synthetic Counterinsurgency (SynCOIN) (Graham 2011) dataset. All examples in this chapter have been drawn from these datasets.

2.2 SNePS 3

We use SNePS 3 (Shapiro 2000) as the KR system for the KBs created by Tractor from the English messages. SNePS 3 is simultaneously a logic-based, frame-based, and graph-based KR system (Schlegel and Shapiro 2012), and is the latest member of the SNePS family of KR systems (Shapiro and Rapaport 1992). In this chapter, we will show SNePS 3 expressions using the logical notation, $(R\ a_1, \ldots, a_n)$, where R is an n-ary relation and $a_1, \ldots, a_n$ are its n arguments. We will refer to such an expression as a "proposition." We will use "assertion" to refer to a proposition that is taken to be true in the KB, and say "assert a proposition" to mean adding the proposition to the KB as an assertion. We will also speak of "unasserting a proposition" to mean removing the assertion from the KB. The arguments of a proposition are terms that could denote words, occurrences of words in the message (called "tokens"), syntactic categories, entities in the domain, events in the domain, classes (also referred to as "categories") of these entities and events, or attributes of these entities and events.

We can classify relations, and the propositions in which they occur, as either: **syntactic**, taking as arguments terms denoting words, tokens, and syntactic categories; or as **semantic**, taking as arguments entities and events in the domain and their categories and properties. A KB is syntactic to the extent that its assertions are syntactic, and is semantic to the extent that its assertions are semantic. The KB first created by Tractor from a message is mostly syntactic. After the syntax-semantics mapping rules have been fired, the KB is mostly semantic. A subtle change that occurs as the mapping rules fire is that terms that originally denote syntactic entities are converted into denoting semantic entities.[2]

2.3 Text Processing

For initial text processing, we use GATE, the General Architecture for Text Engineering (Cunningham et al. 2011), which is a framework for plugging in a sequence of "processing resources" (PRs). The most significant PRs we use, mostly from the ANNIE (a Nearly-New Information Extraction System) suite (Cunningham et al. 2002), are: the ANNIE Gazetteer, for lexicon-based named-entity recognition; the ANNIE NE Transducer, for rule-based named-entity recognition; the ANNIE Orthomatcher, ANNIE Nominal Coreferencer, and ANNIE Pronominal Coreferencer, for co-reference resolution; the GATE rule-based Morphological Analyser for finding the root forms of inflected nouns and verbs; the Stanford Dependency Parser, for part-of-speech tagging and parsing; and the GATE Co-reference Editor, for manual corrections of and additions to the results of the three automatic co-reference resolution PRs. We added to the lexicons, added some rules to the rule-based PRs, added a supplementary part-of-speech tagger, and fixed some program bugs. We did not modify the parser. We can have a person use the Co-reference Editor as part of processing messages, or can process messages completely automatically without using the Co-reference Editor.

The results of GATE processing, with or without the Co-reference Editor, are a set of "annotations," each consisting of an ID, a start and end position within the message's text string, a Type, and a set of attribute-value pairs. Each PR contributes its own set of annotations, with its own IDs, and its own set of attributes and possible values. Only the start and end positions indicate when an annotation of one PR annotates the same text string as an annotation of another PR.

2.4 The Propositionalizer

The Propositionalizer examines the annotations produced by the GATE PRs, and produces a set of SNePS 3 assertions. The stages of the Propositionalizer are: annotation merging; correction of minor errors in syntactic categories; canonicalization

[2]What we call in this chapter the "syntactic KB" and the "semantic KB" were called in other papers the "syntactic propositional graph" and the "semantic propositional graph," respectively. The reason is that, in this chapter, we are exclusively using the logic-based view of SNePS 3, whereas in those papers, we used the graph-based view of SNePS 3. Their equivalence is explained in Schlegel and Shapiro (2012).

of dates, times, weights, and heights; and processing the structured portion of semi-structured messages. Annotations covering the same range of characters are combined into one SNePS 3 token-denoting term. Dates and times are converted into ISO8601 format. Annotation types, subtypes (where they exist), parts-of-speech, and dependency relations are converted into logical assertions about the tokens. The actual text string of an annotation and the root found by the morphological analyzer are converted into terms and related to the annotation-token by the `TextOf` and `RootOf` relations, respectively. Co-reference chains are converted into instances of the SNePS 3 proposition `(Equiv` $t_1, \ldots, t_n$`)`, where $t_1, \ldots, t_n$ are the terms for the co-referring tokens.

Most of the messages we are dealing with have structured headers, generally consisting of a message number and date, and sometimes a time. A message reporting a call intercept generally lists a description or name of the caller and of the recipient, duration, medium (e.g., "cell phone" or "text message"), and intercepting analyst. These are converted into SNePS 3 assertions.

As an example, consider message syn194:

> 194. 03/03/10 - Dhanun Ahmad has been placed into custody by the Iraqi police and transferred to a holding cell in Karkh; news of his detainment is circulated in his neighborhood of Rashid.

The basic information about the word "placed" in SNePS 3 is

```
(TextOf placed n20)
(RootOf place n20)
(TokenRange n20 38 44)
(SyntacticCategoryOf VBN n20)
```

Here, `n20` is a SNePS 3 term denoting the occurrence of the word "placed" in character positions 38–44 of the message text. The last proposition says that the syntactic category (part-of-speech) of that token is VBN, the past participle of a verb (Cunningham et al. 2002, Appendix G).

Some of the dependency information about "placed," with the text to make it understandable is:

```
(nsubjpass n20 n169)
(TextOf Ahmad n169)
(prep n20 n22)
(TextOf into n22)
```

That is, "Ahmad" is the passive subject of "placed," and "placed" is modified by a prepositional phrase using the preposition "into."[3]

Some of the information about "Karkh" is[4]

```
(TextOf Karkh n182)
(SyntacticCategoryOf NNP n182)
(Isa n182 Location)
```

[3]In a dependency parse, each token actually represents the phrase or clause headed by that token.

[4]Note that we are using `Isa` as the instance relation based on sentences like "Fido is a dog." For the subtype (or "subclass") relation we use `Type`.

Notice that in the first two of these assertions, `n182` denotes a token (a word occurrence), but in `(Isa n182 Location)`, it denotes an entity, specifically a location, in the domain. This change in the denotation of individual constants is a necessary outcome of the fact that we form a KB representing the syntactic information in a text, and then gradually, via the syntax-semantics mapping rules, turn the same KB into a semantic representation of the text.

The SNePS 3 KB that results from the Propositionalizer is what we call the syntactic KB. Although it contains some semantic information, such as `(Isa n182 Location)`, most of the information in it is syntactic.

2.5 Enhancement

The syntactic KB is enhanced with relevant information of two kinds: ontological taxonomic information is added above the nouns and verbs occurring in the KB; and geographical information is added to tokens of geographic place names occurring in the message. The information to be added is found by a process called CBIR (Kandefer and Shapiro 2009).

CBIR looks up each noun and verb in ResearchCyc[5] to find the corresponding Cyc concept(s). Then it adds to the KB the terms above those concepts in OpenCyc.[6]

CBIR also looks up proper nouns in the NGA GeoNet Names Server database,[7] and adds information found there to the KB. For example, the information added about Karkh (`n182`) is

```
(Isa n182 SectionOfPopulatedPlace)
(GeoPosition n182 (GeoCoords |33.32174| |44.39384|))
(MGRS n182 38SMB4358187120)
```

Which say that Karkh is a section of a populated place, that its geographic position is 33.32174 N latitude and 44.39384 E longitude, and that its MGRS (Military Grid Reference System) coordinate is 38SMB4358187120.

If CBIR finds MGRS coordinates, but no latitude and longitude, it converts the MGRS coordinates to latitude and longitude using NASA's World Wind software (National Aeronautics and Space Administration 2011).

The information added by CBIR is important to the data association task in deciding when terms from different messages should be considered to be co-referential.

[5] http://research.cyc.com/.

[6] http://www.opencyc.org/.

[7] http://earth-info.nga.mil/gns/html/.

2.6 Major Categories of Entities and Events

The actual message texts determine what categories of entities and events appear in the semantic KBs. For example, in the message, "Owner of a grocery store on Dhubat Street in Adhamiya said …", there is a mention of an entity which is an instance of the category store. So the category of stores is represented in the semantic KB. Nevertheless, there are some categories that play a role in the mapping rules in the sense that there are rules that test whether some term is an instance of one of those categories.

Such major categories of entities include: Person; Organization (a subcategory of Group); company; Location; country; province; city; Date; Time; Phone (the category of phone instruments); PhoneNumber (the category of phone numbers); MGRSToken; JobTitle; Dimension (such as age, height, and cardinality); Group (both groups of instances of some category, such as "mosques," and groups of fillers of some role, such as "residents"); ReligiousGroup (such as "Sunni"); and extensionalGroup (a group explicitly listed in a text, such as, "Dhanun Ahmad Mahmud, Mu'adh Nuri Khalid Jihad, Sattar 'Ayyash Majid, Abd al-Karim, and Ghazi Husayn.")

Major categories of events include: Action (such as "break" and "search"); ActionwithAbsentTheme (such as "denounce" and "report"); actionWithPropositionalTheme (such as "say" and "hear"); Perception (such as "learn" and "recognize"); and Event itself.

2.7 Relations

Relations used in the syntactic and semantic KBs can be categorized as either syntactic relations or semantic relations.

The syntactic relations we use include the following.

- `(TextOf x y)` means that the token y in the message is an occurrence of the word x.
- `(RootOf x y)` means that x is the root form of the word associated with token y.
- `(SyntacticCategoryOf x y)` means that x is the syntactic category (part-of-speech) of the word associated with token y.
- `(r x y)`, where `r` is one of the dependency relations listed in de Marneffe and Manning (2011), for example `nsubj, nsubjpass, dobj, prep`, or `nn`, means that token y is a dependent of token x with dependency relation `r`.

The semantic relations we use include the ones already mentioned (such as `Isa` and `Equiv`), and the following.

- `(Type c1 c2)` means that `c1` is a subcategory of `c2`.
- `(hasName e n)` means that `n` is the proper name of the entity `e`.

- `(GroupOf g c)` means that *g* is a group of instances of the class *c*.
- `(GroupByRoleOf g r)` means that *g* is a group of entities that fill the role, *r*.
- `(MemberOf m g)` means that entity *m* is a member of the group *g*.
- `(hasPart w p)` means that *p* is a part of entity *w*.
- `(hasLocation x y)` means that the location of entity *x* is location *y*.
- `(Before t1 t2)` means that time *t1* occurs before time *t2*.
- `(r x y)`, where *r* is a relation (including `possess`, `knows`, `outside`, `per-country_of_birth`, `org-country_of_headquarters`, `agent`, `experiencer`, `topic`, `theme`, `source`, and `recipient`), means that the entity or event *x* has the relation *r* to the entity or event *y*.
- `(a e v)`, where *a* is an attribute (including `cardinality`, `color`, `Date`, `height`, `GeoPosition`, `sex`, `per-religion`, `per-date_of_birth`, and `per-age`), means that the value of the attribute *a* of the entity or event *e* is *v*.

One relation, although syntactic, is retained in the semantic KB for pedigree purposes: `(TokenRange x i j)` means that the token *x* occurred in the text starting at character position *i*, and ending just before character position *j*. This is retained in the semantic KBs so that semantic information may be tracked to the section of text which it interprets. Two other syntactic relations, `TextOf` and `RootOf`, are retained in the semantic KB at the request of the data association group to provide term labels that they use for comparison purposes.

We believe that the syntactic relations we use are all that we will ever need, unless we change dependency parsers, or the dependency parser we use is upgraded and the upgrade includes new dependency relations. However, we make no similar claim for the semantic relations.

Assertions that use syntactic relations are called “syntactic assertions,” and those that use semantic relations are called “semantic assertions.”

2.8 Representation of Events

To represent events, we use a neo-Davidsonian representation (Parsons 1990), in which the event is reified and semantic roles are binary relations between the event and the semantic role fillers. For suggestions of semantic roles, we have consulted the entries at University of Colorado (2012). For example, in the semantic KB Tractor constructed from message syn064,

> 64. 01/27/10 - BCT forces detained a Sunni munitions trafficker after a search of his car netted IED trigger devices. Ahmad Mahmud was placed in custody after his arrest along the Dour’a Expressway, //MGRSCOORD: 38S MB 47959 80868//, in East Dora.

the information about the detain event includes

```
(Isa n18 detain)
(Date n18 20100127)
(agent n18 n16)
```

```
(GroupOf n16 force)
(Modifier n16 BCT)
(theme n18 n26)
(Equiv n230 n26)
(Isa n230 Person)
(hasName n230 "Ahmad Mahmud")
```

That is, `n18` denotes a detain event that occurred on 27 January 2010, the agent of which was a group of BCT forces, and the theme of which was (co-referential with) a person named Ahmad Mahmud.

2.9 The Syntax-Semantics Mapper

The purpose of the syntax-semantics mapping rules is to convert information expressed as sets of syntactic assertions into information expressed as sets of semantic assertions. The rules were hand-crafted by examining syntactic constructions in subsets of our corpus, and then expressing the rules in general enough terms so that each one should apply to other examples as well.

The rules are tried in order, so that earlier rules may make adjustments that allow later rules to be more general than they otherwise would have to be, and earlier rules may express exceptions to more general later rules. As of this writing, there are 147 mapping rules, that may be divided into several categories:

- *CBIR, supplementary enhancement rules* add ontological assertions that aren't in Cyc, but that relate to terms in the message;
- *SYN, syntactic transformation rules* examine syntactic assertions, unassert some of them, and make other syntactic assertions;
- *SEM, semantic transformation rules* examine semantic assertions, unassert some of them, and make other semantic assertions;
- *SYNSEM, true syntax-semantic mapping rules* examine syntactic assertions and maybe some semantic assertions as well, unassert some of the syntactic assertions, and make new semantic assertions;
- *CLEAN, cleanup rules* unassert some remaining syntactic assertions that do not further contribute to the understanding of the message;
- *INFER, inference rules* make semantic assertions that are implied by other semantic assertions in the KB.

Due to space constraints, only a few rules will be discussed.[8]

An example of a syntactic transformation rule is

```
(defrule passiveToActive
  (nsubjpass ?verb ?passsubj)
  =>
```

[8]The rules are shown using the actual rule syntax.

```
(assert '(dobj ,?verb ,?passsubj))
(unassert
 '(nsubjpass ,?verb ,?passsubj))
(:subrule
 (prep ?verb ?bytok)
 (TextOf by ?bytok)
 (pobj ?bytok ?subj)
 =>
 (assert '(nsubj ,?verb ,?subj))
 (unassert '(prep ,?verb ,?bytok))
 (unassert '(pobj ,?bytok ,?subj))))
```

This rule would transform the parse of "BCT is approached by a man" to the parse of "a man approached BCT." The rule fires even if the "by" prepositional phrase is omitted.

There are also some rules for distribution over conjunctions. One such rule would transform the parse of "They noticed a black SUV and a red car parked near the courthouse" to the parse of "They noticed a black SUV parked near the courthouse and a red car parked near the courthouse" by adding an additional partmod relation, from the token for "car" to the head token of "parked near the courthouse." Then another rule would transform that into the parse of "They noticed a black SUV parked near the courthouse and they noticed a red car parked near the courthouse" by adding a second dobj relation, this one from the token of "noticed" to the token of "car."

Some examples of true syntax-semantics mapping rules operating on noun phrases (presented in the relative order in which they are tried) are:

```
(defrule synsemReligiousGroup
  (Isa ?g relig_group_adj)
  (TextOf ?name ?g)
  =>
  (assert '(Isa ,?g ReligiousGroup))
  (assert '(hasName ,?g ,?name))
  (assert '(Type ReligiousGroup Group))
  (unassert '(Isa ,?g relig_group_adj)))
```

This rule would transform the token for "Sunni," which the GATE named-entity recognizers recognized to name a relig_group_adj, into an entity that is an instance of `ReligiousGroup`, whose name is `Sunni`. It also makes sure that the relevant fact that `ReligiousGroup` is a subcategory of `Group` is included in the semantic KB for the current message.

```
(defrule hasReligion
  (Isa ?religiongrp ReligiousGroup)
  (nn ?per ?religiongrp)
  (hasName ?religiongrp ?religion)
  =>
```

```
    (assert '(MemberOf ,?per ,?religiongrp))
    (assert '(per-religion ,?per ,?religion))
    (unassert '(nn, ?per ,?religiongrp)))
```

This rule would assert about the token of "youth" in the parse of "a Sunni youth" that it is a member of the group named `Sunni`, and that its religion is `Sunni`. It also would unassert the nn dependency of the token of "Sunni" on the token of "youth."

```
(defrule properNounToName
  (SyntacticCategoryOf NNP ?token)
  (TextOf ?text ?token)
  =>
  (assert '(hasName ,?token ,?text))
  (unassert '(SyntacticCategoryOf NNP ,?token))
  (unassert '(TextOf ,?text ,?token)))
```

This rule would transform a token of the proper noun "Khalid Sattar" into a token denoting the entity whose name is `"Khalid Sattar."`

```
(defrule nounPhraseToInstance
  (SyntacticCategoryOf NN ?nn)
  (:when (isNPhead ?nn))
  (RootOf ?root ?nn)
  (:unless (numberTermp ?root))
  =>
  (assert '(Isa ,?nn ,?root))
  (unassert '(SyntacticCategoryOf NN ,?nn))
  (unassert '(RootOf ,?root ,?nn)))
```

This rule would transform the token of "youth" in the parse of "a Sunni youth" into an instance of the category `youth`. The function `isNPhead` returns True if its argument is the head of a noun phrase, recognized by either having a det dependency relation to some token, or by being an nsubj, dobj, pobj, iobj, nsubjpass, xsubj, or agent dependent of some token. (In the corpus we work on, determiners are sometimes omitted.) The `(:unless (numberTermp ?root))` clause prevents a token of a number from being turned into an instance of that number.

Another rule makes the token of a verb an instance of the event category expressed by the root form of the verb. For example, a token of the verb "detained" would become an instance of the event category `detain`, which is a subcategory of `Action`, which is a subcategory of `Event`.

Some examples of syntax-semantics mapping rules that analyze clauses (presented in the relative order in which they are tried) are:

```
(defrule subjAction
  (nsubj ?action ?subj)
  (Isa ?action Action)
  =>
```

```
    (assert '(agent ,?action ,?subj))
    (unassert '(nsubj ,?action ,?subj)))
```

This rule would make the subject of "detained" the agent of a `detain` Action-event.

```
(defrule subjPerception
  (nsubj ?perception ?subj)
  (Isa ?perception Perception)
  =>
  (assert '(experiencer ,?perception ,?subj))
  (unassert '(nsubj ,?perception ,?subj)))
```

This rule would make the subject of "overheard" the experiencer of a `overhear` Perception-event.

Another rule makes the date of an event either the date mentioned in the dependency parse tree below the event token, for example, the date of the capture event in "Dhanun Ahmad Mahmud Ahmad, captured on 01/27/10, was turned over to . . ." is `20100127`, or else the date of the message being analyzed.

A final set of syntax-semantics mapping rules convert remaining syntactic assertions into "generic" semantic assertions. For example, any remaining prepositional phrases, after those that were analyzed as indicating the location of an entity or event, the "by" prepositional phrases of passive sentences, etc., are transformed into an assertion using the preposition as a relation holding between the entity or event the PP was attached to and the object of the preposition.

Cleanup rules unassert syntactic assertions that were already converted into semantic assertions, for example, unasserting `(TextOf` *x y*`)` and `(RootOf` *x y*`)` when `(Isa` *y x*`)` has been asserted. Other cleanup rules unassert remaining syntactic assertions that do not contribute to the semantic KB, such as the `SyntacticCategoryOf` assertions.

The inference rules make certain derivable assertions explicit for the benefit of the data association operation. For example, the agent of an event that occurred at some location on some date was at that location on that date, and the member of a group g_1 that is a subgroup of a group g_2 is a member of g_2.

2.10 Results

In order for a person to get an idea of what is in the semantic KBs, we have implemented a simple natural language generation function that expresses the information in a KB in short formalized sentences. Each relation is associated with a sentence frame whose slots are filled in from the relation's arguments. A term with a proper name, or which is co-referential with one with a proper name, is expressed by its name. Otherwise, terms that are instances of some category are expressed by a symbol constructed from its category. For example, some of the information in the semantic KB that Tractor constructed from syn064, shown and discussed in Sect. 2.8, is

```
detain18
Instance of: detain
detain18's Date is |20100127|.
detain18 has the relation agent to |BCT forces|.
detain18 has the relation theme to |Ahmad Mahmud|.
detain18 has the relation after to search32.

|BCT forces|
Instance of: Organization
detain18 has the relation agent to |BCT forces|.

search32
Instance of: search
search32's Date is |20100127|.
search32 has the relation theme to car108.
detain18 has the relation after to search32.

|Ahmad Mahmud|
Instance of: (setof Person trafficker)
|Ahmad Mahmud|'s sex is male.
|Ahmad Mahmud|'s Religion is Sunni.
|Ahmad Mahmud| has the relation possess to car108.
|Ahmad Mahmud| is located at Expressway.
|Ahmad Mahmud| is located at Expressway's Date is
               |20100127|.
detain18 has the relation theme to |Ahmad Mahmud|.
arrest65 has the relation theme to |Ahmad Mahmud|.

arrest65
Instance of: arrest
arrest65's Date is |20100127|.
arrest65 is located at Expressway.
arrest65 has the relation theme to |Ahmad Mahmud|.
place55 has the relation after to arrest65.

place55
Instance of: place
place55's Date is |20100127|.
place55 is located at |East Dora|.
place55 has the relation in to custody59.
place55 has the relation after to arrest65.
```

```
|East Dora|
Instance of: (setof Location SectionOfPopulatedPlace)
|East Dora|'s GeoPosition is latitude |36.29534|,
                            longitude |44.47780|.
|East Dora|'s Longitude is |44.4091|.
|East Dora|'s MGRS is 38SMB4496078958.
|East Dora|'s MGRSRadius is |0.5|.
place55 is located at |East Dora|.
```

2.11 Evaluation

The mapping rules were developed by testing Tractor on several corpora of messages, examining the resulting semantic KBs, finding cases where we were not happy with the results, examining the initial syntactic KBs, and modifying or adding to the rule set so that an acceptable result was obtained. These "training" messages included: the 100 messages from the Soft Target Exploitation and Fusion (STEF) project (Sambhoos et al. 2008); the 7 Bomber Buster Scenario messages (Gross et al. 2012); the 13 messages of the Bio-Weapons Thread, 84 messages of the Rashid IED Cell Thread, and 115 messages of the Sunni Criminal Thread, of the 595-message SynCOIN dataset (Graham 2011; Graham et al. 2011). None of these messages were actual intelligence messages, but are "a creative representation of military reports, observations and assessments" (Graham et al. 2011). Tractor is still a work in progress. We have not yet finished testing, modifying, and adding to the mapping rules using these training sets.

We are currently developing a "grading rubric" to measure the correctness and completeness of the semantic KBs produced by Tractor against manually produced "gold standard" semantic KBs. We will then have to produce those gold standard KBs, and compare them with those produced by Tractor. We hope to report on this grading rubric, and on Tractor's grades in a future paper.

Nevertheless, we can now evaluate how general the mapping rules are, and whether they are perhaps overly general. The generality of the rules will be tested through examination of how often the mapping rules fire on a "test" dataset, not previously examined. We'll look at the amount of syntactic and semantic data there are in the processed KBs from the messages in our test and training sets. We'll also look at how many mistakes Tractor makes on the test dataset, to test for over-generality. Combined, these three experiments will show that our rules are general, but not overly so, that the amount of semantic data in the resultant semantic KBs is quite high, and that the degree of semantization compares well with that of our training sets. We begin by addressing the question of, given that the mapping rules were developed using the training messages, how general are they? To what extent do they apply to new, unexamined, "test" messages? To answer this question, we used the 57 messages of the Sectarian Conflict Thread (SCT) of the

Table 2.1 The number of mapping rules in each category, the number of those rules that fired on any message in the SCT dataset, the total number of times those rules fired, and the average number of times they fired per message

Rule type	Rule count	Rules fired	Times fired	Firings/msg
CBIR	1	1	474	8.32
SYN	23	13	1596	28.00
SEM	5	5	328	5.75
SYNSEM	99	56	2904	50.95
INFER	9	8	135	2.37
CLEAN	10	8	6492	113.89
TOTAL	147	91	11,929	209.28

Table 2.2 For the total SCT dataset, the number of syntactic assertions, the number of semantic assertions, and the percent of assertions that are semantic in the syntactic KBs, the semantic KBs, and in the semantic KBs without counting the assertions added by CBIR

KB	Syntactic	Semantic	Percent semantic (%)
Syntactic	2469	1149	31.76
Semantic	538	48,561	98.90
without CBIR	538	5646	91.30

SynCOIN dataset. These messages, averaging 46 words per message, contain human intelligence reports, "collected" over a period of about 5 months, which describe a conflict among Christian, Sunni, and Shi'a groups. The messages describe events in detail, and entities usually only through their connection to some group or location.

We divided the rules into the six categories listed in Sect. 2.9, and counted the number of rules used in the SCT corpus, along with the number of rule firings, as seen in Table 2.1. Of the 147 rules currently part of the system, 91 fired during the processing of this corpus for a total of 11,929 rule firings. Sixty-nine rules fired five or more times, and 80 were used in more than one message. Sixty-two percent of all the rules and fifty-seven percent of the true syntax-semantics mapping rules fired on the test messages. We conclude that, even though the rules were developed by looking at specific examples, they are reasonably general.

The purpose of the syntax-semantics mapping rules is to convert syntactic information about the words, phrases, clauses, and sentences in a message into semantic information about the entities and events discussed in the message. We are still in the process of developing the rule set, so it is useful to measure the percentage of each KB that consists of semantic assertions. Table 2.2 shows the

Table 2.3 Percent of the semantic KBs which are semantic for the BBS and STEF training sets, excluding the CBIR enhancements

Dataset	Syntactic	Semantic	Pct semantic (%)
BBS	57	750	92.94
STEF	517	8326	94.15

number of syntactic assertions,[9] the number of semantic assertions, and the percent of assertions that are semantic in the initial syntactic KBs, the final semantic KBs, and the final semantic KBs without counting the semantic assertions added by CBIR (*see* Sect. 2.5). The numbers are the totals over all 57 messages of the SCT dataset. As you can see, before the mapping rules, the KBs are almost 70 % syntactic, whereas after the mapping rules they are more than 90 % semantic. CBIR is purely additive, so it does not reduce the number of syntactic assertions in the KB, but it does increase the semantic content of the KBs to nearly 99 %.

The percentage of the semantic KBs from the test message set that is semantic, 91.30 %, is very similar to that of the training message sets. For example, the semantic content of the semantic KBs of two of these training sets, the BBS and STEF datasets, are 92.94 % and 94.15 %, respectively, as shown in Table 2.3. We conclude that, even though we are still developing the mapping rules, the ones we have so far are converting a large part of the syntactic information into semantic information, and doing so in a way that generalizes from the training sets to test sets.

Since the mapping rules were designed using the training datasets, it is possible that some of the rules that fire in our test dataset (as shown in Table 2.1) are erroneous. That is, the rules may be *too* general. In order to verify that the rules function as expected, we manually verified that the rules were applied only where they should have been.

In order to perform this experiment we ran the mapping rules on each message in the dataset, noting after each rule firing whether the firing was correct or incorrect. Rules which fired due to misparses earlier in the process were not counted as rules used. A rule was counted as firing correctly if its output was semantically valid and in accord with the intent of the rule.

As Table 2.4 shows, very rarely were rules applied overzealously. Therefore we can say with some certainty that the rules are not only general enough to fire when processing messages from corpora other than the training set, but they are not overly general; the firings produce a valid semantization of the messages.

[9]The `TokenRange`, `TextOf`, and `RootOf` assertions, which are syntactic, but are retained in the semantic KB for pedigree information and to assist in the downstream scoring of entities against each other, as explained at the end of Sect. 2.7, have been omitted from the count.

Table 2.4 The number of rules used in each category, along with the number of times rules from each category were used in the SCT dataset, and the number of times they were used correctly

			Fired correctly	
Rule type	Rules used	Times fired	Number	%
CBIR	1	474	474	100
SYN	13	1567	1548	98.79
SEM	5	328	328	100
SYNSEM	56	2651	2431	91.7
INFER	8	85	72	84.7
CLEAN	8	6492	6492	100
Total	91	11,597	11,345	97.8

2.12 Comparison with Other Systems

Our system produces results which are much different from those of the most related system we're aware of—Orbis Technologies' proprietary Cloud Based Text-Analytics (CTA) software. The output of the two systems is not directly comparable. CTA attempts to identify and find relationships among entities, in the process identifying the entities' types as either Person, Organization, Location, Equipment, or Date. Where we identify all the types of entities (and have more types, such as Group and Event), Orbis only seems to identify them when they are in a relation. An Orbis relation is simple—an entity is `associated with` another entity. Tractor uses a large set of relations for representing complex relationships between entities.

Within the 57 SCT messages, Tractor identified (among many other things) 34 entities which were members of specific groups, the religion of 17 entities, 203 locations of events or entities, and 33 persons or groups with specific roles. It additionally identified 102 agents of specific events, 128 themes of events, and over 125 spatial relationships such as "in," "on," and "near."

2.13 Conclusions

Tractor is a system for message understanding within the context of hard and soft information fusion for situation assessment. Tractor's processing is bottom-up—find whatever is in the text, rather than top-down—look for pre-specified entities, events, and relations. Tractor uses GATE Processing Resources (PRs) for text analysis, including named-entity recognition, co-reference resolution, part-of-speech tagging, and dependency parsing. The propositionalizer converts the annotations produced by the GATE PRs into a hybrid syntactic-semantic knowledge base (KB) represented in the SNePS 3 knowledge representation system. Relevant ontological and geographic information is added to the KB, and then hand-crafted syntax-semantics mapping rules convert the syntactic information into semantic information. Although these rules were devised by looking at specific "training" message sets, 62 % of them fired on a separate set of "test" messages. Moreover, not counting syntactic information that is used by later stages of fusion, Tractor, operating on the test messages, was found to convert syntactic KBs that are 68 % syntactic into semantic KBs that are 91 % semantic (99 % semantic when added ontological and geographical information is counted). Not counting rule firings on syntactic assertions that resulted from misparsings, 98 % of the rule firings on the test messages resulted in semantically correct assertions that were in accord with what the rule was designed to do.

Acknowledgements This work has been supported by a Multidisciplinary University Research Initiative (MURI) grant (Number W911NF-09-1-0392) for "Unified Research on Network-based Hard/Soft Information Fusion," issued by the US Army Research Office (ARO) under the program management of Dr. John Lavery.

References

Cunningham H, Maynard D, Bontcheva K, Tablan V (2002) GATE: a Framework and graphical development environment for robust NLP tools and applications. In: Proceedings of the 40th anniversary meeting of the Association for Computational Linguistics (ACL'02)

Cunningham H, Maynard D, Bontcheva K, Tablan V, Aswani N, Roberts I, Gorrell G, Funk A, Roberts A, Damljanovic D, Heitz T, Greenwood MA, Saggion H, Petrak J, Li Y, Peters W (2011) Text processing with GATE (Version 6). The University of Sheffield, Department of Computer Science

de Marneffe M-C, Manning CD (2011) Stanford typed dependencies manual. Stanford University, September 2008. Revised for Stanford Parser v. 1.6.9 in September 2011. http://nlp.stanford.edu/software/dependencies_manual.pdf

Gómez-Romero J, Garcia J, Kandefer M, Llinas J, Molina JM, Patricio MA, Prentice M, Shapiro SC (2010) Strategies and techniques for use and exploitation of contextual information in high-level fusion architectures. In: Proceedings of the 13th international conference on information fusion (Fusion 2010). ISIF

Graham JL (2011) A new synthetic dataset for evaluating soft and hard fusion algorithms. In: Proceedings of the SPIE defense, security, and sensing symposium: defense transformation and net-centric systems 2011, pp 25–29

Graham JL, Rimland J, Hall DL (2011) A COIN-inspired synthetic data set for qualitative evaluation of hard and soft fusion systems. In: Proceedings of the 14th international conference on information fusion (Fusion 2011). ISIF, pp 1000–1007

Grishman R (2011) Information extraction: capabilities and challenges. Notes prepared for the 2011 International Summer School in Language and Speech Technologies, Tarragona

Gross GA, Nagi R, Sambhoos K, Schlegel DR, Shapiro SC, Tauer G (2012) Towards hard+soft data fusion: Processing architecture and implementation for the joint fusion and analysis of hard and soft intelligence data. In: Proceedings of the 15th international conference on information fusion (Fusion 2012). ISIF, pp 955–962

Kandefer M, Shapiro SC (2009) An F-measure for context-based information retrieval. In: Lakemeyer G, Morgenstern L, Williams M-A (eds) Commonsense 2009: Proceedings of the ninth international symposium on logical formalizations of commonsense reasoning. The Fields Institute, Toronto, pp 79–84

Kandefer M, Shapiro SC (2011) Evaluating spreading activation for soft information fusion. In: Proceedings of the 14th international conference on information fusion (Fusion 2011). ISIF, pp 498–505

Lehmann F (ed) (1992) Semantic networks in artificial intelligence. Pergamon Press, Oxford

National Aeronautics and Space Administration. NASA World Wind (2011). http://worldwind.arc.nasa.gov/java/

Parsons T (1990) Events in the semantics of English: a study in subatomic semantics. MIT, Cambridge

Poore AB, Lu S, Suchomel BJ (2009) Data association using multiple frame assignments. In: Liggins M, Hall D, Llinas J (eds) Handbook of multisensor data fusion, Chap. 13, 2nd edn. CRC, Boca Raton, pp 299–318

Prentice M, Shapiro SC (2011) Using propositional graphs for soft information fusion. In: Proceedings of the 14th international conference on information fusion (Fusion 2011). ISIF, pp 522–528

Prentice M, Kandefer M, Shapiro SC (2010) Tractor: a framework for soft information fusion. In: Proceedings of the 13th international conference on information fusion (Fusion 2010), Th3.2.2

Sambhoos K, Llinas J, Little E (2008) Graphical methods for real-time fusion and estimation with soft message data. In: Proceedings of the 11th international conference on information fusion (Fusion 2008). ISIF, pp 1–8

Schlegel DR, Shapiro SC (2012) Visually interacting with a knowledge base using frames, logic, and propositional graphs. In: Croitoru M, Rudolph S, Wilson N, Howse J, Corby O (eds) Graph structures for knowledge representation and reasoning. Lecture notes in artificial intelligence, vol 7205. Springer, Berlin, pp 188–207

Shapiro SC (2000) An introduction to SNePS 3. In: Ganter B, Mineau GW (eds) Conceptual structures: logical, linguistic, and computational issues. Lecture notes in artificial intelligence, vol 1867. Springer, Berlin, pp 510–524

Shapiro SC, Rapaport WJ (1992) The SNePS family. Comput Math Appl 23(2–5):243–275 [Reprinted in Lehmann (1992 pp. 243–275)]

Shapiro SC, Schlegel DR (2013) Natural language understanding for soft information fusion. In: Proceedings of the 16th international conference on information fusion (Fusion 2013) ISIF, 9 pp. (unpaginated)

University of Colorado (2012) Unified Verb Index. http://verbs.colorado.edu/verb-index/

Chapter 3
Cognitive Aspects of Higher Level Fusion

Dale A. Lambert and Kerry Trentelman

Abstract A commander's situation awareness is critical to his or her decision making in a crisis but the ability for a commander to form that situation awareness is often overestimated. Automated data fusion offers a means of supporting a commander's situation awareness, with automated higher level fusion supporting the higher level functions of comprehension and projection. This chapter outlines a software-implemented psychological model that allows a machine to perform comprehension and projection, and interact with a commander.

3.1 Situation Awareness and Data Fusion

3.1.1 Situation Awareness

Crisis management systems typically utilize an assortment of people and equipment to aid a commander's situation awareness of a crisis so that the commander can then decide the course of action that will be taken. Within this framework there is a tendency to assume that: the commander receiving more detailed information is better; the commander receiving more reliable information is better; the commander having more direct control is better; and the commander having more communication is better. However, experiments conducted with teams operating with a bushfire simulator demonstrated that all four of these assumptions may break down (Omodei et al. 2004).

> More detailed information can lead individuals to become information overloaded, with their effectiveness suffering because they fail to realize that they are overloaded. More reliable information can result in less effectiveness when individuals are informed or observe that some information sources are only partially reliable, and as a consequence, pay insufficient attention to the reliable information from those sources. More direct control can be counter productive when individuals seek greater control while they have insufficient situation awareness to support it. More communication impacts negatively when insufficient situation awareness results from communication bottlenecks (Lambert 2009).

D.A. Lambert (✉) • K. Trentelman
Defence Science and Technology Group, Adelaide, SA, Australia
e-mail: Dale.Lambert@dsto.defence.gov.au; Kerry.Trentelman@dsto.defence.gov.au

G. Rogova, P. Scott (eds.), *Fusion Methodologies in Crisis Management*,
DOI 10.1007/978-3-319-22527-2_3

Crisis commanders exhibit the normal human fallibilities, which are often significantly underestimated. An analysis of aviation accidents found that 80–85% of all aviation accidents are attributed to human error (Endsley 1999), highlighting Reason's (Reason 1990) observation that while humans are often heroes, they can also be hazards. Furthermore,

> Problems with SA were found to be the leading causal factor in a review of military mishaps. [and] In a study of accidents among major airlines, 88% of those involving human error could be attributed to problems with situation awareness as opposed to problems with decision making or flights skills (Endsley 1999).

As a second example, Giompapa et al. (2006) observed that human operators tracking air tracks can competently cope with only about 6.8 tracks at any given time, which is suspiciously close to Miller's (1956) demonstrated human short-term memory capacity of 7 ± 2 items, and woefully short of expectations often placed on human operators.

Higher level machine fusion has the potential to substantially alleviate shortcomings in human situation awareness. The dominant understanding of situation awareness comes from Endsley,

> Situation awareness is the perception of the elements in the environment within a volume of time and space, the comprehension of their meaning, and the projection of their status in the near future (Endsley 1988).

As the psychological literature regards sensation as a precursor to perception, situation awareness can be thought of in terms of the four levels: Level 0: Sensation; Level 1: Perception; Level 2: Comprehension; and Level 3: Projection.

3.1.2 Data Fusion

The dominant understanding of Data Fusion comes from the Joint Directors of Laboratories (JDL) Model (e.g., White 1988). It allows for a machine-based assessment of the environment, based on the four levels: Level 0: Sub-Object Assessment; Level 1: Object Assessment; Level 2: Situation Assessment; and Level 3: Impact Assessment. The key observation is that the situation awareness and data fusion levels are fairly direct parallels (Lambert 2001). Thus,

> Situation awareness is the function of fusion performed by people, while machine fusion is the function of situation awareness performed by machines (Lambert 2012a).

This observation rejects the traditional division of human and machine based labor, by allowing both people and machines to operate and interact at Level 0 through to Level 3. Figure 3.1 illustrates the relationship (artwork courtesy of Michael Broughton). This mixed initiative approach offers scope for improved

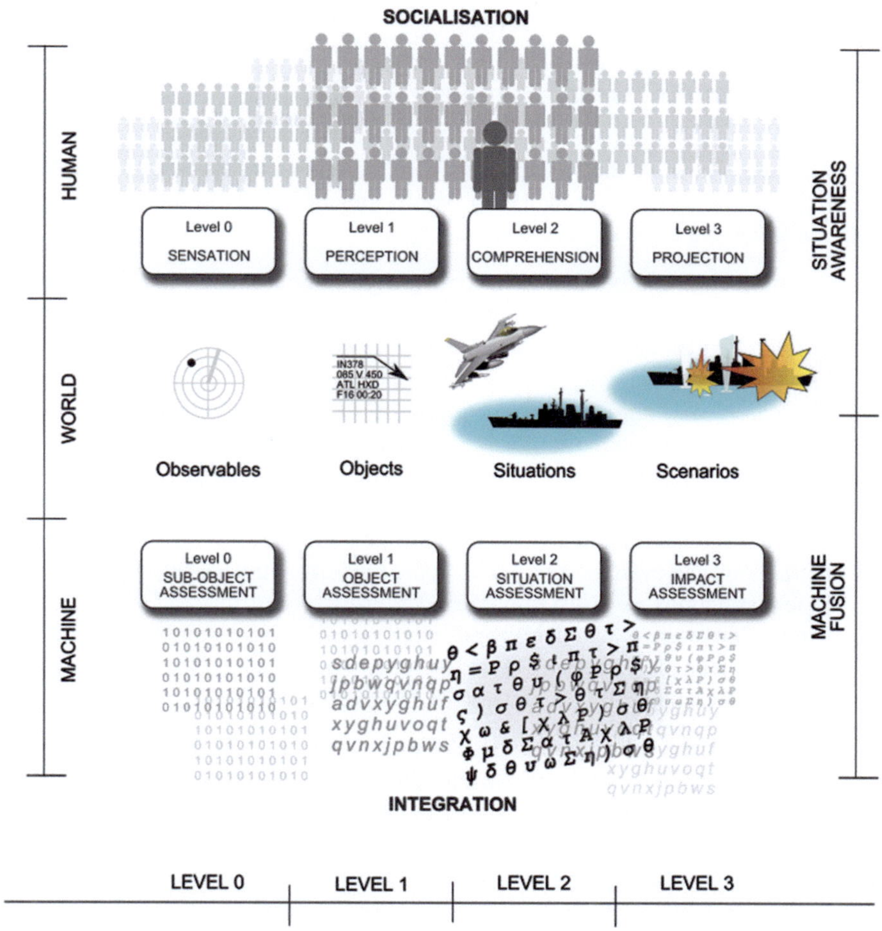

Fig. 3.1 Situation awareness and data fusion

situation awareness, by allowing respective human and machine strengths and weaknesses to be compensated by the complementary strengths and weaknesses of the other.

A distinction is often drawn within the data fusion community between lower level fusion and higher level fusion. Lower level fusion relates to Level 0 and Level 1 and so refers to machine-based sensation and perception. It is relatively mature. Higher level fusion relates to Level 2 and Level 3 and so refers to machine-based comprehension and projection. It is the more challenging. This chapter outlines the cognitive approach to higher level fusion initiated by the first author.

3.2 The Automation Principle

3.2.1 Computer Science Paradigm

Upon announcement of the so-called first-generation computers in a 1946 publication (reproduced as Burks et al. 1971), Von Neumann's team proposed a machine with the following three key elements.

(a) Computers were to be viewed as *logical devices* rather than *electrical devices*, and as a consequence, machine architecture was constructed to reflect an abstract logical structure. This included reference to *organic* concepts like *memory*.
(b) *Instructions and data for the machine were to be stored within the machine's memory*, both in numeric form. The computer could now accept several instructions at a time and carry them out without human intervention, by first storing them in memory. External switching was no longer necessary to perform each instruction and the computer could reference instructions, repeat instructions and even modify instructions, since from the hardware perspective, these instructions were just numerical data values.
(c) *Binary digits* were adopted as the preferred numerical representation. Efficiency could be gained if humans were prepared to accommodate the numerical representation naturally engendered by the machine's architecture.

Proposals (a) and (b) migrated the machine closer to the human by embedding human conceptualization into the machine. Proposal (c) migrated the human closer to the machine by requiring users to think in terms of binary digits.

Adaptation by both human and machine remained but machine adaptation came to dominate over time. Through the autonomy offered by the *stored program concept* in (b), the machine was now in command of the instructions it was to execute and these instructions were unambiguously designed. This led to the formulation of *machine languages* as *formal languages* for communicating instructions to the machine. More sophisticated programming languages were developed to represent more abstract aspects of human conceptualization. The progression of subsequent *higher level language structures* with *graphical and speech interfaces* is now history.

Each of the steps in this progression resulted from a revised form of the Automation Principle originally stated in MacLennan (1983).

> The Automation Principle stipulates that if any of the tasks undertaken by the user in communicating to a computer are mechanical, tedious, error prone or prevalent, then they should be automated within the computer and thereafter interfaced to as if they were primitive operations (Lambert 2009).

The Automation Principle invites machine adaptation by embedding human conceptualization within the machine and then treating those conceptualizations as a primitive interface thereafter. Assembly languages; interpreters; compilers; floating point arithmetic; higher level languages, and higher level language structures;

graphical displays and the desktop metaphor; and speech interfacing are familiar applications of the Automation Principle, which is the defining characteristic of the classical Computer Science paradigm.

Data fusion is equally a product of the Automation Principle. The invention of radar immediately prior to World War II is a familiar example of the automated sensation of objects, thereby alleviating the human user from needing to travel to within visual range of observables. Level 0 signal processing then alleviated the need for a human user having to analyze the return signals to identify features of the observable, by automating that conceptual activity. Level 1 tracking then alleviated the need for a human user having to analyze the detected features to identify the movement of objects, by automating that conceptual activity. Higher level fusion continues this progression. Level 2 situation fusion alleviates the need for a human user having to analyze the relations between detected objects with properties, by automating that conceptual activity. Level 3 scenario fusion alleviates the need for a human user having to analyze the consequences of relations between detected objects, by automating that conceptual activity. *The fusion paradigm expressed by the JDL model in fact arises from an application of the Automation Principle.*

3.2.2 Automation Limit

Continued practice of the Automation Principle logically terminates at the Automation Limit, where *the conceptualization embedded within machines resembles the conceptualization embedded within people.* Lambert (1999) suggested that the Automation Limit will have two consequences. The first consequence is that we *will seek to interface to these machines in a similar way to how we interface to people.* Machines will begin the have the look and feel of people. This prompted Lambert (1999) to promote *virtual advisers* as an appropriate situation awareness interface at the Automation Limit. Lambert established a team to implement virtual advisers (Taplin et al. 2001; Broughton et al. 2002). Those virtual advisers have matured significantly since that time. Figure 3.2 is a recent photograph in which Adam Saulwick is speaking with a virtual adviser in a controlled form of English to obtain situation awareness. The virtual adviser responds verbally in English while exhibiting the appropriate facial emotions as the accompanying synthetic battlespace is automatically manipulated to emphasize the points being made.

The second consequence noted by Lambert (1999) is that we *will seek to predict and explain the behavior of these machines in a similar way to how we predict and explain the behavior of people.* From a behaviorist standpoint, at the Automation Limit our machines become cognitive machines, since we will predict and explain their behavior by ascribing beliefs and volitions, together with a presumption of rationality that they will act to satisfy their volitions, given their beliefs (Dennett 1971). In doing so, these cognitive machines facilitate the transition from lower-level fusion to higher level fusion by enabling the machines to perform comprehension and projection and control the interfaces like the virtual adviser and synthetic battlespaces. This chapter focusses on this second consequence.

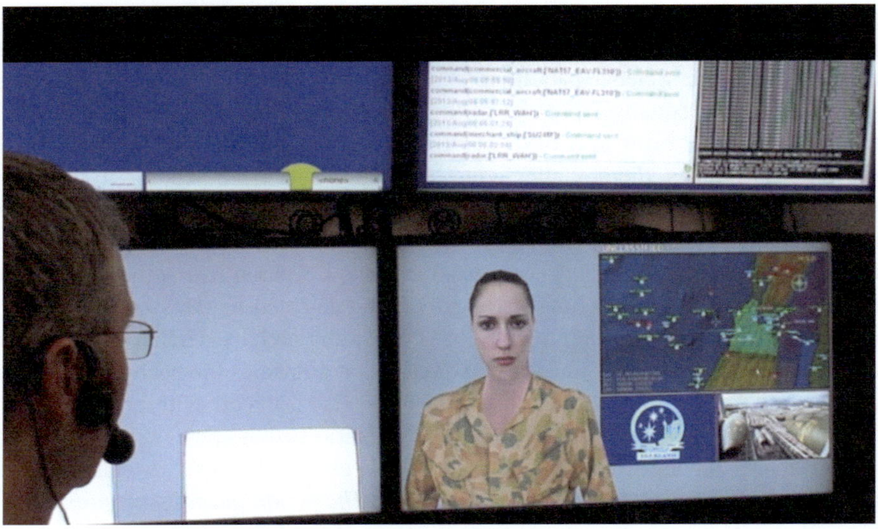

Fig. 3.2 Interaction with a virtual adviser

3.3 Machines with ATTITUDE

3.3.1 Attitude Programming

The first author proposed and implemented a functionalist model (e.g. Block 1980) for cognitive machines (Lambert 2009). More recently that ATTITUDE Cognitive Model has been extended to the ATTITUDE Psychological Model. The term "attitude" was adopted because the model was designed around 'Folk Psychology' propositional attitudes from the Philosophy of Mind discipline within Philosophy (e.g. Churchland 1988), while also providing an integrating framework for the ABC (Affect, Behaviour and Cognition) Model of Attitudes used by the Social Cognition (e.g., Rosenberg and Hovland 1960; Ostrom 1969) community within Psychology.

Folk Psychology ascribes mental states to cognitive individuals as beliefs, expectations, hopes, *et cetera*. These mental states are termed *propositional attitudes* because in everyday language they are represented by *propositional attitude expressions* of the form <subject> <attitude> that <proposition-expression>. Fred believes that the sky is blue; Tom expects that Mary will win lotto; Mary hopes that Tom is insightful are three sample propositional attitude expressions. In a propositional attitude expression: the subject, e.g., Fred, expresses which individual has the propositional attitude; the propositional expression, e.g., the sky is blue, expresses some assertion about the world; and the attitude, e.g., believes, expresses the kind of response the subject has toward the proposition.

Lambert (1999) proposed *propositional attitude instructions* as a programming construct. With subtle modification, the propositional attitude expression Fred believes that the sky is blue becomes an instruction Fred believe that the sky is blue that instructs cognitive machine Fred to believe that the sky is blue. When formalized as believe(Fred, blue(sky)) it then becomes a propositional attitude instruction that instructs Fred to believe that the sky is blue. Propositional attitude programming provides the basis for *attitude programming* at the cognitive level. Attitude programming is a logical step for both Computer Science and Machine Fusion in seeking to reach the Automation Limit. Cognitive level interfaces and cognitive programming constructs offer the same kinds of advantages that assembly languages offer over machine languages or that higher level languages offer over assembly languages.

3.3.2 ATTITUDE *Psychological Model*

Implementation of the ATTITUDE Psychological Model relies on two central ideas. The first, as noted above, is to apply propositional attitude expressions of the form <subject> <attitude> that <proposition-expression> as propositional attitude instructions of the form <attitude> (<subject>, <propositional-expression>). The second is to: associate each <subject> with a different psychologically coded individual; associate each <attitude> with a different type of memory within that psychological individual; and associate each <propositional-expression> with an expression stored within that memory type for that psychological individual. So for example, the propositional attitude instruction believe(Fred, blue(sky)) is seeking to have (computational) psychological individual Fred store the expression blue(sky) within its belief memory. The ATTITUDE Psychological Model operates through the interaction between expressions stored in different kinds of memory. Figure 3.3 presents an example of the ATTITUDE Psychological Model memory structure. It comprises: long-term memory; and the volition, interaction, affectation, awareness, long-term, and attention working memories, with the latter incorporating short-term memory.

Activity for an ATTITUDE agent begins when it *wakes*, resulting in the contents of its long-term memory being loaded into its corresponding long-term working memories. Long-term memory comprises both *assertional long-term memory* and *episodic long-term memory*, with the former housing *declarative* representations and the latter storing *procedural* representations. Assertional long-term memory is further divided into *semantic long-term memory*, which prescribes the meanings of Mephisto formal language expressions, and *epistemic long-term memory*, which expresses knowledge about the world using Mephisto formal language expressions. The conversion to long-term working memory allows the declarative meanings in semantic memory to become executable; the declarative knowledge in epistemic memory to become executable; and the cognitive routine knowledge in episodic memory to become executable. Mephisto (Lambert and Nowak 2008) is a formal semantic representation system that defines a range of concepts across the five layers: metaphysical (e.g., exists, time, space); environmental (e.g., air,

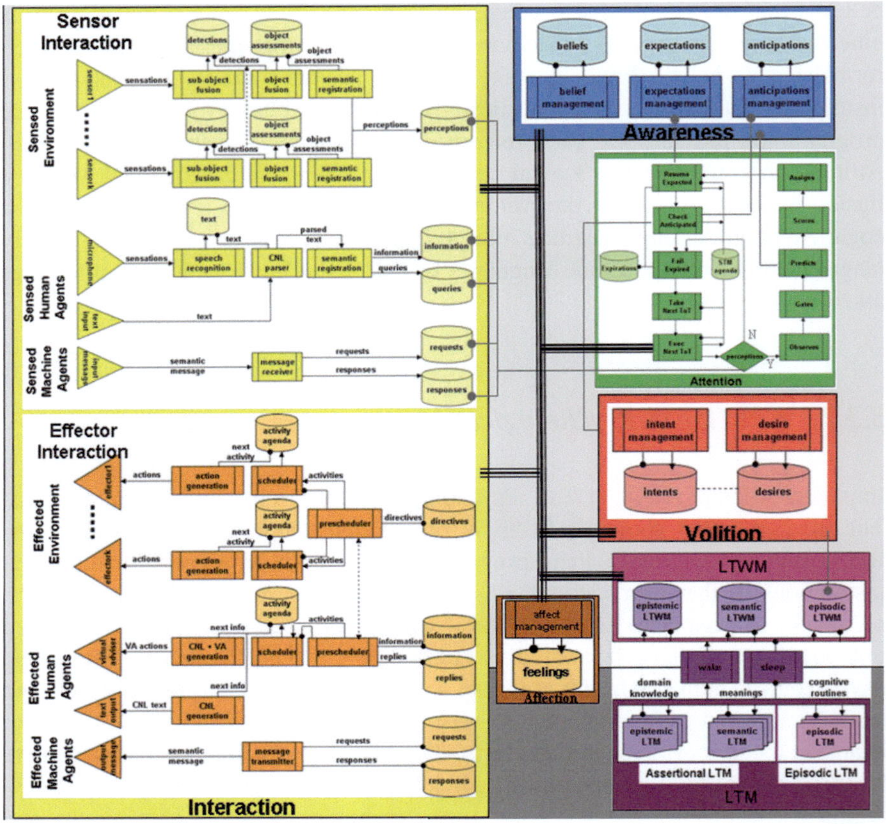

Fig. 3.3 ATTITUDE psychological model

water, land); functional (e.g., senses, moves, informs); psychological (e.g., believes, wants, performs); and social (e.g., agrees, commands, owns). The psychological machine conceptualizes the world through these Mephisto concepts. Mephisto is the machine's language of thought (Fodor 1987).

The waking of an ATTITUDE agent also invokes the *initialization* intention. *Intentions* are *independent volitions* held in volition working memory. Intentions are responsible for the autonomous activity of an ATTITUDE agent, with the agent undertaking an activity only if it has a volition to do so. *Desires* are *dependent volitions* and a tree structure of desires frequently forms through cascading goal decomposition of a root intention. Volitions are able to affect mental behavior within an ATTITUDE agent because whenever a new volition arises, the agent attempts to match that volition with the goals of cognitive routines in episodic working memory, and then executes the matching cognitive routines with that pattern-matched context. This mechanism achieves Dennett's (1971) aforementioned presumption of rationality.

Cognitive routines consist of *a Mephisto goal expression* and *a body expression* that specifies *a recipe for behavior as a transition network of propositional attitude instructions containing Mephisto propositional expressions*. Performance or execution of a cognitive routine involves the agent attempting to successfully transition from the starting propositional attitude instruction of the body through to one of the final propositional attitude instructions of the body, by performing each propositional attitude instruction encountered. Performing a propositional attitude instruction either succeeds or fails, with success or failure determining which propositional attitude instruction in the network is attempted next, and with performance of a propositional attitude instruction resulting in side effects in the other working memories of the agent. For example, performing a believe propositional attitude instruction results in a new belief in the awareness working memory while performing a desire propositional attitude instruction results in a new desire in the volition working memory. Each cognitive routine should achieve the routine's goal if performance to a final propositional attitude instruction of the body is accomplished. Cognitive routines typically interact with one another across multiple levels of abstraction, being clipped together dynamically like pattern-matched "lego blocks" in an attempt to achieve a successful outcome. The first author chose the term "routine" because it conflates the human factors notion of "routine behavior" with the old computer science notion of "subroutine." The initialize intention invokes an initialize routine, which when performed, establishes a number of intentions needed by the agent to properly function in its relevant world domain.

The attention working memory of an ATTITUDE agent is used to manage the *focus of attention* of the agent. An ATTITUDE agent is continually receiving external sensory updates and effecting external actions, while managing the performance of multiple cognitive routines seeking to satisfy a collection of both related and unrelated volitions. The performance of a cognitive routine in context gives rise to mental behavior represented as *trains of thought* stored in short-term memory within attention working memory. Short-term memory therefore stores the competing pending mental activities associated with the agent at any given time, with the highest priority train of thought selected and performed at each opportunity. Machine short-term memory is not bound by the human limitation of 7 ± 2 items (Miller 1956), though very large short-term memories will incur a priority scheduling penalty. Some trains of thought in short-term memory will also be in a non-executable suspended state awaiting satisfaction of a particular Mephisto expression. The attention component must also monitor suspended trains of thought and fail those that have *expired* by surviving beyond their due date.

After an agent wakes, interaction working memory comes into play. Interaction involves either sensing the world through sensors or effecting the world through effectors, and with ATTITUDE, sensors and effectors can be dynamically added on the fly at any time. Sensing involves (1) sensing *signals*, *images* or *text* in the environment; (2) sensing human user input through either *speech* or *text*; or (3) sensing *messages* from other ATTITUDE agents. Signal, image and text sources produce detections that are tracked to identify objects that are semantically registered as *perceptions expressed in Mephisto*. The ATTITUDE Psychological Model has operated with text and signal data sources. The predecessor ATTITUDE Cognitive Model (Lambert 2009) has operated with image sources. User interaction is currently implemented through either direct manipulation devices (Shneiderman 1982) or the Lexpresso-controlled natural language (Saulwick 2014). The latter employs a restricted subset of *English language* sentences to express both information and requests through either speech or text. Agent interaction allows ATTITUDE agents to issue *requests* and *responses* to one another directly *in Mephisto*. Effecting analogously involves: (1) effecting the environment through *actions* undertaken by effector controllers based on *Mephisto directives* to those controllers; (2) effecting the human user with *information* and *replies* through *Mephisto expression control* of interfaces like virtual advisers, virtual environments, *et cetera*; or (3) effecting *messages* to other ATTITUDE agents as *requests* or *responses*. The model allows each of these components to execute through different computational processes potentially on different machines in different parts of the world. This single *distributed cognition* approach was demonstrated at the 12th International Information Fusion Conference in Seattle in 2009 when an ATTITUDE agent with sensors, effectors and cognition running live on different computers, performed higher level fusion on track data for a 15-min demonstration, and reported through a virtual adviser embedded within a synthetic battlespace (Lambert 2012b).

The semantically registered Mephisto perceptions produced from sensory interaction with the external world are processed by the attention component in the first instance. This typically generates side effects in the awareness working memory. Awareness working memory holds beliefs, expectations and anticipations. When a propositional attitude instruction of the form believe(i, α) is performed with self-referential indexical i, a believes(α) expression is inserted into the belief working memory. This can subsequently be queried and so *facilitates an accessible record of what the agent believes*. When a propositional attitude instruction of the form expect(i, α, by(t)) is performed, an expects expression about α is inserted into the expectation working memory, a deadline of t is introduced into the attention expiry working memory for the associated train of thought, and execution of that train of thought is suspended until either the expectation succeeds and is removed through the perception of α before time t, or it expires and is removed by not perceiving α before time t. *This facilitates suspension of the execution of a cognitive routine until a particular contextual perception of the world occurs*. When a propositional attitude instruction of the form anticipate(i, α, with(β), priority(r)) is performed, it succeeds and an expression of anticipation of α with response to intend β is inserted into the anticipation working memory. Whenever an unexpected perception matching the anticipated α occurs, a contextual intention to satisfy β is then invoked. *This facilitates initiating contextual execution of cognitive routines based on perceptions of the world*.

The attention process matches incoming perceptions to waiting expectations and anticipations. Each perception in an incoming batch of perceptions is identified as an *observation* and triaged to one or more observation classes by a *gating* process.

For each perception the predicted expectations and anticipations associated with that observation class are identified. Each perception within the batch of perceptions associated with an observation class is *scored* against the waiting expectations and anticipations for that observation class to determine which perceptions best match with which expectations and anticipations. Based on matched scores, an *assignment* process assigns particular observations to particular expectations. With expectations, this facilitates a contextual pattern match resumption of the suspended trains of thought associated with those expectations. With anticipations, it assigns particular observations to particular anticipations. This results in contextual pattern matched intentions being placed in volition working memory, leading to the contextual invocation of cognitive routines.

Affectation working memory is the final working memory. It encodes the emotional state of an agent, describing an agent as joyful, contented, unhappy, frightened, *et cetera*. It also encodes how an agent feels about other agents, and actual or potential states of the world, e.g. angry with, jealous of, outraged by, concerned about. These formalisms have been influenced by Wierzbicka's work on emotions within the Natural Semantic Meta-language framework (Wierzbicka 1990) and can be used to ensure the right empathy is applied when interacting with other human or machine agents.

3.3.3 Attitude Fusion

Lambert proposed the State Transition Data Fusion (STDF) Model as a companion to the JDL Model (Lambert 2012a). The STDF contention is that each layer of the JDL Model operates with a different notion of state of the world, but that the fusion process is essentially the same, varying only contextually with that notion of state of the world. Figure 3.4 illustrates the STDF Model for Level 2 Situation Assessment.

Through a comparison of Figs. 3.1 and 3.3 it becomes apparent that the ATTITUDE Psychological Model provides an architecture for conducting higher level data fusion in accordance with the STDF Model. The various sensors attached to an ATTITUDE agent facilitate Level 0 Sub-Object Assessment and Level 1 Object assessment being performed in association with the sensory portions of interaction working memory. The transition to Level 2 Situation Assessment occurs when the outputs of Level 1 are semantically registered in symbolic form as perceptions. The executing trains of thought in attention working memory often make predictions about what states of the world might occur, and then store expectations and/or anticipations in awareness working memory to record how these states of the world might be perceived. Processes operating on attention working memory perform propositional association between these predicted observations and the incoming perceptions, potentially resulting in the initiation or resumption of some trains of thought. Trains of thought can be resumed because their expectations have been satisfied, in which case they typically record their updated awareness of the state of the world as beliefs in awareness working memory, or they can have past

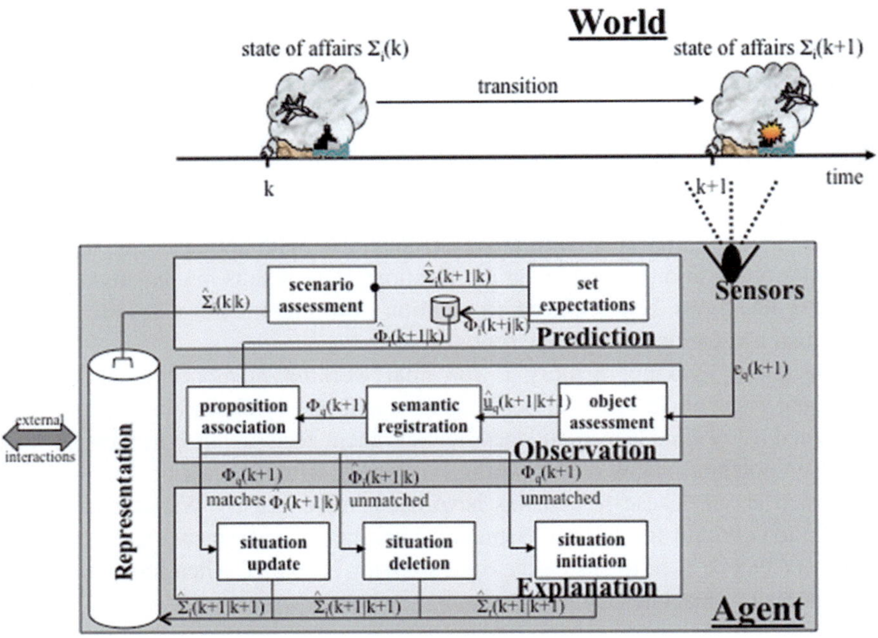

Fig. 3.4 STDF model for level 2 situation assessment

their expectation deadline, in which case they typically record their awareness of expectation failure as beliefs in awareness working memory. Newly matched anticipations produce new volitions in volition working memory, causing routines to be invoked as new trains of thought that typically record updated awareness of the state of the world as beliefs in awareness working memory. Each of these trains of thought may also choose to possibly effect the future state of the world by issuing directives to effectors in interaction working memory.

3.4 Crisis Management

Figure 3.2 comes from an implemented synthetic scenario in which a crisis erupts when a military aircraft makes a missile strike on a cargo ship in a busy air and maritime environment. An implementation of the ATTITUDE Psychological Model senses both textual information and live observations reporting the position and component velocities of objects in the environment in real time. In response the cognitive software performs higher level fusion to evaluate the situations to provide better situation awareness to the commander. The commander is able to verbally ask about the strike situation or any of the potentially affected objects in the region, and the system will generate an automated explanation of the relevant situations.

To achieve this three types of scenario background domain knowledge are required, with each corresponding to a type of long-term memory. The implemented model must understand the meaning of the Mephisto concepts. This is stored in *semantic* memory and comprises both axioms and definitions. For example, the metaphysical level concept of one spatial region S1 connecting to another region S2 is formally given by connects(S1, S2). It is defined by a collection of Boolean Connection Algebra (Stell 2000) axioms, the first of which is entered into semantic memory by tells_front([semantic], all([S], (exists(S) => connects(S, S)))). The social level concept of conflict is instead provided through a definition, which is entered into semantic memory by tells_define([semantic], conflict(@(X, T, SX), @(Y, T, SY), A, B), wants(@(X, T, SX), A) & wants(@(Y, T, SY), B) & (A => ~B)). The @-notation reflects the perdurant stance taken by Mephisto. Here @(X, T, S) is used to represent process X at time T at spatial region S. The second type of background domain knowledge is *epistemic* knowledge which is used to record declarative facts. For example the location of a sea port at Eaglevista is entered into epistemic memory by tells_front([epistemic], sea_port(eaglevista_seaport) & at(eaglevista_seaport, _, coordinate(degrees(56.33333), degrees(−026.16667), metres(0.0)))). The locations of air and sea lanes are also specified. Epistemic memory contains gating conditions. These are used to triage the incoming set of observations to various object classes (missiles, radars, military aircraft, military vessels, commercial aircraft, merchant ships, etc.) in order to process them more effectively. For example a missile gating condition filters out those objects that are believed to have been launched. Epistemic memory also contains definitions for the specific states of an object class. For example the missile object class has four states: approaches target, retreats from target, launched and strikes. The predicate missile_approaches_target is defined in terms of the missile's course with respect to the target and its distance from the target. The predicate launched is defined in terms of a newly detected missile appearing within the vicinity of an object of friendly or unknown allegiance. The predicate strikes is defined in terms of a missile being within some small distance of a target at a given time. The third type of background domain knowledge is *episodic* knowledge which is used to specify procedural processes in terms of propositional attitude instructions. These processes are called cognitive routines and are syntactically defined as routine(<goal>, <body>) where <body> is a recursively defined regular expression of propositional attitude instructions. A regular expression of propositional attitude instructions can be: (a) a propositional attitude instruction, (b) a sequence ^(α_1, α_2, ... α_k), (c) a repetition *(α), or (d) a selection + (α_1, α_2, ... α_k); where α_1, α_2, ... α_k and α are regular expressions. The goal of a cognitive routine should be satisfied when execution of its body succeeds. Note that:

(a) Each propositional attitude instruction either succeeds or fails.
(b) A sequence ^(α_1, α_2, ... α_k) succeeds if α_1 succeeds, then α_2 succeeds, ..., then α_k succeeds; and fails as soon as one of the α_i fails.
(c) A repetition *(α) repeatedly performs α until α fails, at which point *(α) succeeds. *(α) cannot fail.

(d) A selection $+(\alpha_1, \alpha_2, \ldots \alpha_k)$ fails if α_1 fails, then α_2 fails, ..., then α_k fails; and succeeds as soon as one of the α_i succeeds.

The following presents a very simple cognitive routine that is used for monitoring a missile approaching a target.

```
routine(missile_approaching_target(@(Missile, Missile_Time, Missile_Where),
                                    @(Target, Target_Time, Target_Where),
                                    Speed_Missile, Course_Missile),
^(believe_first(i, missile_approaches_target(
                        @(Missile, Missile_Time, Missile_Where),
                        @(Target, Target_Time, Target_Where),
                        Speed_Missile, Course_Missile)),
  (Number_Missed_Updates is 0),
  (Max_Missed_Updates is 3),
  *(
   ^((Number_Missed_Updates =< Max_Missed_Updates),
    now(Now),
    Expiry is Now + 5,
    +(^(expect(i, update(striking_missile,
                        missile_approaches_target(
                       @(Missile, New_Missile_Time, New_Missile_Where),
                       @(Target, New_Target_Time, New_Target_Where),
                       New_Speed_Missile, New_Course_Missile)),
                       by(Expiry)),
      act(i, updated_missile_approaches_target(
                      @(Missile, New_Missile_Time, New_Missile_Where),
                      @(Target, New_Target_Time, New_Target_Where),
                      New_Speed_Missile, New_Course_Missile))),
     (Number_Missed_Updates is Number_Missed_Updates + 1)),
    clear([Now, New_Missile_Time, New_Target_Time, New_Missile_Where,
          New_Target_Where, New_Speed_Missile, New_Course_Missile]))),
+(^((Number_Missed_Updates > Max_Missed_Updates),
   fail),
  succeed))).
```

During initialization, all object states are anticipated. Once an anticipation is met by a matching gated object, a corresponding cognitive routine is called. In the case of the missile object class, when a routine monitoring a particular missile terminates, another waiting anticipation should be met by that same missile. The missile approaching target routine is called when Missile is first detected approaching a target. Thus the initial step is to believe that Missile is in the missile approaches target state. Execution of the cognitive routine continually loops as long as: (1) Missile has not missed three consecutive updates within a currently set maximum update rate of 5s, and (2) Missile remains in the missile approaches target state. If either of these requirements fails, then the cognitive routine terminates. While Missile remains in the missile approaches target state, new temporal and spatial information is continually updated. This information is later used in automated situation reports. For example it can be reported that a particular missile approached a target for some length of time, and at the end of that time, the missile was at some coordinate. The following code shows how information accumulated by the cognitive routine can be converted into English language text which can be spoken by a virtual adviser.

```
english(missile_approaches_target(
                @(Missile,
                        (timestamp(Y1, M1, D1, H1, Mi1, S1),
                         timestamp(Y2, M2, D2, H2, Mi2, S2)),
                        (coordinate(_, _, _), coordinate(Lat2, Long2, _))),
                @(Target, _, _), _, _), E) :-
        atom_codes(Missile, E_Missile),
        atom_codes(Target, E_Target),
        subtract_time(timestamp(Y2, M2, D2, H2, Mi2, S2),
                      timestamp(Y1, M1, D1, H1, Mi1, S1),
                      Timeperiod),
        english(Timeperiod, E_Duration),
        append([E_Missile, " approached ",
                E_Target, " for ", E_Duration, ". "], E1),
        english(coordinate(Lat2, Long2), E_Where),
        append([E1, "At the end of that time it was at ", E_Where, ". "], E), !.
```

It is also possible to provide automated alerts that are triggered when an object is detected in a particular state. For example when a missile is in a strikes state, or when a commercial aircraft veers off an air lane.

3.5 Conclusion

When applied collectively semantic, epistemic and episodic background domain knowledge combine with the implemented ATTITUDE Psychological Model to generate a higher level comprehension of the environment that can then be shared with a commander to improve his or her situation awareness. The commander can interact with the implemented system by asking questions verbally, and the system will generate automated explanations of relevant situations. The system will also trigger alerts in the event of a crisis.

Acknowledgments The authors would like to acknowledge the contributions of Adam Saulwick, Steven Wark, Marcin Nowina-Krowicki, and Michael Broughton.

References

Block NJ (1980) Introduction: what is functionalism? Readings. In: Block NJ (ed) Philosophy of psychology, vol 1. Harvard University Press, Cambridge, MA, pp 171–184

Broughton M, Carr O, Estival D, Taplin P, Wark S, Lambert DA. Conversing with Franco, Focal's virtual adviser, Conversational Character Workshop, Human Factors 2002, Melbourne, Australia, 2002

Burks AW, Goldstine HH, von Neuman J (1971) Preliminary discussion of the logical design of an electronic computing instrument. In: Bell CG, Newell A (eds) Computer structures: readings and examples. McGraw-Hill Book Company, New York, NY, pp 92–119

Churchland PM (1988) Matter and consciousness: a contemporary introduction to the philosophy of mind, a Bradford book. MIT Press, Cambridge

Dennett DC (1971) Intentional systems. In: Dennett DC (ed) Brainstorms: philosophical essays on mind and psychology. The Harvester Press Limited, Brighton, Sussex, pp 3–22

Endsley MR (1988) Design and evaluation for situation awareness enhancement. In: Proceedings of the Human Factors Society 32nd Annual Meeting Santa Monica, CA: Human Factors Society, pp 97–101

Endsley MR (1999) Situation awareness and human error: designing to support human performance. In: Proceedings of the High Consequence Systems Surety Conference, Albuquerque, NM

Fodor JA (1987) Psychosemantics: the problem of meaning in the philosophy of mind. The MIT Press, Cambridge, MA

Giompapa S, Farina A, Gini F, Graziano A, Di Stefano R (2006) A model for a human decision-maker in a command and control radar system: surveillance tracking of multiple targets. In: Proceedings of the 9th International Conference on Multisource Information fusion, Florence, Italy

Lambert DA (1999) Advisers with attitude for situation awareness. In: Proceedings of the 1999 Workshop on Defense Applications of Signal Processing, LaSalle, IL, pp. 113–118

Lambert DA (2001) Situations for situation awareness. In: Proceedings of the 4th International Conference on Information Fusion, Montreal, QC

Lambert DA (2009) A blueprint for higher level fusion systems. J Inf Fusion 10:6–24

Lambert DA (2012a) The state transition data fusion model, Ch 3. In: Blasch E, Bossé É, Lambert DA (eds) High-level information fusion management and system design. Boston, MA, Artech House, pp 33–79

Lambert DA (2012b) Coalition approach to high-level information fusion, Ch 12. In: Blasch E, Bossé É, Lambert DA (eds) High-level information fusion management and system design. Boston, MA, Artech House, pp 251–277

Lambert DA, Nowak C (2008) The Mephisto conceptual framework. DSTO technical report TR-2162, Department of Defence

MacLennan BJ (1983) Principles of programming languages: design, evaluation and implementation. CBS College Publishing, New York, NY

Miller GA (1956) The magical number seven, plus or minus two: some limits on our capacity for processing information. Psychol Rev 63:81–97

Omodei MM, Wearing AJ, McLennan J, Elliott GC, Clancy JM (2004) More is better? Problems of self regulation in naturalistic decision making settings. In: Proceedings of the 5th Naturalistic Decision Making Conference, Stockholm, Sweden

Ostrom T (1969) The relationship between the affective, behavioural and cognitive components of attitude. J Exp Soc Psychol 5(1):12

Reason J (1990) Human error. Cambridge University Press, Cambridge

Rosenberg MJ, Hovland CI (1960) Cognitive, affective, and behavioral components of attitudes. In: Hovland CI, Rosenberg MJ (eds) Attitude organization and change. Yale University Press, New Haven, CT

Saulwick A (2014) Lexpresso: a controlled natural language. In: Proceedings of CNL 2014, 4th International Workshop on Controlled Natural Language, Galway, Ireland, August 20–22, 2014

Shneiderman B (1982) The future of interactive systems and the emergence of direct manipulation. Behav Inf Technol 1(3):237–256

Stell JG (2000) Boolean connection algebras: a new approach to the region-connection calculus. Artif Intell 122:111–136

Taplin P, Fox G, Coleman M, Wark S, Lambert DA (2001). Situation awareness using a virtual adviser. In: Talking Head Workshop, OzCHI 2001, Fremantle, WA, Australia

White FE (1988) A model for data fusion. In: Proceedings of the 1st National Symposium on Sensor Fusion, Vol. 2, Chicago, pp. 143–158

Wierzbicka A (1990) Emotions across languages and cultures. Cambridge University Press, Cambridge

Chapter 4
Information Quality in Information Fusion and Decision Making with Applications to Crisis Management

Galina L. Rogova

Abstract Designing fusion systems for decision support in complex dynamic situations such as crises requires fusion of a large amount of multimedia and multispectral information coming from geographically distributed sources to produce estimates about objects and gain knowledge of the entire domain of interest. Information to be fused and made sense of includes but is not limited to data obtained from physical sensors, surveillance reports, human intelligence reports, operational information, and information obtained from social medial, opportunistic sensors and traditional open sources (internet, radio, TV, etc.). Successful processing of this information may also demand information sharing and dissemination, and action cooperation of multiple stakeholders. Decision making in such environment calls for designing a fusion-based human–machine system characterizing constant information exchange between all nodes of the processing. The quality of decision making strongly depends on the success of being aware of, and compensating for, insufficient information quality at each step of information exchange. Designing the methods of representing and incorporating information quality into such processing is a relatively new and a rather difficult problem. The chapter discusses major challenges and suggests some approaches to address this problem.

4.1 Introduction

Decision making in the complex dynamic environments such as crises requires tracking and monitoring natural and man-made activities for building current and predicted situational picture. Building such situational picture involves gathering and fusing a large amount of heterogeneous multimedia and multispectral

---

This chapter is an extended and revised version of (Rogova and Bosse 2010)

G.L. Rogova (✉)
State University of New York at Buffalo, Buffalo, NY, USA
e-mail: rogova@buffalo.edu

G. Rogova, P. Scott (eds.), *Fusion Methodologies in Crisis Management*,
DOI 10.1007/978-3-319-22527-2_4

information of variable quality and data rates coming from geographically distributed sources to gain knowledge of the entire domain. Data and information[1] to be processed and made sense of includes but is not limited to data obtained from physical sensors (infrared imagers, radars, chemical, etc.), human intelligence reports, operational information, and information obtained from open sources (traditional such as newspapers, radio, TV as well as social media such as Twitter, Facebook, Instagram). Successful processing of this information also demands information sharing and dissemination, and action cooperation of multiple stakeholders such as different national and international authorities, law enforcement and regulatory agencies, and commercial companies.

There are multiple reasons for information deficiency in the crisis environment. Physical sensors can be unreliable or improperly used, fusion models can be imperfect, opportunistic human sensors and human decision makers can have variable expertise, their mental model imperfect and their decisions are usually affected by cognitive biases. Another source of information deficiency may be imperfection of domain knowledge and statistical information about the crisis, which is unavoidable due to the fact that characteristics of crises even of the same type are rarely exactly the same. There is an inevitable delay in data transmission in the dynamic crisis environment, and information entering the system can be obsolete. Information obtained from social networks and opportunistic sensors has additional deficiency stemming from misinformation, possible malicious intent and the pedigree problem. In addition, the lack of proper consideration of context may result in using inadequate or erroneous domain knowledge or inappropriate models and parameters for quality assessment. Such complex disaster environments call for an integrated fusion-based human–machine system, in which some processes are best executed automatically while for others the judgment and guidance of human experts and end-users are critical.

There have been many definitions of information fusion in the literature. For example, Wald in (2002) defined it as "...a formal framework, in which are expressed the means and tools for the alliance of data originating from different sources. It aims at gaining information of greater quality." Varshney (1997) described it as "...a synergistic combination of information, which is performed in order to better understand the phenomenon under consideration, to achieve improved accuracy and obtain information of better quality." Thus obtaining better quality of information as the result of fusion is one of the goals of designing fusion systems. Although the subject of information and data quality has been receiving significant attention in the recent years in many areas including communication, business processes, personal computing, health care, and databases (see for example Lee et al. 2006; Helfert 2001a; Madnick et al. 2009) the problem of information quality in the fusion-based human–machine systems for decision making has just recently begun to attract attention. The main body of the literature on information

[1]In the rest of this chapter, data and information will be used interchangeably.

fusion concerns with building an adequate uncertainty model without paying much attention to the problem of representing and incorporating other quality characteristics into fusion processes.

There are various fusion models considered in the literature. The most used one is the five level JDL[2] data model (White 1988) comprising interrelated lower level fusion (Levels 0/1), and higher level fusion (Levels 2–4); and its extension (Blasch and Plano 2003), which includes Level 5 addressing the issues associated with the human components of the fusion process. Information in these models is not processed sequentially but there is a feedback across and within levels. The successful processing of this information requires being aware of, and compensating for insufficient information quality at each step of information exchange. At the same time, good quality of input information does not, of course, guarantee sufficient quality of the system output and therefore insufficient quality of the information may build-up from one sub process to the other.

Level 4 is called sensor and process refinement, which is considered here as a quality control level. Figure 4.1 shows major points of information exchange and quality considerations in the fusion-based system.

In Fig. 4.1, Q_t^{scene} denotes the quality of situation to be assessed; Q_t^{s}—the quality of information sources, Q_t^{sm}—quality of source model output, Q_t^{fp}—quality of information fusion processes at each level, Q_t^{ip}—quality of presenting information to the user, $Q_t^{\mathrm{ca}}, Q_t^{\mathrm{cc}}$—quality of context estimation and context characteristics, respectively, and $Q_{t,}^{\mathrm{d}} Q_t^{\mathrm{a}}$—quality of decisions and actions, respectively.

It is important to note that the quality assessment and control depends on context and its quality. *Quality of Context* is defined in Rogova et al. (2013) as "any information describing the quality of information that is used as context information." Similar to information utilized in the fusion processes, the information defining a context can be obtained from available databases, observations, the result of sensor fusion, received reports, mining of available information sources (e.g., traditional and social media), or from various levels of information fusion. Of course, the quality of any such information as well as the inference process for obtaining it could be insufficient for a particular use: it might be uncertain, unreliable, irrelevant or conflicting. Knowledge of the quality of this information and its effect on the quality of context characterization can improve contextual knowledge. At the same time, knowledge about a current context can improve the quality of observation and fusion results. Thus there are two interrelated problems concerning both information and context quality: imperfect information used in context estimation and discovery negatively impacts context quality while imperfect context characterization adversely affects the characterization of quality of information used in fusion as well as the fusion results. This interrelationship represents one of the challenges of modeling and evaluating context quality and of

[2]JDL: Joint Directors of Laboratories, a US DoD government committee overseeing US defense technology R&D; the Data Fusion Group of the JDL created the original data fusion 4 level model.

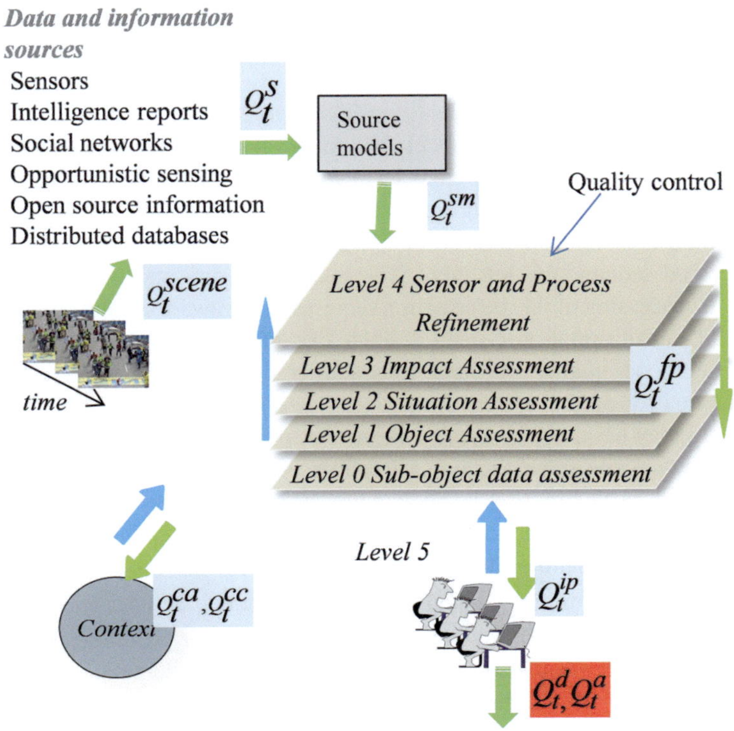

Fig. 4.1 Information flow in a fusion-based system

using context in defining the quality of information used in fusion and the quality of fusion process results.

In order to monitor and control information quality we need to know:

- Ontology of quality characteristics.
- How to assess information quality of incoming heterogeneous data as well as the results of processes and information produced by users (where do the numbers come from?)
- How to evaluate usability of information?
- How to combine quality characteristics into a single quality measure?
- How to evaluate the quality of the quality assessment procedures and results?
- How to compensate for various information deficiencies?
- How do quality and its characteristics depend on context?
- How does subjectivity, i.e., user biases, affect information quality?

The reminder of this chapter is an effort to establish a conceptual framework in which some of these questions may be addressed.

4.2 Information Quality: Definitions and Ontology

There are several definitions of information quality available in the literature:

- "Quality is the totality of characteristics of an entity that bear on its ability to satisfy stated and implied needs" (Standard 8402 1986).
- "Quality is the degree, to which information has content, form, and time characteristics, which give it value to specific end users" (O'Brien and Marakas 2005).
- "Quality is the degree, to which information is meeting user needs according to external, subjective user perceptions" (Wang and Strong 1996).
- "Quality is fitness for use" (Juran and Godfrey 1988).

While having different emphases, all of these definitions point to the fact that information quality is a "user-centric" notion and needs to be measured in terms of the potential or actual value for the users. In the human–system context "users" can be either humans or automated agents and models. However, assessment of information quality as "fitness for use," is based on the "objective" characteristics of information representing inherent properties of information (Fig. 4.2). Here information quality representing "fitness of use" will be defined as *subjective quality. Subjective quality* represents the level of satisfaction of users and process designers in relation to their specific goals, objective, and functions in a specific context. The inherent information characteristics (information about information) considered independently from information users and context, which gives value to subjective quality, will be defined in this chapter as *objective quality* or *meta-data*.

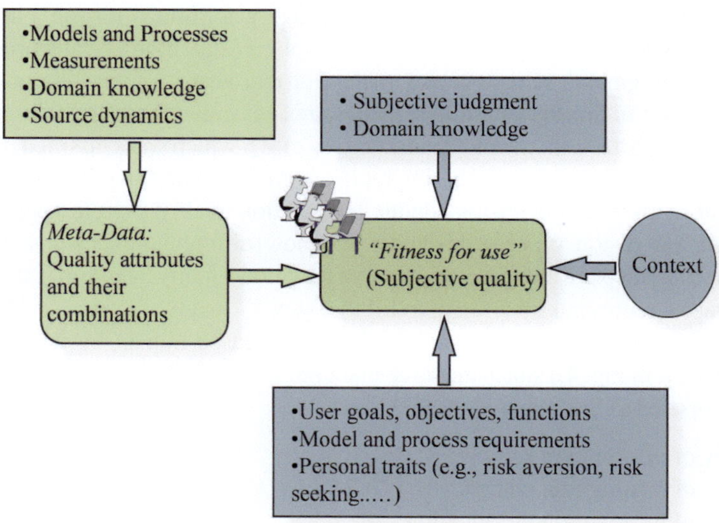

Fig. 4.2 Meta-data and "Fitness for use"

This consideration of two different types of quality is similar to one considered for sensor data in Bisdikian et al. (2009) where subjective information quality is referred to as information quality and objective one—volume of information. Meta-data is represented and measured by its attributes since "without clearly defined attributes and their relationships, we are not just unable to assess Information Quality (IQ); we may be unaware of the problem" (Bovee et al. 2003). Meta-data can be obtained as model and process results, domain knowledge, learning, level of agreement between sources, source dynamic, or direct measurements. It is important to notice that meta-data attributes and the method of their evaluation can differ for hard data produced by physical sensors and numerical models, and soft data coming from human sources.

Due to the fact that all users, whether human, automatic, or hybrid processes, have different data and information requirements, the set of attributes considered and the level of quality considered satisfactory vary with the user's perspective, the type of the models, algorithms, and processes comprising the system. Therefore the general ontology designed to identify possible attributes and relations between them for a human–machine integrated system will require instantiation in every particular case.

There have been multiple information quality ontologies, identifying quality attributes, classifying them into broad categories and relations. In Wang and Strong (1996) quality attributes are obtained by processing the result of a two-stage data consumer survey (so-called empirical approach well established in the marketing disciplines) to capture data consumers' perspectives on data quality. Four classes of quality dimensions (intrinsic, contextual, representational, and accessibility) were identified as most important for data consumers in industry and government. The model in Bovee et al. (2003) is "closely parallel" to Wang and Strong (1996) and considers four attributes of quality as "fitness for use": integrity, accessibility, interpretability, and relevance. In Helfert (2001b) three categories were enumerated: pragmatic, semantic, and syntax for measuring information quality in data warehouse. In Smets (1997) representation of information quality is limited by information imperfection, a limited subcategory of IQ, which was classified into two general categories: uncertainty and imprecision. At the same time there is no clear understanding of what dimensions define information quality from the perspective of information fusion process designers and how different dimensions defining information quality are interrelated. The information quality ontology introduced in this chapter represents an attempt to fill this gap.

The type of information exchange in the fusion-based human–system environment as shown in Fig. 4.1 notes the three main interrelated categories of information quality proposed in this paper (Fig. 4.3):

1. Quality of information source.
2. Quality of information content.
3. Quality of information presentation.

As it can be seen in Fig. 4.1 each node of the fusion process (machine or human) can represent an information source and information recipient at the same time,

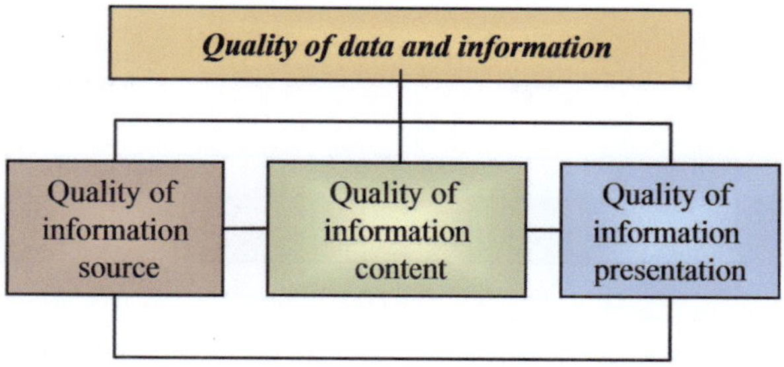

Fig. 4.3 The highest level of information quality ontology

and quality of information/decision/action of such node as a source of information transmitted to the next node depends on the quality of the information and the quality of the process represented by this node.

Quality of information presentation is related to the human nodes. The importance of considering information presentation as a component of the overall quality stems from the fact that the quality of decision and actions produced by this node not only depends on incoming information and human mental model but also the way information is presented.

It is important to notice that these attributes cannot be considered in isolation, for example, quality of a fusion process depends on the quality of the information source producing input information for this process. Each of these quality attributes is further decomposed and each subattribute is discussed in detail in the next section.

4.2.1 Quality of Information Content

There are five major attributes of the quality of information content: *accessibility*, *availability*, *relevance*, *timeliness*, *and integrity* or lack of imperfection, which represents complex notion characterized by many other attributes (Fig. 4.4).

Accessibility and *availability* refer to the ability of users to access this information effectively. Accessibility is related to the cost of obtaining this information, which can be measured by time required for accessing this information. While the cost of obtaining information represents an intrinsic quality of information, its “fitness of use” depends on specific user constraint and specific context. Users have to compare the cost of obtaining this information with the benefits of its utilization. For example, accessibility of a particular piece of information can be considered good if the time required for obtaining this information is much smaller than the time available for making decisions based on this information. *Availability* is an important characteristic, which has a binary value since information can be either available or not. If availability is 0, all other attributes are irrelevant. Both availability and accessibility are time dependent since if information is

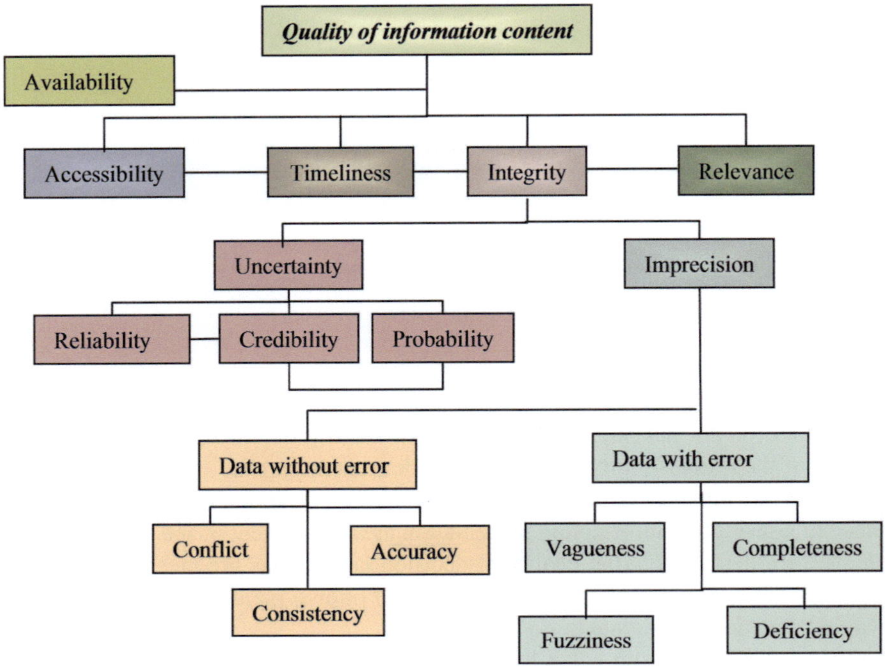

Fig. 4.4 Ontology of quality of information content

unavailable now to an agent it does not mean it will always be unavailable to this or all other agents.

Timeliness is a subjective attribute of information and as an attribute of the content of information is different form *timeliness* of information presentation and can be measured by utility of the information under consideration at the time it becomes available. Timeliness is especially important in a highly dynamic crisis environment when information can quickly become obsolete.

Another important attribute is relevance, which determines which information has to be incorporated into fusion process at a particular time. For example, information about containers arriving in a harbor is not *relevant* when we are evaluating threat presented by a small boat approaching a cruise ship. At the same time, if later we obtain information that the possible attack of a small boat may be a part of a coordinated attack, information about containers may become relevant. While utilizing relevant information can potentially improve the result of any fusion node providing that information is reliable, utilization of irrelevant information can hamper the outcome of the fusion processes. Relevance is defined by many authors as relation to the matter at hand and therefore depends on context as well as goals and functions of decision makers. The dynamics of context, goals, and functions of the decision makers in the dynamic environment make *relevance* a temporal attribute. Thus irrelevant information can become relevant later or relevant information can become obsolete at a certain time (Tawfik and Neufeld 1996).

Defining the relevancy of information is becoming more and more important in the "fog of crisis" when a huge amount of highly dynamic information is produced, and available to be processed and exchanged. Incorporation of irrelevant data into fusion processes not only can skew the quality of the result of this process, and negatively affects the performance of the other fusion nodes utilizing this result but also tie possibly limited crisis response and computational resources. Relevance also depends on the quality of information sources (sensor reliability, truthfulness, expertise, reputation, etc.).

Integrity or lack of imperfection of the content of information is the most studied category of information quality (see for example Bovee et al. 2003; Smets 1997; Rogova and Nimier 2004; Bosc and Prade 1997; Smithson 1989; Jousselme et al. 2003). In the context of a human–system integrated environment imperfection will be defined as something that causes inadequacy or failure of decision making and/or actions. Motivated by Smets (1997) we consider here two major characteristics of imperfection: *uncertainty* and *imprecision*. Uncertainty "is partial knowledge of the true value of the data" and arises from either a lack of information or imperfection of both formal and cognitive models (Smets 1997; Jousselme et al. 2003). It can be either objective and represented by *probabilities* reflecting relative frequencies in repeatable experiments or subjective and represented by *credibility* (believability) describing information which is not completely trustworthy. Uncertainty is the most studied in information fusion component of imperfection (Bosc and Prade 1997; Smithson 1989; Jousselme et al. 2003; Krause and Clark 1993; Klir and Wierman 1999; Dragas 2013). A good review of various types of uncertainty is given in Jousselme et al. (2003).

Another uncertainty characteristic, *reliability* (see for example Rogova and Nimier 2004) can be defined in two different ways. It can be understood as relative stability of physical or human sensor, or model output considered from one time to another under the same environmental characteristics. It can also be understood as a measure of accuracy of *probability* and *credibility* and is usually represented by reliability coefficients, which measure adequacy of each belief model to the reality. Incorporation of reliability coefficients is important due to the fact that the majority of fusion operators presume that information sources are equally reliable, which is not always the case. Therefore it is necessary to account for variable information reliability to avoid decreasing in performance of fusion results.

Imprecision can be possessed by information either with or without error. Thus information without error can be approximate (lacking *accuracy*) or *conflicting* and *inconsistent*. Accuracy represents the degree, to which data corresponds to characteristics of objects or situations. Consistency of a piece of information is usually measured when it is compared with some background knowledge, e.g., databases or knowledge obtained or inferred earlier. *Consistency* of transient and background information is especially important for situation assessment in the dynamic crisis environment since it can lead to discovery of new and unexpected critical situations. *Conflict* assumes several pieces of parallel or temporal information contradicting each other and may occur if either these pieces of information have different reliability or they report on different objects or situations. Information with error can be *incomplete*, *deficient* (lacking important pieces, which may prevent its usage), vague (ambiguous), or fuzzy (not well defined).

Such characteristics of imprecision as *inaccuracy, fuzziness, and vagueness* are inherent to soft data that usually are expressed in natural language, which is ambiguous by its nature (Dragas 2013). Besides, errors in information conveyed by humans can be influenced by cognitive biases and or even intentional. The ontology of quality of information content adopted in this chapter and presented in Fig. 4.4 is inspired by the one introduced in Smets (1997).

There are multiple theories developed to deal with information *uncertainty* and *imprecision*. Thus, *uncertainty*, for example, can be represented and reasoned about by probability, Bayesian probability, belief, and interval probability theories; *vagueness* by rough sets and interval probabilities, *fuzziness* by possibility and fuzzy set theories, *conflict* by belief and possibility theories. Selection of one of these theories depends on many factors, for example, existence of prior probability, type of information (soft, hard, or both), whether the hypotheses about the state of environment under consideration are exhaustive, etc. Thus selection of these theories strictly depends on context. The Transferable Belief Model (TBM) (Smets 2002) is suggested here as the one of the most appropriate for the uncertain dynamic crisis context. The TBM is a two-level model, in which quantified beliefs in hypotheses about an object or state of the environment are represented and combined at the *credal* level while decisions are made based on probabilities obtained from the combined belief by the *pignistic* transformation at the *pignistic level*. Dempster–Shafer beliefs (Shafer 1976), probability and possibility (Dubois and Prade 1988) distributions can be expressed as belief structures represented in the framework of the TBM allowing representing both soft and hard information (Yager 2012). Beliefs are sub-additive, which permits for numerically expressing uncertainty and ignorance. Within the TBM, the unnormalized Dempster's rule can combine basic belief masses based on multiple pieces of evidence, and allows for incorporation of belief reliability. Moreover, the TBM works under the open world assumption, i.e., it does not assume that the set of hypotheses under consideration is exhaustive. It also permits to represent conflict. These properties of the TBM has been successfully exploited in information fusion in general and in the crisis context specifically (see for example Delmotte and Smets 2004; Rogova et al. 2006, 2013; Rogova 2014)

4.2.2 Quality of Information Sources

From the information quality point of view we consider two types of information sources: subjective and objective. Subjective sources such as human observers, intelligence agents, newspaper reporters, experts, and decision makers supply observations, subjective beliefs, hypotheses, and opinions about what they see or learn. These sources use subjective judgment to produce this information, quality

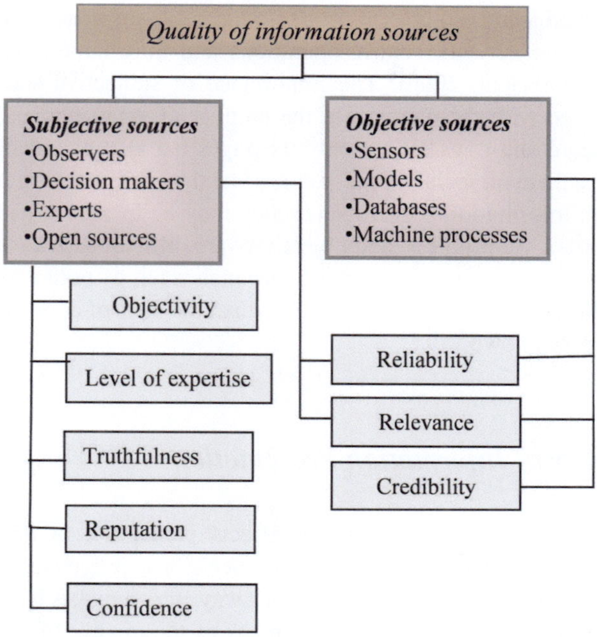

Fig. 4.5 Ontology of quality of information sources

of which is defined by their *level of expertise*, *reputation*, and *objectivity* as well as their intentions defined by the *truthfulness*. Information coming from subjective sources is usually represented in non-numeric form (so-called "soft information"). Quality of objective information sources such as sensors, models, automated processes is free from biases inherent to human judgment and depends only on how well sensors are calibrated and how adequate models are. As opposite to subjective sources, objective sources deliver information in numerical form ("hard information"). Ontology of quality of information sources is presented in Fig. 4.5.

The quality of objective sources comprises *relevance* as well as their *credibility and reliability.* An objective source can be *irrelevant* of it reports information about attributes of an object or events different from one under consideration or it does not work properly or is not designed to deal with the object or event under consideration. Reliability and credibility of objective information source measure the correctness of the choice of a source model or fusion process, their parameters affecting their performance, as well as existence or lack of errors in database. Reliability of objective source can depend on context. For example an optical sensor can be very reliable during a sunny day and not reliable at night.

Relevance and *reliability* can also measure the quality of subjective sources. Confidence of subjective sources is self-evaluated and is similar to credibility of objective sources. *Reliability* is related to quality of beliefs or plausibility assigned to this answer by a human agent or a model (Rogova and Nimier 2004). The notion of *reliability* of subjective sources is similar to the notion of *trust* used, for example, in the literature on network centric operations and information sharing (see for example Hexmoor et al. 2006). The importance of subjective sources in crises recently increased with the increase of the number of smart phones, which allow civilians to record and transmit not only messages but also pictures to emergency centers and post them on social media. At the same time it is very difficult to evaluate quality of such information sources since they may be in many cases anonymous and their expertise, truthfulness, and objectivity are difficult to evaluate. Subjective sources in crises also include trained professionals such as police and ambulance whose reliability, level of expertise are higher than the ones of opportunistic sensors and in many cases known.

4.2.3 Quality of Information Presentation

The quality of information presentation affects perception of decision makers and end users, and influences their actions, decisions, judgments and opinions. Information must be presented on time and in a way, which makes it understandable, complete, and easy to interpret. Thus attributes of the quality of presentation are related to when, which, and how information is presented. The ontology of quality of information presentation is shown in Fig. 4.6.

Timeliness is affected by two factors: whether the information is presented by the time it must be used and whether the presented dynamics information corresponds to the real world information obtained at the time of the presentation. *Completeness* is the ability of an information system to represent every meaningful state of the real world system by representing all necessary information related to this state (Wang and Wang 1996). An important related problem here is whether it is beneficial to present the value of information quality along with information itself (Wang and Wang 1996; Chengalur-Smith et al. 1999). For example, it was found that the benefits of incorporating meta-data depend on the experience of the decision maker (Fisher et al. 2003). Interpretability defines to what extent the users can understand information presented while understandability characterizes the level, to which the user is able to follow the logic of automatic processes producing

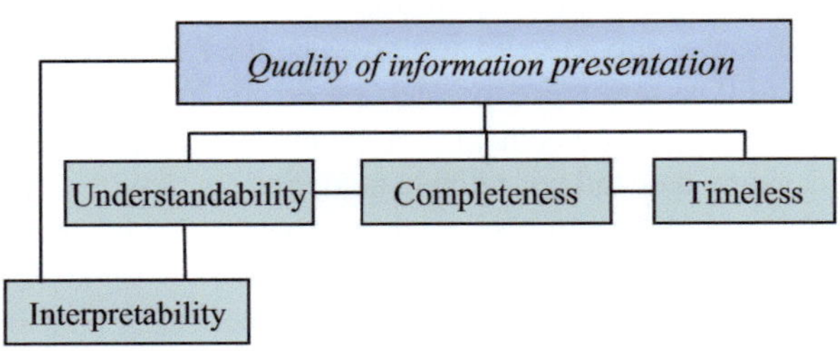

Fig. 4.6 Quality of information representation

this information. It is important to mention that the quality of such information attributes as interpretability and understandability depends on the level of training and expertise of the user and can be high for one use and poor for the others.

4.2.4 Higher Level Quality

Higher level quality measures how well the quality of information is assessed.[3] The needs for considering the higher level quality stem from the fact that the processes of assessing the value of the attributes of information quality have their limitations. These limitations come from wrong or poor quality of context used for quality estimation, imperfect domain knowledge, difficulty of finding an adequate quality model and its parameters, lack or insufficient amount of additional information that may be required for quality assessment, and subjectivity of quality evaluation performed by experts and end-users. Thus, for example, assessment of probability may need more observations or more time than may be available, and therefore the result is not completely reliable. In some cases point-values probabilities cannot be estimated and are presented as intervals, which limit *accuracy* of probability estimation. Another example is wrong context used for assessing sensor reliability. Moreover, assessment of overall information quality may require a combination of quality measures expressed in the framework of different belief theories, obtained from decision makers of different levels of expertise, or a combination of quality values represented in both numeric and symbolic form. The information quality ontology shown in Figs. 4.3, 4.4, 4.5, and 4.6 may also serve as the basis for an ontology for higher level quality.

Since comprehensive assessment of quality of information requires taking into consideration higher level quality, and therefore establishing relations between attributes of quality at different levels. It is important to mention that while quality of source is always a higher level quality for quality of information content, some attributes of information content or information source can serve as higher level quality for other attributes of information source or information content respectively. For example, reliability of a model (a quality attribute characterizing information source) producing plausibility of a hypothesis is a higher level quality for plausibility assessment (an attribute characterizing information content). We can also consider a certain level of belief that a source of information is reliable as a higher level quality for the assessment of credibility of this source.

---

[3]Usually this measure is referred to *uncertainty* only and is called "higher order uncertainty," which is treated without relation to the other quality attributes. Here we define this measure for any quality characteristic and consider it with relation to other attributes.

4.3 Assessing the Values of Information Quality

In order to compensate for insufficient information quality in a fusion-based human–machine system it is necessary to be able to assess the values of quality attributes and combine these values into an overall quality measure.

Information quality of meta-data represented by single attributes can be assessed by utilizing:

- A priori domain knowledge
- Measurements
- Outputs of models and processes
- Learning from past experience
- A level of agreement between sources
- Source dynamics, i.e., the degree of consistency with prior reports for human observers or prior readings for sensors.

Selection of a particular method for defining the value of information quality depends on the attribute under consideration and information available. Thus a priori domain knowledge about the context of the problem under consideration can provide quality values for many attributes, such as source reputation, expertise, level of training; information availability and accessibility, or sensor reliability and credibility, which will be difficult or even impossible to estimate otherwise. The methods utilizing contextual information have proven themselves very successful. In Rogova et al. (2013), context-dependent credibility, reliability, and timeliness have been introduced for sequential decision making for threat assessment. The context there was represented by the time-dependent distance between an observed target and a sensor, and a time-dependent threshold on credibility. A case study involving an evidential learning system for sequential decision making designed for a quantitative evaluation of the described method showed the benefits of context considerations. Context as a source of reliability estimation has been reported in many publications (see for example Fabre et al. 2001; Appriou 2001; Delmotte and Borne 1998; Jiang et al. 2008; Cholvy and Nimier 2003). The authors of Fabre et al. (2001) present two methods of defining source reliability based on contextual information. Both methods utilize subjective contextual information modeled by the theory of fuzzy events and used in connection with probability theory. In the Local Contextual Combination Method (LCCM), reliability of each source is defined by associating each sensor and each hypothesis to the context, and is computed as a Bayesian probability mass on the frame of discernment (defined by validity domain of each source). In another global contextual processing method (GCPM), inspired by the method suggested in Appriou (2001), reliability of one (or of several sensors) is computed as the probability of conjunction of the fuzzy subsets corresponding to each source for each contextual variable. In Delmotte and Borne (1998), expert knowledge is used to represent reliability by a possibility distribution defined on the sensor domain.

In many cases a priori domain knowledge does not directly contain values of quality attributes but includes certain information such as training examples, which can be exploited for learning. Measurements usually produce *accuracy*.

Models and processed outputs can serve as a source of assessing integrity of data obtained with these models (relevance, level of conflict, fuzziness, ambiguity, credibility). In general, information can be considered *relevant* for situation assessment if a change of its value affects the set of hypotheses about objects or situations under consideration, the levels of belief assigned to these hypotheses (e.g., reduces ignorance, strife or reduces/increases conflict), increase of information, or values of utilities of a set of possible courses of action. Subjective judgment of human experts is used when there is no a priori domain knowledge of, e. g., probability of non-repeatable events, or when it is important to know the quality of information from the subjective point of view of an expert, e.g., the level of understandability of information. As it was mentioned above, *consistency* measures the correspondence of observed or inferred information with expected information obtained from databases and domain knowledge. A method of defining relative reliability of model outputs expressed as beliefs and based on relative distance between bodies of evidence is proposed in Jiang et al. (2008). In Cholvy and Nimier (2003) a mathematical model of a discrete classification of credibility and reliability, and examples of a source reliability evaluation in a fusion process using symbolic data are presented.

While the value of any information quality attribute may be defined in various ways, the users and process designers are mostly interested in how good the quality is. Thus the subjective quality scores usually measure the level of satisfaction with the information under consideration in relation to the decision makers' goals and objectives or the purpose of the models or processes. Such level of satisfaction with objective quality attributes is based on the value of these attributes and can be expressed either in a linguistic form (e.g., good, fair, bad) or numerically by a number from the interval [0 1]. Some attributes, e.g., availability, have binary values (0 or 1) only. The level of satisfaction then can be represented and treated within a certain uncertainty theory (e.g., fuzzy, belief, or possibility).

One of the ways proposed here is to measure the level of satisfaction by measuring a particular quality attribute in relation to a certain context specific, and in many cases subjective, threshold (threshold satisfaction). The relation between a threshold and the value of the quality attribute can be transformed into an uncertainty measure within an uncertainty theory under consideration, e.g., beliefs or possibility. In some cases, the quality score of a particular attribute may be defined by comparing a different attribute with a context specific threshold. Thus, for example, if the reliability of source is lower than a certain threshold, information produced by this source may be considered irrelevant. In this case the degree of information relevance can be defined by the function of the distance between the threshold and the source reliability. In the human system environment relevance can be defined by a human-in-the-loop and be represented either by a number between 0 and 1 or in linguistic form (relevant, maybe relevant, irrelevant). It is important to mention that subjective quality scores should be considered along with the quality of presentation and the quality of the users and experts.

One of the most difficult for evaluation and at the same time very important in the disaster environment quality attribute is *trust*, which is defined here as measure of reliability and intent of a subjective information source. The problem is that most often the information is coming from open source information or opportunistic sensors such as the bystanders or family members whose identity, intentions, and source of information may not be known to the emergency responders. Another problem is that crisis response involves communications between multiple agencies and multiple crisis responders. They do not usually have a shared history and they may interact for the first time during the disaster, especially during large scale disasters. In such situation, responders cannot rely on their past experience to assess the reliability and reputation of others and have to dynamically establish a measure of trust (Kostoulas et al. 2006). Trust can be partially based on the valuation of the information that the source provides (Chopra and Haimson 2005); if any portion of that information is believed to be unlikely or inaccurate, the value of trust of the information sources can be reduced (Pichon et al. 2012). The believability of information usually evaluated by information recipient and is in many cases subjective.

Depending on the context and user requirements the overall quality may relate to a single attribute, a combination of several or all the attributes. Evaluation of the overall quality measure requires a method of combing different quality attributes into a combined measure. The subset of attributes considered depends on user goals, objectives, and functions as well as the purpose of model or process of interest. For example, a combination of credibility, reliability, and timeliness, and a method of incorporating their combination into sequential decision making threat assessment have been introduced in Rogova et al. (2013). In Rogova (2014) reliability and credibility were combined; relevance and truthfulness as an overall quality measure was considered in Uddin et al. (2012). Other combinations as completeness and understandability, completeness and timeliness can be also considered. In many cases, the overall quality represents a trade-off between different attributes, for example, between completeness and understandability, or credibility and timeliness. While designing an overall quality measure one has to take into account the hierarchy of quality attributes, their possible different importance under different contexts and for different users, and the quality of the values assigned to them (higher level quality). The problem of combing several quality attributes into an overall quality measure is similar to the problem of multi-criteria decision making, which requires comparing several alternatives based on the values of the criteria under consideration. Different quality attributes can be considered and different criteria while alternatives are different values of an overall quality measure.

One of the possible ways of representing such unified quality is to consider a *t-norm* or weighted average of the quality scores defined for each component while the weights are non-negative and their sum is normalized to unity. The weights representing a trade-off between the attributes under consideration are context specific and can be assigned by the users based of their needs and preferences.

A more general representation of an overall quality measure can be obtained by training a neural network, which can serve as a tool for transforming vector of individual quality scores into a subjective unified quality score. Another way of representing a unified quality of information is to consider utility of decisions or actions or *total information* based on this information. For example, *relevance and timeliness* can be measured by the increase of utility of decision or actions after the incorporation this piece of information.

If the quality attribute values are represented within the framework of an uncertainty theory their combination can be obtained by the combination rule defined in this theory. For example, it is possible to use conjunctive combination of reliable quality attributes expressed within the framework of the possibility theory. It is also possible to represent the unified quality measure as a belief network, in which quality of single attributes is expressed within a belief theory (Bovee et al. 2003). This method of attribute combination is especially appropriate when values of single attributes are heterogeneous, i.e., expressed in different forms e.g., point-value numbers, intervals, and linguistic values.

4.4 Quality Control

To ensure successful decision making and actions in the fusion base machine-human system environment it is important to deliver the right information at the right time. This requires not only awareness of the quality values but also employing quality control measures. Quality control should be implemented at each step of information exchange. Thus it is necessary to account for such quality attributes as reliability, credibility, relevance, timeliness when raw data and information enter the fusion processes and delivered to the human users; quality of models and fusion results (credibility, reliability, timeliness, etc.) when information is transferred between and within fusion levels; quality of judgment such as trust, confidence when information is transferred between humans, reliability, credibility, timeliness, etc. when information is transferred between human and automatic module of the system.

Quality control can include:

- Eliminating information of insufficient quality from consideration
- Incorporating information quality into models and processing by
  - Employing formal methods, e.g., methods of belief change to deal with inconsistency
  - Modifying the fusion processes to account for data and information quality.
- Modifying the data and information by compensating for its quality before processing or presenting to the users
- Delaying transmission of information to the next processing level or to decision makers until it has matured as a result of additional observations and/or computations improving its associated information quality
- Combination of strategies mentioned above.

Selection of a particular quality control method depends on the type of quality attributes under consideration. For example, *irrelevant* information is usually eliminated from consideration while *credibility* is often explicitly incorporated into models and processes.

Elimination of information of insufficient quality requires criterion to be used for deciding when the quality of information is not sufficient. In the majority of cases such criterion is related to "fitness of use" and depends on context, user objective goals, task, and cognitive model. One of such criteria is a threshold satisfaction when the value of a quality attribute or a combination of several attributes is compared with a certain dynamic context and user specific threshold. For example, one can consider a time varying threshold on credibility of hypotheses about the state of the environment or accessibility of information measured by the difference between the cost of getting this information and its utility.

Elimination of information can be applied to a subset of multiple sources in order to avoid or mitigate conflict and possible decrease of confidence of the fusion result or to deal with computational complexity to decrease error caused by approximation. Elimination of information can be also used to deal with pedigree problem to avoid double counting and decrease credibility of the result. The problem with pedigree of sources producing hard data (e.g., physical sensors in distributed fusion) received significant attention in the past and is usually approached by employing dynamic network analysis. Elimination of information and pedigree problem has become more complicated when more and more information is coming from social sensing, e.g., bystanders, Twitter, Facebook, etc. because of often unknown dependence between sources, rumors, lack of *prior* information about quality of sources, and possible outdated information. One of the methods of selection of least dependent sources is suggested in Uddin et al. (2012). This method exploits the notion of "close" and "non-close" source in a social graph to suppress rumors and chain messaging.

Methods of modifying information to make it totally or at least equally reliable before fusion processing often employ reliability coefficients, which measure the adequacy of the source of incoming information (both soft and hard), and model and fusion processes outputs to the reality. As it was described in section 4.2.1, there are two kinds of reliability coefficients: the one, which measures historical correctness of the source, e.g., fusion result and usually applied to "hard results" (correct/not correct). Another kind of reliability coefficients represents a second level of information quality for measurement accuracy, human agent confidence, or sensor model and process output credibility expressed within an uncertainty theory used. Reliability as a second-level uncertainty is important in fusion applications since the majority of the fusion operators are symmetrical and based on optimistic assumption that the sources are equally reliable. This assumption is often incorrect and ignoring reliability (absolute or relative) of information can lead to contra intuitive results (Haenni 2005). Information modification to account for its reliability before fusion is usually performed by discounting credibility and using the discounted values in the fusion model (see for example Smets 2002; Shafer 1976; Dubois and Prade 1988; Fabre et al. 2001; Appriou 2001; Delmotte and Borne 1998; Jiang et al. 2008;

Pichon et al. 2012; Schubert 2008). Reliability coefficients are also employed for modifying fusion processes. Methods of compensating for *reliability* are reviewed in for example Rogova and Nimier (2004). Modification of possibility fusion by incorporating reliability in the case when (1) a subset of sources is reliable but we do not know which one; (2) only an order of the reliabilities of the sources is known but no precise values are described in Dubois and Prade (1992).

Delaying the transferral of information demands a balance between timeliness and other characteristics of information quality, e.g., credibility, reliability, accessibility, relevance. This method of quality control is based on assumption that the quality may be improved over time when more information becomes available. At the same time, in the crisis situation waiting may result in unacceptable decision latency and unwanted, even catastrophic consequences. Therefore, the cost of waiting for additional information has to be justified by obtaining the results of better quality and ultimately better decisions and actions. Thus, the timeliness of the decision is defined by a context-dependent balance between the waiting time and improved information quality. There may be several criteria to consider for dealing with the trade-off between decision latency and improved decision outcome. One of these criteria is the Maximum Expected Utility Principle (von Neuman and Morgenstern 1947). According to this criterion, a new observation is justified if the difference between maximum expected utility with the new observation and without the new observation is greater than the cost of obtaining this new observation. The core difficulty of utilizing the Maximum Expected Utility Principle is a problem of finding the utility for each decision and utilizing this principle in the highly dynamic uncertain crisis environment. A more appropriate criterion in such environment is threshold satisfaction, which compares a predefined situation, user goals, and domain knowledge based threshold with the quality of information obtained from the sensors (physical or human) or fusion processes. It is important to note that when the users of this information are human the quality threshold also depends on their personal traits, for example, of their level of risk tolerance. The selection of quality attributes to be combined with timeliness and a method of measuring the value of quality depend on a situation. Sensor management (e.g., physical sensors or members of a crisis management team) should be employed when it is either impossible or very costly to get information of acceptable quality with current set of sensor or sensor configuration.

4.5 Conclusions

This chapter has discussed major challenges and some possible approaches addressing the problem of data and information quality in the fusion-based human–machine information environment. In particular, this chapter presents an ontology of quality of information and identifies potential methods of assessing the values of quality attributes, combining these values into an overall quality measure as well as possible approaches to quality control. Designing the methods of representing and incorporating information quality into fusion systems in general and in the crisis environment in particular is a relatively new and a rather difficult problem and more research is needed to confront all its challenges. One of the main challenges is the growing possibilities of communications resulting in a growing number of opportunistic human sensors, a huge amount of information coming from social networks, and growing importance of human–centric fusion and related problem of modeling quality of information resulting from this.

References

Appriou A (2001) Situation assessment based on spatially ambiguous multisensor measurements. Intl J Intell Syst 16(10):1135–1166

Bisdikian C, Kaplan L, Srivastava M, Thornley D, Verma D, Young R (2009) Building principles for a quality of information, specification for sensor information. In: Proc. of the 12th International conference on information fusion, Seattle, WA, USA, 6–9 July, pp 1370–1377

Blasch EP, Plano S (2003) Level 5: user refinement to aid the fusion process. In: Dasarathy B (ed) Multisensor, multisource information fusion: architectures, algorithms, and applications. Proc of the SPIE, vol 5099

Bosc P, Prade H (1997) An Introduction to the fuzzy set and possibility theory-based treatment of flexible queries and uncertain or imprecise databases. In: Motro A, Smets P (eds) Uncertainty in information systems: from needs to solutions. Kluwer, Dordrecht, The Netherlands, pp 285–324

Bovee M, Srivastava RP, Mak B (2003) A conceptual framework and belief-function approach to assessing overall information quality. Int J Intell Syst 18:51–74

Chengalur-Smith I, Ballou D, Pazer H (1999) The impact of data quality information on decision making: an exploratory analysis. IEEE Trans Knowl Data Eng 11(6):853–864

Cholvy L, Nimier V (2003) Information evaluation: discussion about STANAG 2022 Recommendations. In: Proc. of the RTO IST Symposium on military data and information fusion

Chopra K, Haimson C (2005) Information fusion for intelligence analysis. In: The Proceedings of the 38th Hawaii international conference on system sciences

Delmotte F, Borne P (1998) Context-dependent trust in data fusion within the possibility theory. In: Proceedings of the IEEE Intl. conference on systems, man and cybernetics, pp 78–88

Delmotte F, Smets P (2004) Target identification based on the transferable belief model interpretation of dempster-shafer model. IEEE Trans Syst Man Cybern A 34:457–471

Dragas V (2013) An ontological analysis of uncertainty in soft data. In: Proc. of the 16th international conference on information fusion, Istanbul, Turkey, pp 1566–1573

Dubois D, Prade H (1988) Possibility theory: an approach to computerized processing of uncertainty. Plenum, New York

Dubois D, Prade H (1992) Combination of fuzzy information in the framework of possibility theory. In: Abidi MA, Gonzalez RC (eds) Data fusion in robotics and machine intelligence. Academic, San Diego, CA, pp 481–505

Fabre S, Appriou A, Briottet X (2001) Presentation and description of two classification methods using data fusion based on sensor management. Inf Fusion 2:47–71

Fisher CW, Chengalur-Smith I, Ballou DP (2003) The impact of experience and time on the use of data quality information in decision making. Inf Syst Res 14(2):170–188

Haenni R (2005) Shedding new light on Zadeh's criticism of Dempster's rule. In: 7th International conference on information fusion (FUSION), pp 879–884

Helfert M (2001) Managing and measuring data quality in data warehousing. In: Proc. of the world multiconference on systemics, cybernetics and informatics, pp 55–65

Helfert M (2001) Managing and measuring data quality in data warehousing. In: Proc. of the World multiconference on systemics, cybernetics and informatics, pp 55–65
Hexmoor H, Wilson S, Bhattaram S (2006) A theoretical inter-organizational trust-based security model. Knowl Eng Rev 21(2):127–161
Jiang W, Zhang A, Yang Q (2008) A new method to determine evidence discounting coefficient. In: Huang D-S et al (eds) ICIC 2008, vol 5226, LNCS. Springer, Heidelberg, pp 882–887
Jousselme A-L, Maupin P, Bosse E (2003) Uncertainty in a situation analysis perspective. In: Proc. of the FUSION'2003-6th Conference on multisource-information fusion, pp 1207–1214
Juran JB, Godfrey AB (1988) Juran's quality handbook, 5th edn. McGraw-Hill, New York
Klir GJ, Wierman MJ (1999) Uncertainty-based lnformation, vol 15, 2nd edn, Studies in fuzziness in Soft Contputing. Physica, Heidelberg
Kostoulas D, Aldunate R, Pena-Mora F, Lakhera S (2006) A decentralized trust model to reduce information unreliability in complex disaster relief operations, intelligent computing in engineering and architecture. Lect Notes Comput Sci 4200:383–407
Krause P, Clark D (1993) Representing uncertain Knowledge: an Artificial Intelligence Approach. Kluwer Academic Publishers, Dordrecht, The Netherlands
Lee Y, Pipino L, Frank J, Wang R (2006) Journey to data quality. MIT Press, Cambridge, MA
Madnick SE, Lee YW, Wang RY, Zhu H (2009) Overview and framework for data and information quality research. ACM J Data Inf Qual 1(1): Article 2
O'Brien JA, Marakas G (2005) Introduction to information systems. McGraw-Hill/Irwin, New York
Pichon F, Dubois D, Denoeux T (2012) Relevance and truthfulness in information correction and fusion. Int J Approx Reason 53(2):159–175
Rogova G (2014) Adaptive real-time threat assessment under uncertainty and conflict. In: Proc. 4th IEEE conference on cognitive methods in situation awareness and decision support, San Antonio, TX
Rogova G, Bosse E (2010) Information quality in information fusion. In: Proc. of the 13th International conference on information fusion, Edinburg, Scotland, July 2010
Rogova G, Nimier V (2004) Reliability in information fusion: literature survey. In: Proc. of the FUSION'2004-7th Conference on multisource-information fusion, pp 1158–1165
Rogova G, Scott P, Lollett C, Mudiyanur R (2006) Reasoning about situations in the early post-disaster response environment. In: Proc. of the FUSION'2006-9th Conference on multisource information fusion
Rogova G, Hadrazagic M, St-Hilaire M-O, Florea M, Valin P (2013) Context-based information quality for sequential decision making. In: Proc. of the 2013 IEEE international multi-disciplinary conference on cognitive methods in situation awareness and decision support (CogSIMA)
Schubert J (2008) Conflict management in Dempster-Shafer theory by sequential discounting using the degree of falsity. In: Magdalena L, Ojeda-Aciego M, Verdegay JL (eds), Proceedings of IPMU'08, Torremolinos (Malaga), 22–27 June 2008, pp 298–305
Shafer G (1976) A mathematical theory of evidence. Princeton University Press, Princeton, NJ
Smets P (1997) Imperfect information: imprecision—uncertainty. In: Motro A, Smets P (eds) Uncertainty management in information systems: from needs to solutions. Kluwer, Dordrecht, The Netherlands, pp 225–254
Smets P (2002) Data fusion in the transferable belief model. In: Proc. of the FUSION'2000-Third conference on multisource-multisensor information fusion, pp 21–33
Smithson M (1989) Ignorance and uncertainty: emerging paradigms. Springer, New York
Standard 8402, 3. I (1986) International organization of standards
Tawfik AY, Neufeld EM (1996) Irrelevance in uncertain temporal reasoning. In: Proc of the 3rd Intl. IEEE workshop on temporal representation and reasoning, pp 196–202
Uddin M, Amin M, Le H, Abdelzaher T, Szymanski B, Nguyen T (2012) On diversifying source selection in social sensing. In: Proc. of the 9th International conference on networked sensing systems (INSS), pp 1–8
Varshney P (1997) Multisensor data fusion. Electron Commun Eng J 9(6):245–253

von Neuman J, Morgenstern O (1947) Theory of games and economic behavior. Princeton University Press, Princeton, NJ
Wald L (2002) Data fusion: definitions and architectures: fusion of images of different spatial resolution. Les Presses, Ecole des Mines de Paris, Paris, France
Wang R, Strong D (1996) Beyond accuracy: what data quality means to data consumers. J Manage Inf Syst 12:5–34
Wang Y, Wang R (1996) Anchoring data quality dimensions in ontological foundations. Commun ACM 39(11):86–95
White FE (1988) A model for data fusion. In: Proc. 1st National symposium on sensor fusion
Yager R (2012) Conditional approach to possibility-probability fusion. IEEE Trans Fuzzy Syst 20(1):46–55

Chapter 5
Uncertainty Representations for Information Retrieval with Missing Data

Anne-Laure Jousselme and Patrick Maupin

Abstract Retrieving items such as similar past events, or vessels with a specific characteristic of interest, is a critical task for crisis management support. The problem of information retrieval from incomplete databases is addressed in this paper. In particular, we assess the impact of the uncertainty representation about missing data for retrieving the corresponding items. After a brief survey on the problem of missing data with an emphasis on the information retrieval application, we propose a novel approach for retrieving records with missing data. The general idea of the proposed data-driven approach is to model the uncertainty pertaining to this missing data. We chose the general model of belief functions as it encompasses as special cases both classical set and probability models. Several uncertainty models are then compared based on (1) an expressiveness criterion (non-specificity or randomness) and (2) objective measures of performance typical to the Information Retrieval domain. The results are illustrated on a real dataset and a simulation controlled missing data mechanism.

5.1 Introduction

In crisis situations, an effective response highly depends on the ability of the decision maker to reach rapidly the adequate level of situation awareness and thus on his/her ability to quickly access and adequately process information (Chen and Dahanayake 2007). In particular, relying on past experiences helps not only

---

This work is an updated version of Jousselme and Maupin (2013).

A.-L. Jousselme (✉)
NATO Science and Technology Organization, Centre for Maritime Research and Experimentation (CMRE), Viale San Bartolomeo 400, 19126 La Spezia, Italy
e-mail: anne-laure.jousselme@cmre.nato.int

P. Maupin
Command, Control and Intelligence (C2I) Section, Defence R & D Canada - Valcartier, Quebec, QC, Canada G3J 1X5
e-mail: patrick.maupin@drdc-rddc.gc.ca

G. Rogova, P. Scott (eds.), *Fusion Methodologies in Crisis Management*,
DOI 10.1007/978-3-319-22527-2_5

saving time not reinventing the wheel, but also avoiding mistakes. The hypothesis that similar events will lead to similar consequences and can thus be processed in an equivalent manner is the principle of case-based reasoning (CBR) systems (Aamodt and Plaza 1994). For instance, Improvised Explosive Device (IED) incidents have been recorded since between 2004 and 2008 by the US National Counter Terrorism Center (NCTC), made available through the Worldwide Incidents Tracking System (WITS) for research purposes.[1] In the maritime domain, the National Geospatial-Intelligence Agency compiles and publishes weekly the hostile acts against commercial shipping worldwide.[2] Piracy events can then be retrieved and further analyzed for an anticipated response.

Besides CBR approaches and still in the aim of better answering user's needs, many decision support systems use a "query-answer" principle. Indeed, in complement to data-driven approaches, a user-driven approach appears to be very efficient to overcome the overwhelming amount of information from diverse sources that operators under pressure need to process in a crisis situation. Guided by the mission at hand and the context in general, the user is the one knowing his/her information needs and actively searching for information across the different systems at his/her disposal. As an example, to analyze the Automatic Identification System (AIS) data and look for anomalies, the Vessel Traffic Services (VTS) operator executes a series of queries on the ships contacts database to retrieve all tankers not emitting the Estimated Time of Arrival in a given area, all ferries at a certain distance of their usual routes, etc.

Unfortunately, more than often, the databases are incomplete, uncertain, and imprecise mostly because the perpetrators identities, intents, procedures, and tactics are only partially known: Events are partially recorded, and vessels do not transmit the mandatory information. As a result the operators are left with databases plagued with a data quality problem known as *missing data* [see Kim et al. (2003) for a taxonomy of such *dirty data*]. However, observations containing missing data should not be rejected because valuable information can be derived from the remaining data and, in fact, from the data fields missing, which can give clues on adversary strategies of deception.

The problem tackled in this work is the retrieval of items previously stored and structured in a database degraded by missing data. The items could be, for example, records of past events (e.g., IED or piracy attacks) or timely events as represented by AIS information. The question here is how to efficiently retrieve items from incomplete databases. Only a few works directly address the problem of missing data in information retrieval [see for instance Yi (2011) or Schmidt and Vorobieva (2007)]. We identified however relevant research areas in which the problem of missing data is addressed. The standard (statistical) methods for dealing with missing data (Lynch 2003; Farhangfar et al. 2007; Yassir and Nayak 2012) either completely delete the records or database attribute with missing data (case or

---

[1] http://www.nctc.gov.

[2] http://msi.nga.mil/NGAPortal/MSI.portal?_nfpb=true&_pageLabel=msi_portal_page_64.

pairwise deletion), or replace the missing data by one (simple imputation) or several (multiple imputation) estimations of the missing data. Although deleting records could be acceptable in some applications such as classification where an abstraction of the database content is sought for, it is more problematic in information retrieval since (1) items with empty fields are frequent and (2) they may contain very precious information. In other words, in those kinds of applications, we do not want to miss any relevant record even if it is incomplete. Another drawback of statistical approaches is their inadequacy to deal with categorical variables.

Contrary to probabilistic information retrieval approaches (Fuhr 1992; Crestani et al. 1998) and their evidential extensions (e.g., Chen 1988; da Silva and Milidiú 1993; Chowdhary and Bansal 2011; Lalmas 1998) where the uncertainty is used to model the probability of relevance of items with respect to user's information need (query), we conceive here the uncertainty representation rather at the attribute value level as it is done in probabilistic databases (Dalvi et al. 2009; McClean et al. 2001; Lee 1992). Probabilistic databases directly address the problem of uncertain data, and thus that of missing data: Each (imperfect) attribute value of the database is replaced by a probability distribution over the set of possible values for that attribute. This results in a series of possible instances of the database named as *possible worlds*, an idea very close to the multiple imputation techniques from statistical processing of missing data. The interest of such an approach is that it works not only for real-valued data but also categorical variables or any other type of variable. For a richer representation of uncertainty, possibilistic and evidential database models have been proposed in which the uncertain and imprecise attribute value for a given item is represented by a belief function (Lee 1992).

Evidential databases are extensions of probabilistic databases designed to deal with imprecise and uncertain data (Telmoudi and Chakhar 2004; Hewawasam et al. 2005; Bach Tobji et al. 2008; Zaffalon 2002). In this work, we adopt the uncertainty model of evidential databases and propose a retrieval approach based on a distance calculation between belief functions. The items are then ordered according to their distance to the query. Our specific objective is to compare different uncertainty representations (at the attribute level) and highlight their impact on the quality of the retrieval algorithm. As uncertainty representations, we consider the models of classical sets (which corresponds to the standard multiple imputation model), probabilities, and belief functions. The interest of these models is their consistency with the more general model of belief functions. We show that a distance between belief functions as a criterion for retrieval reduces to standard distances measures between sets or probabilities. In Sect. 5.2, we exemplify the retrieval problem on two cases of interest: (1) retrieving vessels within AIS dataset and (2) retrieving IED events. We describe briefly the datasets and introduce the information retrieval formulation together with some standard measures of performance. In Sect. 5.3, after setting the bases of belief function theory we present the proposed model for information retrieval with missing data. Section 5.4 is dedicated to some experimental comparisons of the different models, on synthetic data and on real data. Some conclusions are drawn in Sect. 5.5 together with some intent for future works.

5.2 Retrieving Items from Imperfect Databases

As an example of records of past events, the WITS is a US government's database on terror attacks accessible on the NCTC website.[3] These IED incidents have been recorded since between 2004 and 2008 by the US National Counterterrorism Center, made available through the WITS for research purposes, and processed here by Geocommons.[4] Figure 5.1 is an excerpt of the terrorist attacks in Afghanistan ranging between January 2004 and March 2008.

The AIS used by VTS to track, locate, and identify vessels is based on a cooperative transmission of information by the different types of vessels across the world. It is mandatory for the vessels (above specific length) to report their AIS information which is classically separated into a static and a dynamic part. The static information contains in particular the Name, MMSI,[5] IMO[6] Number, Call Sign, Width, Length, Expected Time of Arrival (ETA), Destination, Ship Type, Cargo ID, as illustrated in Fig. 5.2a. The dynamic information which is updated frequently contains in particular the Time, MMSI, Latitude, Longitude, Course Over Ground, True Heading, Speed Over Ground, Navigation Status, AIS Message Type as illustrated in Fig. 5.2b. The unique identity of the ship is represented by its MMSI, appearing in both parts.

WHEN					WHERE						WHO			WHAT							DAMAGE							ID
Date of Incid	Year	M	Day	Ap	Region of A	Country	Afghanista	City	Latitud	Longit	Nu	Ch	Nationality	Ev	M	As	Su	IEI	Discription of Attack	Summary	Nu	Facility Type	Nu	Oi	Oil	En	dar	ICN Code
07/26/2004	2004	7	26	0	South Asia	Afghanis	Kandahar	Kanda	31.61	65.7	1	3	Unknown	4	0	0	0	1	Police chief escapes assassination attempt	On 26 July 2004, in Kandahar, Afghanistan, an improvised explosive device exploded near a police vehicle carrying Kandahar's police chief, causing no injuries. No group claimed responsibility.	0		0	0	0	0	99	200464138
04/03/2005	2005	4	3	0	South Asia	Afghanis	Nangarhar	Jalalal	36.76	65.9	1	1	Afghanista	4	0	0	0	1	Government office building damaged in IED attack by Taliban in Jalalabad, Nangarhar, Afghanistan	################################	0	Government	0	0	0	0	99	200566091
04/04/2005	2005	4	4	0	South Asia	Afghanis	Kandahar		31.61	65.7	1	3	Unknown	4	0	0	0	1	1 vehicle damaged in IED attack in Kandahar Province, Afghanistan	################################	0	Vehicle	0	0	0	0	99	200566138
04/10/2005	2005	4	10	0	South Asia	Afghanis	Kandahar	Dama	34.35	67.8	1	3	Unknown	4	0	0	0	1	Unknown facility targeted in IED explosion in Daman, Kandahar Province, Afghanistan	On 10 April 2005, in Daman, Kandahar Province, Afghanistan, an improvised explosive device (IED) exploded, causing no injuries or damage. No group claimed responsibility.	0	NA	0	0	0	0	99	200566610
06/20/2005	2005	6	20	0	South Asia	Afghanis	Khowst		35.21	70.8	1	3	Unknown	4	0	0	0	1	UN vehicle badly damaged in IED attack in Khowst Province, Afghanistan	On 20 June 2005, in Khowst Province, Afghanistan, assailants detonated an improvised explosive device (IED) near a UN vehicle, causing significant damage, but no casualties. No group claimed responsibility.	0	Vehicle	0	0	0	0	99	200567649
04/26/2005	2005	4	26	1	South Asia	Afghanis	Kandahar	Unkno	31.6	65.8	1	3	Unknown	4	0	0	0	1	14 police officers killed in bombing in Kandahar Province, Afghanistan	On 26 April 2005, in Kandahar Province, Afghanistan, an improvised explosive device (IED) exploded as a police vehicle passed over it, killing 14 police officers and destroying the vehicle. No group claimed responsibility.	0	Vehicle	14	0	0	0	99	200570780
05/01/2006	2006	5	1	0	South Asia	Afghanis	Oruzgan	Tarin	32.62	65.9	1	1	Afghanista	4	0	0	0	1	1 civilian, 1 soldier wounded in VBIED attack by suspected Taliban in Tarin Kowt, Oruzgan, Afghanistan	################################	2	Military	0	0	0	0	99	200689522

Unknown Empty Restricted access Not Applicable

Fig. 5.1 Terrorist Attacks, Afghanistan, Jan. 2004–Mar. 2008 from the WITS database (National Counterterrorism Center (NCTC) 2010). Different types of missing data are *highlighted*

[3]www.nctc.gov.

[4]www.geocommons.com.

[5]Maritime Mobile Service Identity.

[6]International Maritime Organization.

a

	Time	MMSI	Name	IMO	CallSign	Width	Length	AntennaFromBow	AntennaFromPort	ETAMonth	ETADay	ETAHour	ETAMinute	Draught	Destination	ShipTypeId	CargoId
1	1325376000	211389230	NULL	NULL	DBXM	5	16	5	2	NULL	NULL	NULL	NULL	NULL	NULL	37	NULL
2	1325376000	211404810	ESCHWEGE	9252644	DBGZ	10	66	29	5	1	2	15	0	3.6	BALTIC SEA	55	NULL
3	1325376000	212076000	NEFTEGAZ-66	8418564	5BYL2	16	81	12	14	11	5	15	0	5	THYBORON	9	9
4	1325376000	215729000	VICTOR HENSEN	7360655	9HYZ7	10	40	17	5	NULL	NULL	NULL	NULL	3.8	WORKING AREA RAUMA	9	0
5	1325376000	219002827	MIE MALENE	NULL	5QRJ	5	15	2	2	3	3	11	4	2.8	SKAGEN	30	NULL
6	1325376000	219013001	HG306 TOBIS	NULL	OUPS	10	40	9	5	NULL	NULL	NULL	NULL	NULL	NULL	30	NULL
7	1325376000	238151640	SVETI KRSEVAN	9326562	9AA2126	18	88	44	8	12	31	0	0	2.4	DOMINCE OREBIC	6	0
8	1325376000	255804540	PONTA SAO LOURENCO	7716763	CSEO	16	106	85	9	NULL	NULL	NULL	NULL	4.7	SETUBAL	7	3
9	1325376000	304879000	SAFMARINE SALOUM	9327578	V2BO4	24	148	NULL	NULL	12	21	12	0	6	FOR ORDERS	0	0
10	1325376000	312275000	ARGONAUT	7639642	V3ON2	NULL	NULL	NULL	NULL	1	1	9	0	NULL	BATUMI	7	0
11	1325376000	338125699	LA BELLA VITA	NULL	NULL	NULL	NULL	NULL	NULL	NULL	NULL	NULL	NULL	NULL	NULL	NULL	NULL
12	1325376000	356683000	TIMBER TRADER VIII	9054767	3EVK9	18	96	82	12	12	30	12	0	6	KAOHSIUNG	7	0
13	1325376000	366921820	MARS	7338327	WDD7627	8	28	7	6	NULL	NULL	0	0	4.5	NULL	52	NULL
14	1325376000	366964110	MISS DANA	NULL	WAD8305	7	20	15	4	5	24	10	0	2	BBI	9	0
15	1325376000	367029020	BOSTON PILOT MYSTIC	NULL	VHF 20	NULL	NULL	NULL	NULL	NULL	NULL	NULL	NULL	2	OWN-SHIP	50	NULL
16	1325376000	367177370	HELEN D COPPEDGE	7400510	WCA5274	9	30	10	2	NULL	NULL	NULL	NULL	4	NULL	52	NULL
17	1325376000	367410780	DETROIT	NULL	WDE8881	14	43	10	7	11	17	12	0	2.9	CATLETTSBURG KY	31	NULL
18	1325376000	370270000	OCEANIC HAWK	9561899	3FQJ3	27	170	150	16	12	27	10	0	5.2	VANCOUVER	7	0
19	1325376000	371885000	DURBAN HIGHWAY	9536961	H9CR	32	199	40	24	1	4	8	0	9.4	BEZEE	7	0
20	1325376000	563831000	APL QATAR	9400148	9VMJ5	32	275	219	16	1	3	6	0	9.9	SAVANNAH	7	9
21	1325376000	564719000	AFFINITY	9289776	9VDT5	32	228	189	7	NULL	NULL	NULL	NULL	13.1	NEWYORK,USA	8	0
22	1325376000	579000000	DREDGE POTTER	NULL	AEMC	14	82	28	4	NULL	NULL	0	0	NULL	NULL	33	NULL

b

	Time	Id	Latitude	Longitude	CourseOverGround	TrueHeading	SpeedOverGround	RateOfTurn	PositionAccuracy	NavigationalStatus	AISMessageId
1	1325376000	212777000	29.6119766666667	-15.3124866666667	34	32	16.7	0	1	0	1
2	1325376000	224637000	36.1329666666667	-5.43786666666667	231.6	55	0	0	0	0	1
3	1325376000	232704000	56.7590833333333	8.104295	235	201	1.2	0	0	0	1
4	1325376000	235050584	35.8876666666667	14.5151666666667	305	NULL	0	NULL	0	5	3
5	1325376000	236568000	45.6430333333333	13.2876216666667	20	96	0.2	0	1	1	3
6	1325376000	245843000	43.5261666666667	-1.497	179	120	0	0	0	5	3
7	1325376000	246612000	54.673315	14.9523516666667	260.7	259	13.1	0	1	0	1
8	1325376000	304080889	60.5271666666667	27.1661666666667	333	169	0	0	0	5	3
9	1325376000	352698000	42.8602666666667	-10.0942	178	180	19	0	1	0	1
10	1325376000	366951140	30.087985	-90.91302	271	NULL	0.1	NULL	0	0	1
11	1325376000	367050250	30.3515183333333	-88.506225	358.6	NULL	0	NULL	0	0	1
12	1325376000	367129670	29.909705	-89.9807033333333	116.4	46	0	0	1	0	1
13	1325376000	367189580	30.3179516666667	-87.3098083333333	92.9	NULL	1.9	NULL	0	5	3
14	1325376000	367339380	26.090655	-80.1753883333333	208	NULL	0	NULL	0	NULL	18
15	1325376000	367362230	31.1488566666667	-87.947575	247.4	244	3.8	0	1	0	1
16	1325376000	367390240	29.5554383333333	-90.39145	10.5	NULL	0	NULL	0	0	1
17	1325376000	367397510	29.5735616666667	-90.3833116666667	263.3	NULL	0	NULL	0	NULL	1
18	1325376000	367416070	35.8291666666667	-7.83483333333333	272	272	8.7	0	1	0	1
19	1325376000	367448060	29.9367316666667	-90.3511633333333	108.7	NULL	6.1	NULL	0	0	1
20	1325376000	367459210	30.2895083333333	-91.2207733333333	28.4	NULL	0.1	NULL	0	0	1
21	1325376000	367505090	30.0676383333333	-93.3451783333333	154	41	0	0	1	0	1
22	1325376000	370912000	44.4443133333333	12.2464166666667	334	149	0	0	0	5	3

Fig. 5.2 AIS information with missing data. (**a**) Example of the static information from an AIS dataset, (**b**) example of the dynamic information from an AIS dataset

5.2.1 Characteristics of the Dataset

The database is a table of *n* items (events, cases, vessels, etc.) described by *m* features (descriptors, attributes, etc.) in which several kinds of imperfections can arise:

- data may be missing [(?) see Schafer and John 2004 for a survey]: (1) They exist but have not been gathered yet (Blank cell), (2) the value is really unknown with no possibility of retrieving it (Unknown), (3) the field does not apply to the particular case of that event (NA—Not Applicable) or (4) the information is classified and not accessible (#—Restricted Area);
- the attributes describing the vessels are of different types, defined over heterogeneous domains (e.g., categorical, ordered, binary, real, interval, free text, etc.);
- inconsistencies may arise between several dependent (correlated) variables;
- some values may be erroneous;

- some values may be approximated. In the example of the dynamic AIS database, a field "Position Accuracy" is added to specify that the express some uncertainty regarding the position values. In the example of the WITS database, a field "Approximated date" is added to specify that the date is indeed only an approximation of the real event date;
- some variables are correlated either in a conjunction form (e.g., MMSI and Name, Width, Length and Type, Type and Cargo ID) or a hierarchical form (e.g., When: Date (most specific) = Year & Month & Day; Where: Region $\subset$ Country $\subset$ Province $\subset$ City $\subset$ LatLong coordinates).

In the example of the specific AIS dataset above, a preliminary data cleanup has been performed and erroneous or inconsistent data or have been deleted and replaced by NULL. To such an operation, we could have preferred an adequate uncertainty representation.

Most of the retrieval techniques simply ignore the items with missing data since empty fields do not match queries of the standard tools. In the work presented here, we aim at improving the performances of Information Retrieval (IR) techniques so more relevant items can be retrieved, while less irrelevant items would not be retrieved. The idea is to first replace the missing value by an adequate uncertainty representation so it can be, in a second step, considered by the retrieval task. Beyond the performances of the IR technique, the main purpose of this work is to highlight the impact of the uncertainty representation on the IR performances. Ultimately, a deeper analysis will come up with an optimal uncertainty representation (guided by the context of use) which improves the IR algorithm so that relevant items are not ignored. As performance measures, we consider the tradeoff between the *precision* (proportion of retrieved items that are relevant) and the *recall* (the proportion of relevant items that are retrieved) as it will be detailed in Sect. 5.2.4.

5.2.2 Missing Data Categories (Mechanisms)

Several causes for missing data are identified, for instance (Schafer and John 2004): "Don't know", "Refuse", "Unintelligible", "Restricted area", "Does not apply", etc. Depending on the cause of missingness, different missing data categories are distinguished in the machine learning and statistics literature which are linked to the mechanism which provided the missing data. We partition the dataset $\mathcal{D}$ into two subsets of items $\mathcal{D}_{\text{obs}}$ and $\mathcal{D}_{\text{miss}}$ representing the items with completely observed values and missing values, respectively. Let us denote by R the binary random variable of missingness: If $R = 1$ then the value is observed, if $R = 0$, the value is missing. Three missing data mechanisms are generally distinguished: (1) Missing At Random (MAR), in which the mechanism is conditionally independent of the missing values given the observed variables, i.e. $P(R|\mathcal{D}) = P(R|\mathcal{D}_{\text{obs}})$; (2) Missing Completely At Random (MCAR), a special case of MAR, in which the missing data mechanism is completely independent of the domain variables, i.e. $P(R|\mathcal{D}) = P(R)$;

and (3) Non Missing At Random (NMAR) if the missing data depends on either the unobserved values, i.e. $P(R|\mathcal{D}) \neq P(R|\mathcal{D}_{\text{obs}})$. The MAR mechanism for missing data will be used for simulation purposes in Sect. 5.4. Note however that using this hypothesis as a basis for further processing is controversial since it is a strong assumption that cannot be tested (Zaffalon 2002).

5.2.3 *Information Retrieval Problem Formulation*

Let $\mathcal{D}$ be a set of n items $\mathbf{x}$ described by a set of m descriptors[7] $\{x_i\}$. We can note $\mathcal{D} = \{\mathbf{x}^{(i)}\}_{i=1}^{n}$. The aim of an Information Retrieval task is to retrieve an ordered list of relevant items in response to a query q by means of an information retrieval function ir_d defined as:

$$ir_d : \mathcal{Q} \to \mathcal{P}(\mathcal{D}) \tag{5.1}$$

$$q \mapsto r_q \subseteq \mathcal{D} \tag{5.2}$$

$\mathcal{Q}$ is the set of all possible queries, $\mathcal{P}(\mathcal{D})$ is the power set of $\mathcal{D}$ (the set of all subsets of $\mathcal{D}$), and r_q is the set of items retrieved by the ir_d function and thus identified as relevant to the query q. The items of r_q are ranked in order of increasing distance d from q. The top ranked items are the most relevant to the query. The distance quantifies then the inverse of a degree of relevance of an item to a query. Indeed, in the case of a normalized distance measure: If $d(\mathbf{x}, q) = 0$, then x is totally relevant to q; if $d(\mathbf{x}, q) = 1$, then x is totally irrelevant to q. The set of retrieved items (or relevant items) according to ir_d may be thus defined according to that distance measure as:

$$r_q = \{\mathbf{x} \in \mathcal{D} | d(\mathbf{x}, q) \leq \tau\} \tag{5.3}$$

where τ is a threshold over d. If each feature is represented by a random variable X_j whose domain is denoted as $\mathcal{X}_j$, then an item $\mathbf{x}$ of $\mathcal{D}$ is a realization of the joint random variable X whose domain is $\mathcal{X} = \mathcal{X}_1 \times \ldots \mathcal{X}_m$. The domains maybe of different natures (as discussed in Sect. 5.2.1), and of different sizes. We consider here the case of categorical variables hence $\mathcal{X}_j$ is a discrete finite and unordered set of possible N_j values, i.e. $\mathcal{X}_j = \{x_{j,1}, \ldots, x_{j,N_j}\}$. In the case of a binary variable ($N_j = 2$), we will note $\mathcal{X}_j = \{x, \neg x\}$.

Following Burkhard (2004), let us define an Information Entity (IE) as $X_j = x_{j,l}$ meaning that the variable X_j is assigned the particular value $x_{j,l}$ of its domain $\mathcal{X}_j$. A case $\mathbf{x}$ is a subset of IEs, i.e. $\mathbf{x} = \{(X_j = x_{j,l})\}_{j \in [1,m]}$. As it will be detailed

---

[7]Descriptors are also called *terms*, *features*, *attributes*, etc.

in Sect. 5.3.3, an IE can be either precise, $X_j = x_{j,l}$, either non-specific, $X_j = (x_{j,1}, \ldots, x_{j,g})$, either uncertain, $p(X_j = x_{j,g}) < 1$, or both non-specific and uncertain.

A query is also a subset of IEs and can be denoted as $\mathbf{q} = \{(X_j = x_{j,l})\}_{j\in[1,m]}$. The query could itself be thus either precise, non-specific, or uncertain. Note that uncertain queries will not be considered.

The key element of the information retrieval task is the distance measure d. Depending on the data (i.e. their domain of definition, their inter-dependence, existence of empty fields, etc.), different distance measures are used. If we consider m possibly dependent and heterogeneous (with domains of different types) descriptors X_j, then a cautious way of computing the distance between a query and the database items is by a weighted sum of the individual distances computed feature by feature:

$$d(\mathbf{x}, \mathbf{q}) = \sum_{j=1}^{m} w_j d_j(x_j, q_j) \tag{5.4}$$

This is one of the most used distance measure for IR problems which naturally accounts for heterogeneous features.

5.2.4 IR Efficiency

Let us denote by R_q the set of relevant items, that is the ideal set of items to be retrieved by the ir_d function. As mentioned in Brini et al. (2005), "for an efficient Information Retrieval System (IRS) these two sets must be equal as often as possible" (Brini et al. 2005). The efficiency of an IR system is classically assessed by measures such as the *precision*, quantifying the proportion of retrieved items that are relevant, i.e. $\text{Prec} = \frac{|R_q \cap r_q|}{|r_q|}$, and the *recall* quantifying the proportion of relevant items that are retrieved, i.e. $\text{Rec} = \frac{|R_q \cap r_q|}{|R_q|}$. Like in classification systems, a tradeoff between these two measures is needed. An aggregated measure based on the two previous ones is the F_β-measure defined as $F_\beta = (1+\beta^2)\frac{\text{Prec}\cdot\text{Rec}}{(\beta^2\cdot\text{Prec}+\text{Rec})}$ where β is a non-negative real-valued constant allowing to balance between Prec and Rec in different contexts. The latter will be used in the upcoming experiments. Note however that other more powerful measures exist and will be used in future works. Having this ground truth information is not trivial and in many cases it is not completely available. When all relevant items of a dataset cannot be identified, alternative measures of performance are proposed which overcome that issue, such as in Buckley and Voorhees (2004), Wu and McClean (2006), and Ahlgren and Grönqvist (2006).

5.3 Handling Uncertainty for Missing Data in Information Retrieval

As proposed in evidential databases, we adopt the belief function model as a general model for representation of missing attribute values as it encompasses as special cases both the classical set and probability models.

5.3.1 Belief Functions as a General Model for Uncertainty Representation

Let X be a frame of discernment, i.e. the set of all possible values for an unknown variable x, and let it denote by A any subset of X. Let $\mathcal{P}(X)$ be the power set of X and let $\mathcal{P}(X)'$ be the power set of X bereft of the empty set, $\mathcal{P}(X)\backslash\varnothing$. A Basic Probability Assignment (BPA) m is a mapping from $\mathcal{P}(X)$ to $[0, 1]$ satisfying:

$$\sum_{A\subseteq X} m(A) = 1 \quad \text{and} \quad m(\varnothing) = 0 \tag{5.5}$$

A subset A of X such that $m(A) > 0$ is called a *focal element*. Belief and plausibility functions are mappings from $\mathcal{P}(X)$ to $[0, 1]$ defined, respectively, as $\mathrm{Bel}(A) = \sum_{B\subseteq A} m(B)$ and $\mathrm{Pl}(A) = \sum_{B\cap A\neq\varnothing} m(B)$. We also have $\mathrm{Bel}(A) = 1 - \mathrm{Pl}(\overline{A})$ where $\overline{A}$ is the complement of A.

Whenever the focal elements of m are all singletons (i.e., $m(A) = 0$ iff $|A| > 1$), then m is said to be a Bayesian BPA and defines a probability distribution over X. A categorical BPA is such that the mass is focused on a single focal element (i.e., $\exists A$ such that $m(A) = 1$). A simple support belief function (SSBF) is such that $m(A) = \alpha$ and $m(X) = 1 - \alpha$). If we are dealing only with categorical belief functions, then the framework is simply that of classical (crisp) sets. Figure 5.3 illustrates the links between the three models as both non-specificity and randomness are canceled.[8] Ultimately, we obtain a representation with no uncertainty, being both precise and certain (i.e., focused on a single singleton).

In Lee (1992), similar but also other cases of uncertainty representations are considered: we restrict to exhaustive frames of discernment (i.e. the type *inapplicable* of Lee (1992) is not considered here) but we consider the case of Bayesian belief functions (as in McClean et al. 2001).

[8]The typology of uncertainty types referred here is the one of Klir and Yuan (1995), in which fuzziness is omitted. Note that randomness is called *discord* in Klir and Yuan (1995).

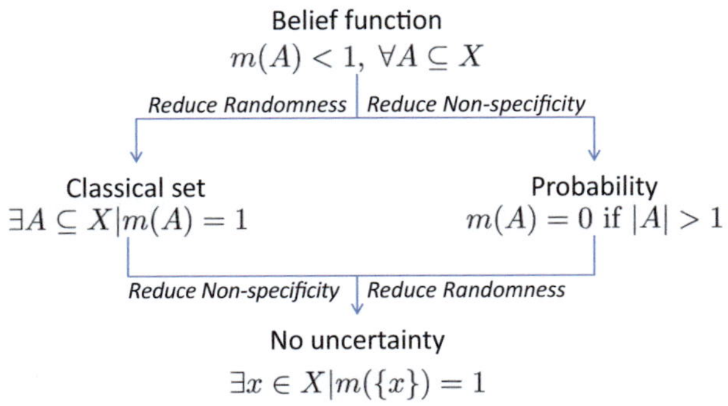

Fig. 5.3 Three models for uncertainty representation. Two types of uncertainty are represented (non-specificity and Randomness) by belief functions. Each of the two types of uncertainty is canceled as we change from belief functions to classical sets or to probabilities

5.3.2 Distance Measures

For the sake of consistency, we need a distance measure which reduces to classical measures within the different models so that the results can be compared to traditional approaches. We use the distance defined in Jousselme et al. (2001) which satisfies these consistency properties. It is defined as:

$$d_J^{(2)}(m_1, m_2) = \sqrt{(\mathbf{m}_1 - \mathbf{m}_2)'\mathbf{Jac}(\mathbf{m}_1 - \mathbf{m}_2)} \tag{5.6}$$

where

$$\mathbf{Jac}(A, B) = \frac{|A \cap B|}{|A \cup B|} \text{for all } A, B \in \mathcal{P}(X)' \tag{5.7}$$

with $\mathbf{Jac}(A, B)$ being the Jaccard Euclidean distance between two belief functions defined over X, $\mathcal{P}(X)'$ is the power set of X bereft of the empty set $\varnothing$. It appears that:

- Probability (randomness only): $d_J(p_1, p_2) = \sqrt{(\mathbf{p}_1 - \mathbf{p}_2)'(\mathbf{p}_1 - \mathbf{p}_2)}$ is the Euclidean distance between probability distributions.
- Classical set (non-specificity only): $d_J(A, B) = \sqrt{1 - \frac{|A \cap B|}{|A \cup B|}}$ which turns to be proportional to the classical Jaccard index for quantifying similarity between sets. That means that $d_J(A, B) = 0$ if $A \subseteq B$. So an item will be retrieved iff $x \in B$.
- No uncertainty: $d_J(x, x') = 0$ if $x = x'$ and $= 1$ if $x \neq x'$. If the uncertainty relatively to the missing value x is not modeled, and if the query does not contain itself some uncertainty, and if a wrong value had been assigned to x, then the item will not be retrieved (based on this descriptor).

5.3.3 *Proposed Approach*

Each missing value of row i and column j of an item $\mathbf{x}$ of the set $\mathcal{D}$ is denoted by $?_{i,j}$ and is replaced by a belief function. In the standard imputation model the estimated value is either correct in which case the item will be retrieved or wrong in which case the item will not be retrieved. Within the multiple imputation models, K tables of data are processed in parallel (as K distinct simple imputation models) and the results are combined for an improved accuracy.

We see the multiple imputation model as the classical set model for uncertainty representation: The uncertainty regarding the true value of $?_{i,j}$ is modeled by a classical set $A \subseteq \mathcal{X}_j$ meaning that all we know is that $?_{i,j} \in A$. Equivalently to simple and multiple imputation approaches, the best way to identify A could be an issue but this is out of the scope of this work.

In the case of probability model, $?_{i,j}$ is replaced by a probability distribution p defined over $\mathcal{X}_j$. Then, $p(x_{i,j})$ represents the probability that the missing value $?_{i,j}$ be equal to $x_{i,j}$, for each $x_{i,j} \in \mathcal{X}_j$. Several way to obtain the values of p are possible: p can be deduced from the prior probabilities of X_j from the database, or assigned a more subjective value, or output by a regression method and in that case accounts for the dependence between the different variables (ability to predict $?_{i,j}$ from the other values of the database), etc.

In the case of a belief function model, $?_{i,j}$ is replaced by a BPA m defined over $\mathcal{X}_j$. Then, $m(A)$ represents the uncertainty that the missing value $?_{i,j}$ belongs to A, for all $A \subseteq \mathcal{X}_j$.

The approach proposed here differs from Lee (1992) mainly by the query process. In Lee (1992), the query (or select operation) is based on the evaluation of pairs $[\mathrm{Bel}(q|x_j^{(i)}); \mathrm{Pl}(q|x_j^{(i)})]$ of belief and plausibility values of a query q knowing that the value of attribute j of item (tuple) $\mathbf{x}^{(i)}$ is $x_j^{(i)}$. Here, the relevant items will be retrieved by means of a distance measure d between q and any item $\mathbf{x}$ of $\mathcal{D}$.

5.4 Comparing Uncertainty Representations

We can summarize our hypothesis as follows: Items with missing data are not retrieved by classical query tools (or they could all be retrieved by a simple modification of the query). Replacing the missing value by a fixed one is known as the simple imputation method. No uncertainty is expressed then. We would like to know if explicitly expressing some uncertainty about the value of the missing attribute improves the retrieval process. To such a purpose, we need to make this uncertainty representation varying and compare different instantiations of it.

We use in this section a formalization of their main constructs together with a general model of distance measures and operators introduced in Sect. 5.3.3.

5.4.1 Elements Under Evaluation

In an attempt to answer the question "What can be compared when comparing uncertainty representations?", we distinguished in Jousselme and Maupin (2012) between two kinds of elements that are Uncertainty *calculus* elements on the one hand and Atomic Decision Procedures on the other hand. This distinction is roughly the one between a theory and its use in a specific fusion algorithm. We then identified several elements related to uncertainty representations which can be compared:

1. Uncertainty *calculus* elements among which (UC-1) the *mathematical model* for uncertainty representation, (UC-2) the *uncertainty measures* understood as uncertainty-based information (Shannon entropy-like) measures, (UC-3) *inference rules and combination operators*, (UC-4) *Transformation functions*.
2. Atomic Decision Procedures among which (ADP-1) *Universe of discourse*, (ADP-2) *Instantiated uncertainty representation*, (ADP-3) *Reasoning step*, (ADP-4) *Decision step*.

Relatively to these elements, we will compare in this paper (UC-1) mathematical models, i.e. classical sets vs probabilities vs belief functions, and (ADP-2) instantiated uncertainty representations, i.e. several probability or belief assignments.

Moreover, we distinguish between two kinds of complementary evaluation methods namely *formal evaluations* and *empirical evaluations*: Formal evaluations are often focusing on the axiomatic structure of uncertainty *calculi* and typically rely on theorem proving techniques; Empirical evaluations rely as the name says on experiments, the gathering of observations, and in general (but not necessarily) on ground truth information used to evaluate the quality of predictions made when using a given uncertainty *calculi*.

5.4.2 Atomic Decision Procedure

To guarantee that only the uncertainty representation is assessed, we must ensure that it is the only variable in the evaluation process. We keep thus the decision procedure at its simplest expression and call it Atomic Decision Procedure. In Fig. 5.4, the ADP is represented in which h is a variable uncertainty representation. Each attribute value x_j of each item $\mathbf{x}$ is replaced by a belief function representing its imperfection: If the attribute is not missing, it is replaced by a categorical belief function focused on a singleton corresponding to the attribute value. If the attribute value is missing, it is replaced by a variable belief function covering the range of uncertainty representations of interest. The query is represented as a vector of binary values $\mathbf{q} = \{(X_j = x_{j,l})\}_{j \in [1,m]}$. Each information entity $(X_j = x_{j,l})$ is replaced by either (1) a categorical belief function focused on a single value to be retrieved or (2) a vacuous belief function if no query concerns that particular attribute. For

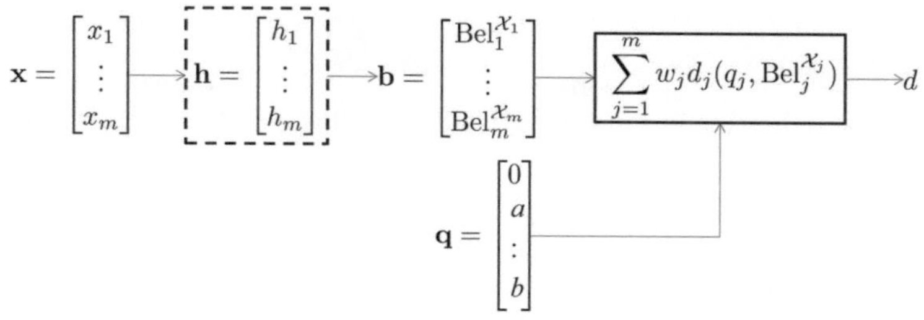

Fig. 5.4 Atomic decision procedure for handling uncertainty for missing data in IR

each attribute j of the item i, the distance between the belief function $\mathrm{Bel}_j^{\mathcal{X}_j}$ defined over the frames $\mathcal{X}_j$ and the query component $q_j$corresponding to the attribute j is computed. The individual distance values so obtained are then aggregated by a weighted average.[9]

5.4.3 *Criteria for Evaluation*

Regarding the different criteria for uncertainty representation evaluation that have been identified in Costa et al. (2012) and that are still currently discussed and refined within the Evaluation of Techniques for Uncertainty Representation (ETUR) working group,[10] we consider the criteria of (a) expressiveness and (b) precision and accuracy. Whereas the first one will be assessed through a formal method, the second one will be assessed through an empirical method requiring thus experimentation by means of IR measures of performance as introduced in Sect. 5.2.4.

Expressiveness

According to the models described in Sect. 5.3.1, uncertainty representations can express either:

1. Non-specificity only: Only the range of the value x, i.e., $x \in A$ is known and no uncertainty about that proposition exists, i.e. $m(A) = 1$. This is the classical set (CS) model;
2. Randomness only: The value of x is precise but its value among a subset A of possible values is uncertain: $\sum_{x \in A} p(\{x\}) = 1$. This is the probability (PT) model.

[9]We used here equal weights.

[10]http://eturwg.c4i.gmu.edu.

3. Non-specificity and Randomness: Several ranges for x are uncertain: $m(A) < 1$ for several $A \subseteq X$. This is the belief function (BF) model.

The three models can be thus be ordered according to their expressiveness as: $CS \prec_{\text{expr}} PT \prec_{\text{expr}} BF$.

Precision and Recall

Contrary to the above criteria which can be assessed through formal evaluation methods, the criteria of precision and recall need to be assessed through empirical evaluations requiring some experiments and involving a dataset and an ADP. In these experiments, we considered 3 models: (1) A classical set model where $|A|$ is the only parameter and ranges from 1 to $|\mathcal{X}_j|$, (2) a probability model where p, the probability distribution over $\mathcal{X}_j$ is the only parameter and ranges from a categorical distribution to a uniform one through a random degradation (six steps $P1$ to $P6$ are shown below), and (3) a SSBF model where $|A|$ and α are the two parameters ranging from 1 to $|\mathcal{X}_j|$ and from 0 to 1, respectively. Note that the case $\alpha = 1$ corresponds to the classical set model.

In order to evaluate the performances of the IR approach, we need a complete dataset (without missing data) to be used as the reference dataset on which the set of relevant items R_q is unambiguously identified. The dataset selected is the Solar Flares dataset taken from the UCI Machine Learning Repository.[11] This dataset contains 1389 samples described by 10 categorical features. We artificially decimated the dataset following a MAR mechanism implemented as a coin tossing based scheme degrading successively the dataset with proportions of missing data ranging from 0.1 to 0.7, with an increment of 0.1 % at each step. For each proportion of missing data, the Atomic Decision Procedure described in Sect. 5.4.1 (Fig. 5.4) is applied with 3 types of belief functions degenerated to represent different models of uncertainty representation, namely a classical set model, a probability model, and a SSBF model. For each model, the F-measure is computed: The higher the measure, the better the retrieval methods, hence the better the uncertainty representation (since it is the only variable within the ADP). Figure 5.5 shows the results obtained for several uncertainty representation models. A threshold of 0.71 has been selected for retrieving the relevant items according to Eq. (5.3).[12] In light of these figures, we firstly observe (Fig. 5.5a) that introducing an amount of uncertainty indeed improves the performances compared to one standard simple imputation method (replacement by frequency histogram mode). Figure 5.5b–d illustrates the impact of the uncertainty representation on the overall retrieval results. In particular, one probability model ($P1$ of Fig. 5.5b) provides good performances.

---

[11] http://archive.ics.uci.edu/ml/index.html.

[12] Among the series of results obtained for different values of τ we selected these ones as they were amongst those with (1) a clear difference between the models and (2) good performances. Further results will be provided in an extended version of our work.

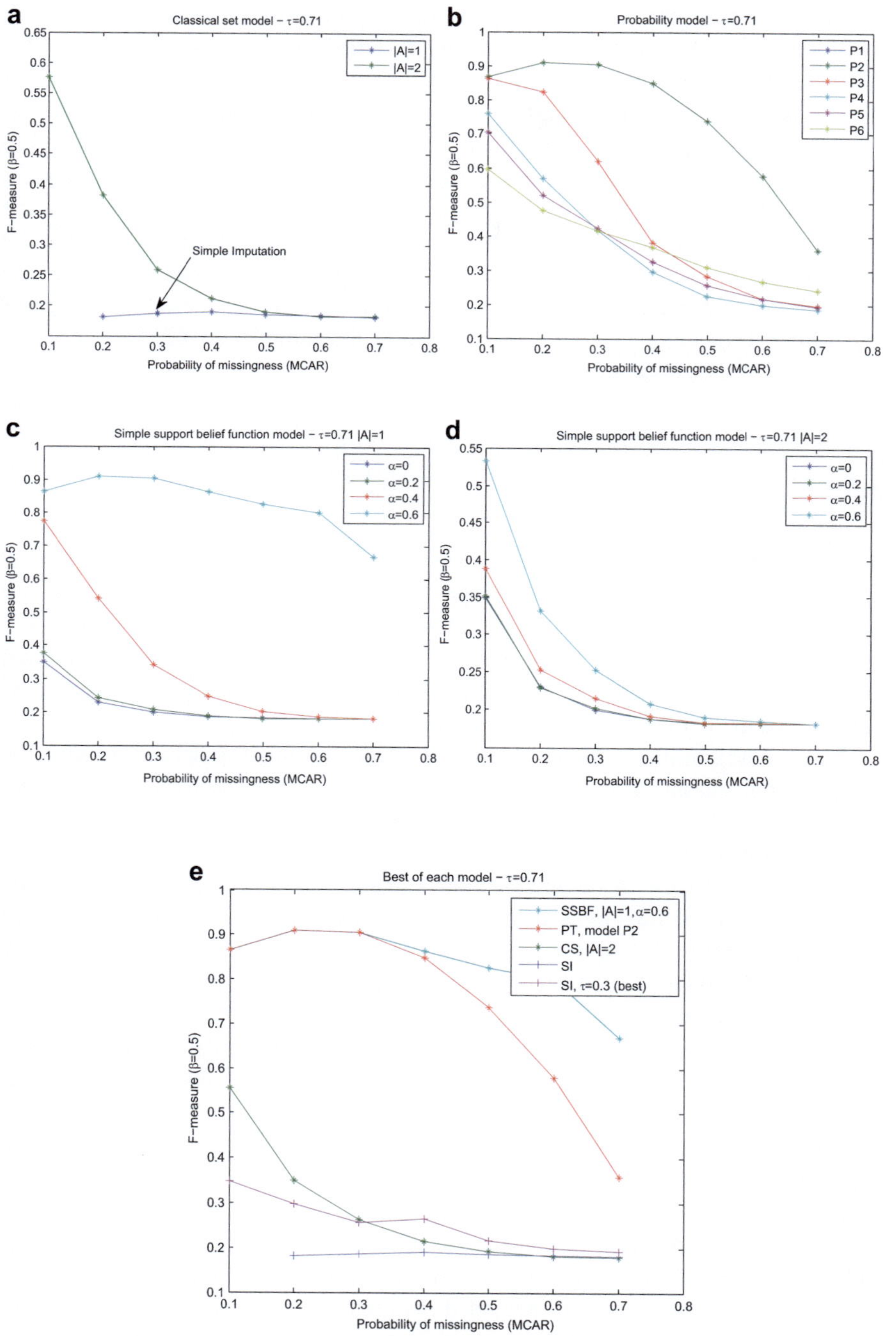

Fig. 5.5 Retrieval results according to the F-measure obtained on the Solar Flares dataset for a threshold $\tau = 0.71$ over the distance values. (**a**) Classical set model, (**b**) probability model, (**c**) simple support belief function model, $|A| = 1$, (**d**) simple support belief function model, $|A| = 2$, (**e**) best of each model

This probability distribution has a high value over the most frequent element of $|\mathcal{X}_j|$ and the remaining probability is randomly assigned to the other elements of $|\mathcal{X}_j|$. As the distribution tends to a uniform one, the performances tend toward that of the classical set model. Figure 5.5c, d shows several variations of the SSBF model for $|A| = 1$ and $|A| = 2$, respectively. A SSBF with a mass of 0.6 assigned to the singleton with the most frequent element of $\mathcal{X}_j$ provides the best performances over all the tested models. For the sake of comparison, Fig. 5.5e summarizes the best results of each model. In particular, we see that the SSBF model seems to be more robust to an increased proportion of missing data compared for instance to the probability model. In this figure, we also added the best result obtained for the Simple Imputation model, with a different threshold τ of 0.3. Due to space restrictions, all the results cannot be shown.

5.5 Conclusions

Retrieving items such as similar past events, or vessels with a specific characteristic of interest, for crisis management support is often a critical task especially in incomplete databases. A lot of the retrieving tools simply ignore items with empty fields although these can be very relevant to the query. The quality of the situation awareness thus depends on the quality of the retrieving method to adequately consider all the items whether they are complete or not.

We proposed in this work an approach to deal with missing data in information retrieval applications. We showed in this work that accounting for the uncertainty about the missing has an impact on the retrieval performances.

In accordance to evidential databases, we considered the general uncertainty representation scheme of belief functions and we replaced the missing values by a belief function defined over the domain of features, as an extension to standard imputation techniques. Among the different types of variables which possibly constitute a database, we considered the particular case of categorical variables. The retrieval task is performed, through an aggregated measure of distance within the framework of belief functions. On a specific example of a real dataset, we observed the impact of different uncertainty representations for missing data on the retrieval task. The simulation results show that introducing a small amount of uncertainty, i.e. simply considering two values instead of a single one, indeed increases the performance: More relevant items are retrieved while less irrelevant items are not retrieved. Moreover, additional uncertainty (e.g., probability or belief function) yet increases the performances for some models. In particular, the belief function model shows a higher robustness to high missing data proportions. Other tests on other datasets must be performed to extend the validity of these results. In particular, from these preliminary results, not surprisingly, it seems that the definition of the focal elements (i.e. cardinality of the set in the case of a classical set) has a higher impact than the numerical values of the BPA (i.e. probability distribution). Further tests are required to better identify the role of these numerical assignments.

The main purpose of the study was to highlight the impact of uncertainty representation on the information retrieval performances. Now, a next step would be to identify the optimal uncertainty model for a given application. The proposed approach allows to account for the possible correlation between uncertainty representations across features, but this has not been exploited yet. The analysis of other criteria of performances such as the computational complexity or the interpretation (i.e. meaning of numerical values) appear also to be of primary importance for an efficient user interaction in crisis situations.

References

Aamodt A, Plaza E (1994) Case-based reasoning: foundational issues, methodological variations, and system approaches. AI Commun 7(1):39–59

Ahlgren P, Grönqvist L (2006) Retrieval evaluation with incomplete relevance data: a comparative study of three measures. In: 15th ACM international conference on information and knowledge management, Arlington

Bach Tobji MA, Ben Yaghlane B, Mellouli K (2008) A new algorithm for mining frequent itemsets from evidential databases. In: Magdalena JVL, Ojeda-Aciego M (ed) Proceedings of IPMU, pp 1535–1542

Brini A, Boughanem M, Dubois D (2005) A model for information retrieval based on possibilistic networks. In: String processing and information retrieval (SPIRE 2005), Buenos Aires. Lecture notes in computer sciences. Springer, New York, pp 271–282

Buckley C, Voorhees EM (2004) Retrieval evaluation with incomplete information. In: Proceedings of the 27th annual international ACM SIGIR conference on research and development in information retrieval (SIGIR '04), Sheffield, pp 25–32

Burkhard H-D (2004) Case completion and similarity in case-based reasoning. Comput Sci Inf Syst 1(2):27–55

Chen LA (1988) On information retrieval and evidential reasoning. Tech. Rep. UCB/CSD-88-429, EECS Department, University of California, Berkeley

Chen N, Dahanayake A (2007) Role-based situation-aware information seeking and retrieval for crisis response. Int J Intell Control Syst 12:186–197

Chowdhary KR, Bansal VS (2011) Information retrieval using probability and belief theory. In: International conference on emerging trends in networks and computer communications (ETNCC), pp 188–191

Costa PCG, Laskey K, Blasch E, Jousselme A-L (2012) Towards unbiased evaluation of uncertainty reasoning: The URREF Ontology. In: Proceedings of the 15th International Conference on Information Fusion, Singapore

Crestani F, Lalmas M, Van Rijsbergen CJ, Campbell I (1998) Is this document relevant? ... probably: a survey of probabilistic models in information retrieval. ACM Comput Surv 30(4):528–552

Dalvi N, Re C, Suciu D (2009) Probabilistic databases: diamonds in the dirt (extended version). Commun ACM 52:86–94

da Silva WT, Milidiú RL (1993) Belief function model for information retrieval. J Am Soc Inf Sci 44(2):10–18

Farhangfar A, Kurgan L, Pedrycz W (2007) A novel framework for imputation of missing values in databases. IEEE Trans Syst Man Cybern - A: Syst and Humans 37(5):692–708

Fuhr N (1992) Probabilistic models in information retrieval. Comput J 35:243–255

Hewawasam GK, Premaratne K, Subasingha M-L, Shyu SP (2005) Rule mining and classification in imperfect databases. In: Proceedings of the 7th international conference on information fusion

Jousselme A-L, Grenier D, Bossé E (2001) A new distance between two bodies of evidence. Inf Fusion 2:91–101

Joussselme A-L, Maupin P (2012) A brief survey of comparative elements for uncertainty calculi and decision procedures assessment. In: Proceedings of the 15th international conference on information fusion, 2012. Panel Uncertainty Evaluation: Current Status and Major Challenges

Jousselme A-L, Maupin P (2013) Comparison of uncertainty representations for missing data in information retrieval. In: Proceedings of the international conference of information fusion, Istanbul

Kim W, Choi B-J, Hong E-K, Kim S-K, Lee D (2003) A taxonomy of dirty data. Data Min Knowl Discov 7(1):81–99

Klir GJ, Yuan B (1995) Fuzzy sets and fuzzy logic: theory and applications. Prentice Hall International, Upper Saddle River

Lalmas M (1998) Information retrieval and Dempster-Shafer's theory of evidence. In: Applications of uncertainty formalisms. Lecture notes in computer science, Chap. B. Springer Berlin/Heidelberg, pp 157–176

Lee SK (1992) Imprecise and uncertain information in databases: an evidential approach. In: Proceedings of the 8th international conference data engineering, pp 614–621

Lynch SM (2003) Missing data. www.princeton.edu/~slynch/soc504/missingdata.pdf

McClean S, Scotney B, Shapcott M (2001) Aggregation of imprecise and uncertain information in databases. IEEE Trans Knowl Data Eng 13:902

National Counterterrorism Center (NCTC) (2010) Worldwide Incidents Tracking System (WITS) report on terrorism. http://www.nctc.gov/, April 2011

Schafer JL, John WG (2004) Missing data: our view of the state of the art. Psychol Methods 7(2):147–177

Schmidt R, Vorobieva O (2007) Applying case-based reasoning for missing medical data in ISOR. In: LWA 07, pp 275–280

Telmoudi A, Chakhar S (2004) Data fusion application from evidential databases as a support for decision making. Inf Softw Technol 46:547–555

Wu S, McClean S (2006) Evaluation of system measures for incomplete relevance judgment in IR. In: Flexible query answering systems. Lecture notes in computer sciences, vol 4027. Springer, New York, pp 245–256

Yassir A, Nayak S (2012) Issues in data mining and information retrieval. Int J Comput Sci Commun Netw 2:93–98

Yi X (2011) Discovering and using implicit data for information retrieval. Ph.D. thesis, University of Massachusetts Amherst

Zaffalon M (2002) Exact credal treatment of missing data. J Stat Plann Inference 105(1):105–122

Part II
Context in Crisis Management

Chapter 6
Crisis Management and Context

Galina L. Rogova

Abstract Successful management of critical situations created by major natural and man-made activities requires monitoring, recognizing, fusing, and making sense of these activities in order to support decision makers in either preventing a crisis or acting effectively to mitigate its adverse impact. Context plays an important role in crisis management since it provides decision makers with important knowledge about current situations and situation dynamics in relation to their goals, functions, and information needs, to enable them to appropriately adapt their decisions and actions. Efficient context exploitation for crisis management requires a clear understanding of what context is, how to represent it and use it. The chapter provides a brief discussion of the key issues of the problem of context definition, representation, discovery, and utilization in crisis management.

6.1 Introduction

The Merriam-Webster dictionary defines crisis as "unstable or crucial time or state of affairs, in which a decisive change is impending; *especially*: one with the distinct possibility of a highly undesirable outcome." Crisis situations can be caused by natural and/or man-made activities and usually result in dangerous social, political, economic, and large-scale environmental events. It is important to monitor, recognize, and make sense of these activities as early as possible in order to either prevent a crisis or mitigate its outcome. There are two types of critical situations to be considered: declared or discovered and underlying or hidden crises. Often a crisis situation results in different crisis situations (cascading crises) and it is necessary to monitor the environment during a declared crisis for a timely discovery of these "cascading crises" while they are still evolving.

This chapter is an extended and revised version of (Rogova 2009)

G.L. Rogova (✉)
State University of New York at Buffalo, Buffalo, NY, USA
e-mail: rogova@buffalo.edu

G. Rogova, P. Scott (eds.), *Fusion Methodologies in Crisis Management*,
DOI 10.1007/978-3-319-22527-2_6

Predicting and recognizing crises, and effectively responding to them require systematical gathering and analyzing a large amount of heterogeneous data and information of varying significance and accuracy, often redundant, unreliable, or contradictory. This information has to be fused and reasoned about to produce knowledge about current and predicted situations, which is a task of higher level fusion (Situation and Threat assessment), (see for example Liggins et al. 2009). This knowledge is essential for providing crisis management actors with situation awareness necessary for successful control, prevention, and mitigating crises.

The success of this effort strongly depends on the quality of situation understanding, communication, and actions, and requires (Rogova 2009):

- Contextual understanding of the characteristics and behaviors of the dynamic environment of the declared crisis to support:
  - damage assessment,
  - public health response,
  - search and rescue,
  - evacuation,
  - resource allocation, identification of obstacles and interruptions to emergency operations.
- Reasoning leading to detection, discovery, and recognition of evolving underlying crises.
- Prediction of consequences of current situations to support preventive and mitigating measures dealing with:
  - public safety and security,
  - property loss, possible adverse environmental, economic, and political consequences.
- Effective and secure communication between all actors involved.

Crisis responders and decision makers have to be aware of current context and context dynamics to appropriately adapt their decisions and actions. Declared crises afford decision maker and crisis responders context by providing essential information about the current and future states of the environment, which is required to perform their functions and achieve their goals. The hierarchical structure of crisis management demands context consideration at the different levels of granularity to provide crisis management actors with the information tailored to their specific tasks. It is important to know that decisions and actions of crisis management actors can affect the states of the environment and therefore context at a certain level of granularity.

For evolving undeclared crises, context may be utilized for detecting deviations from usual, which can trigger abductive reasoning aimed at explaining the observations uncharacteristic of the context under consideration and possible discovering of new possibly unexpected or even unimaginable situations and threat. Awareness of discovered new situations allows crisis management actors to adjust their plans and actions and act efficiently in a new context. Context also simplifies communications between the crisis management actors since it "has the effect of narrowing down communicative possibilities of the message as it exists in abstraction from context" (Leech 1981).

In general, context exploitation can offer benefits to higher level fusion and crisis management by:

- transforming data into information and information into knowledge
- acquiring knowledge
- eliminating or reducing ambiguity
- detecting inconsistencies
- explaining observations
- constraining processing
- facilitating effective communications.

Efficient context exploitation requires a clear understanding of what context is, how to represent it, and use for processing information about the environment for providing knowledge to the crisis management actors to support effective decisions and action. Utilization of context in crisis management raises further questions concerning:

- the relations between situations, context, and knowledge
- the way context representation, contextual reasoning, and decision making are affected by the uncertainty inherent to the crisis environment

The remainder of this chapter provides a discussion of key issues of definition, representation, discovery, and utilization of context in crisis management.

6.2 Context

The notion of context has been used in diverse research areas for a long time but "while most people tacitly understand what context is they find it hard to elucidate" (Dey 2001). Context has many facets that sometime lead to defining it based on certain narrow characteristics of the particular problem being addressed. Such definitions are too specific to the problem under consideration making it difficult, even sometime impossible, to understand which characteristics of context should be taken into account in general.

A definition that does allow for better understanding of context and its relation to situations—and therefore one that is more appropriate for formalizing and utilizing context in building higher level fusion and crisis management processes—was proposed in Gong (2005) and further discussed in Steinberg and Rogova (2008) and Rogova and Steinberg (in press).

This definition assumes two different context paradigms: "*Context of X*" (CO) and "*Context for X*" (CF), which correspond to two basic meanings conveyed by dictionaries (Gong 2005):

"To knit or bind together; to unite closely" (CO) and
"That which surrounds, and gives meaning to, something else" (CF).

A reference item X, for which context is considered, is a topic of interest represented by any physical or conceptual entity, for example, a situation and event of interest such as a natural phenomenon or terrorist threat. Reference items are represented by a set of *problem variables* describing situational items, their relationships, and behavior that an agent wishes to evaluate and characterize. They are evaluated on the basis of the characteristics of both CO and CF (*context variables*). A CO is a part of the environment corresponding to a set of items and relationships of interest "grouped or contained by X". We have certain expectations about X based on a given CO, e.g., in the context of earthquake we can expect damaged roads and bridges; or in the context of wildfire we can expect burn victims. Alternatively, a CF defines the contextual space of items externally related to and referenced by X: the weather provides a *context for* search and rescue after an earthquake (*context of*). Reasoning about entities and relationships while considering them as problem variables within a certain context corresponds to reasoning about situations. Such reasoning produces an answer to the question: "what is going on in the part of the environment corresponding to the reference item(s) characterized by a set of problem variables within a specific context characterized by a specific set of contextual variables." Context variables can serve as problem variables when they represent reference items for a different context.

A set of items and relationships defining a CF corresponds to auxiliary variables determined (by some process) to be relevant to a given problem. They affect knowledge about problem variables contained in a CO, reasoning about them and, therefore, affect decisions and actions based on the values and behaviors of problem variables. For example, the time of the day (CF) affects the possible distribution of casualties considered in the context of earthquake (CO).

Situation assessment processes use CF to better understand reference items and relations between them. Crisis responders and decision makers have to take CF into consideration to optimize decisions and actions. CF is equally necessary for both crisis response in the case of declared crises and discovery of evolving crises. For example, evacuation measures during a wildfire have to take into account the strength and the direction of the wind, which represent CF for evacuation and rescue. At the same time CF for observed and/or inferred reference items and relations between them may support recognition of an evolving hidden crisis representing an undiscovered CO of these reference items. For example, an epidemic of flu in one country may be considered as CF for flu-like illnesses in another country. Consideration of the contextual variables of this CF will provide a better understanding of the nature of this illness and help to discover a possible evolving flu epidemic.

While COs can be declared, inferred, or discovered as the result of reasoning, CFs can be either given, obtained as the result of direct observations, or inferred. A CO is a background context, which usually provides a more general and stable environment. Consideration of CO and CF also offers a clear understanding of relations between context and situations. Reasoning about entities and relationships while considering them as problem variables within a certain context corresponds to reasoning about situations. Such reasoning produces an answer to the question "what is going on in the part of the environment corresponding to the reference item(s) within a specific context?" Thus we can define a *problem (crisis) context* as a meta-situation (situation of higher level of granularity), comprising a set of relationships involving context variables: ($C = \mathrm{PV}_i$, CV_i, R_i), where PV_i and CV_i are problem and context variables respectively; and R_i are relationships between various problem variables, various context variables, and between problem and context variables. Modeling problem context for crisis management is generally reduced to the following problem (Gong 2005): "Given an entity of interest (a physical object, a situational item, and an event) what context or a sequence of contexts can be formed, such that a task about this entity can be accomplished."

Since the crisis environment inherently involves uncertainty, we can say that the value of a problem variable PV_i is pv_i with plausibility pl_i in a certain context CF_j for which this problem variable is a reference item: $\mathrm{pl}(\mathrm{PV}_i = \mathrm{cV}_i)_{|\mathrm{CF}j} = \mathrm{pl}_i$ Plausibility here can be replaced by another uncertainty metric—belief, possibility, probability, etc.—depending of the uncertainty theory considered.

Crisis management comprises a complex distributed hierarchical organization comprising actors with different goals, functions and information needs, in which decisions at certain levels of hierarchy may have a tactical character (e.g., activities in direct response to casualties reported at a specific location), while decisions at higher levels may have a more strategic character such as understanding evolving crisis situations related to a larger region. Complex hierarchical relations between decision makers and their goals, objectives, functions, and information requirements define specific sets of problem and context variables and relations between corresponding problem contexts. Thus consideration of context at different levels of granularity in crisis management is necessary for partitioning of the knowledge space to provide specific knowledge to various decision makers and crisis responders.

The authors of Brezzilon (1999) and Brézillon and Pomerol (1999) provide a relationship between knowledge and context as applied to decision making and actions. They consider context as "a shared knowledge space that is explored and exploited by participants" and introduce the notions of external knowledge, contextual knowledge, and proceduralized context. Thus external knowledge represents context under consideration at a highest level of granularity. It is background knowledge, which is known to each actor of situation management but does not directly affect their decisions and actions. *Contextual knowledge* is a part of the context, which is relevant to a given problem at hand at a given time for a given agent (human or automatic). Contextual knowledge is the part of knowledge that can have several realizations ("proceduralized context"), which dynamically instantiates contextual knowledge based on current information and transforms it into "functional knowledge" used for reasoning, decision making, or actions according to a specific focus of a specific time. In the terms of CO and CF, we can think about external knowledge as CO at a certain level of granularity. For example, earthquake in a certain area (CO), characterized by a vast knowledge of different types about general earthquake characteristics related to damage, casualties, the road conditions, hospital capacities, etc. in this and surrounding areas represents external knowledge for crisis management actors with their goals, objectives, and tasks.

At the same time each actor has to know specific situations related to his specific goals and objective, and functions. For example, for an ambulance dispatcher who has to know the distribution of casualties and distribution of ambulances (CO at a lower level of granularity as related to earthquake in general), which represent his contextual knowledge. In the dynamic environment contextual knowledge has many realizations and in order to send ambulances he needs to know current casualty and ambulance situation (CO) along with the weather and time of the day (CF), which represents the instantiation of his contextual knowledge (proceduralized context) to fulfil his task.

6.3 Context Representation

As defined in the previous section, context is a meta-situation for a set of problem variables under consideration. It should provide contextual knowledge to the crisis management actors and allow for complex dynamic representation of their particular goals and functions as well as objects, their characteristics, and interrelationships characterizing the crisis environment. Thus a context representation should (Steinberg and Rogova 2008):

- offer a formal and uniform way for specifying CO and CF at different levels of granularity,
- be flexible enough to easily adapt to the dynamics of the crisis situation,
- be broad enough to include possible and unknown cascading crises,
- provide a high level of formality in description of contextual information,
- be easily understandable by the users and support information sharing and reuse,
- provide for dealing with uncertain situation variables and relations.

Models with these characteristics are similar to ones used for modeling situations and include *ontology-based* and *logic-based* models (see for example Steinberg and Rogova 2008; Rogova and Steinberg in press; Stang and Linnhoff-Popien 2004; Bettini et al. 2010; McCarthy 1993).

Ontology is an established framework for knowledge representation and for reasoning about situations (Kokar et al. 2009; Little and Rogova 2009). It can provide a high degree of rigor in specifying core concepts, sub-concepts, facts and their inter-relationships to enable realistic representation of contextual knowledge for reasoning, information sharing and reuse.

Logic-based models define contexts in terms of facts, expressions and rules. McCarthy (1993) introduces a formalization of logic-based models built on a relation *ist*(*c*,*p*), read as "proposition *p* holds in the context *c*." The *ist* concept can be interpreted as validity: *ist*(*c*; *p*) is true if and only if the proposition *p* is true in context *c*. Information fusion systems generally deal with uncertain data and therefore it is necessary that context representation in logic-based models allows for uncertain statements, rules, and beliefs assigned to them. To use *ist*(*c*,*p*) concept for making assertions about uncertain situational items in this framework, it is necessary to expand McCarthy's definition of *ist(c,p)* by incorporation of an uncertainty measure—expressed as probability, possibility, or belief—in place of McCarthy's binary one. For example, *bel (a,ist(c,p))* can be used to represent an agent *a*'s belief that proposition *p* is true in the context *c*. Since an agent's knowledge about context can be uncertain, *bel* may represent a combination of belief in the validity of proposition *p* and belief associated with context characteristics.

Other logic based models are *Situation Theory* (Akman and Surav 1997) and *Extended Situation Theory* (Devlin 1991). *Extended Situation Theory* expands Situation Theory by modeling context with situation types corresponding to objects of Situation Theory. The variety of different contexts is addressed in the form of rules and beliefs related to a particular point of view. Situation Theory represents units of information as *infons*, which are denoted as $\sigma = (R, a_1, \ldots, a_n, i)$, where R is an n-place relation and $a_1, \ldots, a_n$ are state variables that range over entities of types appropriate for a given relation R. In "classical" Situation Theory, i is a binary variable, which is equal to 1 if a relationship $R(a_1, \ldots, a_n,)$ holds, 0 otherwise. In Extended Situation Theory (Akman and Surav 1997) context is modeled by situation types corresponding to objects of Situation Theory that supports two kinds of infons: (1) factual infons to state facts, and (2) constraints (which correspond to parametric conditionals) to capture the if–then relations holding within a context. To capture uncertain if-then relations holding within the context representing a part of the uncertain environment, it is necessary to consider *uncertain* infons. Theory incorporates uncertain infons simply by redefining the binary variable i of classical Situation Theory as a continuous variable $i \in [0,1]$ to represent the belief that R holds. An equivalent probabilistic extension of Situation Theory was derived independently in Steinberg (2008).

The context representations mentioned above can provide a high level of formality in the description of contextual information, while each of them has its advantages and drawbacks. Therefore, hybrid models may better serve for modeling context for crisis management. Such hybrid models may include a combination of ontology-based and situation theory (Kalyan et al. 2005) or contextual graphs (Thagard and Shelley 1997), incorporating both action and context nodes (variables and relationships) as well as paths through them. Although the contextual graph is not free of weaknesses, e.g., there may be problems with time representation and complexity, it offers certain advantages over other approaches since "it allows a representation of knowledge and reasoning in a way that is directly comprehensible by users. At the action level, making explicit contextual elements allows to explain the reasons for the choice of an action on another one. Thus, information in contextual graphs is useful and usable for operators" (Brézillon 2003).

6.4 Selection of Context Variables

Context representation and exploitation requires methods of selecting context variables describing CF and CO. Selection of context variables is one of the major challenges of modeling and exploiting problem contexts. In general, such selection should be relevance based for constraining domain ontology of the context under consideration and relations of context variables with problem variables as well as goals and functions of crisis management actors.

A context variable can be called relevant if the values of problem variables, decision, or action under consideration change with the value of the context variable. If the problem variables are objects, object attributes, or relations, this relevance can be defined in terms of mutual information between problem and the candidate context characteristic. If the problem variables are situational items, we can call a contextual characteristic relevant if a change of its value affects the uncertainty distribution of hypotheses about these situational items, and therefore decisions of crisis managers and actions of crisis responders. Relevance of a context variable can be also defined by decision makers and responders based on their information needs.

Another criterion for selecting a particular context variable is the increase in information or achieving a higher action utility as the result of using that characteristic for estimation and/or inferencing. Action utility includes consideration of benefits of a particular action and its cost that may involve costs of data collection, communications and processing, as well as the cost of lost opportunity.

The problem of selecting relevant context variables is complicated by the fact that relevance is often time-dependent. The crisis environment and therefore information needs and fusion processes are often dynamic, which makes the utility of information time-dependent. Situations of interest are also often dynamic, such that the availability of any sought data may be time-dependent. Furthermore, the cost of data acquisition and processing varies with resource and situation states.

6.5 Context Aware Crisis Management

Figure 6.1 presents a framework for designing context aware crisis management systems and shows the interrelations between context, fusion, and crisis management processes. As it can be seen from Fig. 6.1, a context engine interacts with and supports interrelated fusion and crisis management processes. Context plays an essential role in higher level fusion, in which variables of interest including relation—and situational variables are being evaluated and reasoned over to afford situation understanding. Generally, the larger context in which a problem is considered, the more fully will it be understood being conditioned on a large number of mutually independent context variables. Context consideration can improve the results of fusion products and therefore situation awareness, decisions, and actions of crisis management actors by taking into account the quality of input information (e.g., reliability of observations and reports) and interim results of the processes involved in fusion. For instance, a CO can serve for selecting relevant observations and provide expectations for sensor and process management. A CF can, for example, be used to improve fusion results by incorporating context based reliability into sources' predictive uncertainty models such as probability or belief.

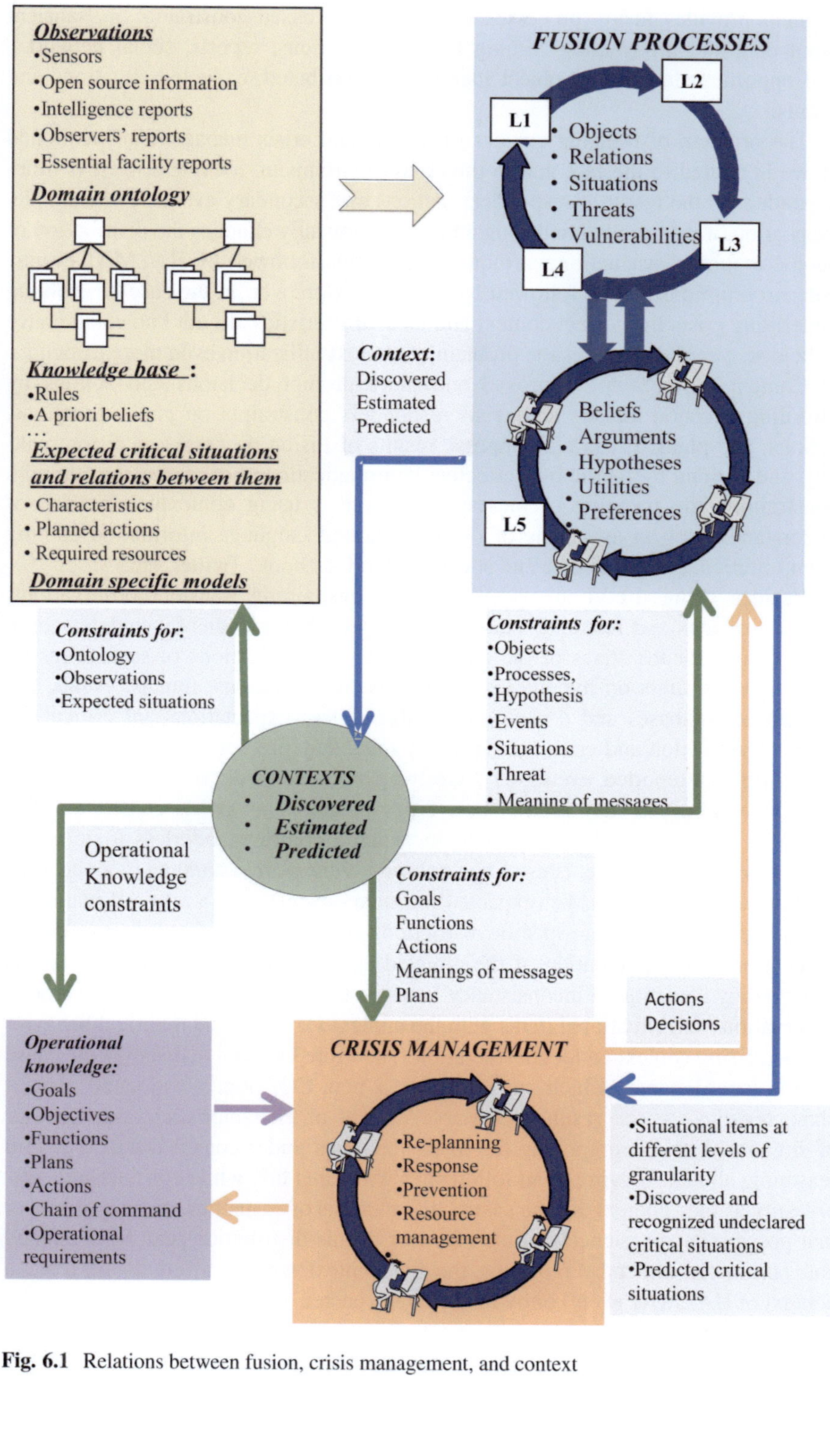

Fig. 6.1 Relations between fusion, crisis management, and context

Context provides fusion processes with relevance based constraints on transient input data and information coming from observations, reports, social networks, and opportunistic sensors; object identity and attributed, expected situations and events.

The problem of utilizing context for fusion and crisis management mentioned above is related to the fact that in the crisis environment, a context itself is time-dependent as the result of responders' actions and secondary events. For example, evacuation of the population during a flood dynamically changes the distribution of people requiring evacuation; earthquakes are often followed by HAZMAT events, volcano eruptions can be followed by fires, etc. Moreover, in the case of evolving underlying crisis the correct context and its characteristics are not known and have to be discovered that makes the problem of context utilization even more difficult.

Context provides for improved and more prompt decisions and actions by affording decision makers and crisis responders constraints on goals, functions, actions, and plans. It can also improve results of fusion processes, decision making, and actions by facilitating effective communications between actors through constraining the meaning of messages as well as using contextual filtering of understanding of an avalanche of dynamic natural language information coming from traditional media source and social media (Facebook, Twitter, etc.)

Besides being useful in evaluating and constraining transient observations, object attributes and relational states, context provides a means for understanding expectations for the states of the environment and implications of such states by offering constraints on hypotheses about expected situations, threats, events, the content of databases and domain knowledge. These expectations are critical for higher level fusion and crisis management since they present means for discovery of possible unreported events or underlying non-declared crises representing a new context, which may require new types of action or action constraints. For example, discovery of Hazmat event, in which a colorless, odorless toxic gas is vented to the atmosphere, caused by unobserved/unreported damage to a chemical facility during an earthquake (expected secondary event) or as a result of malicious human activity can require specific restrictions on the routing of ambulances or/and evacuation of the population of the affected area. Discovery of a new context is triggered by detection of inconsistency between the characteristics and behavior of observations and situational items with the context variables and their and behavior characterizing context under consideration. Therefore we need to discover the cause of this inconsistency, which can be either a new CO, poorly estimated context characteristics, or the result of variable quality of characteristics and behavior of situational items comprising the reference items under consideration. Thus, in reasoning about an environment following a wide-area fire, while considering *wide-area-fire* as the "context of" the spatial distributions of casualties, we may discover that possible inconsistencies of the behavior of this distributions can be explained and eliminated either by replacing the fire context with a different one (e.g., a context of HAZMAT event) or by modifying expected distributions in context under consideration.

An effective way of solving the problem of discovery of a new context explaining this inconsistency is to use abductive reasoning or "inference for best explanations" (Thagard and Shelley 1997; Josephson 1990). The abductive process of reasoning from effect to cause requires:

- constructing or postulating hypotheses about possible context,
- computing plausibility of these hypotheses,
- selecting the most plausible hypothesis.

Hypothesis construction can be either implemented by the analysis of possible cascading crises, simulated scenarios, and data mining, or provided by domain experts. As it was mentioned above, context represents a meta-situation for problem variables and therefore the problem of hypothesis plausibility computations and hypothesis selection can be solved by employing various inference methods designed for situation assessment such as belief-based argumentation (Rogova et al. 2006) or belief networks (Jensen 1996). Abductive reasoning for context discovery for crisis management has to take into account the lack of complete knowledge and insufficient and perhaps even non-existent statistical information. Moreover, the set of hypotheses about possible context may not be exhaustive, and an open world assumption has to be included in the process of hypotheses evaluation. This requirement imposes certain constraints on a possible framework for uncertainty representation, which has to deal with the lack of statistical data and must allow for reasoning and fusion under the open world assumption. One of the most appropriate frameworks for uncertainty representation in such environment seems to be the Transferable Belief Model (Smets and Kennes 1994), which assumes an existence of not probabilities but sub-additive basic belief masses existing independently of any probabilistic model, can deal with the lack of statistical data, and allows for reasoning under the open world assumption.

The process of hypothesis selection has to take into account the following considerations (Josephson 1990):

- whether the degree, to which the hypothesis to be selected is better than alternatives,
- the degree of credibility of the hypothesis by itself, independently of considering the alternatives; one should be cautious about accepting a hypothesis even if it is clearly the best one we have if it is not sufficiently plausible in itself.

Abduction reasoning can result in discovering a new context or poor quality of the estimation of context and/or problem variables. A very important consideration here is the quality (credibility, reliability, etc.) of context variables, assessed reference items as well as the quality of incoming information. One unavoidable problem related to the quality of incoming information hampering the problem of context discovery is the problem of time delay affecting the timeliness of decisions and actions. The time delay is a combination of time required for discovery of an event, communication and information processing, and response time. To account for the time delay it might be necessary to consider projected in time characteristics and behavior of situational items and characteristics of CF. At the same time incorporation of the time delay into fusion processes and actions represents one of the major challenges of higher level fusion and context exploitation.

6.6 Conclusions

Context plays an important role in crisis management by constraining observations, domain knowledge, and a set of crisis specific models and decisions of crisis management actors as well as expectations about situational items and relations for situation and impact assessment processes. Context considerations provide the crisis responders and decision makers with necessary information for effective management of a detected and recognized crisis as well as crisis prevention or consequence mitigation. The chapter has discussed the notion of context in relation to crisis management and situation assessment, methods of context representation and discovery as well as presented a general framework for utilizing context in crisis management. Context exploitation in the uncertain crisis environment is a difficult problem, and designing a context aware crisis management system requires more research of such unsolved problems as efficient representation of dynamic context, selection of context characteristics, context quality control, scalability, and context discovery.

References

Akman V, Surav M (1997) The use of situation theory in context modeling. Comp Intell 13:3

Bettini C, Brdiczka O, Henricksen K, Indulska J, Nicklas D, Ranganathan A, Riboni D (2010) A survey of context modelling and reasoning techniques. Perv Mob Comput 6(2):161–180

Brézillon P (2003) Context dynamic and explanation in contextual graphs. In: Blackburn P et al (eds) CONTEXT 2003, vol LNAI 2680. Springer, New York, NY, pp 94–106

Brézillon P, Pomerol J-C (1999) Contextual knowledge sharing and cooperation in intelligent assistant systems. Le Travail Humain 62(3):223–246

Brezzilon P (1999) Context in problem solving: a survey. Knowl Eng Rev 14(1):47–80

Devlin K (1991) Logic and information. Press Syndicate of the University of Cambridge, Cambridge

Dey A (2001) Understanding and using context. Pers Ubiquit Comput 5(1):4

Gong L (2005) Contextual modeling and applications. In: Proc IEEE International Conference on SMC, vol V1. IEEE, Washington, DC

Jensen FV (1996) An introduction to Bayesian networks. Springer, New York, NY

Josephson J (1990) On the logical form of abduction. In: AAAI Spring symposium series: automated abduction. AAAI, Menlo Park, CA, pp 140–144

Kalyan A, Gopalan S, Sridhar V (2005) Hybrid context model based on multilevel situation theory and ontology for contact centers. In: Proc of the 3rd Int'l Conf on Pervasive computing and communications workshops (PerCom 2005 Workshops). IEEE, Washington, DC, pp 3–7

Kokar M, Matheus CJ, Baclawski K (2009) Ontology-based situation awareness. Inf Fusion 10:83–98

Leech G (1981) Semantics: the study of meaning, 2nd edn. Penguin books, Harmondsworth
Liggins ME, Hall DL, Llinas J (eds) (2009) Handbook of multisensor data fusion. CRC Press, London
Little EG, Rogova G (2009) Designing ontologies for higher level fusion. Inf Fusion 10:70–82
McCarthy J (1993) Notes on formalizing context. In: Proceedings of the Thirteenth International Joint Conference in Artificial intelligence, Chambery, France, pp 555–560
Rogova G (2009) Context awareness in crisis management. In: Proc of the IEEE Military Communications Conference (MILCOM 2009), 18–21 October 2009, Boston, MA
Rogova G, Steinberg A (in press) Formalization of context in information fusion. In: Snidaro L, Garcia J, Llinas J, Blasch E (eds.), Context enhanced information fusion, New York, NY: Springer
Rogova G, Scott P, Lollett C, Mudiyanur R (2006) Reasoning about situations in the early post-disaster response environment. In: Proc 9th International Conference on Information fusion, FUSION 2006, July 10–13, 2006
Smets P, Kennes R (1994) The transferable belief model. Artif Intell 66:191–243
Stang T, Linnhoff-Popien C (2004) A context modeling survey. In: First International Workshop on Advanced context modelling, reasoning and management, Nottingham, UK
Steinberg A (2008) Foundations of situation and threat assessment, Chapter 26. In: Liggins M, Hall D, Llinas J (eds) Handbook of multisensor data fusion. CRC Press, Boca Raton, FL
Steinberg AN, Rogova GL (2008) Situation and context in data fusion and natural language understanding. In: Proc Eleventh International Conference on Information fusion, Cologne
Thagard P, Shelley CP (1997) Abductive reasoning: logic, visual thinking, and coherence. In: Dalla Chiara M-L et al (eds) Logic and scientific methods. Kluwer, Dordrecht, pp 413–427

Chapter 7
A Multi-Agent Context-Management System for RECON Intelligence Analysis

Alexis Morris, William Ross, and Mihaela Ulieru

Abstract Adaptive systems require technologies to enable high synchronicity between its users and their unfolding situation dynamics, in concert with system response actions. To be effective, a multi-dimensional view of context must be considered and incorporated. This work advances the development of such a system for RECON, an initiative to support intelligence analysts with a novel context-management and case-based recommendation capability. The central concepts involved in the management of explicit and implicit contexts are presented and are developed into a novel multi-agent approach. In particular a new context-sensitive cognitive model and a community of expert service-oriented agents are proposed to facilitate and improve system adaptations to user-specific, situational, and system states. These designs pave the way towards future developments and experiments in improving human–machine interaction with adaptive context-management systems.

7.1 Introduction

Information overload presents a critical challenge to the development and success of advanced socio-technical systems, where humans-in-the-loop must make sense of an ever-increasing inflow of data in order to perform their tasks. Information generation enables unique perspectives of the world to be derived from raw data and facts, and with increases in data volume there is a need for better management of how this information is processed, presented, and consumed, especially in time-sensitive situations. With advances in technology, it has become increasingly important for decision-makers and strategists to make sense of world dynamics quickly, in order to stay ahead of unfolding situations. Failure to maintain control of dynamic situations (e.g., due to poor information management) can be the cause

A. Morris (✉) • W. Ross
Faculty of Computer Science, University of New Brunswick, Fredericton, NB, Canada
e-mail: alexis.morris@unb.ca; william.ross@unb.ca

M. Ulieru
School of Information Technology, Carleton University, Ottawa, ON, Canada
e-mail: mihaela@theimpactinstitute.org

G. Rogova, P. Scott (eds.), *Fusion Methodologies in Crisis Management*,
DOI 10.1007/978-3-319-22527-2_7

of small annoyances like reduced inefficiency, or even larger catastrophes (Vicente 2004). In both cases efficient management is vital, particularly as the trend towards a more information-dependent society continues.

To manage the dynamics of real-world information monitoring and sense-making, there are many different explicit and implicit contexts that can be considered by an adaptive system. However, the challenge is in finding the "right" context so that the system, in turn, can act as an aid (rather than a hindrance) to the expert human user. As in Ross et al. (2013), *context* is considered as anything that can be used to correctly identify the situation of a user. Explicit context can be provided directly by the user or generated based on the user's actions, such as system tasks recently performed (based on system logs) and current location data (based on mobile global-positioning systems) (Yılmaz and Erdur 2012). Implicit context, on the other hand, provides a less concretized notion, describing users' psycho-physiological states such as their current cognitive mood and stress level. These can be obtained through active and passive sensing of users via bio-metric sensors, although such states can also be deduced from explicit sources such as camera monitoring of facial expressions (Schmidt et al. 1999).

The successful management of both kinds of context is important. Systems that are adaptive to the dynamics of a wide range of contexts can increasingly support properties favouring the "5 Rights" (Fischer 2012)—i.e., providing the right information to the right person in the right place, at the right time, and in the right way (e.g., based on the preferences of the user). Such highly adaptive systems are becoming a focus at present (Hong et al. 2009), and this work aims to contribute a possible path forward in their development and practical usage, especially in the intelligence analysis domain.

7.1.1 RECON Intelligence Analysis

The RECON (REcommending Cases based on cONtext) system is a recent initiative aimed at providing a capability for intelligence analysts that takes into account their need for relevant information consumption in a time-sensitive environment. As a part of Defence Research & Development Canada's iVAC (Intelligent Virtual Analyst Capability) project (Gouin et al. 2012), RECON uses an adaptive-systems approach for information offloading and filtering to assist intelligence analysts. Its architecture has been outlined previously in Ross et al. (2013) and combines the following components: human–computer interface (HCI), data collector, case-based recommendation (CBR), brain–computer interface (BCI), and context manager. It is envisioned that such a unique combination of components, enhanced through the use of explicit and implicit context management, can better support analysts in performing their tasks by satisfying the different information "rights" mentioned above.

Intelligence analysis provides an interesting domain for the study of how to develop a flexible and adaptive technological solution to the problem of information overload, and the present work focuses on extending the current RECON system by introducing a multi-agent, context-management architecture. This approach, designed with flexibility and decentralization in mind, has at its core a community of service-oriented expert agents that can be extended towards a host of applications and user domains. This paper contributes to the discussion of adaptive context-management systems with a proposed architecture based on human factors and a context-sensitive cognitive model.

The remainder of the paper is organized as follows. Section 7.2 provides a human-factors perspective to context management, which is used in both Sects. 7.3 and 7.4, respectively, to present the context-management architecture in detail and to highlight the use of the cognitive modelling paradigm in context management. Section 7.5 then discusses related work in the field and highlights the novelty of the proposed context-management approach, while Sect. 7.6 presents the conclusion and offers direction for future work.

7.2 Human Factors for Context Management

The "human factor" is considered to be a significant contributor to effective system design and run-time operation (Vicente 2004). However, it is difficult to incorporate human behaviour into system designs due to the inherent multi-dimensional nature of socio-technical systems (Baxter and Sommerville 2011; Morris 2009). Such systems, as presented in Vicente (2004), can be considered according to the following five levels: physical, psychological, social, organizational, and political. Each level presents important considerations for context-management systems, as they must make their response adaptations fit both the unfolding real-world and human-factor dynamics.

At the *physical* level an adaptive system should fit the context related to the user's physical body, including its location and functional bio-states (e.g., muscle signals, heart rate, and galvanic skin response). At the *psychological* level the system should fit the internal cognitive view of the user, including the user's mental states, moods, intentions (as far as can be deduced), and particular (or known) cognitive limitations (e.g., working memory load, stress responses, and alertness levels). At the *social* level the system should fit the social-team dynamics of the user's personal network of both devices and independent agent actors (whether other humans or autonomous software entities). At the *organizational* level the system should fit the formal structures in place within the user's organization and respond to the dynamics within that structure, in addition to the role of the organization with other organizations (e.g., when information must be shared between two organizations the goals of these organizations must be considered). Lastly, at the *political* level the system should fit the current policies in place, enabling specific measures to be taken by the system in response to the dynamics at this level (such as the cascading effect of the introduction or removal of communications policies).

A modular approach to account for the multi-dimensionality of human factors is proposed for the RECON system, in the next sections, and is based on the notion of expert agents, each targeting a unique perspective for system adaptation.

7.3 A Multi-Agent, Context-Management Architecture for RECON

The RECON context-management architecture, while also being concerned with both the explicit and implicit contextual landscape of the user and system, emphasizes flexibility (i.e., being able to use different service providers to achieve the same function) and adaptability (i.e., being able to respond to changes in order to support the user in reducing information overload). To achieve these goals, a four-layer multi-agent systems approach is proposed as a foundation for the context-management architecture. The following four layers, identified in Fig. 7.1, are described below: the agent communication middleware; the generic, expert service-oriented agent community; the agent-oriented interaction logic; and the application-specific interfaces. Together, these layers create a system in which a community of experts—consisting of both humans and agents—can work together synergistically to reduce information overload and improve organizational effectiveness through emergent adaptation, which is discussed in the final subsection.

Communication is an essential requirement for a community of flexible agents (Bellifemine et al. 2001; Gregori et al. 2006). Software agents must be able to communicate dynamically with other agents without hard-coding specific agent communication at design time, thereby necessitating the need for agent discovery at run-time. Such a mechanism is greatly facilitated by existing agent communication middleware, such as can be found in JADE (Bellifemine et al. 2001), a FIPA-compliant, Java-based multi-agent platform, and SPADE (Gregori et al. 2006), a FIPA-compliant, Python-based multi-agent platform.

A communication channel is provided in this middleware to enable agent-to-agent interaction. In such communication middleware, an agent directory is present in which agents can register to become part of the community. The specific service(s) provided by the agent is also tracked using the service directory. Agents requiring a specific service need only search the directory to find the corresponding agent(s) that provides the service. It is then left to the agent's internal logic to determine which agent service to select.

7.3.1 A Generic Expert Agent Community

RECON envisions a multi-solutions approach towards reducing the problem of information overload, where each independent solution can be combined to form a

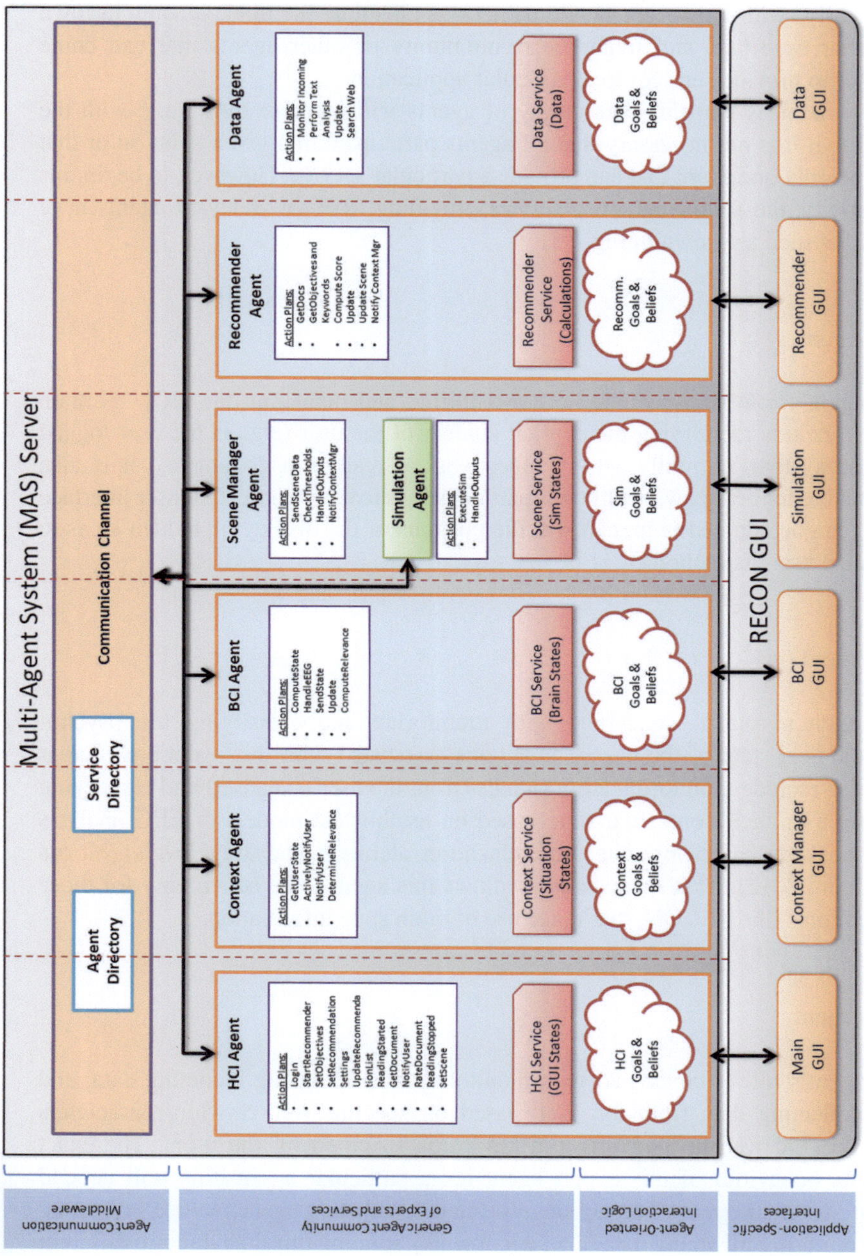

Fig. 7.1 A four-layer, multi-agent context-management middleware architecture in which a community of generic, expert service-oriented agents provides services to a domain-specific application, RECON, concerned with reducing a user's information overload

single, more comprehensive solution. These various solutions, including improved HCI adaptation, BCI monitoring, and data filtering, can be considered systems in their own right, each with a particular "expertise." Each agent can be created generically, without a specific domain or application in mind, having its own particular expertise, and living in a community of other agents that can come together to provide services to a particular application.

The vision is that this community of agents will increase over time, with the caveat that it is not necessary that all agents participate in a given solution or that there be only one agent that can provide a particular service. However, to begin this community, the following seven service-providing, RECON-relevant agents have been identified as shown in Fig. 7.1.

HCI Agent

This agent's role is concerned with monitoring and managing the HCI. Its main actions include identifying the explicit context of the user (e.g., is the user logged into the system, currently setting objectives, or reading a document). It is also responsible for being aware of how a particular application's graphical user interface (GUI) can be adapted (e.g., can specific portions of the display be hidden so as to minimize user distraction).

BCI Agent

This agent's role is concerned with monitoring and classifying the psycho-physiological state of the user. The actions associated with this agent's particular expertise include monitoring EEG signals (from the user's headset) and classifying the user's implicit contextual state based on established models of EEG analysis (e.g., measures based on excitement, relaxation, alertness, and stress levels) (Morris and Ulieru 2012). This specialization allows this agent to be active only for those applications, like RECON, that make use of brain state information.

Data Agent

This agent's role is concerned with monitoring and analyzing incoming data, and with collecting data from the Web based on specific Web-crawling parameters (e.g., specific website domains and keywords specified by the user). The exact run-time behaviour of this expert agent for a particular application will depend on the specified needs. The actions associated with this agent include collecting and monitoring incoming data (e.g., documents), running this data through different text-analyzing engines [such as AlchemyAPI (2015) and OpenCalais (2015)], and storing this processed data in a database for later use by other agents.

Scene Agent

This agent's role is concerned with the creation and monitoring of *scenes*. A scene refers to a particular aspect of a situation that a user may wish to offload to the system (e.g., perhaps the user is interested in being notified if $> X$ documents are found containing a particular set of keywords). The actions associated with this agent include storing scenes created by the user, monitoring incoming processed data to determine if specific scene conditions have been met, and issuing a notification if a scene's condition threshold has been reached.

Simulation Agent

This agent's role is concerned with the creation and execution of simulations, whose results act as particular scene conditions. The actions of this agent include storing the location of external simulation models or the models of internal simulations supported directly by the agent, as well as the input parameters that are passed to these simulations (e.g., a real number in the range [0,1] specifying the system's confidence in the relevance of a particular keyword based on the number of recent documents containing this keyword). Other actions include executing the simulations and storing the results in a database, which can be accessed by other interested agents.

Recommender Agent

This agent's role is concerned with ranking processed data (i.e., tagged documents) based on specific recommendation criteria so as to present the user with a recommendation of the most relevant system-data available. The actions of this expert agent include storing the user-specific recommendation criteria (e.g., relevant keywords, preferred website domains, and user rating history), and updating the recommendation list based on newly processed documents and user action and feedback (e.g., which documents have been read and what ratings were provided by the user).

Context Agent

This agent's role is concerned with assessing the current overall context of the user. The actions of this agent include acquiring all available implicit and explicit context from the other agents within the community, determining the current context of the user, managing what information is sent to the user (e.g., from the Recommender and Scene Agents), and initiating available GUI interventions (via the HCI Agent) to reduce experienced information overload on the part of the user. It is from these other agents that the context agent will collate and make sense of this information based on the active agents (i.e., those which form part of the multi-solutions approach for the specific application under consideration).

7.3.2 Agent-Oriented Interaction Logic

As discussed above, an agent will attempt to find other agent services it needs within the community proactively and, thus, self-organize around its goals without any explicit direction within the code. This is accomplished through the use of goal-oriented, belief-based agents. The Belief-Desire-Intention (BDI) software model enables an agent to be deliberative, i.e., the agent can select the action it will perform next based on the current state of the system (Carbo et al. 2001). This state includes all of the facts within the agent "world" (i.e., system)—incoming documents, HCI logs, BCI data, and simulation results—but that each agent, because of its expertise, restricts its consideration of these facts to those relevant to the completion of its goals. This model also ensures that these agents have bounded rationality so that the specific beliefs of a particular agent are not known to other agents unless they are explicitly shared. Such features enable flexible agent-to-agent interaction. Ultimately, the present architecture allows several agents to offer the same expert service, and an agent can choose which other agent(s) to interact with based on criteria such as past experience and recommendations from trusted "friends" (Carbo et al. 2001; Morris and Ulieru 2012). In this way, agents can be highly flexible, while still maintaining distinct expertise.

7.3.3 Application-Specific Interfaces

To link the generic agents within the expert community to a specific domain application an interface is required. This interface acts like a mechanism for articulating the needs of the application while respecting the expert services provided by the agents. Because the agents are both belief-based and goal-based, the articulation of application-specific requirements must necessarily take the form of world facts and agent beliefs. For example, for the Data Agent, beliefs can be provided by the application interface specifying the location of the data repository to monitor, or a list of important keywords to search for, and specific beliefs (flags) related to the agent's text analyzers.

There need not be a graphical interface associated with each belief-based agent at the application layer (e.g., in the case of the Data Agent, particular beliefs can be specified in code by the application developer, thus disallowing user-specified modifications). However, for the RECON application, a GUI has been created for each of the agents to increase the ability to tailor the system to a specific user, as the application targets the problem of reducing information overload, and this may require high levels of user-interface customization (Lafond et al. 2014).

7.3.4 Towards Emergent Adaptation in RECON

Rather than by considering the effects of any one agent, it is the combination of the interaction and particular expertise of each agent that enables system-level adaptation to occur (Ross et al. 2012). Specifically, in RECON, adaptation is primarily a combination of the following: the HCI Agent, i.e., the interaction log of what the user has done and is currently doing, as well as the explicit situational context and the possible GUI adaptations that have been enabled in code; the BCI Agent, i.e., what is the current psycho-physiological state of the user and how does this compare to the user's "normal" state; the Scene and Recommender Agents, which provide recommendations/notifications to the user; the Context Agent, which must collate this disparate information into action(s) that help reduce analyst overload; and the Data Agent, which influences the number of documents in the system, via its filter, and the meta-information associated with each document, via the text analysis mechanisms it employs. Thus, the adaptation of the system can be said to emerge as a result of the total interactions between each agent, based on their perspective of the unfolding situation, the user dynamics, and their particular action plans (response behaviors) (Bar-Yam 1997).

7.4 Cognitive Modelling for Context Management in RECON

As discussed in Sect. 7.2, the human psychological-level context represents an important challenge in the design of adaptive systems. However, in addition to detecting the user's psychological states, it is important to be able to operationalize this information in order to improve adaptive system behaviour. In this section, a human-performance cognitive-behavioural model is considered for the purposes of context management in RECON, mapping states to behaviours and system adaptation. Such an approach allows for system-level decisions about possible adaptations that could be beneficial based on the user's particular context.

The COntextual COntrol Model (COCOM), described in Gore et al. (2002) and based on the work of Hollnagel (1998), is a foundational model outlining four different control states that can be in effect for an individual based on the amount of time remaining for decision-making. These states—strategic control, tactical control, opportunistic control, and scrambled control—represent a continuum from strategic control, where the decision-maker has sufficient time to plan, to scrambled control, where the decision-maker is faced with very limited (to potentially no time) to plan. These control states result in set parameters that can be considered in calculations, such as the estimation of available time to an event horizon (i.e., the time until a significant event occurs) (Morris et al. 2012).

Similar control-specific notions are also applicable to context management and, in the case of RECON, a natural mapping can be made for the intelligence

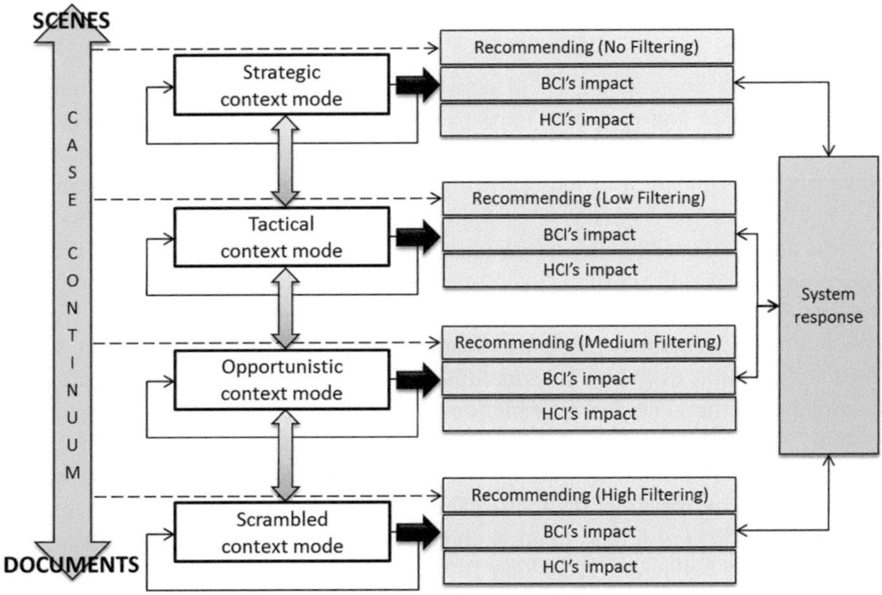

Fig. 7.2 The COCOM model (Gore et al. 2002) applied to context management

analysis domain. As shown in Fig. 7.2, the COCOM model has been fitted to support RECON's context-management approach. In RECON, two classes of recommendation exist: (1) cases (or scenes), which can include things such as newly simulated situation projections; and (2) documents, which consist mainly of new input data from the Web. Cases tend to reflect higher-level, strategic and tactical outlooks and would generally be created only when the user has sufficient time.

A user will be in one of four (possibly overlapping) context modes, determined by the context manager based on both explicit (e.g., HCI logs) and implicit (e.g., BCI state classification) context. These context modes, ordered according to decreasing time-to-plan and directly based on the mapping from COCOM states, are as follows: the *strategic context mode*, where there is a significant amount of time remaining before a decision is required; the *tactical context mode*, where there is sufficient time remaining to consider alternate avenues; the *opportunistic context mode*, where time is limited; and the *scrambled context mode*, where time is very limited (or has run out) and a decision must be made as soon as possible.

These four modes map to system adaptation and recommendation. When the strategic context mode has been identified, the system performs no special filtering of recommendations and alerts, allowing the user to view a wide range of information and case projections that may not be "on-task." Likewise, when the tactical context mode is deduced by the context manager, the system makes use of low filtering and hence the more off-task alerts and case projections are not directly presented to the user. In the opportunistic context mode, the system uses a medium level of filtering for recommendations and alerts, allowing near-task information to be shown to the user. Lastly, in the scrambled contextual mode, the system adapts with high filtering of incoming recommendations and alerts, presenting only on-task information to the user. The determination of on-task recommendations and alerts is based to a large extent on user preferences, such as the ranking of current objectives, keywords, and scenes, while the determination of the current context mode, as discussed earlier, is based on a combination of user-context data from both explicit and implicit sources. The effectiveness of the adaptation is adjustable based on direct feedback from the user following a system response, thereby allowing for improved adaptation in the future.

7.5 Related Work on Multi-Agent Systems for Context Management

In this section, a selection of related work on the use of multi-agent systems in context management is presented. These efforts reflect the importance of the context-management problem and the applicability of agent-based approaches.

In Yılmaz and Erdur (2012), the authors describe a multi-agent mobile context-aware system, where humans with client agents on their mobile phones navigate around a city, while the client agent connects to the context agent to determine nearby points-of-interest based on a generic points-of-interest ontology, in addition to finding people with similar interests close by. Such an approach, while multi-agent, does not use multiple agents in determining context, nor does it use a cognitive model for guiding context classification. Instead, it defines context strictly as being a location-based entity.

In Wei et al. (2008), the authors propose a multi-agent architecture for achieving user requests in a pervasive computing environment. At its core, it uses three agents to achieve context-awareness: the context collection agent, the ontology agent, and a reasoner agent. In some respects this is similar to the current work, as multiple agents are involved in context management; however, the specific agents used, as well as the mechanism for ultimately determining context are quite different (e.g., they do not use a cognitive model to guide context classification).

In Qureshi and Perini (2008), the authors propose a context-management system comprising four agents: a context monitor agent, a configuration selector agent, an enforcer agent, and a visualizer agent. Although a multi-agent approach is employed, this research focuses on the problem of improving software robustness in the face of viruses and malware and does not emphasize a community of generic expert agents.

In Song et al. (2010), the authors propose an ontology-based decision support for military information systems. As part of the approach, a multi-agent methodology is described in which five agents work together to support decision-making: a sensor agent, a context-management agent, a decision-support agent, a user agent, and an information-service agent. While similar in terms of domain, this approach does not share the same agents as RECON, nor is the emphasis on reducing information overload, although it does target improved decision-making. In addition, context appears to be derived only from rules pertaining to the sensor data and not a combination of the data (and, thus, context) arriving from several agents.

These works have shown that, while agent-oriented approaches have had a long and prominent history in context-aware systems literature (Hong et al. 2009), the use of a multi-agent approach to improve a system's context-awareness has received relatively less consideration and remains a promising avenue of exploration.

7.6 Conclusion

The RECON vision targets the adaptive context-aware systems domain for intelligence analysts, and has an explicit human-factors view of context management. Moreover, systems that employ HCI, BCI, and simulation in their determination of context are rare, as are systems that rely on a cognitive model as the basis for their context classification. These advanced domains have been combined in the RECON architecture for intelligence analysis and are extended with the early multi-agent context-management foundations introduced in this paper.

As part of future work, the proposed extension will be integrated within the existing RECON implementation. In addition, a human-in-the-loop, serious gaming experiment is being developed to investigate the effectiveness of this approach in reducing information overload based on the principles of adaptive context management. It is envisioned that such an approach will be suitable not only for the intelligence analysis domain, but also for a broader user group. This will further the goal of futuristic, and adaptive software-intensive systems.

Acknowledgements This work was funded in part by Defence R&D Canada, Thales Research and Technology Canada, and a research partnership grant from the Department of National Defence and the Natural Sciences and Engineering Research Council of Canada.

References

Alchemyapi (2015) http://www.alchemyapi.com

Bar-Yam Y (1997) Dynamics of complex systems, vol 213. Addison-Wesley, Reading, MA

Baxter G, Sommerville I (2011) Socio-technical systems: from design methods to systems engineering. Interact Comput, Elsevier, 23(1):4–17

Bellifemine F, Poggi A, Rimassa G (2001) JADE: a FIPA2000 compliant agent development environment. In: Proceedings of the fifth international conference on autonomous agents. ACM, New York, pp 216–217

Carbo J, Molina JM, Davila J (2001) A BDI agent architecture for reasoning about reputation. In: 2001 IEEE international conference on systems, man, and cybernetics, vol 2. IEEE, New York, pp 817–822

Fischer G (2012) Context-aware systems: the 'right' information, at the'right' time, in the 'right' place, in the 'right' way, to the 'right' person. In: Proceedings of the international working conference on advanced visual interfaces. ACM, New York, pp 287–294

Gore BF et al (2002) Human performance cognitive-behavioral modeling: a benefit for occupational safety. Int J Occup Saf Ergon, Taylor & Francis, 8(3):339–351

Gouin D, Lavigne V, Bergeron-Guyard A (2012) Human-computer interaction with an intelligence virtual analyst. In: Proceedings of knowledge systems for coalition operations. IHMC, Pensacola, FL

Gregori ME, Cámara JP, Bada GA (2006) A jabber-based multi-agent system platform. In: Proceedings of the fifth international joint conference on autonomous agents and multiagent systems. ACM, New York, pp 1282–1284

Hollnagel E (1998) Context, cognition, and control. In: Waern Y (ed) Co-operative process management. Taylor & Francis, London, pp 27–52

Hong JY, Suh EH, Kim SJ (2009) Context-aware systems: a literature review and classification. Expert Syst Appl, Elsevier, 36(4):8509–8522

Lafond D, Proulx R, Morris A, Ross W, Bergeron-Guyard A, Ulieru M (2014) Hci dilemmas for context-aware support in intelligence analysis. In: The sixth international conference on adaptive and self-adaptive systems and applications, ADAPTIVE 2014. IARIA pp. 68–72, 2014

Morris A (2009) Socio-technical systems in ICT: a comprehensive survey. Technical Report. University of Trento. http://eprints.biblio.unitn.it/archive/00001670/

Morris A, Ulieru M (2012) FRIENDs: brain-monitoring agents for adaptive socio-technical systems. Multiagent Grid Syst 8(4):329–347

Morris A, Ross W, Ulieru M, Chouinard P (2012) Modelling public security operations: evaluation of the holistic security ecosystem (HSE) proof-of-concept. Technical Report, DTIC Document

Opencalais (2015) http://www.opencalais.com

Qureshi NA, Perini A (2008) An agent-based middleware for adaptive systems. In: The eighth international conference on quality software, 2008. IEEE, New York, pp 423–428

Ross W, Morris A, Ulieru M (2012) NEXUS: a synergistic human-service ecosystems approach. In: 2012 IEEE sixth international conference on self-adaptive and self-organizing systems workshops (SASOW), IEEE, New York, pp 175–180

Ross W, Morris A, Ulieru M, Guyard AB (2013) RECON: an adaptive human-machine system for supporting intelligence analysis. In: 2013 IEEE international conference on systems, man, and cybernetics (SMC). IEEE, New York, pp 782–787

Schmidt A, Beigl M, Gellersen HW (1999) There is more to context than location. Comput Graph 23(6):893–901

Song S, Ryu K, Kim M (2010) Ontology-based decision support for military information systems. In: 2010 Long island systems, applications and technology conference (LISAT). IEEE, New York, pp 1–5

Vicente K (2004) The human factor: revolutionizing the way people live with technology. Routledge, London

Wei Z, Wang N, Kang M, Zhou W (2008) An agent-based context-aware middleware for pervasive computing. In: International symposium on information science and engineering, 2008. ISISE'08. vol 2. IEEE, New York, pp 116–119

Yılmaz Ö, Erdur RC (2012) ICONAWA–an intelligent context-aware system. Expert Syst Appl 39(3):2907–2918

Part III
Social Media and Crisis Management

Chapter 8
On the Challenges of Using Social Media for Crisis Management

Thomas Delavallade, Simon Fossier, Claire Laudy, and Gaëlle Lortal

Abstract In crisis situations, the challenge of understanding the current situation is tightly linked to the ability to process the variety and the amount of information provided by the multiple sources. In particular, social media can provide additional insight on real-time events, providing that the information that they relay is accurately retrieved, evaluated, and fused. In this chapter, we describe various mechanisms and functions necessary for information fusion and understanding, starting from social media exploration and retrieval, then describing the fusion process and the associated management of information uncertainty, concluding with a description of the methodology and experiments we use to tackle the intrinsic big volume of data and processing required for social media information analysis.

8.1 Context

8.1.1 Crisis Management

The accumulation of major natural disasters since the beginning of the new millennium with tremendous human and financial impacts has lead the society to gradually accept the predominance of resilient policies over preventative ones. Prevention and prediction are of course still deeply needed, but even with the most sophisticated measures large-scale disasters will still regularly occur. This is exactly what the authors of Roth et al. (2013) argue in their report commissioned by the Swiss Federal Office for Civil Protection: "Probably the most pervasive trend has been a shift from the protection paradigm, focused on disaster prevention through

T. Delavallade (✉)
Thales, 4 av. des Louvresses, 92622 Gennevilliers, France
e-mail: thomas.delavallade@thalesgroup.com

S. Fossier • C. Laudy • G. Lortal
Thales, 1 av. Augustin Fresnel, 91767 Palaiseau, France
e-mail: simon.fossier@thalesgroup.com; claire.laudy@thalesgroup.com; gaelle.lortal@thalesgroup.com

G. Rogova, P. Scott (eds.), *Fusion Methodologies in Crisis Management*,
DOI 10.1007/978-3-319-22527-2_8

strategic planning and effective command and control structures, to the resilience paradigm where disaster management is about adaption during and after a disaster."

Crisis management is a wide field theorized under several aspects. Lerbinger (1997) categorizes seven types of crises according to their nature: natural, technological, of confrontation, malevolence, of skewed management values, deception, and management misconduct. It is necessary to be able to categorize a crisis as different crisis management strategies should be involved in order to deal with or resolve a crisis. In Coombs (2007), a clustering of crisis types was proposed, to underline the kind of impact a crisis has, into: victim-centered, accidental, or preventable. For instance, the victim cluster encompasses:

- Natural disaster: acts of nature such as tornadoes or earthquakes;
- Rumor: false and damaging information being circulated about you organization;
- Workplace violence: attack by former or current employee on current employees on-site;
- Product tampering/Malevolence: external agent causes damage to the organization.

Several characteristics of crises can be drawn from the literature. In Aligne and Mattioli (2010), the analysis adopts Dr. Gal Crocq (2007) criteria for characterizing crisis:

- a break (disorder, disturbance);
- a bifurcation (transition period, decisive moment, choice);
- a threat (danger, severity);
- an important stake (large number, large scale);
- the urgency (surprise, miss time);
- a degraded, unordered situation (means shortage);
- a difficulty, a psychological tension.

It leads them to adopt the following crisis definition (Aligne 2009): "A situation of break with regard to the previous events, which threatens the functioning and the values of the individual and/or the community, and where appears an urgent necessity to act despite the degradation of the means, the information and the control. The characteristic of a crisis is that it is unpredictable, out of frame, that it exceeds the existing means, and that it cannot be anticipated by scenarios".

8.1.2 Social Media in Crisis Management

Mobile means of communication explosion (as smartphones) combined with online social media accessible via mobile apps brought new behaviors in our societies. In events involving a great number of people, behaviors have changed, even in vital situations (Deparis 2013).

Having in mind the examples[1] of the 2004 Indian Ocean tsunami, the 2010 Haïti earthquake, the 2011 tsunami in Japan, the 2012 Sandy hurricane, or the 2013 typhoon in the Philippines, the lessons learnt clearly reveal the importance of local, flexible, collaborative, and self-organized processes for crisis management and reconstruction, as is pointed by Meier (2013). Such a trend has been undeniably favored by the development and spread of mobile technology in a first step and of social media in a second step.

The aforementioned crises have shown the very common and widespread use of these technologies by citizens to report real-time information. For instance, in Castillo et al. (2010), the authors report having collected more than 4.7 million tweets about the 2010 Chilean earthquake published between 27/02/2010 and 02/03/2010. And during the Sandy Hurricane, Twitter officially reported that more than 20 million tweets had been published between 27/10/2012 and 01/11/2012.

One of the usual behaviors is the quasi-systematic sending of snapshots, photos, opinions, and alerts on social media. Sometimes before thinking of their own security, stakeholders local to the events broadcast their information (Starbird and Palen 2010). When an unexpected event happens, potentially unpleasant or dangerous, local social media users use their means of online communication (mainly Twitter) in order to quickly spread a piece of news to a large amount of people, hopefully including emergency responders (Huang et al. 2010). Several studies (Sakaki et al. 2010; Vieweg et al. 2010; Crooks et al. 2013) show that social media could be used to analyse the human behaviors during a crisis but also to support the responders on the field and crisis situation assessment (American Red Cross 2010). Several methodologies (Singh et al. 2010; Starbird and Palen 2010) have been proposed to enable social activity-based event detection.

8.1.3 Tools for Crisis Management

These technology-driven behavioral changes have nurtured the development of collaborative and crowd-sourcing applications which aim at exploiting the huge amount of information reported by citizens, and at the same time at encouraging such spontaneous contributions through dedicated tools, in order to enhance situation awareness.

Following the presidential election in Kenya in 2007, riots broke out in the country. Then, local journalists set up a platform, which was able to gather text messages or emails. This platform enabled to collect the anonymous evidences of abuses suffered by people. These evidences have been synthesized in a cartographic application displaying the chain of events and the state of atrocities perpetrated all over the country—this process is often referenced as *crisis mapping*. This application has greatly simplified the coordination task of Organisation Non

[1] The list does not intend to be exhaustive.

Gouvernementale (ONG) on the spot. Further to this event, and according to its successful use, the platform has been generalized and named *Ushahidi* (Swahili for *witness*).[2] Under the sponsorship of the eponym non-profit organization, other solutions have emerged, purposing a set of applications enabling coordination during a crisis.

In the same perspective of a global involvement of the society in a resilience-oriented process, it is interesting to note that these platforms have been initially developed in the open source community by volunteers who spontaneously gathered their efforts to provide concrete and operational tools to support ongoing crisis management efforts. Sahana was indeed developed in the aftermath of the Indian Ocean tsunami, while Ushahidi was created to monitor the situation in Kenya after the 2007 elections which were followed by violent ethnic confrontations.

Such crowd-sourcing platforms are commonly used since the 2010 Haïti earthquake. For instance, the emergency teams from the city of New York use the Sahana platform. This illustrates that Public Protection and Disaster Relief organizations (PPDRs) have acknowledged the importance of social media for crisis management. Craig Fulgate, FEMA (US Federal Emergency Management Agency) administrator, has explicitly characterized the issue: Social media is imperative to emergency management because the public uses these communication tools regularly (San Su et al. 2013). Citizens use social media as a daily communication means and they do so also and maybe even more during crisis events. A huge amount of information is therefore available and updated in real-time, some pieces of which being potentially relevant and still unknown to PPDRs. Besides these social media may temporarily be the sole communication means available during a crisis, because telephone communications are down or emergency call centers are saturated as is reported in Peary et al. (2012) for the Japanese tsunami case. So the potential benefits of a better use of social media to enhance bi-directional communications between citizens and PPDRs are obvious. The OCHA (United Nations Office for the Coordination of Humanitarian Affairs) has released a technical report in 2011 expressing clearly this need (Initive 2011). In the same perspective the European Commission has launched several projects, like COSMIC,[3] ATHENA[4] or iSAR+,[5] to tackle precisely this issue and develop guidelines but also technical platforms allowing PPDRs to better integrate social media in their crisis management processes.

Other initiatives took advantage of the Web to aggregate a large volunteers' community and propose a solution for crisis monitoring. That way, the CrisisCommons community[6] proposes Internet users to contribute to their Wiki in adding resources as software, applications, documentation, advices, etc. to provide a larger knowledge base.

---

[2]Official Website of the foundation: http://ushahidi.com/.

[3]http://www.cosmic-project.eu.

[4]http://research.shu.ac.uk/centric/index.php/research-projects/athena.

[5]http://isar.i112.eu.

[6]http://crisiscommons.org/.

8.1.4 Need for Uncertain Information Fusion

Because of the underlying uncertainty of the human-generated information feeding such platforms, manual validation processes are necessary. Information may not be accurate enough to be directly exploited by PPDRs (Public Protection and Disaster Relief organizations) or it may even be false. Social media are indeed a place where rumours and misinformation thrive, even during crisis events. This issue is actually a major concern for PPDRs. Different surveys have indeed highlighted that trustworthiness was the major barrier for the adoption of social media monitoring tools by the PPDRs who do not consider information coming from social media as reliable (San Su et al. 2013; Flizikowski et al. 2014).

In order to address this issue, networks of volunteers like *HOT—humanitarian OpenStreetMap team*, the *SBTF—Standby Task Force*, and *Humanity Road* have risen. Such organizations have even decided to federate themselves under the *Digital Humanitarian Network*. These networks are activated in case of a major crisis in order to set up Virtual Operations Support Teams (VOST), teams of volunteers spread all over the world which take care of data collection, validation, and mapping.

Given the amount of data to process, tools to automate, at least partly, or to ease the validation stage are deeply needed. In Imran et al. (2014), the authors have chosen a micro-tasking approach. They designed a set of tools to address the different tasks composing the validation process. For each of these tasks a specific tool has been developed to simplify user interactions, identifying the specific information that needs to be manually labeled and the right time to ask for such a label. Such tools represent a great step towards faster, less resource-consuming, and more efficient crowd-sourcing processes which should help to provide more reliable information to PPDRs.

If such processes seem very well adapted to label videos, images, to provide accurate geolocation information or translations, they do not, by themselves, offer any support to volunteers in the assessment of the credibility of the events reported by the citizens. Furthermore we believe that such an assessment can hardly be performed using such a distributed approach as the one used in micro-tasking. Indeed, as it is mentioned throughout (Capet and Delavallade 2014), evaluating information implies comparing and fusing several pieces of information which may corroborate or contradict to a certain extent the pieces of information to be evaluated.

We have therefore adopted a different approach, focusing on decision support for the evaluation of information. In the next section we will present OsintLab, the social media monitoring platform we have developed, focusing on data exploration and situation awareness capabilities which are rather limited on existing platforms used for crisis management. Then we describe a generic methodology we have developed to tackle the information evaluation issue and its limitations. The rest of the paper is dedicated to the new approach we propose in order to cope with these limitations. More precisely Sect. 8.3 details the method we have designed to identify and fuse the relevant pieces of information, with regard to the target event that needs to be evaluated. Depending on the degree of conflict between those pieces of information, different hypotheses may be generated by the fusion process. In Sect. 8.4 we describe the algorithm we have developed to allow the fusion method to deal with uncertain information and provide a global assessment of the certainty assigned to the different fusion hypotheses. Finally, Sect. 8.5 describes the specific implementation of the proposed fusion method which allows to deal with the Big Data nature of social media data. Experimental results are also presented in the second part of this section.

8.2 Information Management Challenges

8.2.1 Social Media Monitoring

Existing crisis management platforms, like Ushahidi, Sahana, or those described in Kumar et al. (2011), MacEachren et al. (2011), Stollberg and de Groeve (2012), and Cameron et al. (2012), which integrate a social media monitoring component, mainly focus on crisis mapping. Therefore they provide a timeline and a geographic projection of messages collected from the social media plus collaborative means to annotate the messages, their geolocation, their severity, etc. A more comprehensive analysis of the messages to highlight the trends, identify the main topics of interest is often lacking and so are the tools which could help analysts to assess the credibility of the mapped information and detect global informational patterns such as rumors. Besides, almost all of these platforms are specifically tuned for Twitter. If it is indeed a very important communication channel with regard to crisis management, others need also to be considered. In time of crisis it is indeed very common to observe the creation and intense activity of dedicated blogs, forums, and Facebook pages.

To address the weaknesses of these tools we have developed a dedicated platform for social media mining and monitoring called OsintLab (Gouttas 2013). It provides an end-to-end processing chain to collect, analyze, exploit and visualize in a near real-time process, information published in the social media like blogs, forums, Twitter, or Facebook. Semi-automated, relying on highly interactive man-machine interfaces to allow analysts investigate large amounts of textual information, it aims at helping them to identify hot topics, uncover potential threats, and understand information flows patterns. Through a search engine, a reporting engine providing statistical summaries, dashboards, and a graph visualization interface, all interconnected, it offers various ways to dig in the data.

Because maps play a key role in crisis management, which is illustrated by the growing importance of *crisis mapping* tools, a GIS (geographic information system) capability has been integrated in OsintLab. Combined with a search engine it allows not only to visualize the locations on a map of the collected social media data but also to specify geographic regions of interest as an additional search criterion. Figure 8.1 gives a snapshot of the HMI which ensures this *crisis mapping* capability within OsintLab.

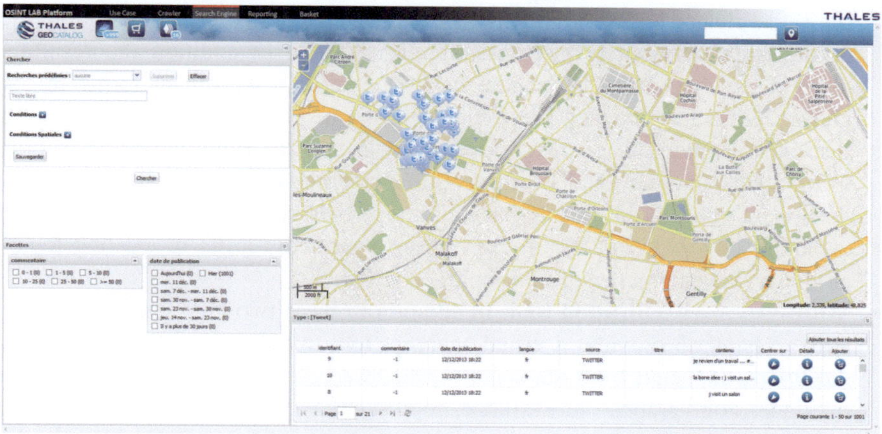

Fig. 8.1 Crisis mapping view within OsintLab

The OsintLab platform integrates various analytical components dedicated to situation awareness: a text clustering engine, a pattern-based classifier, and a multi-graph visualization tool.

- **Pattern-based text classifier**: automatically classifying texts collected in social media in classes defined by users is an important tool for analysts. It helps them sort the huge amount of incoming texts according to criteria they have selected. For situation awareness, users can classify the texts in categories of interest, so that each category contains all the texts matching a given pattern. This way, users can easily and quickly retrieve all those texts which deal with a known topic of interest. Patterns are defined through a combination of criteria available in a search engine:

  – Boolean request combining various keywords, applied on the text content;
  – Criteria applied on text meta-data, such as the nickname of the author of the text, the publication source of the text, its publication date, and its geolocation.

  In addition, an alert mechanism has been integrated on top of this classification process. Alert patterns can be specified by the analyst so that he is warned as soon as the frequency of texts being assigned to a category exceeds a given threshold on a given time window.

- **Text clustering**: the automatic pattern-based classification process is useful only to sort texts in categories which are known in advance but it does not enable to automatically identify emerging and unexpected topics, which is of key importance when it comes to the management of very complex crisis situations. To do so OsintLab integrates a text clustering engine. Based on relational analysis theory, this engine builds clusters of texts sharing similar terms, without having to predefine the number of clusters to be built. This number is automatically identified through an optimization process: the similarity between all texts of the same cluster has to be maximized while maximizing at the same time the dissimilarity between texts from different clusters. In addition the resulting clusters are automatically labeled using the most discriminant terms of the cluster: those terms which are frequent in the texts of the cluster and rare in the texts of other clusters. More details about this algorithm are given in Pine et al. (2005). In order to cope with real-time information updates, an incremental version of the algorithm has been developed. It enables to follow the time evolution of topics and therefore to observe semantic drifts of existing topics as well as the emergence of new topics.

- **Multi-graph visualization tool**: In addition to these two algorithmic functionalities, OsintLab provides a visual analytics tool to ensure situation awareness. All that is collected from the social media is accessible through a multi-graph interface. Four graphs are available, representing the relations between texts, authors, sources, and topics (extracted by the clustering engine). Besides, thanks to a graph clustering engine, based on the Louvain algorithm, described in Blondel et al. (2008), communities of authors and communities of topics are automatically identified and highlighted. These four graphs are interconnected so that a selection of any text, author, source, or topic leads to the selection of the associated entities in the three other graphs. One of the most interesting interconnections is the one between topics and authors. Indeed it allows to quickly understand what the main topics are and who discuss them. In addition to these four analytical dimensions, a timeline allows to take into account the temporal dimension. Besides, several filters are available to ease the navigation process and to enable users to progressively refine their analysis. Figure 8.2 provides a screenshot of this multi-dimensional visualization tool. In this figure, the three graphs, from left to right are representing relations between texts,

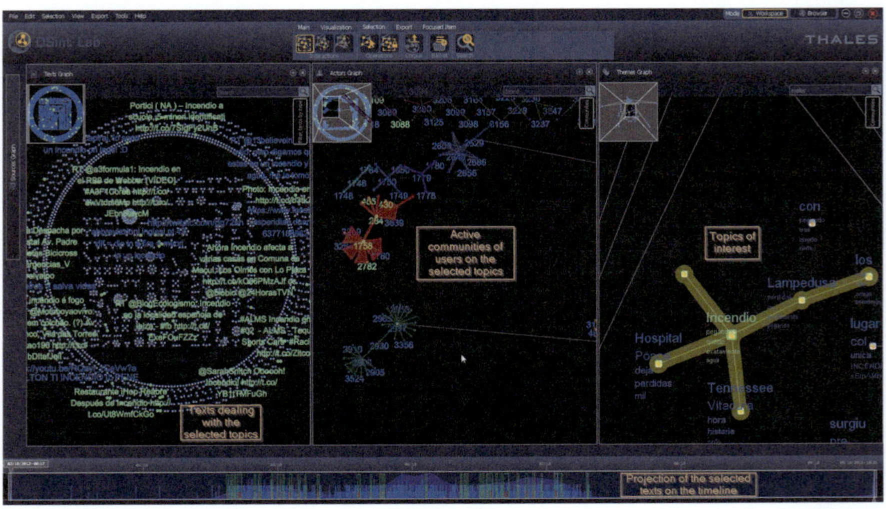

Fig. 8.2 Multi-graph visualization interface within OsintLab

authors & topics, sources have been omitted. The timeline is at the bottom. A topic has been selected and the texts dealing with this topic and their authors have been automatically highlighted.

Through this multi-graph view, analysis of the contents of the collected messages can be coupled with the analysis of the relations between their authors. This is a unique feature which is key to understand and deeply analyze information propagation dynamics. For example, for a given topic we can not only follow in time the different messages dealing with this topic, but we can also identify the main involved communities on the author graph and observe the relations between these authors. In turn this can be used when cross-checking the validity of a given information by taking into account the dependencies between the different people reporting similar or contradictory information. For instance, if an event is confirmed by 10 different persons, its credibility should be reinforced. But if these persons are all belonging to the same community and often communicate together, there is a high chance that they simply report a piece of information they hold from the same source. The 10 reports should then be considered as redundancy rather than corroboration.

This is a first way to address the need for information evaluation tools. However, this is still too manual an approach which can hardly be used in real crisis situations where a huge amount of information needs to be validated as fast as possible. To tackle this issue we have developed a new generic methodology which is presented in the next section, decomposing the overall information evaluation process in smaller and simpler processes. Such decomposition is important as it enables to identify those sub-processes which can be automated.

8.2.2 Information Evaluation

As illustrated in the previous sections, the rise of crowd-sourcing-based approaches offers renewed situation awareness capabilities for crisis management. It is actually a double-edged sword. Due to the inherently imprecise and uncertain nature of the information reported by citizens, such approaches are indeed critically requiring information evaluation mechanisms.

Focusing on Twitter, some researchers have shown that machine learning approaches could be used to build models able to predict the credibility of tweets, based on a set of features including both meta-data about tweet authors and variables extracted from an analysis of the tweet contents (Castillo et al. 2011). Even with standard machine learning techniques, they managed to build predictive models with a good accuracy with respect to human evaluations, the key issue being the selection and extraction of relevant features.

Their approach is promising and it is tempting to adapt it to handle any kind of social media data and not only Twitter, but it also has some important drawbacks which should not be ignored. First of all, to train their models it is required to have a labeled dataset, i.e., a set of tweets for which we know if it is credible or not. This is usually a very demanding and costly task. To circumvent the problem they used a crowd-sourcing approach, asking different people to assess the credibility of different tweets, with at least 3 different assessments for each tweet; the final label of each tweet was then picked automatically according to a majority vote. This is fine if we want to model how an ordinary person perceives the credibility of a tweet taken in isolation, but this is completely different from what we would like to achieve: model the reasoning process of an information analyst who most probably will judge the credibility of an information by cross-checking it with other information he may collect on the same subject. Building a suited labeled dataset and identifying the relevant features which could help to model such a cross-checking analysis still remains a very difficult and costly task. That is one of the main reasons which led us to look for alternative models.

This cross-checking activity is in fact very common for journalists or intelligence analysts, may they work in the defense or civil sector. It is probably the military domain which has best studied and conceptualized information evaluation. Actually a specific step of the intelligence cycle is dedicated to information evaluation and an official doctrinal corpus has been published over the years by the different defense organizations to make explicit the underlying concepts and standardize the operationalization of this process. In the NATO doctrine, NATO (2003) defines the degree of confidence that can be placed in a piece of information as a two-dimensional variable, distinguishing two components: the reliability of the information source and the credibility of the informational content, understood as its confirmation by other sources.

This definition suffers from certain weaknesses, which are discussed in Capet and d'Allonnes (2014). To overcome the main limitations we have proposed a 3-step, semi-automatic, information scoring process. It is described in Lesot et al. (2011) and illustrated in Fig. 8.3.

The proposed scoring system takes as inputs structured events, reported in textual documents, which are automatically extracted by natural language processing (NLP) tools. We assume that the information evaluation process is initiated by a user request: an information analyst interested by a specific event, queries the proposed system about this event in order to assess its credibility.

This query initiates the scoring process and in a first step all the relevant events are retrieved, i.e., those which are similar enough to the evaluated event. The analyst is then asked to check which ones are indeed relevant and if they corroborate or invalidate the queried event.

In a second step all these relevant events, plus the one to be evaluated, are scored individually to determine for each of them a confidence level, taking into account both the reliability and uncertainty of the source which reported the event. Uncertainty of the source is measured according to the linguistic tags used to report the event, such as *maybe*, *probably*, *could*, etc. These tags are identified during the information extraction stage performed by the aforementioned NLP tools. More details about this stage are given in Goujon (2009). There are several ways to assess

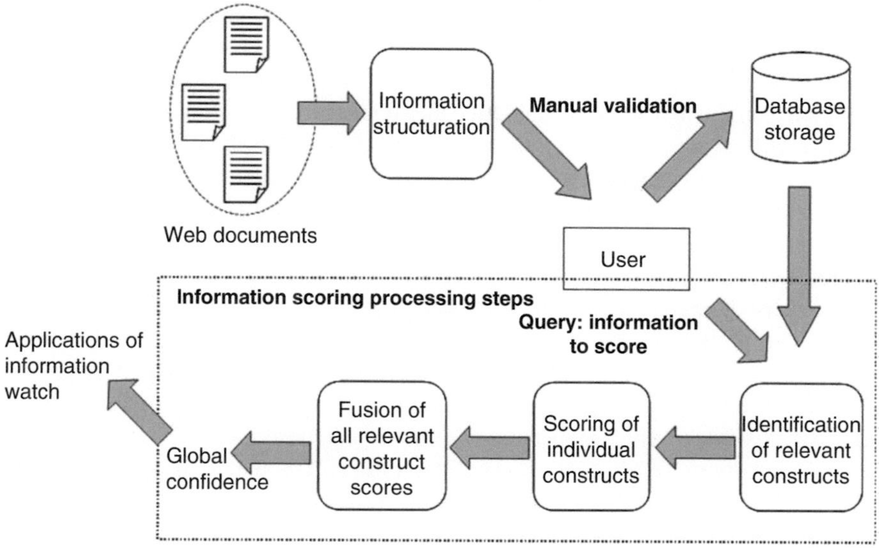

Fig. 8.3 Global architecture of the proposed information scoring process

the source reliability and the interested reader can refer to Pichon et al. (2014) for a review of this field and the description of a multi-criteria aggregation approach.

In the third step, the confidence scores of the different relevant events are aggregated in a single credibility score, taking into account the dependencies between the different sources which have reported the considered events. In order to have more flexibility in the choice of the aggregation measures while having enough expressivity to model uncertainty we decided to choose the possibility theory as our uncertainty management framework.

In Lesot et al. (2014) we present the results of an experimental evaluation of this information scoring system. As mentioned previously, it is very costly to build a labeled dataset to properly evaluate such a system. To circumvent the problem we built a synthetic dataset on which our system showed good performances. However, because our system is semi-automatic, another interesting assessment would be provided by end users. We did not perform such usability tests in a large scale and formalized experimental framework. The first qualitative feedback we managed to collect has nonetheless highlighted some limitations in our approach.

Firstly, identifying manually if an extracted event confirms or invalidates another one turns out to be quite cumbersome. Secondly, aggregating all the available pieces of information to build a single credibility measure can be misleading. Having a single credibility measure for an event is not perfectly in line with the military doctrine since it directly integrates the source reliability which is traditionally left aside as a second quality measure. However the source reliability is also made available in our system with other intermediate measures in order to provide explanations of how the global credibility measure was obtained. Actually the cognitive gap was higher for those analysts with no military intelligence background, the main reason being that such a global aggregation, recommended in the military doctrine, was perceived as hiding the complexity of the situation. In some cases, especially when the reliability of the different information sources is unknown and when some conservative measures are taken, if contradictory views co-exist with regard to an event, our approach tend to lead to situations of ignorance. Without other knowledge, this sounds rational as there is no clue to decide whether the event is true or false. However this is not very satisfying as we would like to have, for instance, more details about the competing hypotheses and their respective credibility.

The work described in the following sections precisely intends to address these issues. The proposed fusion mechanism allows to automatically identify similar events and to assess which events are compatible or inconsistent. When events are compatible they are subsequently fused so that a richer event is created, which was not possible in the framework we have just described. When events are inconsistent, they are all kept as rival hypotheses. The uncertainty management framework proposed hereafter, based on the evidence theory, is furthermore well suited to combine individual uncertainties as defined in the framework described in this section. As mentioned, these uncertainties are derived from a combination of the linguistic tags used to report the different events and the reliabilities of the information sources. Evidence theory is a generalization of the possibility theory which was proposed in Lesot et al. (2011) to compute these individual uncertainties and a transposition in the evidence theory framework of the Yager's discounting operator (Yager 1984) used in Lesot et al. (2011), is straightforward.

8.3 Information Fusion

8.3.1 *High-Level Information Fusion*

Context

Soft data fusion is an ever growing trend in the information fusion community. More and more tracks dedicated to soft information are organized within the International Conference on Information Fusion over the years and numerous authors stress the need for soft data management and fusion. For instance, in the detection of people and complex activities, the use of soft information sources is critical. The authors of Llinas et al. (2010) describe a 5-year program involving several academic actors that aim at addressing major stakes of soft data fusion. It includes the development of a framework as well as evaluation methods. In addition, many authors consider new issues raised by soft data fusion. Among them Gross et al. (2012) and Date et al. (2013) quote NLP, transformation of data into comprehensive and semantic data structures, soft data association and graph matching. Studies such as Belov and Gerken (2009) and Kohler et al. (2013) emphasize the importance of integrating soft data in situation awareness support systems. As the automation of soft data fusion is a very challenging issue, due to, e.g., error estimation, normalization, and context extraction for information interpretation, the authors propose a mixed approach that embeds the participation of a human analyst into the fusion process.

As seen before, the use of social media during crisis is valuable but may also lead to a huge amount of information to be processed in order to manage the crisis. As information items provided through social media relate different aspects of the same situation, the pieces of information are linked through a large information network. One of the issues for using such an amount of information is to be able to access relevant parts of it efficiently. For example, this enables to highlight repetitive schemes in crisis as well as deviation from usual behavior.

Three Fusion Functions

The InSyTo Synthesis soft data fusion framework relies on the use of semantic graph structures to store soft data and uses a graph algorithm to carry out the fusion process. It enables three different operations on networks of information that are depicted on Fig. 8.4 and described hereafter.

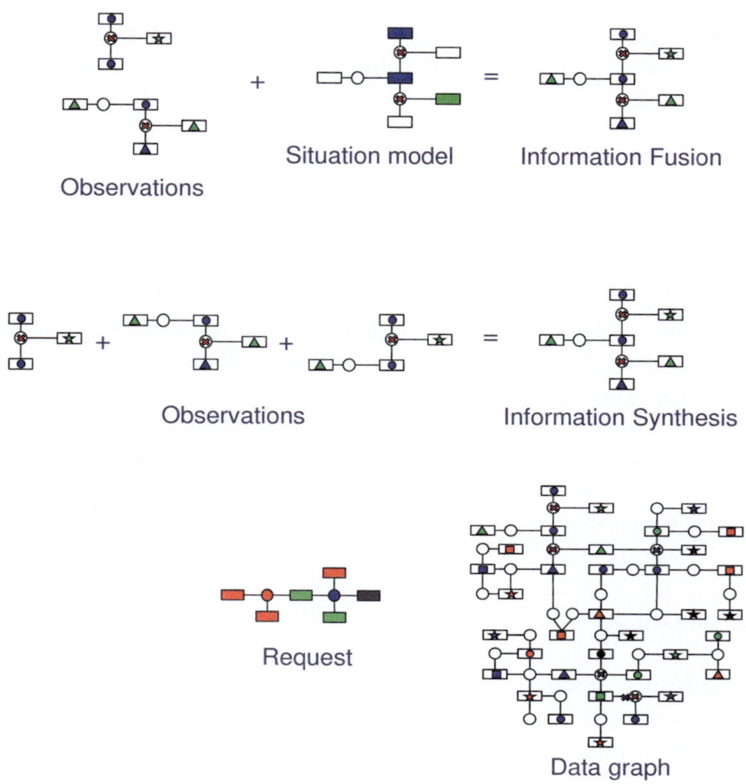

Fig. 8.4 InSyTo three functions

When a model of a situation of interest (e.g., an incident in the subway, at a specific station, with a specific cause and casualties) is available, one may want to monitor the situation and raise alarms if an instance of such a situation is happening. Therefore, different observations, coming potentially from different sources, are filtered out in order to keep observations of interest only. They are then assembled through **information fusion** in order to provide a representation of the ongoing situation of interest, as precise as possible.

Information synthesis enables one to collect and organize information about a specific subject. Through information synthesis, all the gathered information items are organized into a network. The redundant part of the information items is detected and eliminated.

All the instances of information corresponding to a specified graph pattern may be found within a network of information, through the **information request** function.

In the remaining part of this chapter, we will focus first on information fusion, which allows to combine several observations of an event, and provide the most reliable "summary" of this event given the provided information. In particular, a method to incorporate uncertainty associated to source reliability is presented in Sect. 8.4. Second, given an already built information graph, we will focus on the information request function, with the aim to support first responders and public protection organizations by providing means to efficiently query this information graph, and show how distributed computing can help process big volumes of data with the associated requests.

8.3.2 Formal Graph-Based Information Representation

Graph-based representations appear to be naturally well adapted to soft data. Our approach relies on the use of bipartite graphs, more specifically a subset of the conceptual graphs (Sowa 1984; Chein and Mugnier 2008) to represent soft data and knowledge. The conceptual graphs formalism is a model that encompasses a basic ontology (called *vocabulary*), graph structures, and operations on the graphs. The vocabulary defines the different types of concepts and relations that exist in the modeled application domain, while the graphs provide a representation of the observations which are provided by the information sources.

Conceptual Graphs

Basic **conceptual graphs** are bipartite graphs containing concept and relation nodes. Figure 8.5 gives an example of a conceptual graph. The rectangular boxes represent concept nodes and the ovals represent relation nodes.

Fig. 8.5 Example of a conceptual graph

Graphical form

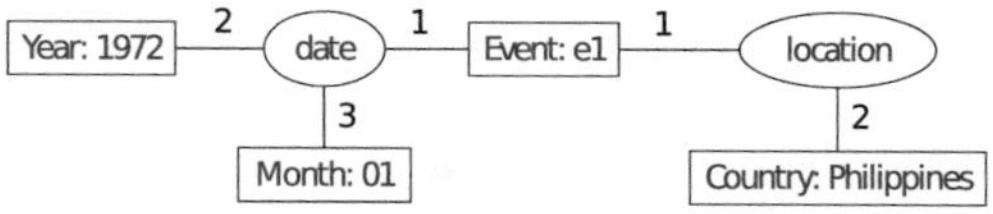

Linear form

date([Event: e1@e1], [Year: 1972 @y1972], [Month: 01 @january]), location([Event: e1 @e1], [Country: Philippines @Philippines]).

The term **concept** is used to refer to a concept node. The concepts represent the "things" or entities that exist. A concept is labeled with two components: the conceptual type and the individual marker.

The **conceptual type** defines the category to which the entity belongs. For instance, in Fig. 8.5 the concept `[Country:Philippines]` is an instance of the category `Country`, i.e., its conceptual type is `Country`.

The **individual marker** relates a concept to a specific object of the world. The object represented by `[Country:Philippines]` has the name (or value) `Philippines`. The individual markers may also be undefined. An undefined or generic individual marker is either blank or noted with a star $\star$, if the individual object referred to is unknown.

The term **relation** is used to refer to a relation node. The relation nodes of a conceptual graph indicate the relations that hold between the different entities of the situation that is represented. Each relation node is labeled with a relation type that points out the kind of relation that is represented.

The notion of **vocabulary** was defined in Chein and Mugnier (2008). The concept types and the conceptual relation types, which are used to label the concept and relation nodes, are organized in hierarchies.

We restrict our approach to relation types that are unordered in order to manage only one hierarchy, the concept types hierarchy.

Formally, we denote the set of concept types as T_C, the set of relation types as T_R, and the set of individual markers that are used to label the concept nodes as markers, which defines a vocabulary $\mathcal{V} = (T_C, T_R, \text{markers})$. A basic conceptual graph G is then defined by a 4-uple $G = (C_G, R_G, E_G, l_G)$, where

- (C_G, R_G, E_G) is a finite undirected and bipartite multigraph. C_G is the set of concept nodes. R_G is the set of relation nodes, and E_G is the set of edges.
- l_G is a naming function of the nodes and edges of the graph G which satisfies:

  1. A concept node c is labeled with a pair $l_G(c) = (type(c), marker(c))$, where $type(c) \in T_C$ and $marker(c) \in \text{markers} \cup \{\star\}$.
  2. A relation node r is labeled by $l_G(r) \in T_R$. $l_G(r)$ is also called the type of r.

Generalization/Specialization

Specialization/generalization relationships can naturally be defined on the graphs, and be used for the fusion, synthesis, and request functions, allowing to define formally the notion of subgraphs, which are specializations of another graph.

Relationships Between Conceptual Types

Given the hierarchical nature of the vocabulary, a partial order holds among the set of conceptual types T_C, interpreted as a relation of specialization: $t_1 \leq t_2$ means that t_1 is a specialization of t_2, that is to say that any instance of the class denoted by t_1 is also an instance of the class denoted by t_2.

Relationships Between Concepts

Given the order on T_C, we can also partially order the concepts that are defined on $T_C \times \{\text{markers} \cup \{\star\}\}$, by a specialization relation as follows. Let $c_1 = [T_1 : m_1]$ and $c_2 = [T_2 : m_2]$ be two concept nodes, we define:

$$c_1 \leq c_2 \quad \text{iff} \quad \begin{cases} T_1 \leq T_2 \\ m_2 = * \quad \text{or} \quad sim(m_1, m_2) \geq thres \end{cases} \tag{8.1}$$

where *sim* is a similarity function and *thres* a user-defined threshold.

For instance, if we consider that the conceptual type `Event` is greater (i.e., more general) than the conceptual type `TerroristEvent`, we have:

`[Event:*]` $\geq$ `[Event:e1]` $\geq$ `[TerroristEvent:e1]`

but `[Event:e1]` and `[TerroristEvent:*]` are not comparable.

Relationships Between Graphs

We also define a specialization relation between graphs. This relation is denoted by $\sqsubseteq$ (in order to avoid confusion with the specialization relation $\leq$ between concepts). Let A and B be two basic conceptual graphs. $\mathscr{C}_A$ and $\mathscr{R}_A$ denote the set of concepts and relations of the graph A, defined over the vocabulary $\mathscr{V}$. Denoting as P_{AB} the set of graph isomorphisms between A and B, we have:

$$A \sqsubseteq B \Leftrightarrow \exists p \in P_{AB}, \begin{cases} p : \mathscr{C}_A, \mathscr{R}_A \to \mathscr{C}_B, \mathscr{R}_B \\ \qquad c_A, r_A \mapsto c_B, r_B \\ \forall\, c_A \in \mathscr{C}_A,\; c_B \leq c_A \\ \forall\, r_A \in \mathscr{R}_A,\; r_B = r_A \end{cases}$$

By extension of the notation, $G_A|_{r_A} \sqsubseteq G_B|_{r_B}$ denotes that the subgraph of G_A restricted to the relation node r_A and its linked concept nodes is a generalization of the subgraph of G_B restricted to the relation node r_B and its linked concept nodes.

Specialization Set

In addition, we define an operator Π that provides the set of all more specialized graphs of a graph: given two graphs A and B, we have

$$B \in \Pi(A) \Leftrightarrow B \sqsubseteq A$$

This set Π may be extended with an element $\diamond$ representing an element of "incoherent information" (this will be useful when considering the result of the fusion of two incompatible graphs). It will be denoted as $\Pi^* = \Pi \cup \{\diamond\}$

8.3.3 Graph Fusion Algorithm

Our approach to soft data fusion is inspired from the maximal join operation on conceptual graphs (Laudy and Ganascia 2009; Laudy 2011). Given two observations expressed as graphs, the fusion process relies on the search of graph isomorphisms linking these two graphs to a new graph, and a selection of the most adequate couple of isomorphisms to determine the actual fused graph, i.e., the graph which describes best the knowledge we have on the situation.

Knowledge Model and Observations

Within the fusion function, we aim at processing (through fusion) a set of observations provided by different sources of information. Not all the information present in an observation is of interest to an operator who has a specific goal of analysis, i.e., who is interested by a particular event. There is, therefore, the need to define the notion of **knowledge model**, which is a reference piece of information which will define the scope of the analysis, i.e., *what we want to know* about the situation.

Formally, this model is expressed as a conceptual graph, generally with all concept nodes filled in the form `[ConceptType:*]`. In the remaining of this paper, the graph which represents the knowledge model will be denoted as $\mathcal{M}$.

Our previous comment on the fact that not all information present in an observation is of interest to the user leads to the following rule for the structuring of an observation into a conceptual graph, such that this conceptual graph is compliant with the knowledge model: *the concepts and relations that are part of an observation but that do not appear in the knowledge model, should not be included in the graph representing the observation.*

To facilitate the exposition of our fusion approach, we further impose that when an observation does not explicitly provide one of the concepts which appears in the knowledge model, its associated graph will anyway include this concept, and the model value will be attributed to said concept. A non-formal way to justify this would be "just because an observation doesn't provide the gender of an person doesn't mean that it doesn't have any".

Formally, this means that the conceptual graph A associated with an observation is necessarily such that $A \sqsubseteq \mathscr{M}$, i.e., $A \in \Pi(\mathscr{M})$. Besides, as will be seen next, the fusion operator that performs the fusion of two observations, or pieces of soft information, is a mapping from $\Pi(\mathscr{M})^2$ to $\Pi(\mathscr{M})$.

Generic Fusion Algorithm

With all the previously defined elements, we can formally define the fusion operator performing soft information fusion. The general definition of a fusion operator $\otimes$ compliant with a model $\mathscr{M}$ is an operator such that:

$$\begin{aligned} \otimes : \Pi^*(\mathscr{M}) \times \Pi^*(\mathscr{M}) &\to \Pi^*(\mathscr{M}) \\ (A, B) &\mapsto A \otimes B \end{aligned}$$

It is possible to define different fusion processes, which therefore lead to different fusion operators, according to the context of use of the fusion (application domain, user preferences, level of importance of the different information items,...). For instance, an example of such operator will be provided in Sect. 8.4.4 on a case of management of incident in the subway.

The fusion algorithm is made of two interrelated components (see Fig. 8.6). The first component is a generic subgraph matching algorithm, which itself relies on the use of fusion strategies (parameters of the algorithm). The generic graph-based fusion algorithm that is used for the three functions (information fusion, information synthesis, and information request). The usage of the algorithm (parameters and launch mode) determines the fusion operator that is used, and thus the function that is realized.

The graph matching component takes care of the overall structures of the initial and fused observations. It is in charge of the structural consistency of the fused information, regarding the structures of the initial observations, within the fusion process.

Fusion Strategies

The fusion strategies are made of similarity, compatibility, and functions over elements of the graphs to be fused [see Eq. (8.1) for instance]. They are used in order to adapt the fusion algorithm to specific context and user requirements. The context

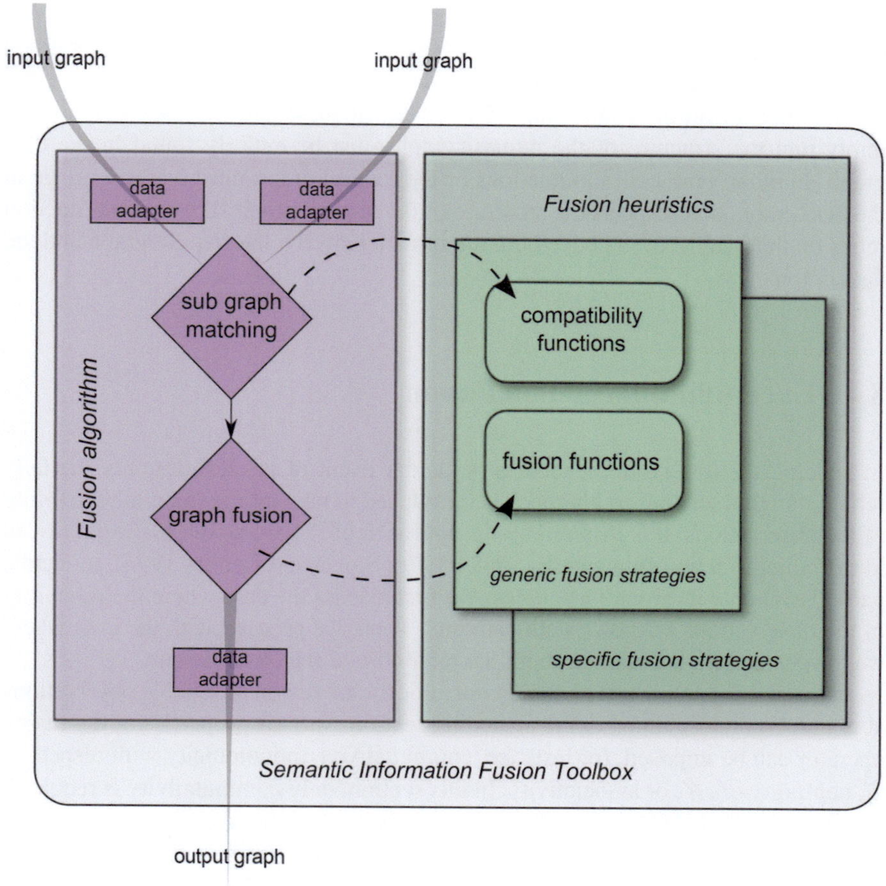

Fig. 8.6 InSyTo algorithm

encompasses the application domain, the semantics of the information items and user preferences. The fusion strategies enable to manage the discrepancies that may be observed in observations of the same situation by different sources (see Laudy 2011).

Within the fusion and synthesis functions, the strategies are adapted to each context of application and rely on user-defined similarity measure and data specific fusion functions.

Within the request function, the strategies that are used in the algorithm follow the specialization/generalization relationships defined before section "Generalization/Specialization". We use a *subsumption* strategy and a whole-structure conservation mode.

The subsumption strategy is used within the nodes to nodes comparison between the request and the data graphs. That is to say that the concepts of the request graph must be more general than the ones of the data graph in order to be fused.

The specialization relationship between the request and the data graphs also imply that the structure of the request graph must be entirely found in the data graph. In other words, all the relations of the request graph must have an image in the data graph, that respects the structure of the request graph. The request function relies on the search for injective homomorphism between the request graph and the data graph.

8.4 Uncertain Information Fusion

In general, the observations about a particular event of interest are only partially reliable. In this section, an approach is introduced to take into account the reliability of the observations. It is inspired by the approach of Pichon et al. (2012) developed in the belief function framework (Shafer 1976; Smets and Kennes 1994), to handle partially reliable information sources. We first discuss the case where the reliability of the observations is known with certainty. Then, we proceed with the case where we have uncertain knowledge about the reliability of the observations.

In this section, $\otimes$ denotes a fusion operator for the fusion of reliable observations about an event, as defined in the preceding section. Various properties of the fusion operator can be imposed, for instance, commutativity, monotonicity with respect to $\sqsubseteq$, neutrality of $\mathcal{M}$, or associativity. In this section, only commutativity is required, i.e., $\forall (A, B) \in \Pi^*(M)^2, A \otimes B = B \otimes A$. Note that the use of the operator $\sqsubseteq$ in the monotony constraint implies that we allow the fusion of observations that are not strictly equal but very similar (see Laudy 2010, 2011 for more details). The similarity process, as well as the level of similarity that should be reached, correspond to the *sim* and *thres* in Eq. (8.1) and are parameters of the global fusion system. They are the core of the notion of "fusion strategy" that is adapted according to the domain of application and level of precision requested.

8.4.1 Certain Case

Let $\mathcal{M}$ be the knowledge model associated to a particular event of interest. Suppose we receive an observation $A \sqsubseteq \mathcal{M}$ about this event. Similarly as in Smets (1993) and Pichon et al. (2012), reliability in the present work means the following: if this observation can be assumed to be reliable, then our knowledge about the event becomes A, and if this observation is assumed to be not reliable, then it is not useful and must be discarded, which amounts to knowing nothing about the event. Note that in our approach, knowing nothing and knowing $\mathcal{M}$ about the event of interest are considered equivalent.

This classical view of the notion of reliability can be extended to the situation where we receive two observations A and B, as follows. There are four elementary cases to consider with respect to the reliability of these observations:

1. If they are both not reliable, then we discard both of them and we only know $\mathcal{M}$ about the event;
2. If observation A is reliable and observation B is not reliable, then we discard observation B and our knowledge about the event is A;
3. If observation A is not reliable and observation B is reliable, then we discard observation A and our knowledge about the event is B;
4. If they are both reliable, then we know $A \otimes B$ about the event.

The reasoning described in the previous paragraph can be formalized as follows. Let $\mathcal{H}_A = \{h_A, \neg h_A\}$ be the assumption space on the reliability of observation A, where h_A (respectively $\neg h_A$) denotes that observation A is reliable (respectively unreliable). Similarly, let $\mathcal{H}_B = \{h_B, \neg h_B\}$ be the assumption space on the reliability of observation B. The set of possible elementary assumptions on the reliability of these two observations is denoted by $\mathcal{H}_{A\times B}$ and defined by $\mathcal{H}_{A\times B} = \mathcal{H}_A \times \mathcal{H}_B = \{(h_A, h_B), (h_A, \neg h_B), (\neg h_A, h_B), (\neg h_A, \neg h_B)\}$. We can define a mapping $\Gamma_{A,B}$ from $\mathcal{H}_{A\times B}$ to $\Pi^*(\mathcal{M})$, which assigns to each elementary hypothesis $h \in \mathcal{H}_{A\times B}$, the result of the fusion of the two observations A and B. $\Gamma_{A,B}(h)$ indicates how to interpret these observations in each of their configuration $h \in \mathcal{H}_{A\times B}$. We have:

$$\Gamma_{A,B}(h_A, h_B) = A \otimes B$$
$$\Gamma_{A,B}(h_A, \neg h_B) = A$$
$$\Gamma_{A,B}(\neg h_A, h_B) = B$$
$$\Gamma_{A,B}(\neg h_A, \neg h_B) = \mathcal{M}$$

8.4.2 Uncertain Case

The difficulty is that in general we have uncertain knowledge about the reliability of the observations. We consider in this paper that this uncertainty is represented by a probability distribution $\text{prob}^{\mathcal{H}_{A\times B}}$ defined on space $\mathcal{H}_{A\times B}$. Following Pichon et al. (2012), this uncertainty is transferred through $\Gamma_{A,B}$ onto space $\Pi^*(\mathcal{M})$ in the form of a probability distribution $\text{prob}^{\Pi^*(\mathcal{M})}$ defined on $\Pi^*(\mathcal{M})$ by:

$$\text{prob}^{\Pi^*(\mathcal{M})}(C) = \sum_{h:\Gamma_{A,B}(h)=C} \text{prob}^{\mathcal{H}_{A\times B}}(h), \quad \forall C \in \Pi^*(\mathcal{M})$$

In this context, $\text{prob}^{\Pi^*(\mathcal{M})}(C)$ is the probability that our knowledge about the event of interest be in the form of conceptual graph $C \sqsubseteq \mathcal{M}$. In short, it is the probability of knowing C. For instance, we may assume that observations A and B

have independent probabilities q_A and q_B, respectively, of being reliable, in which case we obtain:

$$\begin{aligned}\text{prob}^{\Pi^*(\mathcal{M})}(A \otimes B) &= q_A \cdot q_B\\ \text{prob}^{\Pi^*(\mathcal{M})}(A) &= q_A \cdot (1 - q_B)\\ \text{prob}^{\Pi^*(\mathcal{M})}(B) &= (1 - q_A) \cdot q_B\\ \text{prob}^{\Pi^*(\mathcal{M})}(\mathcal{M}) &= (1 - q_A) \cdot (1 - q_B)\end{aligned}$$

The extension of this approach to the case of more than two partially reliable observations does not raise any theoretical issue. One should only be aware that the order in which observations are handled may matter depending on whether $\otimes$ is associative.

8.4.3 Extension of TBM

Let $\text{prob}^{\Pi^*(\mathcal{M})}$ represent our uncertain knowledge about an event of interest. It may, for instance, be the result of the merging of several partially reliable observations according to the scheme above.

It is insightful to remark that $\text{prob}^{\Pi^*(\mathcal{M})}$ is quite close formally to a mass function (Shafer 1976; Smets and Kennes 1994), since a mass function on a finite set Ω is formally a probability distribution on the power set of Ω. Indeed, this comparison can be used to define some concepts inspired from belief function theory, in the present knowledge representation framework, which deals with uncertain and soft knowledge. In particular, we may define the degree of support (or degree of certainty) Sup(A) of a conceptual graph $A \sqsubseteq \mathcal{M}$ as:

$$\text{Sup}(A) = \sum_{B \sqsubseteq A, B \neq \diamond} \text{prob}^{\Pi^*(\mathcal{M})}(B).$$

This definition is directly inspired from the definition of the belief function associated with a mass function (Shafer 1976; Smets and Kennes 1994). Its introduction is motivated by the fact that in some problems, such as the one of query answering studied in Sect. 8.4.4, we may only be interested in a given graph $A \sqsubseteq \mathcal{M}$ and in particular by how much the knowledge derived from the available observations supports this graph.

8.4.4 An Example on Underground Railway Crisis Management

In this section, we will exemplify the approach described previously on a small use case involving an abnormal event in a subway network.

We consider the very simple model $\mathcal{M}$ given in Fig. 8.7 that describes a situation where an incident occurs in the subway network, at a certain location (station or train tunnels), has a certain main cause (human or technical), and can have an impact on the safety of people around the incident location. Part of the concept hierarchy is described in Fig. 8.8, and could of course be completed for a realistic use case.

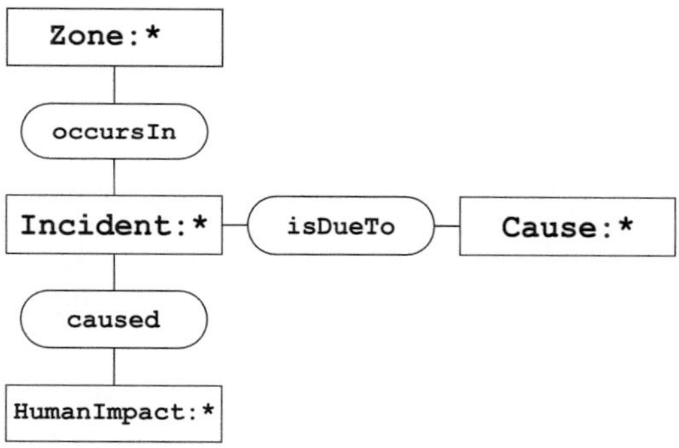

Fig. 8.7 Model of situation of interest

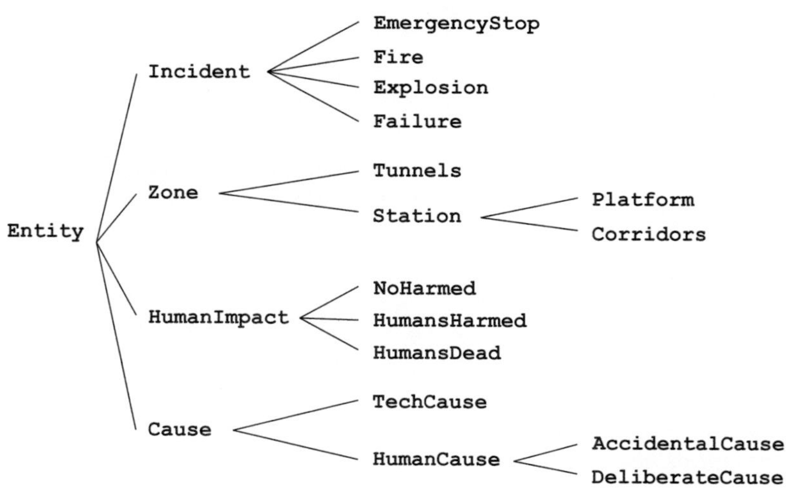

Fig. 8.8 Entities types hierarchy

Observations

In this example, four observations are provided successively to the system, from various sources of information. For illustration purposes, we describe them both in their initial format and in the conceptual graph formalism, though the transformation for the latter to the former is not in the scope of this chapter.

Initial Observations

- OBSERVATION 1, 12:41, Twitter, @paulbismutx (average user, but geolocalized near the station, reliability: 0.6) `Wow, just heard some kind of explosion at Monparnasse stn, people are shocked but everyone seems alrig -ht. Dunno what's up...`
- OBSERVATION 2, 12:42, Twitter, @martin_libe (verified journalist account, reliability: 0.7) `Was next to some kind of RATP technical box which just went BANG! at the end of Montparnasse platform. Now I'm half de af for the day...`
- OBSERVATION 3, 12:46, Twitter, @demonguy75 (average user, geolocalized 15 km from the station,reliability: 0.2) `Heard some ppl got harmed at Montparnasse, bf taking line 4 hope he's ok :'(`
- OBSERVATION 4, 12:51, Twitter, official train company account, reliability: 0.9) `A sound explosion was heard in Montparnasse station minutes ago, due to an electrical relay which malfunctioned and switch ed off abruptly. Nobody was harmed. Traffic will resume shortly.`

Conceptual Graphs

The four graphs corresponding to the observations are shown in Fig. 8.9.

For more conciseness, we will note the observations as ordered lists of their concept nodes, e.g., in the form $G_1 = \{$`Station:Monparnasse, Explosion:*, Noharmed:*, Cause:*`$\}$.

Fusion of Observations

Over time, the observations get to our system and we fuse them. The following illustrates the process.

Situation at 12:41 We only have Observation 1 at our disposal. The reliability of G_1 is 0.6. Therefore, our knowledge about the situation reads:

$$\text{prob}^{\Pi^*(\mathscr{M})}(G_1) = 0.6$$

$$\text{prob}^{\Pi^*(\mathscr{M})}(\mathscr{M}) = 0.4$$

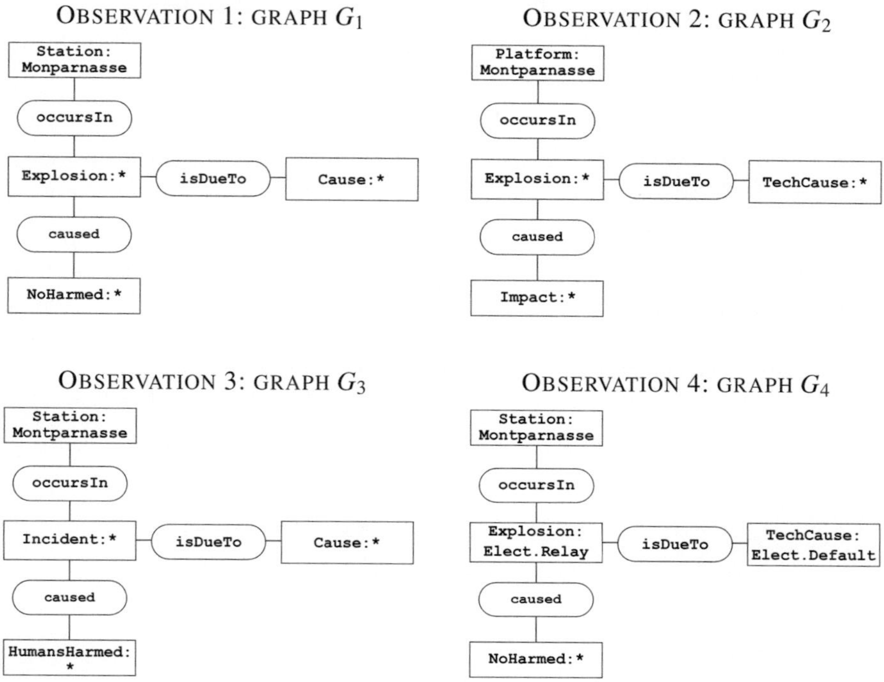

Fig. 8.9 Conceptual graph corresponding to the observations

Situation at 12:42 Observation 2 has been input into the system, with a reliability of 0.7. Following the reasoning of Sect. 8.4.2, we consider four elementary cases and the knowledge we have on the event of interest is then:

$$
\begin{aligned}
\text{prob}^{\Pi^*(\mathscr{M})}(G_1 \otimes G_2) &= 0.6 \quad \times 0.7 \quad = 0.42\\
\text{prob}^{\Pi^*(\mathscr{M})}(G_1) &= 0.6 \quad \times (1-0.7) = 0.18\\
\text{prob}^{\Pi^*(\mathscr{M})}(G_2) &= (1-0.6) \times 0.7 \quad = 0.28\\
\text{prob}^{\Pi^*(\mathscr{M})}(\mathscr{M}) &= (1-0.6) \times (1-0.7) = 0.12
\end{aligned}
$$

The explicit value of $G_1 \otimes G_2$ depends on the fusion operator used. In this case, we will define a simple fusion process, in which the concepts and relations nodes of two subgraphs are fused to form the new fused graph by respecting the following rules:

- For each conceptual type: select the most generic common subtype of the two initial concepts;
- For markers: two markers are compatible iff
  1. one of the markers is $\star$ (and the other is kept), or
  2. for strings, one marker is included in the other (and the longest is kept).

With this fusion operator, the graph that depicts the fusion result is more specific than the two initial ones, and contains all the information that was carried out by the initial observations. For instance:

- `[Platform:Monparnasse]` and `[Station:Montparnasse]` are not comparable by $\leq$, but they are compatible and their fusion reads `Platform`Montparnasse.
- $G_1 \otimes G_2$ is therefore written {`Platform:Montparnasse, Explosion:*, NoHarmed:*, TechCause:*`}.

Situation at 12:46 Observation 3 is compatible with G_2 but incompatible with G_1 or $G_1 \otimes G_2$ since the `NoHarmed` conceptual type is incompatible with `HumansHarmed`. Therefore, the fusion result of G_3 with G_1 or $G_1 \otimes G_2$ corresponds to the incoherent knowledge $\diamond$. The mapping $\Gamma_{1,2,3}$ therefore reads:

$$\begin{aligned}
\Gamma_{1,2,3}((h_1, \quad h_2), h_3) &= \diamond \\
\Gamma_{1,2,3}((h_1, \quad h_2), \neg h_3) &= G_1 \otimes G_2 \\
\Gamma_{1,2,3}((h_1, \neg h_2), h_3) &= \diamond \\
\Gamma_{1,2,3}((h_1, \neg h_2), \neg h_3) &= G_1 \\
\Gamma_{1,2,3}((\neg h_1, \quad h_2), h_3) &= G_2 \otimes G_3 \\
\Gamma_{1,2,3}((\neg h_1, \quad h_2), \neg h_3) &= G_2 \\
\Gamma_{1,2,3}((\neg h_1, \neg h_2), h_3) &= G_3 \\
\Gamma_{1,2,3}((\neg h_1, \neg h_2), \neg h_3) &= \mathcal{M}
\end{aligned}$$

and therefore, the knowledge about the event reads:

$$\begin{aligned}
\text{prob}^{\Pi^*(\mathcal{M})}(G_1 \otimes G_2) &= 0.42 \times (1 - 0.2) = 0.336 \\
\text{prob}^{\Pi^*(\mathcal{M})}(G_1) &= 0.18 \times (1 - 0.2) = 0.144 \\
\text{prob}^{\Pi^*(\mathcal{M})}(G_2 \otimes G_3) &= 0.28 \times \quad 0.2 = 0.056 \\
\text{prob}^{\Pi^*(\mathcal{M})}(G_2) &= 0.28 \times (1 - 0.2) = 0.224 \\
\text{prob}^{\Pi^*(\mathcal{M})}(G_3) &= 0.12 \times \quad 0.2 = 0.024 \\
\text{prob}^{\Pi^*(\mathcal{M})}(\diamond) &= 0.42 \times \quad 0.2 \\
&\quad + 0.18 \times \quad 0.2 = 0.120 \\
\text{prob}^{\Pi^*(\mathcal{M})}(\mathcal{M}) &= 0.12 \times (1 - 0.2) = 0.096
\end{aligned}$$

The result of this calculation shows without surprise that the probability of knowing G_3 about the situation is very small (2.4 %), while the probabilities associated to G_1 and G_2 remain high. The probability associated to the incoherent knowledge provides an insight on the level of inconsistency in our set of observations.

Situation at 12:51 The fourth observation is taken into account, it is fully compatible with $G_1 \otimes G_2$, G_1 and G_2, but not with $G_2 \otimes G_3$ or G_3. We repeat the same process, which yields:

$$
\begin{aligned}
\mathrm{prob}^{\Pi^*(\mathcal{M})}((G_1 \otimes G_2) \otimes G_4) &= 0.336 \times 0.9 = 0.3024\\
\mathrm{prob}^{\Pi^*(\mathcal{M})}(G_1 \otimes G_2) &= 0.336 \times (1 - 0.9) = 0.0336\\
\mathrm{prob}^{\Pi^*(\mathcal{M})}(G_1 \otimes G_4) &= 0.144 \times 0.9 = 0.1296\\
\mathrm{prob}^{\Pi^*(\mathcal{M})}(G_1) &= 0.144 \times (1 - 0.9) = 0.0144\\
\mathrm{prob}^{\Pi^*(\mathcal{M})}(G_2 \otimes G_4) &= 0.224 \times 0.9 = 0.2016\\
\mathrm{prob}^{\Pi^*(\mathcal{M})}(G_2) &= 0.224 \times (1 - 0.9) = 0.0224\\
\mathrm{prob}^{\Pi^*(\mathcal{M})}(G_2 \otimes G_3) &= 0.056 \times (1 - 0.9) = 0.0056\\
\mathrm{prob}^{\Pi^*(\mathcal{M})}(G_3) &= 0.024 \times (1 - 0.9) = 0.0024\\
\mathrm{prob}^{\Pi^*(\mathcal{M})}(\diamond) &= 0.056 \times 0.9\\
&\quad + 0.024 \times 0.9\\
&\quad + 0.120 = 0.1920\\
\mathrm{prob}^{\Pi^*(\mathcal{M})}(G_4) &= 0.096 \times 0.9 = 0.0864\\
\mathrm{prob}^{\Pi^*(\mathcal{M})}(\mathcal{M}) &= 0.096 \times (1 - 0.9) = 0.0096
\end{aligned}
$$

Note that when a new observation needs to be taken into account, we use the knowledge obtained from previous observations, i.e., there is no need to restart the fusion process from the beginning.

Degree of Support of the Knowledge

The probability distribution $\mathrm{prob}^{\Pi^*(\mathcal{M})}$ obtained previously provides a full description of the uncertain and soft knowledge that can be derived from the received observations. Another interesting feature of our approach is to enable the evaluation of the degree of support of any graph relatively to the received observations.

The example of request which can be executed by an operator would be:

How much of the information received supports the fact that something happened on a station platform but without any harm to passengers?

This question can be answered from the probability distribution $\mathrm{prob}^{\Pi^*(\mathcal{M})}$ and the notion of degree of support defined in Sect. 8.4.3.

The request defines an "hypothesis graph" G_h, which in this example is the graph `{Platform:*, Incident:*, NoHarmed:*, Cause:*}`. The degree of support to G_h is then defined by $\text{Sup}(G_h)$, which is computed by taking into account the probabilities of all graphs more precise than G_h.

Here, these graphs are $G_2 \otimes G_4$, $(G_1 \otimes G_2) \otimes G_4$ and $G_1 \otimes G_2$, both more specific than G_h. Therefore:

$$\begin{aligned}\text{Sup}(G_h) &= \text{prob}^{\Pi^*(\mathcal{M})}(G_2 \otimes G_4) + \text{prob}^{\Pi^*(\mathcal{M})}(G_1 \otimes G_2)\\ &\quad + \text{prob}^{\Pi^*(\mathcal{M})}((G_1 \otimes G_2) \otimes G_4)\\ &= \mathbf{0.5376}\end{aligned}$$

The answer to the question is thus:

The fact that an incident happened on a platform without any harm to passengers is explicitly supported at 53.76 % by the information received.

Let us remark that $\text{prob}^{\Pi^*(\mathcal{M})}(\diamond)$, the incoherent knowledge, affects all the graphs by reducing their own support. If $\text{prob}^{\Pi^*(\mathcal{M})}(\diamond)$ is too high, this can be a good indication that either one of the observations did not relate to the event of interest, or that the model chosen for observation representation should be reassessed.

8.5 Information Request on Social Media

Because of the large amount of information available through social media, the information fusion operations have to be adapted to so-called "big data" issues. This section presents the adaptation of the **request function**, so that it can be used on a large information network.

8.5.1 Map/Reduce Fusion Algorithm

Subgraph Isomorphism

The graph fusion algorithm relies on the search for matches between subgraphs and more precisely for a maximal matching subgraph. Given two graphs G and H, subgraph isomorphism enables to determine whether G contains a subgraph that is isomorphic to H (i.e., G contains a subgraph that has the same structure as H and which nodes are pairwise compatible). The maximum subgraph isomorphism problems for two graphs G and H return the largest subgraph SG of G that is isomorphic to H.

Maximal subgraph matching is used in order to determine where to add information in an information graph, and which parts of two information graphs are redundant and should thus be fused rather than be repeated twice in the graph resulting from the fusion.

The decision problem of whether two *graphs* matches is well studied and is in NP (McKay and Piperno 2014). But the problem of finding matching *subgraphs*, known as the *subgraph isomorphism* problem, is known to be NP-complete and difficult to solve in parallel (Plantenga 2013). This problem is well studied and has led to a lot of algorithms, among which several have been parallelized (Plantenga 2013; Zhao et al. 2012; Liu et al. 2009).

The most difficult graph matching problem in our case is to find the set of subgraphs that *maximize* the matching. This *maximum common subgraph* isomorphism problem (MCS) is known to be NP-hard (Ullmann 1976), which makes it intractable on large graphs (Conte et al. 2007).

Because of the computational complexity of the MCS problem, the InSyTo Synthesis framework was not able to manage very big graphs. First, the underlying algorithm had to upload the whole graph in memory, which is not possible for very big graphs. Then, the processing time could be too long regarding operational needs.

Secondly, the InSyTo Synthesis algorithms rely on subgraph isomorphism and maximum subgraph isomorphism algorithms. Both those algorithms must solve highly combinatoric problems, on which the execution time may become too long to satisfy user requirements.

The MCS problem is often solved with tools from the combinatorial optimization domain (Bunke et al. 2002; Raymond and Willett 2002) like constraint programming (Ndiaye and Solnon 2011; Solnon 2010). The parallelization of MCS algorithms remains to be studied. We describe hereafter an approach based on the MapReduce principle for the partial parallelization of the fusion algorithm. The implementation use the Hadoop framework (White 2009). Using Hadoop enables the management of big graphs by avoiding the load of the graphs in memory and distributing the computations over several processing nodes.

Problem Decomposition

In this work, we focus on the request functionality of InSyTo Synthesis. The input data are two graphs. The *data graph* is a big network (more than 20,000,000 nodes). The *request graph* is a relatively small graph. Indeed, the request is "written" by the analyst and contains most of the time 20–50 nodes. The request graph is the equivalent of the model $\mathcal{M}$ defined in Sect. 8.3.3 and used in the (uncertain) fusion function.

The main issue of the subgraph isomorphism algorithm is the combinatoric explosion and the big graph structures. The management of the structural part of the graphs (i.e., which node is linked to which one) is a difficult problem particularly when looking for candidate images for the nodes of one graph in the other graph at the same time. Therefore, within the search for candidate images for the nodes of the request graph in the information graph, our approach is to split up the process into a first phase that will manage the values of the request and information nodes and a second phase in charge of the management of the structural aspects of the graphs.

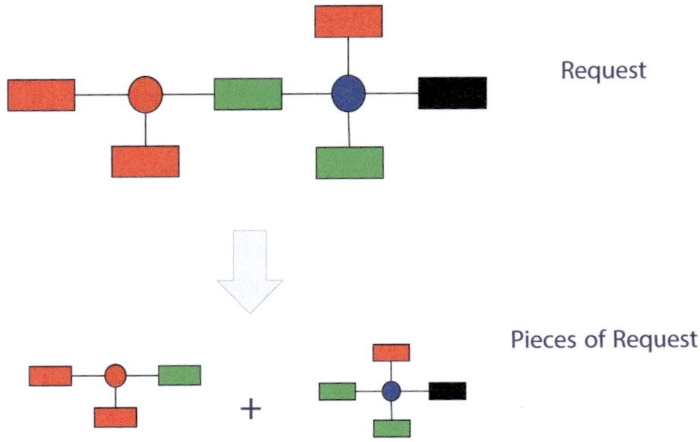

Fig. 8.10 Decomposition of a "request" graph

In order to follow the MapReduce model, we cut the two graphs (data and request) into small pieces that will then be pairwise compared. The comparison take into account the value of the nodes themselves, without taking care of the overall structures of the graphs. Therefore, the comparisons are independent from one another, and the MapReduce model can be applied.

Figure 8.10 shows the decomposition of a request graph into two small graphs. Each small graph corresponds to one of the relation nodes of the request graph, linked to a copy of its neighbor concept nodes.

Once the two graphs are cut around the relations nodes, each relation node and the ordered list of concepts nodes it is linked to constitutes an entry for the MapReduce algorithm. To be readable by the Hadoop framework, the set of pieces of request are written into their linear form (see Fig. 8.5) into a file. Each line of the file is one of the subgraph and will be processed by the MapReduce algorithm.

The MapReduce algorithm looks for all the candidate images for each piece of the request (from the request graph) in the set of pieces of data (from the data graph) (see Fig. 8.11).

The master node of the Hadoop cluster divides the data graph into a set of relation nodes, each being linked a set of concepts. It then distributes the compatibility test of the data relations against the request graph to the worker nodes of the cluster. The worker nodes process the compatibility test and produce, if the test is achieved, output records R as pairs of compatible request and data nodes.

The map function on a request graph $G_r = (C_r, R_r, E_r, l_r)$ and a data graph $G_d = (C_d, R_d, E_d, l_d)$ is detailed hereafter. The compatibility between a subgraph G_{r,r_i} of the request G_r and a subgraph G_{d,d_i} of the data graph G_d follows the specialization relationship $\sqsubseteq$ defined on subgraphs restricted to one relation (see section "Generalization/Specialization").

Once the map step is achieved, the reduce step takes place. The master node of the processing cluster collects the records $R = \{(G_{r,r_i}, G_{d,d_i}) \forall G_{r,r_i} \in G_r | G_{r,r_i} \sqsubseteq G_{d,d_i}\}$ produced during the map phase and combine them so as to associate each subgraph G_{r,r_i} of the request with the list of all compatible subgraphs G_{d,d_i} of the data graph.

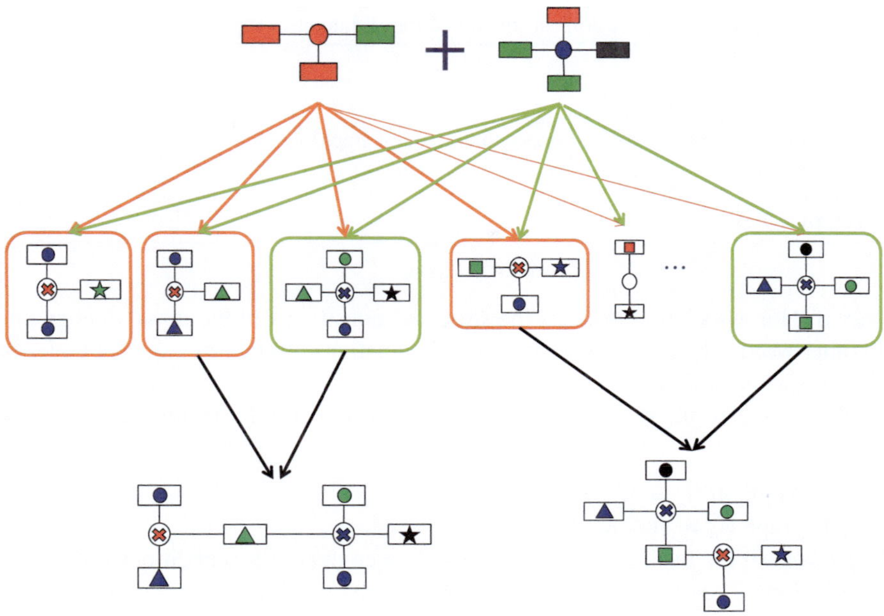

Fig. 8.11 Finding candidate images and reconstructing answer graphs

The reduce function aggregates the set of candidate subgraphs G_{d,d_j} as a list of images of G_{r,r_i}. The produced associative array $r\_img(G_{r,r_i}) = \{G_{d,d_j} \forall G_{d,d_j} \in G_d | G_{r,r_i} \sqsubseteq G_{d,d_j}\}$ has the same set properties than R but permits to improve further query times.

Result Graphs Build-Up

Once the candidate answers to each piece of request are processed, we need to build up answer graphs from these candidates. The different combinations of pieces of answers are provided, taking care of the original structure of the data graph (see Fig. 8.11). The candidate pieces for answers are assemble one with another, if and

Algorithm 1 Map function for a request graph and a data graph

function map(G_r, G_{d,d_i}):
$R \leftarrow \emptyset$
for all $G_{r,r_i} \in G_r$ **do**
  if $G_{r,r_i} \sqsubseteq G_{d,d_i}$ **then**
    {append output record}
    $R \leftarrow R \| (G_{r,r_i}, G_{d,d_i})$
  end if
end for
return R

Algorithm 2 Reduce function on the set of candidate subgraphs

function reduce(R):
$r\_img \leftarrow \emptyset$
for all $(G_{r,r_i}, G_{d,d_i}) \in R$ **do**
  {append subgraph to the list mapped to the request graph}
  $r\_img(G_{r,r_i}) \leftarrow r\_img(G_{r,r_i}) \| G_{d,d_i}$
end for
return r_img

only if their association respects the structural constraints of the initial data graph. In other words, the connectivity between the relations through the concept nodes is checked while building the answer graphs.

Within the result graph reconstruction algorithm, the following notations are used:

- request graph $G_r = (C_r, R_r, E_r, l_r)$;
- data graph $G_d = (C_d, R_d, E_d, l_d)$;
- *fstrategy* is the fusion strategy that is used for compatibility conditions and fusion functions. The fusion strategy used within the request function is the subsumption of concept nodes.
- $r\_img = \{(G_{r,r_i}, G_{d,r_j}) | G_{r,r_i} \sqsubseteq G_{d,r_j}\}$ is the associative array containing the records generated by the MapReduce algorithm.
- G_{r,r_i} is the subgraph of G_r restricted to the relation r_i and its linked concept nodes $c_j \in C_r$;
- $r\_mods = \{G_{r,r_i} | r_i \in R_r\}$ is the set of subgraphs of G_r restricted to each relation r_i of G_r.
- $ans = \{G_{ans_i} | G_{ans_i} = subgraph(G_d) \wedge G_r \sqsubseteq G_{ans_i}\}$ the set of subgraphs of G_d that are answers to the request (i.e., more specific than) G_r.
- $asso_i = \{(G_{r,r_i}, img_i)\}$ is an associative array that associates each subgraph restricted to a relation of the request graph G_r to a subgraph restricted to a compatible relation of the data graph G_d that is its image in the current association. An image of a relation node can be added in an association if and only if it is compatible (regarding node values and graph structures) with the already associated relations.
- $assos = \{asso_i\}$ is the set of all association tables $asso_i$ containing relations nodes of the data graph G_d associated to the relation nodes of request graph G_r and that respect the conditions given by G_r and the fusion strategy *fstrategy*
- $graph(asso_i) \rightarrow G_{answer_i}$ is a function that transforms an association table $asso_i$ into a graph by reconstructing the graph structure.

The input of the algorithm are r_img, the table of potential relation images of the data graph for each relation node of the request and the fusion strategy *fstrategy* that relies on the subsumption between concept nodes and graphs (see section "Generalization/Specialization").

8.5.2 Experimentations

This section presents an illustrative experimentation. The experimentation was conducted on a big information network containing terrorist attacks related information.

Hadoop: A Framework for Processing "Big Data"

Hadoop is a implementation of a distributed file system along with a model for performing distributed data processing on this file system.

Hadoop's file system is called HDFS (Hadoop Distributed File System). It allows the partitioning of data across many nodes. A Hadoop cluster scales computation capacity, storage capacity, and I/O bandwidth by simply adding commodity servers.

The data processing model is called MapReduce, and allows processing of large data sets. Users specify a map function that processes a key/value pair to generate a set of intermediate key/value pairs, and a reduce function that merges all intermediate values associated with the same intermediate key. The advantage

Algorithm 3 Result graph reconstruction algorithm

associateModel(*r_img*, *fstrategy*):
$assos \leftarrow \emptyset$
for all $img \in r\_img(r\_mods_0)$ **do**
  {append an element to the associations:}
  $assos \leftarrow assos \| (r\_mod_0, img)$
end for
for $r\_mod \in \{r\_mods_1 \ldots r\_mods_n\}$ **do**
  $i \leftarrow 0$
  while $i + 1 < |assos|$ **do**
    {iterate over *assos* while modifying it}
    $i \leftarrow i + 1$
    $found \leftarrow \bot$
    for all $img \in r\_img(r\_mod)$ **do**
      if $compatible(assos_i, (r\_mod, img, fstrategy))$ **then**
        $associate(assos_i, (r\_mod, img, fstrategy))$
        $found \leftarrow \top$
      end if
    end for
    if $\neg found$ **then**
      {remove the ith association:}
      $assos \leftarrow \{assos_0, \ldots, \widehat{assos_i}, \ldots, assos_n\}$
    end if
  end while
end for
$ans \leftarrow \emptyset$
for all $asso \in assos$ **do**
  $ans \leftarrow ans \| graph(asso)$
end for
return *ans*

of MapReduce is that the processing can be performed in parallel on multiple processing nodes (multiple servers) so it is a system that can scale very well.

Since it's based on a functional programming model, the map and reduce steps each do not have any side effects (the state and results from each subsection of a map process do not depend on another), so the data set being mapped and reduced can each be separated over multiple processing nodes.

The use of Hadoop for the parallelization of our algorithms has two advantages. The first one is that the graphs are not loaded in memory, but sub-parts of them are passed along the processing nodes, thanks to the distributed file system. The second one is that parts of the isomorphism search process itself can be distributed on the different processing nodes, enabling a speed-up of the overall process.

The Global Terrorism Database

The Global Terrorism Database[7] (GTD) is an open-source database that contains the descriptions of all the reported terrorist events that occurs worldwide between 1970 and 2012.

The GTD has the following characteristics:

- It contains information on over 113,000 terrorist attacks;
- It includes information on more than 47,000 bombings, 14,400 assassinations, and 5600 kidnappings since 1970;
- It includes information on 45 to more than 120 variables for each case;

The GTD is supervised by an advisory panel of 12 terrorism research experts and over 4,000,000 news articles and 25,000 news sources were reviewed to collect incident data

For each GTD incident, information is available on the date and location (region, country, state, city) of the incident, perpetrator group name, tactic used in attack, the weapons used and nature of the target, the number of casualties, etc.

Other variables provide information unique to specific types of cases, including kidnappings, hostage incidents, and hijackings.

This database represents a huge work and is particularly valuable for intelligence analysts. Therefore smart functionalities that would enable the recognition of modus operandi, for instance, are eagerly awaited.

Within our experimentations, we imported the data contained in the GTD into an information network. The importation of the data is agnostic to the contents of the database.

The GTD database is a classical "table". Each line of the table contains the description of an event, each column of the table corresponds to a characteristic

[7] The National Consortium for the Study of Terrorism and Responses to Terrorism (START) makes the GTD available via an online interface on the website of the project: http://www.start.umd.edu/gtd/.

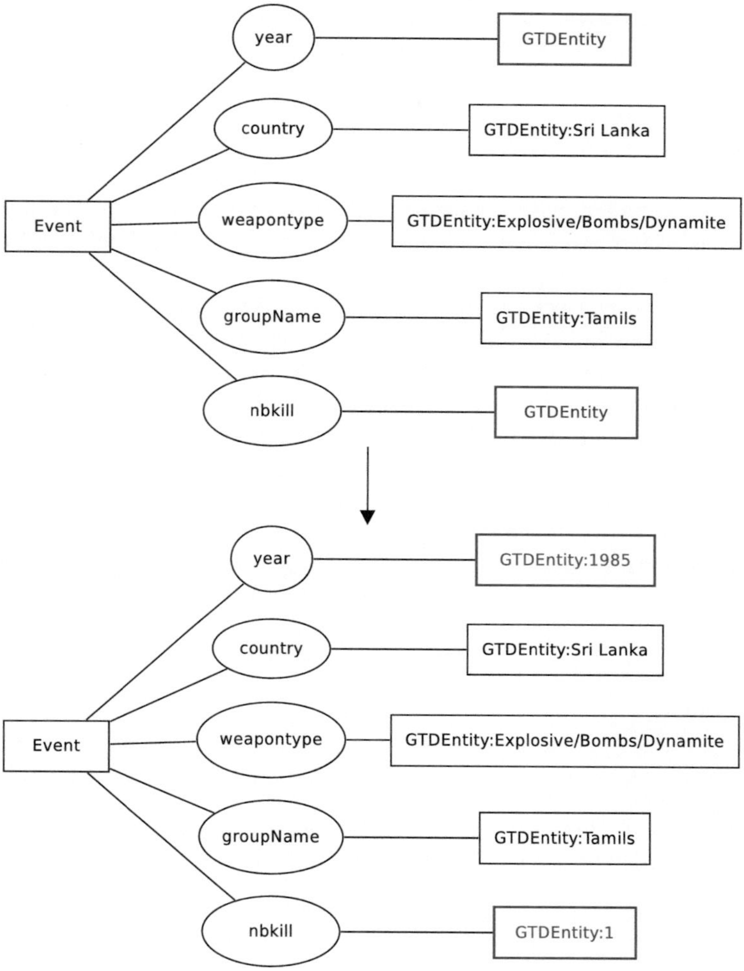

Fig. 8.12 Example of request and answer on the information network

of the events (date, location, attack type, etc.). To import the GTD data into the conceptual graphs format, we first automatically built the type hierarchy described in section in "Conceptual Graphs". Therefore, two concept types were defined, `GTDEvent` and `GTDEntity` that are directly attached to the root type of the hierarchy. The types of relations are extracted from the column labels of the table.

The data itself is then imported into a graph structure by creating a `GTDEvent` which value is the identifier of the event contained in the first column of the table. This concept is then linked to concepts of type `GTDEntity` and value to values of the other columns through relation nodes of the types corresponding to the labels of the columns.

The resulting network of information contains more than 10,000,000 concept nodes linked through 13,560,000 relation nodes.

Results

We launched several requests on the network made of the GTD data. These requests are patterns of terrorist events containing some known and unknown values. The Fig. 8.12 depicts an example of such a request.

It requests the network for information on years, types of weapons used, and number of persons killed in all the terrorist events perpetrated by Tamils and that occurred in Sri Lanka. This request gave us 2935 answers under the form of graphs describing these terrorists events (see Fig. 8.12 for one of the answers).

The whole data network can be processed through the parallelized version of the fusion algorithm. This was not worth considering, using a classical algorithm for the maximum subgraph isomorphism problem.

Acknowledgements We would like to thank Johann Dreo, Christophe Gouguenheim, Florence Aligne, and Juliette Mattioli for fruitful discussions and collaborative work.

References

Aligne F (2009) Which information and decision support system for the crisis management? In: NATO RTO-MPIST-086

Aligne F, Mattioli J (2010) The role of context for crisis management cycle. In: Burstein F, Brézillon P, Zaslavsky AB (eds) Supporting real time decision-making. Annals of information systems, vol 13. Springer, New York, pp 113–132

American Red Cross (2010) The case for integrating crisis response with social media. http://www.scribd.com/doc/35737608/White-Paper-The-Case-for-Integrating-Crisis-Response-With-Social-Media

Belov N, Gerken P (2009) Mixed initiative soft data fusion associate. In: 12th international conference on information fusion, pp 1897–1902

Blondel V, Guillaume J-L, Lambiotte R, Lefebvre E (2008) Fast unfolding of communities in large networks. J Stat Mech: Theory Exp 10:P10008

Bunke H, Foggia P, Guidobaldi C, Sansone C, Vento M (2002) A comparison of algorithms for maximum common subgraph on randomly connected graphs. In: Lecture notes on computer science, vol 2396. Springer, Heidelberg, pp 123–132

Cameron M, Power R, Robinson B, Yin J (2012) Emergency situation awareness from twitter for crisis management. In: Proceedings of the 21st international conference companion on World Wide Web. ACM, Lyon, pp 695–698

Capet F, d'Allonnes AR (2014) Information evaluation in the military domain: doctrines, practices and shortcomings. In: Capet P, Delavallade T (eds) Information evaluation. Wiley-ISTE, Hoboken, p 103

Capet P, Delavallade T (2014) Information evaluation. Wiley-ISTE, Hoboken

Castillo C, Mendoza M, Poblete B (2010) Twitter under crisis: can you really trust what we RT? InL 1st workshop on social media analytics, SOM'10

Castillo C, Mendoza M, Poblete B (2011) Information credibility on Twitter. In: Proceedings of the 20th international conference on World Wide Web, number 11. ACM, Hyderabad
Chein M, Mugnier M-L (2008) Graph-based knowledge representation: computational foundations of conceptual graphs. Springer, New York
Conte D, Foggia P, Vento M (2007) Challenging complexity of maximum common subgraph detection algorithms: a performance analysis of three algorithms on a wide database of graphs. J Graph Algorithms Appl 11(1):99–143
Coombs WT (2007) Crisis management and communications. IPR. http://www.instituteforpr.org/crisis-management-communications/
Crocq L (2007) Médecin général des armées, Formation et sensibilisation NRBC, Haut Comité Français pour la défense civile
Crooks A, Croitoru A, Stefanidis A, Radzikowski J (2013) #Earthquake: twitter as a distributed sensor system. T GIS 17(1):124–147
Date K, Gross GA, Khopkar SS, Nagi R, Sambhoos K (2013) Data association and graph analytical processing of hard and soft intelligence data. In: Proceedings of the 16th international conference on information fusion. IEEE, New York, pp 404–411
Deparis É (2013) Création de nouvelles connaissances décisionnelles pour une organisation via ses ressources sociales et documentaires. Ph.D. thesis, Université de Technologie de Compiégne, Compiégne
Flizikowski A, Holubowicz W, Stachowicz A, Hokkanen L, Kurki T, Paivinen N, Delavallade T (2014) Social media in crisis management the iSAR+ project survey. In: Hiltz S, Pfaff M, Plotnick L, Shih P, Tapia A (eds) Proceedings of the international ISCRAM conference
Goujon B (2009) Uncertainty detection for information extraction. In: Proceedings of the international conference on recent advances in natural language processing, pp 118–122
Gouttas C, de Maupeou S, Mabiala Y, Delavallade T, Huyot B (2013) Visual analytics innovative services for cyber-security. In: NATO symposium on visual analytics, IST-116
Gross G, Nagi R, Sambhoos K, Schlegel D, Shapiro S, Tauer G (2012) Towards hard+soft data fusion: processing architecture and implementation for the joint fusion and analysis of hard and soft intelligence data. In: 15th international conference on information fusion, pp 955–962
Huang C-M, Chan E, Hyder A (2010) Web 2.0 and internet social networking: a new tool for disaster management? Lessons from Taiwan. BMC Med Inform Decis Mak 10(1):57. doi:10.1186/1472-6947-10-57. http://www.biomedcentral.com/1472-6947/10/57.ISSN:1472-6947
Imran M, Castillo C, Lucas J, Meier P, Rogstadius J (2014) Coordinating human and machine intelligence to classify microblog communications in crises. In: Hiltz S, Pfaff M, Plotnick L, Shih P, Tapia A (eds) Proceedings of the international ISCRAM conference
Initive HH (2011) Disaster relief 2.0: the future of information sharing in humanitarian emergencies. Technical report, UN Foundation and Vodafone Foundation Technology Partnership
Kohler H, Lambert DA, Richter J, Burgess G, Cawley T (2013) Implementing soft fusion. In: Proceedings of the 16th international conference on information fusion, FUSION 2013, Istanbul. IEEE, New York, pp 389–396
Kumar S, Barbier G, Abbasi M, Liu H (2011) Tweettracker: an analysis tool for humanitarian and disaster relief. In: International conference on weblogs and social media (ICWSM), pp 1–1
Laudy C (2010) Introducing semantic knowledge in high level information fusion. Ph.D. thesis, Paris 6 University
Laudy C (2011) Semantic knowledge representations for soft data fusion — efficient decision support systems - practice and challenges from current to future. INTECH Open Access Publisher
Laudy C, Ganascia J-G (2009) Using maximal join for information fusion. In: Graph structures for knowledge representation and reasoning (GKR 2009) collocated with the twenty-first international joint conference on artificial intelligence (IJCAI-09)
Lerbinger O (1997) The crisis manager: facing risk and responsibility/Otto Lerbinger. Erlbaum, Mahwah

Lesot M-J, Delavallade T, Pichon F, Akdag H, Bouchon-Meunier B, Capet P (2011) Proposition of a semi-automatic possibilistic information scoring process. In: 7th conference of the European Society for Fuzzy Logic and Technology, pp 949–956

Lesot M, Pichon F, Delavallade T (2014) Quantitative information evaluation: modeling and experimental evaluation. In: Capet P, Delavallade T (eds) Information evaluation. Wiley-ISTE, Hoboken, p 187

Liu Y, Jiang X, Chen H, Ma J, Zhang X (2009) MapReduce-based pattern finding algorithm applied in motif detection for prescription compatibility network. In: Advanced parallel processing technologies, vol 5737. Springer, Berlin/Heidelberg, pp 341–355

Llinas J, Nagi R, Hall D, Lavery J (2010) A multi-disciplinary university research initiative in hard and soft information fusion: overview, research strategies and initial results. In: 13th international conference on information fusion, pp 1–7

MacEachren A, Robinson A, Jaiswal A, Pezanowski S, Savelyev A, Blanford J, Mitra P (2011) Geo-Twitter analytics: applications in crisis management. In: Proceedings of 25th international cartographic conference, p 1

McKay BD, Piperno A (2014) Practical graph isomorphism, II. J Symb Comput 60:94–112

Meier P (2013) Two signs of things to come in disaster response: learning from typhoon Yolanda. Opening Keynote Address at CrisisMappers 2013 http://www.engagingcities.com/article/two-signs-things-come-disaster-response-learning-typhoon-yolanda

NATO (2003) STANAG2511. Intelligence Reports

Ndiaye SN, Solnon C (2011) CP Models for Maximum Common Subgraph Problems. In: 17th international conference on principles and practice of constraint programming (CP). Lecture notes in computer science. Springer, New York, pp 637–644

Peary B, Shaw R, Takeuchi Y (2012) Utilization of social media in the East Japan earthquake and tsunami and its effectiveness. J Nat Disaster Sci 34(1):3–18

Pichon F, Dubois D, Denoeux T (2012) Relevance and truthfulness in information correction and fusion. Int J Approx Reason 53(2):159–175

Pichon F, Labreuche C, Duqueroie B, Delavallade T (2014) Multidimensional approach to reliability evaluation of information sources. In: Capet P, Delavallade T (eds) Information evaluation. Wiley-ISTE, Hoboken

Pine JA, Lemoine J, Benhadda H (2005) Un nouvel outil de classification non supervisée de documents pour la découverte de connaissances et la détection de signaux faibles: Rares text. Journées Sur Les Systémes D'Information Elaborée, 1–15 (Online)

Plantenga T (2013) Inexact subgraph isomorphism in MapReduce. J Parallel Distrib Comput 73:164–175

Raymond J, Willett P (2002) Maximum common subgraph isomorphism algorithms for the matching of chemical structures. J Comput Aided Mol Des 16(7):521–533

Roth F, Giroux J, Herzog M (2013) Crisis mapping in Switzerland: a stakeholder analysis. 3rg special report, Risk and Resilience Research Group, Center for Security Studies (CSS), ETH

Sakaki T, Okazaki M, Matsuo Y (2010) Earthquake shakes Twitter users: real-time event detection by social sensors. In: Proceedings of the 19th international conference on World Wide Web, pp 851–860

Shafer G (1976) A mathematical theory of evidence. Princeton University Press, Princeton

Singh VK, Gao M, Jain R (2010) Situation detection and control using spatio-temporal analysis of microblogs. In: Rappa M, Jones P, Freire J, Chakrabarti S (eds) Proceedings of the 19th international conference on World Wide Web. ACM, pp 1181–1182

Smets P (1993) Belief functions: the disjunctive rule of combination and the generalized Bayesian theorem. Int J Approx Reason 9(1):633–664

Smets P, Kennes R (1994) The transferable belief model. Artif Intell 66(2):191–234

Solnon C (2010) Alldifferent-based filtering for subgraph isomorphism. Artif Intell 174(12–13):850–864

Sowa J-F (1984) Conceptual structures: information processing in mind and machine. Addison-Wesley, Reading

Starbird K, Palen L (2010) Pass it on: retweeting in mass emergency. In: Proceedings of the 7th international ISCRAM conference

Stollberg B, de Groeve T (2012) The use of social media within the global disaster alert and coordination system (gdacs). In: Proceedings of the 21st international conference companion on World Wide Web. ACM, pp 703–706

Ullmann JR (1976) An algorithm for subgraph isomorphism. J ACM 23(1):31–42

Vieweg S, Hughes AL, Starbird K, Palen L (2010) Microblogging during two natural hazards events: what Twitter may contribute to situational awareness. In: Proceedings of the SIGCHI conference on human factors in computing systems, pp 1079–1088

White T (2009) Hadoop: the definitive guide, 1st edn. O'Reilly Media, Sebastopol

Yager R (1984) Approximate reasoning as a basis for rule-based expert systems. IEEE Trans Syst Man Cybern 14:636–643

San Su Y, Wardell C III, Thorkildsen Z (2013) Social media in the emergency management field. Technical report, National Emergency Management Association (NEMA), CNA Analysis and Solutions

Zhao Z, Wang G, Butt AR, Khan M, Anil Kumar VS, Marathe MV (2012) SAHAD: subgraph analysis in massive networks using Hadoop. In: Proceedings of the 2012 IEEE 26th international parallel and distributed processing symposium. IPDPS '12. IEEE Computer Society, Washington, DC, pp 390–401. doi:10.1109/IPDPS.2012.44. http://dx.doi.org/10.1109/IPDPS.2012.44. ISBN: 978-0-7695-4675-9

Chapter 9
Towards a Crowd-Sensing Enhanced Situation Awareness System for Crisis Management

Andrea Salfinger, Werner Retschitzegger, Wieland Schwinger, and Birgit Pröll

Abstract Natural and man-made crises pose severe challenges on emergency responders, as they need to gain timely *Situation Awareness* (SAW) in order to decide upon adequate rescue actions. Computational SAW systems aim at supporting humans in rapidly achieving SAW by means of Information Fusion (IF), thus reduce information overload by fusing data stemming from various sensors to situation-level information. Recently, the increasing popularity of social media on mobile devices has enabled humans to act as *crowd sensors*, who broadcast their observations on the unfolding crisis situation over social media channels. Consequently, SAW systems for crisis management would benefit from exploiting social media as additional data source. Therefore, the aim of this chapter is to investigate upon how crowd-sensing can be incorporated into SAW systems for crisis management, by elaborating on the following issues: How can the SAW system seek and retrieve additional information from social media that may complement the situational picture obtained with other types of sensors? How can the SAW system adapt this crowd-sensing alongside the monitored situation, to keep pace with the underlying real-world incidents? We attempt at illustrating potential solutions towards these questions, by examining how crowd-sensing can deliver input data for SAW systems, elaborating on the challenges such systems need to overcome in order to identify and extract relevant information from social media, and finally, discussing the architecture of a situation-adaptive SAW system capable of exploiting both conventionally sensed data and unstructured social media content.

A. Salfinger (✉) • W. Retschitzegger • W. Schwinger
Department of Cooperative Information Systems, Johannes Kepler University Linz, Altenbergerstr. 69, 4040 Linz, Austria
e-mail: andrea.salfinger@cis.jku.at; werner.retschitzegger@cis.jku.at; wieland.schwinger@cis.jku.at

B. Pröll
Institute for Application Oriented Knowledge Processing, Johannes Kepler University Linz, Altenbergerstr. 69, 4040 Linz, Austria
e-mail: bproell@faw.jku.at

G. Rogova, P. Scott (eds.), *Fusion Methodologies in Crisis Management*,
DOI 10.1007/978-3-319-22527-2_9

9.1 Introduction

9.1.1 The Role of Information Fusion for Achieving Situation Awareness

Natural and man-made crises pose severe challenges on emergency responders, as they need to gain timely *Situation Awareness* (SAW), i.e., an appropriate understanding of the situation(s) at hand, in order to decide upon adequate rescue actions. However, humans' situation assessment (i.e., the mental process leading to SAW), which consists of the *perception* of elements in the environment, the *comprehension* of their meaning (i.e., understanding their interrelations to infer the overall situation), and their *projection* into the future (i.e., anticipating how the situation will develop, as a basis to undertake adequate counteractions and measures) (Endsley 1995), is hampered by information overload, time pressure, and incomplete or missing information. To mitigate this problem and support humans' situation assessment, systems capable of computational situation assessment have been developed. These *SAW systems* mimic a human's situation assessment process by means of Information Fusion (IF)—thus, observe the environment through various sensors (analogous to a human's physical sensing, e.g., vision), fuse the sensors' measurements to infer the underlying real-world objects (corresponding to a human's perception), assess the relationships between these objects to detect the overall ongoing situation(s) (corresponding to a human's comprehension), and finally, analyze these situations' impact and likely development (corresponding to a human's projection) (Niklasson et al. 2008). Hence, the outcome of these computational IF steps represents already aggregated, high-level situational information the SAW system's human operators can be provided with. Consequently, the SAW system's human operators do not need to sift through and interpret the massive amount of incoming data themselves, which reduces their information overload and response time towards critical situations.

9.1.2 Peculiarities of Crisis Management

Whereas such SAW systems have already proven their value in a range of control center domains [e.g., road-traffic (Baumgartner et al. 2010, 2014; Meyer-Delius et al. 2009) and maritime monitoring (Edlund et al. 2006; Rhodes et al. 2007)], their adoption for emergency response and crisis management is hampered by the following peculiarities: First, unlike domains that can be comprehensively monitored by sensors, crises, and their effects typically cannot be extensively captured due to their unpredictability, uniqueness, and large-scale dimensions

(Walle and Turoff 2008). Even for crises that can be forecast to some extent (e.g., meteorological phenomena like hurricanes), it is largely unknown what effects they will cause where and when. Therefore, the monitoring SAW systems may face *limited sensor coverage* (and thus, limited data) on the unfolding crisis situation and its impacts. Second, as crisis situations themselves are dynamic, crisis management organizations' information need substantially differs across the monitored *crisis situation's evolution* [stretching across preparation, mitigation, response, and recovery (Walle and Turoff 2008)]. Especially in the aftermath of a crisis, a substantial amount of situational information is delivered by on-the-ground human observers (e.g., reports on injuries and damage), and thus often not easily accessible for machine-based (real-time) processing.

9.1.3 Crowd-Sensing to the Rescue

However, technological progress and associated societal changes started to ameliorate this *sensor coverage problem* encountered in the crisis management domain. Nowadays ubiquitous availability of social media on mobile devices enables humans to act as *crowd sensors* or *citizen sensors* (Sheth 2009): In matters of seconds, a human bystander observing a specific event can directly broadcast his/her observation to the entire world. As social media platforms often provide dedicated APIs,[1] their plethora of human communication is also readily accessible for machines and thus can be computationally monitored in (near) real time. This so-called *crowd-sensing* can also be employed for collecting observations on unfolding real-world crisis situations shared via social media messages (such as Facebook status updates and microblogging), as has been confirmed in several case studies (e.g., Cameron et al. 2012; Dashti et al. 2014; Dugdale et al. 2012; Shaw et al. 2013). Although social media platforms are primarily intended for human communication, various crowd-sensing systems have already demonstrated that they can be used to automatically retrieve *situational update information* for crisis management (e.g., Ulicny et al. 2013; Yin et al. 2012; Purohit and Sheth 2013; Abel et al. 2012; MacEachren et al. 2011; Rogstadius et al. 2013; Kumar et al. 2011; Sakaki et al. 2010; Smid et al. 2011). However, these approaches have been focused on retrieving and aggregating crowd-sensed information, but have not yet examined on a situation-level fusion of these information with data obtained from other sensors, nor on adapting towards the specific information needs of their human operators throughout the unfolding crisis situation.

[1] Application programming interface.

9.1.4 Situation-Adaptive Crowd-Sensing

Therefore, the aim of the present chapter is to elaborate on the following issues: First, to address the *sensor coverage problem*, the SAW system should automatically retrieve information from social media that complements the current situational picture obtained with other types of sensors. However, due to the *dynamics* of the monitored real-world crisis situations, this crowd-sensing needs to be dynamic itself: For instance, during crisis preparation, the crowd will provide different types of information than in the aftermath of a crisis, and also the information need of the SAW system's human operator will differ. Thus, second, we need to address the question how the SAW system could adapt this crowd-sensing alongside the monitored situation, to keep pace with the underlying real-world incidents and the implied varying information provision (by the human crowd sensors) and information needs (of its human operators). Once the SAW system has detected a specific real-world situation, it should therefore be capable of adjusting its crowd-sensing to retrieve additional information on this situation. While monitoring the unfolding situation, the SAW system could project the crisis' further evolution, and adjust its crowd-sensing to proactively look out for the expected information, to allow for a timely detection of situational updates. Conversely, the (already known) situation that has driven crowd-sensing can also be exploited in the interpretation of the crowd-sensed content, and the extraction and fusion of relevant information. Summarizing, to address these issues, in this chapter we introduce *situation-adaptive crowd-sensing* for crisis SAW: We discuss the architecture of a crowd-sensing enhanced SAW system, which keeps track of the monitored real-world situation's evolution and reuses the detected (and projected) situational context to optimize its crowd-sensing configuration and processing.

9.1.5 Structure of this Chapter

In the next section, we will examine the prerequisites of our work, by outlining the fundamental background of social media usage in crisis situations and discussing which types of information potentially of value for crisis management organizations are provided within social media. Consequently, Sect. 9.3 reviews the state of the art of computational systems for making use of such social media content in crisis management, and analyzes the systems' open issues. To tackle the identified open issues, we sketch the architecture of a situation-adaptive, crowd-sensing enhanced SAW system for crisis management in Sect. 9.4. Based on a case-study on a real-world social media data set, we will outline the challenges its different processing blocks need to address, and point out potential technical solutions based on our prototypical implementation. Finally, we summarize our approach in Sect. 9.5.

9.2 Situational Update Information from Social Media

9.2.1 Social Media Usage in Crisis Situations

The popularity of social media is still on the rise,[2] and its users generate billions of messages every day: According to Twitter's usage statistics, roughly 9000 tweets are generated per second.[3] Social networks thus more and more are becoming vital "information hubs" for disseminating all types of information in (near) real-time, which is also the case in crisis and disaster situations: Such exogenous events (i.e., events happening outside the social network which trigger an increased social media usage) like crisis events even reinforce people's turn to social media (Imran et al. 2014a), who then employ social networks to actively broadcast or seek timely situational information. In the light of this, various case studies examined on how people make use of social media during and after crisis events [e.g., studies on social media activities triggered by flooding events in Queensland, Australia (Cameron et al. 2012; Shaw et al. 2013), in Colorado (Dashti et al. 2014), and Germany (Fuchs et al. 2013), flooding and wildfires in the USA (Vieweg et al. 2010), earthquakes in Japan (Sakaki et al. 2011), or the Haiti earthquake in 2010 (Dugdale et al. 2012)], which confirmed that crisis-relevant information is broadcast via social media. It could be shown that fine-grained *situational update information* is posted from on-the-ground observers (Vieweg et al. 2010), sometimes even before this information reaches the news (Ulicny et al. 2013) or authorities responsible for taking appropriate counteractions.

9.2.2 Types of Crisis Information in Social Media

Consequently, the question arises of what types of crisis-related information are published in social media, and whether these would represent valuable information for crisis management organizations. Therefore, Olteanu et al. (2015) analyzed social media messages regarding various crisis events with respect to crisis management organizations' specific *information needs*. In terms of the different types of information relevant for crisis management, messages on *caution and advice*, *affected individuals*, *infrastructure and utilities*, *donations and volunteering*, *sympathy* and *other topics* could be identified. Furthermore, they also categorized the different encountered sources to analyze the role of the messages' authors (i.e., whether these messages have been posted by eyewitnesses, governments, NGOs, businesses, media or outsiders). Using this categorization, social media messages on 26 different crisis events, involving both natural disasters (such as

[2]http://www.pewinternet.org/2015/01/09/social-media-update-2014/, accessed on 22 April 2015.

[3]www.internetlivestats.com/twitter-statistics, accessed on 22 April 2015.

flooding, earthquakes, meteors, wildfires, and typhoons), as well as human-induced emergencies (such as an explosions, bombings, fires, shootings, train crashes, and building collapses), have been labeled manually by human crowd-workers,[4] which revealed that these information types of relevance for crisis management can be found in social media. It could be shown that the proportions of the different types of information exhibit considerable variations between different crisis events, but are generally stable across comparable crisis events (Olteanu et al. 2015). Thus, for future crises, we can expect to observe similar types of information in social media, but need to expect that some types of information may be sparse (e.g., on-the-ground observations are frequently dominated by news, sympathy and comments).

9.3 Related Work: Crowd-Sensing Systems for Crisis Management

Based on these promising case studies revealing the potential of social media for supporting crisis management (cf. Sect. 9.2), several crowd-sensing systems have been proposed, with the aim of retrieving crisis-relevant social media content and processing it to actionable information for emergency managers and operators. An extensive survey reviewing the state of the art has been presented in Imran et al. (2014a), with the aim of elaborating on the different processing steps employed by recent crowd-sensing approaches and pointing towards interesting techniques developed in related fields, but not investigating upon the situation-level fusion of crowd-sensed information. Whereas existing, elaborate IF architectures for the application domain of crisis management (such as Scott et al. 2004; Jakobson et al. 2006) already allow for the inclusion of unstructured textual content provided by human observers [i.e., so-called *soft sensors* (Llinas 2010)], these do not elaborate on means for addressing the peculiarities of crowd-sensing. We therefore conducted a survey evaluating current crowd-sensing systems with a focus on these systems' capabilities for supporting a human operator's SAW (Salfinger et al. 2015) by means of situation-level fusion, in which we investigated the following nine advanced approaches: HADRian (Ulicny et al. 2013), ESA (Yin et al. 2012), Twitris (Purohit and Sheth 2013), Twitcident (Abel et al. 2012), SensePlace2 (MacEachren et al. 2011), CrisisTracker (Rogstadius et al. 2013), TweetTracker (Kumar et al. 2011), Toretter (Sakaki et al. 2010), and CIACM (Smid et al. 2011). The conclusions of this evaluation are summarized in the following.

In order to assess these systems' provided level of SAW support in a structured fashion, we compared their abilities in terms of the JDL data fusion model (Llinas et al. 2004), which represents a functional model of the IF levels SAW systems should provide to enhance their operators' SAW, ranging from low-level fusion (L0–L1) to high-level fusion (L2–L3).

[4]The resulting labeled CrisisLex26 data set is available for research purposes under www.crisislex.org.

In particular, these levels comprise JDL Level 0 (L0), responsible for observing the environment through various *sensors*, JDL Level 1 (L1), fusing the measurements retrieved from L0 *within* and *across* sensors in order to infer relevant real-world *objects* (thus, performs object assessment), JDL Level 2 (L2), analyzing the relationships between these objects in order to assess *situations*, JDL Level 3 (L3), projecting these situations' development and impact, JDL Level 4 (L4), realizing resource management or *process refinement*, and JDL Level 5 (L5), *user refinement*, comprising the user's interactions and his/her potential adaptations at all processing levels. Figure 9.1 presents an extension of the JDL model, which maps the processing steps of crowd-sensing systems onto their corresponding JDL levels. Levels L0–L3 base upon each other, by aggregating and interpreting data from the precedent level. Contrastingly, levels L4 and L5 influence other levels, forming feedback loops orthogonal to the functional levels. Thus, both levels, L4 and L5, allow for an adaptive SAW system. Thereby, L4 may comprise feedback loops from object fusion, situation assessment or projection which automatically adapt the lower fusion levels, for instance, the sensing level, by actively triggering the sensors to retrieve more information on identified objects or situations (i.e., perform sensor tuning). L5 enables the user to interact upon any of the levels' components in that, for example, he/she may inspect the sensing output and reconfigure the sensors accordingly.

Based upon comparing the different systems in terms of this evaluation framework, we identified the following open issues.

9.3.1 High-Level Fusion Rarely Attempted

When comparing their functionality with respect to a comprehensive SAW architecture stretching across all IF levels, it becomes apparent that current systems rarely support high-level fusion, thus, do not provide situation assessment [which is, except for HADRian (Ulicny et al. 2013), deferred to the human operator] and projection (which is not targeted in any of the above-mentioned systems).

9.3.2 Lack of Self-adaptivity

Although providing valuable first steps towards crowd-sensing enhanced SAW systems, these systems are designed to be steered by a human operator and his/her ad-hoc information needs and insights, but provide limited (self-)adaptivity towards the *dynamics* of social media (e.g., detecting and reacting to emerging social media trends, such as the adoption of crisis-specific hashtags in Twitter), and system-based, continuous situation monitoring functionality that adapts alongside the evolving real-world situation.

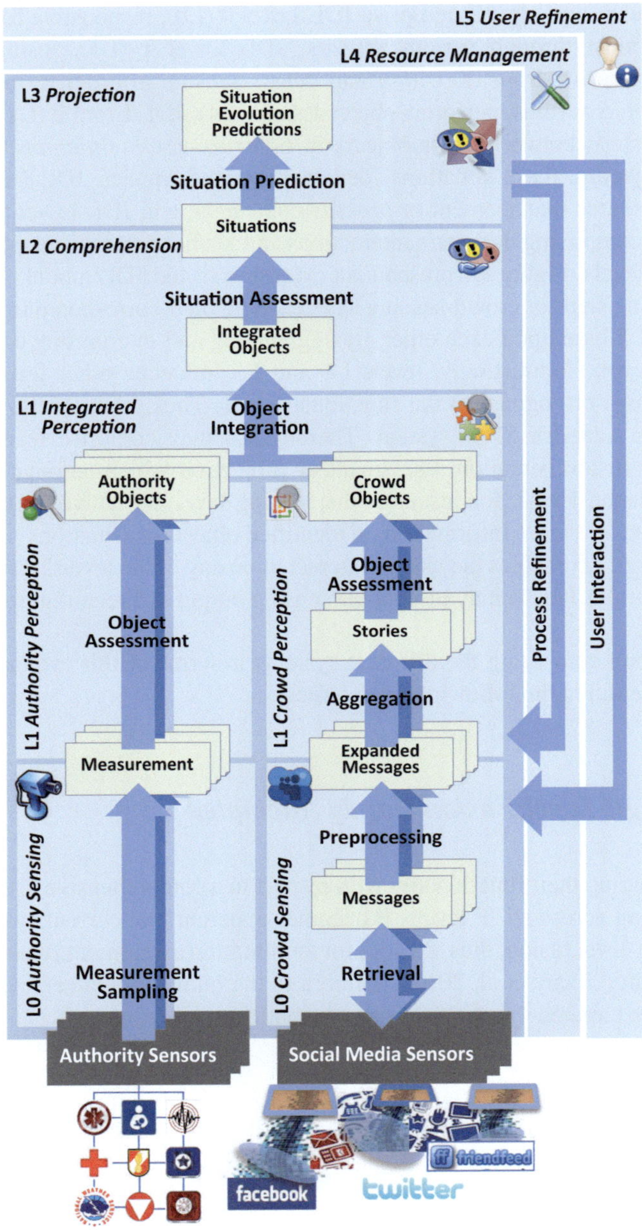

Fig. 9.1 Integrating crowd-sensing into SAW systems conforming to the JDL Information Fusion model

9.3.3 Sparsity Not Addressed

Besides that, many systems base on clustering approaches to infer real-world events from social media content (Yin et al. 2012; Purohit and Sheth 2013; Abel et al. 2012; Rogstadius et al. 2013) or probabilistic event detection (Purohit and Sheth 2013; Sakaki et al. 2010; Smid et al. 2011), thus, summarize frequently posted information, which is assumed to represent a certain degree of confidence. However, actually *sparse* situational update information (Vieweg et al. 2010) may be dominated by general news and comments [as is frequently the case, cf. our discussion of the findings presented in Olteanu et al. (2015) in Sect. 9.2], and therefore may not be brought to the human operator's attention.

9.3.4 Limited Semantic Processing

This is aggravated by the fact that most current systems do not employ a natural language processing (NLP) approach for an in-depth analysis of the actual *semantics* contained in the messages, but pursue a bag-of-words approach (thus, compare textual similarity based on common words), nor incorporate already established *situational context* in this processing.

9.3.5 Specifying the Information Need Rarely Supported

We further note that present crowd-sensing systems rarely provide their users with means of specifying their *information needs* a priori, apart from formulating queries (typically keywords) for the retrieval of messages over the social media platform's API. This has also been one major finding presented in Imran et al. (2014a), who conclude that current crowd-sensing approaches focus on the bottom-up, i.e., data-driven, processing of social media messages (by aggregating related content), but do not provide means for human operators to drive this process by their current information needs [e.g., by specifying which of the *information types* enlisted in Olteanu et al. (2015) would be currently of interest for the human operator]. This, however, represents the key advantage of so-called *template-based* SAW systems (Laxhammar 2008), which allow its users to specify templates (i.e., *situation types*) describing the situations of interest that should be detected by the system.

Therefore, in the next section, we will elaborate on how to integrate social media-based crowd-sensing into such a template-based SAW system, and investigate upon how this allows us to address the identified challenges.

9.4 A Framework for Crowd-Sensing Enhanced SAW Systems

9.4.1 Goals

In the present section, the architecture of a SAW system will be derived which is capable of exploiting both, data delivered from the SAW systems conventional sensors (which we term *authority sensors*), as well as information obtained from social media platforms, i.e., *crowd-sensed* information. Stretching beyond existing crowd-sensing systems' objectives (cf. Sect. 9.3), the focus of our architecture will be on providing means for *situation-adaptive crowd-sensing*, in order to provide the following functionality: Once the SAW system has detected a specific real-world situation, it should be capable of adjusting its crowd-sensing to retrieve additional information on this situation (thus, the SAW system should adapt its *sensing configuration*, i.e., JDL L0, accordingly). While monitoring the unfolding situation, the SAW system could project the crisis' further evolution, and adjust its crowd-sensing to proactively look out for the expected types of information, which aids the timely detection of situational updates. Conversely, the already established situational context that has driven the crowd-sensing can also be exploited in the interpretation of the crowd-sensed content, and the extraction and fusion of relevant information (thus, the SAW system should adapt its *perception configuration*, i.e., JDL L1, accordingly).

9.4.2 Approach

As we will show, *situation-adaptive crowd-sensing* can be realized by including *situational feedback-loops* between the different processing levels (i.e., performing JDL L4 process refinement), which adapt the lower-level IF steps and provide additional situational context for the retrieval and interpretation of social media content based on the currently monitored (or projected) real-world crisis situation. Thus, we will outline how the integration of crowd-sensing into a SAW system implementing the JDL IF principles not only allows to enrich the assessed situational picture by incorporating additional data. It complementary also allows to refine and enhance the crowd-sensing process itself, by evolving it alongside the monitored situation and providing situational context for the interpretation of the retrieved messages. For realizing the required functionality, we will specifically elaborate on concepts for the following challenges: (1) coping with the dynamics of social media by providing an *adaptive crowd-sensing* platform, (2) employing a *semantic analysis* of textual content based on domain ontologies in order to infer the reported events and match them with ontological concepts that can be used for automated reasoning in the situation assessment phase, (3) specifically seeking *sparse situational update information* posted by on-the-ground observers by proposing an aggregation-segregation-based approach, and (4) employing additional context in order to infer the semantics of social media content and refine the crowd-sensing process by exploiting already assessed or projected situations, allowing for an informed, situation-aware processing. We will back the conceptual discussion with pointers towards technical solutions for implementing the sketched building blocks. These are based on the on-going prototype development in the course of our research project *crowd*^{SA} (Pröll et al. 2013),[5] which aims at leveraging social media content for rescue organizations such as the Austrian Red Cross. Therefore, we will subsequently refer to our architecture as the *crowd*^{SA} system.

9.4.3 Illustrative Scenario

To demonstrate the challenges that need to be addressed within such a crowd-sensing enhanced SAW system, we will provide real-world examples from our case studies on the microblogging service Twitter.[6] We will center our discussion on a concrete real-world scenario from the *HawaiiHurricanes* data set (cf. Table 9.1). In the collection of this data set, we initially aimed at monitoring the aftermath of the hurricanes Iselle and Julio affecting Hawai'i in August 2014. However, our keyword set—intentionally very broad, in order to maximize recall—retrieved tweets referring to concurrent, similar crises as well, which makes it a suitable data set for highlighting the challenges faced in real-world crisis situation monitoring settings. Therefore, we will underpin the various crowd-sensing challenges by presenting real-world tweets from this data set, which encompasses observations covering the following three different real-world crisis events (and their aftermath) happening between 9th and 15th August 2014:

- the hurricanes *Iselle* and *Julio* threatening the Hawai'ian islands,[7] which caused severe power outages on the East side of the island of Hawai'i,
- the flooding caused by the remnants of hurricane Bertha in UK,[8] and
- a serious flooding event on Long Island, New York.[9]

Thus, the devised SAW system needs to be capable of discriminating these three crisis situations.

[5] crowdsa.situation-awareness.net.

[6] Therefore, from now on we will use the terms *social media message* and its Twitter-specific equivalent *tweet* interchangeably.

[7] www.latimes.com/nation/nationnow/la-na-nn-hawaii-storm-iselle-juliio-20140808-story.html.

[8] www.bbc.com/news/uk-scotland-28739164.

[9] http://newyork.cbslocal.com/2014/08/13/flash-flood-watches-warnings-in-effect-as-heavy-rain-drenches-parts-of-tri-state/.

Table 9.1 Overview of recorded real-world data sets

Data set	Purpose event	Time period	Service	Remarks
HawaiiHurricanes	Hurricanes *Iselle* and *Julio* and the tropical storm *Genevieve* approaching Hawai'i[a]	08/09/2014–08/21/2014	Twitter Streaming API & Twitter4J	$\sim$ 212,600 tweets
GeneralDisasters	General twitter stream monitoring for English crisis-related keywords[b]	07/31/2014–10/31/2014	Twitter Streaming API & Twitter4J	$> 7.3 \times 10^6$ tweets

[a]By filtering tweets according to the following keywords and hashtags: Hurricane, #HurricaneIselle, #HurricanePrep, #updatehurricaneiselle, #hiwx, #HIGov, Iselle, #Genevieve, #Iselle, #Julio, #HIWX, #HIWx

[b]Tracked by a filter query leaving language and location deliberately unspecified and the following English keywords: Typhoon, Hurricane, Flooding, #Typhoon, #storm, #typhoon, #flood, flood, spring tide, windstorm, disaster

9.4.4 *Outline*

We start our discussion of the different processing layers by illustrating the functionality of the so-called *Core SAW system*, which consists of the essential building blocks that implement the JDL IF levels, but is confined to *authority-sensing*. Thus, it solely receives data specifically directed to this authority's (e.g., a specific rescue organization) SAW system. Thereby, it can be assumed that this information will be relevant for the monitoring task at hand, and thus, its informativeness need not be questioned. Commencing with the discussion of the *Core SAW system* is necessary to understand the implications on incorporating a crowd-sensing layer—notably, what kind of output needs to be provided by the crowd-sensing process in order to serve as input processable by the SAW system. After laying out the functionality and output of the *Core SAW system*, we are then enabled to elaborate on the challenges and solutions for including JDL L0 and L1 components realizing *crowd-sensing*, i.e., the retrieval and fusion of content from social media, and will discuss on how the *Core SAW system* and the *crowd-sensing* components can intelligently complement each other to realize *situation-adaptive* sensing and processing.

9.4.5 *The Core SAW System for Monitoring Evolving Crisis Situations*

In the present section, we will devise the core architecture of a SAW system supporting crisis management. As already mentioned, crisis management organizations have specific *information needs* that should be fulfilled by the SAW system, which vary according the human operator's specific role and across the crisis situation's evolution (e.g., in the aftermath of an earthquake, emergency doctors may want to know the locations of injured people, whereas infrastructure reconstruction units need to know the locations and types of damaged buildings). Therefore, the SAW system should allow for a specification of templates describing the situations that should be detected by the system (i.e., specific event constellations of interest). Furthermore, it should be capable of monitoring crisis situations across the entire crisis management life-cycle (i.e., across the evolutionary phases of situation detection, mitigation, response, and recovery), requiring that the system is capable of tracking the crisis situation across these phases.

Thus, our core SAW system needs to provide the following functionality, which we will discuss in the subsequent sections: (1) allowing for the specification of templates for the situations of interest, (2) detecting ongoing situations, and (3) monitoring a detected situation across its different phases, i.e., tracking the situation across its different evolutionary states. This can be realized by employing a *knowledge-based* situation assessment technique: Event constellations that should be detected by the system (i.e., *situations*) are described by a set of rules (i.e., the *situation templates*). *Situation Assessment* conforms to matching data observed from the environment against these rules, whereby a matched rule corresponds to the detection of a specific real-world *situation instance*. In the following, we will provide more details on this functionality by describing our prototypical implementation.

9.4.6 Implementation

For implementing a template-based SAW system, which essentially requires a means to describe domain knowledge and a mechanism for detecting situations (such as a reasoning component or rule engine), a broad set of technologies can employed, for which we sketch a few exemplary configurations from on our own developments: Whereas we originally employed semantic web technologies for our SAW framework BeAware (Baumgartner et al. 2010) (namely RDF, the graph database Allegrograph, Lisp, and Prolog), in our current version of the SAW framework we favor object-relational technologies for implementing the *crowd*^{SA} prototype, motivated by the comparative performance evaluation of the two approaches in Baumgartner et al. (2014).

For the *crowd*^{SA} prototype, we pursue a UML-based approach for ontology engineering (Kogut et al. 2002), using the model-driven development tool Visual Paradigm[10] as ***OntologyDesigner***. Our SAW Core Ontology is based on Matheus et al. (2003), whereby new application domains can be incorporated easily by simply inheriting from our base classes, such as *Object*, *Situation*, and *Action*. After ontology engineering, in which domain experts model the domain knowledge for describing the monitored environment, being the requirement for template-based situation modeling and assessment, the corresponding database schemes, Data

[10]http://www.visual-paradigm.com.

Access Layers, and persistence mappings can be automatically generated by Visual Paradigm. For our current implementation, we employ a postgreSQL database with the spatial extension postGIS, Java for implementing the framework logic, the object-relational-mapping framework Hibernate,[11] and the business rules platform JBoss Drools Rules.[12]

9.4.7 Configuration: Specifying Situation Templates

After modeling the application domain in terms of an appropriate domain ontology, domain experts need to provide the system with their *information needs*, thus, specify templates describing the situations the SAW system should detect at runtime. As enabler for tracking crisis situations across their different potential evolutionary phases and for situation-adaptive crowd-sensing, we consequently require a template allowing to specify these different *evolutionary phases*, and a technique to track real-world situation instances across these phases accordingly. Therefore, we introduced *Situation Evolution Types* (SETs) (Salfinger et al. 2014b) for specifying potential crisis situation evolutions.

A SET models the different potential evolutionary states of a situation, i.e., it allows to track a crisis situation from its emergence (e.g., the situation *Hurricane threatens inhabited area* is triggered by the formation of a hurricane moving towards inhabited landmass) through its climax (e.g., the hurricane makes landfall and causes damage, such as power outages and flooded roads) to its clearance (e.g., power is restored). As underlying formalism, SETs base on Finite State Machines (FSM) describing the different *states* of a situation, so-called Situation State Types (SSTs). A SST corresponds to a set of *Event* and *Object Types* (e.g., *PowerOutage*, *City*) in specific relations (e.g., the spatial relations *Overlapping*, *Close*, the temporal relation *Before*), and the possible transitions between these states (Salfinger et al. 2014b; Kokar et al. 2008). In order to facilitate SET specification, domain experts need to be provided with appropriate modeling tools. Figure 9.2 shows a screenshot of ***SEM*² *Suite***, a tool suite supporting an interactive and incremental specification of such SETs (Salfinger et al 2014a). The depicted SET captures potential SSTs encountered in expectation of, during and after a hurricane crisis (exemplarily showing also the specification of the SST "PowerOutageInCity", and a composite SST based thereupon, "PowerOutageInArea").

In order to enable ***Situation Assessment***, ***SEM*² *Suite*** compiles each of these SSTs to a rule supplied to the ***Situation State Assessor***'s rule engine (cf. Fig. 9.3 for an outline of the different architecture components), and stores the specified SETs in the ***Situation Evolution Type KB***. The latter is used at runtime by the ***Situation Evolution Assessor*** to reason upon their defined FSMs in order to track evolving situations.

[11] http://hibernate.org.

[12] http://www.drools.org.

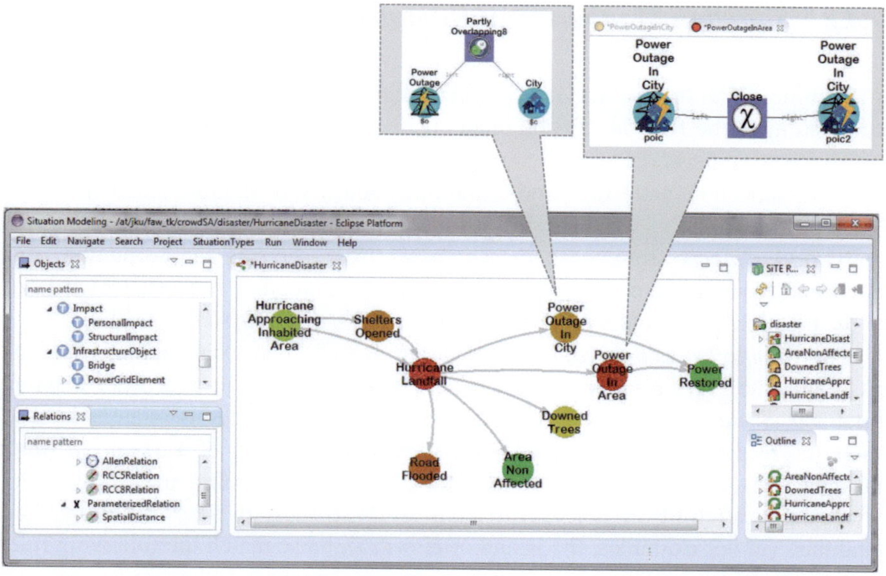

Fig. 9.2 A screenshot of SEM^2 Suite, showing a SET depicting the evolutionary states potentially encountered in the course of a hurricane situation, and a zoom into two of its modeled SSTs

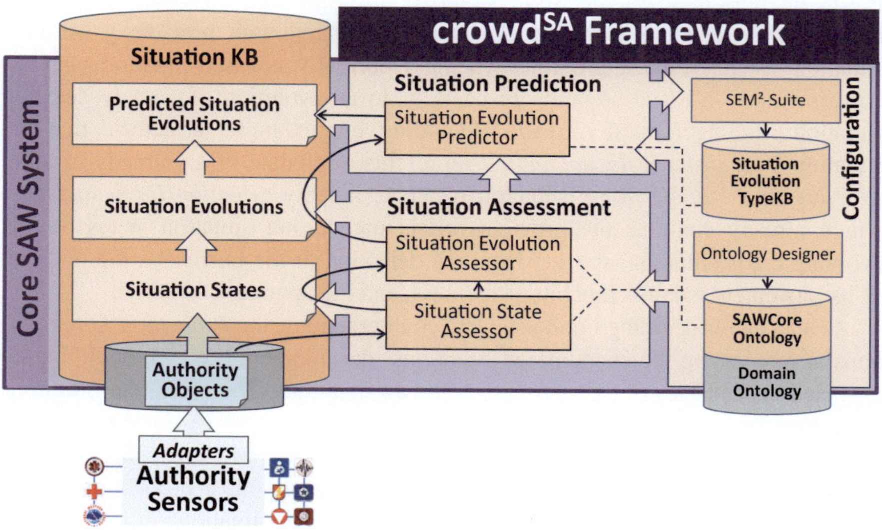

Fig. 9.3 The essential building blocks of the *Core SAW system*

9.4.8 Situation Monitoring at Runtime

At runtime, situation monitoring is performed in the following fashion (cf. Fig. 9.3), conforming to the JDL data fusion model (Llinas et al. 2004).

The *Sensing Level* (L0) retrieves various observations from the environment (e.g., measurements obtained from wind or humidity sensors, or satellite images).

The *Perception Level* (L1) performs object assessment, i.e., infers the monitored real-world ***Objects*** from these measurements (e.g., detects the formation of a hurricane from satellite data and different types of weather sensors, which leads to the creation of an object of the class "Hurricane", e.g., *Hurricane Iselle*).

The *Comprehension Level* (L2) infers the overall situational picture based on analyzing the objects interrelations, which is implemented in a two-tier fashion: First, the ***Situation State Assessor*** detects the current situations' state: Matched rules (corresponding to a specific SST) trigger the creation of a ***Situation State*** instance, i.e., a snapshot of a real-world situation. According to our scenario, when the location of the object *Hurricane Iselle* spatially overlaps with inhabited landmass (in our examples, the island of *Hawai'i*), these match the following rule: PartiallyOverlapping(Hurricane, PopulatedArea). Therefore, following this rule, an instance of the SST "HurricaneLandfall" is created (in our example, the situation state instance *IselleLandfall*).

Second, the ***Situation Evolution Assessor*** performs situation evolution tracking. Based on reasoning on the SETs stored in the ***Situation Evolution Type KB***, the currently detected ***Situation States*** are compared with previously assessed situation instances, in order to infer whether a currently assessed situation snapshot corresponds to a novel situation instance or to an evolution of already detected situation (Salfinger et al. 2014b). Regarding our example, it detects that the situation state instance *IselleLandfall* represents an update of an already existing evolving *HurricaneDisaster* situation instance, notably *SituationHurricaneIselle*, which already contains the previous, crisis preparation situation states of this evolving crisis situation, starting from the detection of the hurricane, the tracking of its movement towards inhabited landmass, and the opening of shelters.

As a basis for deciding upon adequate counteractions, the *Projection Level* (L3) aims at forecasting the encountered situations' development (i.e., what will be the expected next phases of the encountered crisis situation). Therefore, the ***Situation Evolution Predictor*** reasons upon currently ongoing ***Situation Evolutions***, forecast data (e.g., weather forecasts), and potentially historic ***Situation Evolutions*** (i.e., historic situation instances are employed to compute the most probable situation evolution). For instance, regarding the depicted crisis situation *SituationHurricaneIselle*, which is in the situation state where the crisis reaches its climax during the landfall of the hurricane (*IselleLandfall*), the next expected phases are the different situation states that may be encountered in the crisis aftermath phase, such as flooded roads, power outages, and downed trees. Thus, the SAW system can anticipate these states, based on determining which states are reachable from its current state in the SET.

9.4.9 Incorporating Situation-Adaptive Crowd-Sensing

The aim of including a crowd-sensing and -perception pipeline (i.e., including a dedicated L0 and L1 level for sensing and fusing social media content) is to gain potentially relevant information from diverse social media platforms, posted by a *crowd* of human observers [each regarded as a single *crowd sensor* (Nagarajan et al. 2009; Sakaki et al. 2010)], to complement or enhance the situational picture obtained from *authority sensors*. Consequently, the crowd-sensing pipeline's input consists of *social media messages* (e.g., tweets, Facebook status updates), from which relevant observations on evolving events, objects or situations should be extracted for further processing.

Whereas, in the illustrated hurricane scenario, some SSTs can be assessed from authority-specific hardware sensors (e.g., the detection of a hurricane formation and the projection of its likely movement, can be accurately computed from various weather sensors and satellite sources), the assessment of actually encountered damage (as well as the determination of non-affected areas and needed resources) is largely based on human reports and observations, which we can attempt to gather from social media, as will be illustrated in the present section. We will thus elaborate on how relevant observations can be retrieved from social media and can serve in the inference of the situational picture.

As outlined in Fig. 9.4, which sketches the high-level overview of our proposed architecture, this requires the following processing blocks:

- The SAW system requires a means to retrieve messages via the social media platforms' provided APIs, for which we propose an ***Adaptive Crowd-Sensing Platform***.
- The ***Message Pre-Processing*** block attempts at making sense of the retrieved messages and extracting relevant information.
- The ***Message Aggregation*** block groups related observations in order to infer the underlying real-world events, and finally,
- the ***Object Extraction*** block aims at extracting object-level information from these, which can be processed by the core SAW system.

Figure 9.5 shows an expansion of these processing blocks, detailing on their functionality and required components, and highlights the different processing steps based upon the illustrative scenario. In the following sections, we will explicate each of the required processing blocks.

How to Retrieve Data from Social Media? Adaptive Crowd-Sensing

As a first step, messages need to be retrieved from social media platforms via their provided APIs, which typically involves an initial filtering by specifying keywords or geographical regions of interest (Salfinger et al. 2015). However, the question of what to filter for remains challenging due to the dynamically changing and self-organizing nature of social media.

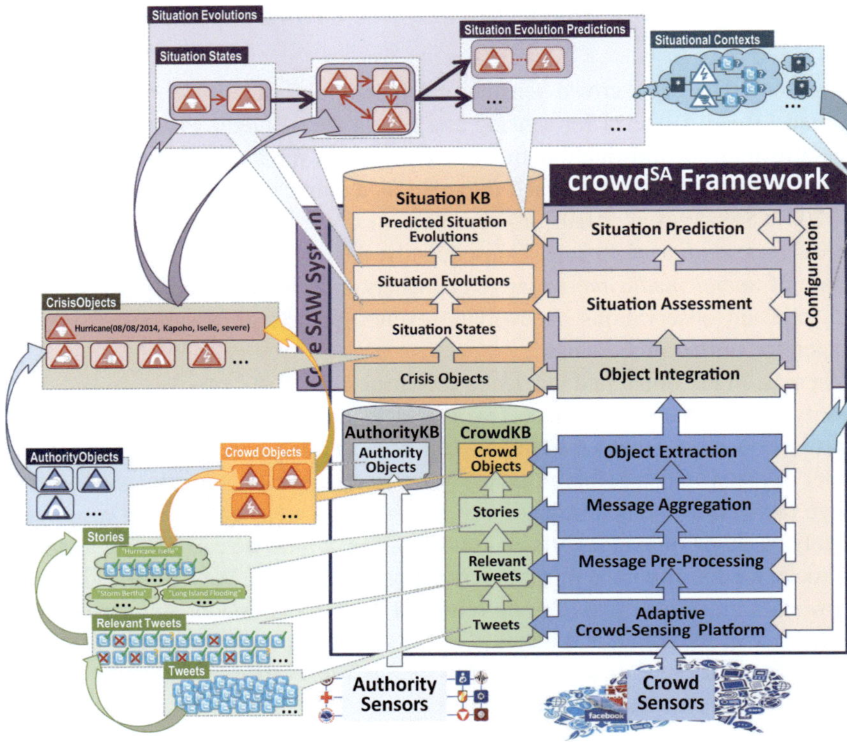

Fig. 9.4 Overview of the *crowd*^{SA} architecture and processing

9.4.10 Coping with the Dynamics of Social Media

Crises often trigger the formation of dedicated user communities, who typically adopt a specialized vocabulary for the crisis at hand (e.g., crisis-specific hashtags), which, however, may vary between communities and organizations. In some cases, this process is even guided by authorities or emergency organizations, who encourage users to post situational update information marked with these crisis hashtags (the Philippine Red Cross, for example, pursues an official strategy on promoting the use of dedicated crisis hashtags since 2012[13]). In our *HawaiiHurricanes* data set, for instance, 4728 tweets contained the crisis-specific hashtag *#Iselle*, but not the general hashtag *#hurricane* (98 % of these did not even comprise the word "hurricane" at all). Thus, such specifically marked situational update information may be missed by tracking general crisis-related keywords (e.g., by just filtering for messages comprising "typhoon" or "hurricane"). However, the burden of identifying

[13]https://irevolution.files.wordpress.com/2014/06/pcdspo-report-ondisaster-response-unified-hashtags.pdf.

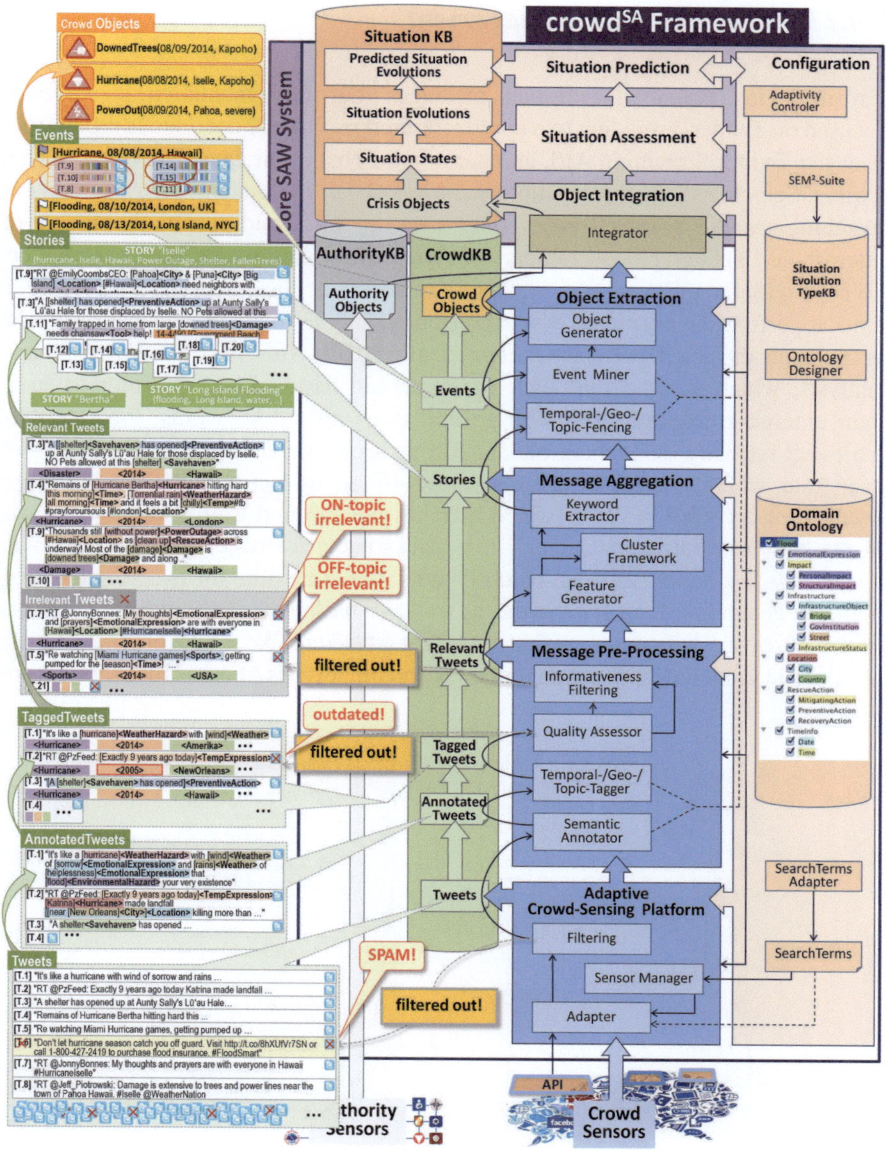

Fig. 9.5 The different crowd-sensing and processing steps illustrated

such novel hashtags and adapting the keyword queries accordingly should not be put solely on the human operator - in order to support an operator under time pressure, an intelligent system should be capable of detecting and adapting towards such emerging social media trends.

9.4.11 Crowd-Sensor Management

For retrieving potentially relevant messages, our implementation is currently focused on Twitter, providing two kinds of ***Adapters***: One for continuous monitoring over the Twitter Streaming API, another one for post-event data gathering over the Twitter Search API (using the library Twitter4J[14]). The adapters retrieve content matching specific ***Search Terms*** (comprising keywords and potentially an area of interest, i.e., a geographical boundary box), and are steered by the ***Sensor Manager***. To cope with the quickly changing nature of social media, a ***SearchTerms Adapter*** aims at performing adaptive keyword selection, similar to the approaches proposed in Kenneth et al. (2014) and Olteanu et al. (2014). This is backed by our data analysis, which showed that such crisis-specific hashtags can be inferred from their emerging co-occurrence with general crisis-related keywords: Fig. 9.6 plots the tweet intensity on hashtag-filtered subsets from the *GeneralDisasters* data set over time, whereby this plot actually allows to reconstruct the different hurricane events that occurred in 2014 (which matches the "ground truth", i.e., the records of the National Hurricane Center[15]): As peaks occur for tweets using the hashtag *#hurricane*, a smaller peak can be observed at the same time for tweets containing both, the general hashtag *#hurricane*, as well as a hashtag assembled from the name of the hurricane happening at that time (e.g., *#Iselle*). We could observe similar behavior in other data subsets on typhoons and other types of crisis (not shown here

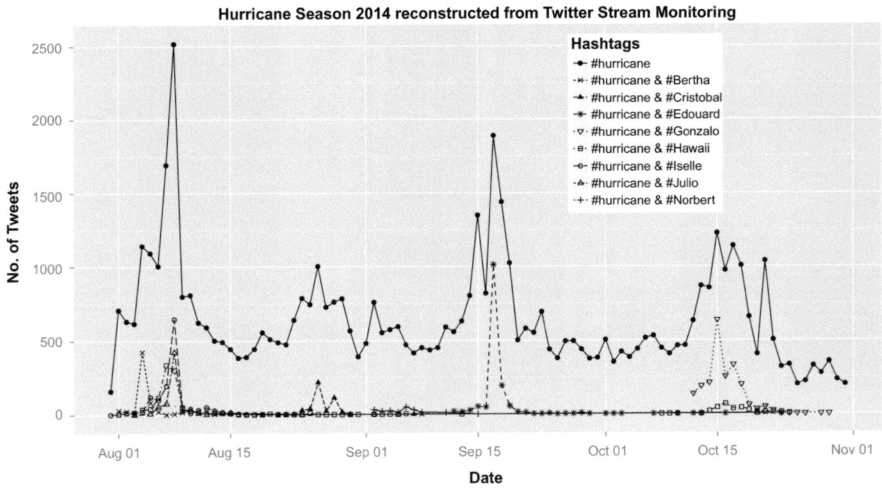

Fig. 9.6 Emerging hashtag correlations between general crisis-related hashtags and crisis-specific hashtags found in the general *GeneralDisasters* data set

[14]http://twitter4j.org.

[15]http://www.nhc.noaa.gov/data/tcr/index.php?season=2014.

due to space constraints). When the ***SearchTerms Adapter*** detects such emerging hashtag correlations, the ***Sensor Manager*** may start new adapter instances tracking for these crisis-specific hashtags.

How to Extract Information from Social Media Data? Message Pre-processing

Once messages are retrieved from social media platforms, the SAW system needs to infer meaningful information from them. Thus, the next challenge revolves around the *unstructuredness* of social media messages, as most of the relevant information (potentially containing spatio-temporal-thematic characteristics (MacEachren et al. 2011; Nagarajan et al. 2009)) is concealed in free-form text in the messages' content, potentially in conjunction with some structured meta-data (depending on the social media platform).

The processing of natural language represents a crucial issue for social media due to its informal nature and the brevity of its content (e.g., tweets are restricted to a length of at most 140 characters). Thereby, contextual information is frequently missing (as opposed, for instance, to news articles, which are written in a self-contained style), as we will illustrate in the subsequent examples.

The following tweets from the *HawaiiHurricanes* data set contain situational update information on the damage caused by *Iselle*, and address the *need* and *provision* of resources within self-organized citizen response actions:

> "There are still people trapped in their homes in Puna HI, the devastation is widespread with trees taking out power poles and roofs #Iselle"

> "#kokua #hilo Puna Needs Ice #iselle power outage - drop off ice at Ben Franklin Sun 8/10 10am to 11am https://t.co/lLc0F1qSQX"

> "Pahoa & Puna Big Island #Hawaii if your freezer is full & power is out contact me. Neighbors with electricity will volunteer to help #Iselle"

However, when retrieving tweets matching crisis-related keywords, a substantial amount of retrieved messages will simply consist of tweets in which these keywords are used in non-crisis contexts, such as proverbs, statements, metaphors, songs, poems, and other citations, for instance:

"RT @ladygaga7561: I am wind and hurricane, the stormy sky and rain when you run dry I'll flood your pain."

Negations need to be correctly resolved, as well as colloquial language:

"hurricane Julio was a non-event here didn't make landfall.So friking hot n humid here tho.hurricane did effect my diving with small surf"

Historic events, that are often posted on anniversary dates and frequently re-tweeted, need to be detected:

"#hurricanecharley made landfall #PuntaGorda #Florida #10yrs ago this week, #remember http://t.co/qnb7aDOSnu #itonlytakesone storm #Hurricane"

Furthermore, ironic tweets pose severe challenges for automated content analysis:

"Hurricane Bertha made landfall in my living room. I wouldn't care but she's dripping on the TV."

9.4.12 Decoding Missing or Unmarked Context

In order to correctly interpret a message, its specific context is key. However, humans often tend to omit explicit contextual information if it is assumed to be shared among the different correspondents, a finding which has also been corroborated in other case studies: Vieweg et al. (2010) identified cases on a tweet data set concerned with the Colorado river flooding in 2010, where people simply mentioned "the river" when actually referring to the Colorado river, which was clear to them from the specific situational and communicative context, a phenomenon Vieweg et al. coined the *decoding of unmarked context*. Humans automatically put the following tweet in context if they are aware of the fact that "Iselle" refers to the hurricane event—but would not be able to interpret it in the same way if they did not know what "Iselle" refers to:

> "Photo: Iselle cleanup continues in the Puna district on Hawaii's Big Island -... http://t.co/M5YICpOLW0 #breaking #news #headlines"

Since this phenomenon is especially pronounced in tweets due to their real-time focus and brevity, the SAW system can establish this missing, situational context by means of situation-adaptive perception: Based on the already assessed situation that Hurricane Iselle has made landfall in Hawai'i, the circumstance the tweet is referring to is already known to the system, allowing for an informed processing by incorporating this situational context in the processing of the message's content.

9.4.13 Semantic Preprocessing

Thus, after storing the retrieved social media messages in a central ***Tweet Repository***, the *crowd*^{SA} prototype's next processing step comprises the ***Semantic Annotation Pipeline***, to perform NLP on the actual message text. For this, we employ the popular NLP software framework GATE (Cunningham et al. 2002), which allows for a custom configuration of various annotation pipelines, and its extension fine-tuned towards the linguistic peculiarities of Twitter content, TwitIE (Bontcheva et al. 2013). Our basic annotation pipeline consists of the classical NLP steps of tokenization, stopword elimination, Part-of-Speech (POS) tagging, stemming (e.g., reducing "flooded" and "flooding" to "flood"), Named Entity Recognition and Classification (NERC) based on ontological and Gazetteer-based lookups, resulting in domain-grounded annotations. NERC refers to the process of annotating textual entities with respect to the *semantic categories* they refer to (e.g., "Steve Jobs" refers to a *person*, "Apple" may refer to *food* or *company*). Whereas automated NERC methods have been extensively studied on longer documents, such as news articles, *entity span identification*, and *entity class annotation* on social media content have proven to be difficult. Even human annotators vary considerably in their annotation results (Corvey et al. 2010), especially in case of location entities, which, however, represent important information for crisis management (MacEachren et al. 2011).

However, as opposed to open-domain Information Extraction (IE) tasks, for the application domain of crisis management, a priori knowledge of the different information types of interest can be employed [as categorized in Olteanu et al. (2015)]. The potential of such a *domain-dependent* analysis of message content for crisis management has been demonstrated in Purohit et al. (2014), which identified resource seeker and supplier behavior in Twitter during crisis situations. Thus, for crisis management, the following categories of information are of primary interest (Scott et al. 2004; Vieweg et al. 2010; Olteanu et al. 2015): (1) reports on the current state of infrastructure and (2) on different types of hazards, and (3) requests for and supplies of resources or help. In order to extract these information, *crowd*^{SA} utilizes an *ontology-based semantic annotation* approach, by employing a domain ontology for annotating information entities in tweets, such as infrastructure entities (e.g., mentions of bridges, buildings etc.), or natural phenomena (e.g., environmental hazards such as flooding or hurricanes). This is motivated by the lexical diversity of natural language, since social media users may employ different terminology to describe the same, as in the following examples:

> "#Iselle may be gone, but many in Hawaii living with no electricity are still feeling the pain. http://t.co/zGCF3OZi1U http://t.co/B6KPPnwPrR"

> "Thousands still without power across #Hawaii as clean up is underway! Most of the damage is downed trees and along the coast #Iselle #hiwx"

> "28 homes destroyed or damaged. 8 thousand w/o power. Hundreds with no running water. Hurricane #Iselle is still hurting us."

> "RT @Jeff_Piotrowski: Damage is extensive to trees and power lines near the town of Pahoa Hawaii .#Iselle @WeatherNation http://t.co/G8Eqg9T..."

In these example tweets, the phrases "no electricity", "without power", "w/o power", and "damage … to … power lines" basically refer to the same problem (notably power outages), which can be overcome by mapping these phrases to the ontological concept *PowerOutage*. Thus, a semantic expansion by mapping from plain text to ontological concepts allows to account for synonyms (e.g., as in Ozdikis et al. 2012). Moreover, this also allows to tackle language heterogeneity, an important issue in social media: As (from a worldwide perspective) only 50 % of Twitter messages are written in English (Hong et al. 2011), this represents a crucial issue in emergency situations like the Haiti earthquake (Dugdale et al. 2012). To adapt the processing pipeline towards different languages, the text preprocessing components (stopword removal, POS tagging, stemming) need to be configured towards the language at hand. Furthermore, the ontological classes need to be annotated with textual labels of the corresponding language, whereas the overall annotation pipeline remains unchanged. For the inference and annotation of proper nouns and proper names (e.g., the Everglades, Microsoft), the annotation pipeline can also incorporate general open-domain knowledge from online Linked Data repositories. For instance, DBpedia[16] can be queried from the annotation pipeline by including the DBpedia Spotlight plugin.[17]

Upon this initial semantic interpretation, the overall *topic*, *time*, and *location* of each tweet need to be determined, i.e., it needs to be inferred *what* has happened *where* and *when*, which may be supported by a *situational context* obtained from a feedback loop from the comprehension or projection level. This *spatio-temporal-thematic* grounding is performed by three dedicated components, i.e., a ***Topic Tagger***, a ***Temporal Tagger***, and a ***Geo-Tagger***. The latter two are motivated by the fact that a tweet's metadata on its creation timestamp and its user's location need not necessarily correlate with its content. Therefore, the ***Temporal Tagger*** aims at resolving the tweet content's time span, i.e., performs *temporal grounding*, which is a key requirement in order to correctly interpret (and discard) messages such as:

> "RT @PzFeed: Exactly 9 years ago today, Hurricane Katrina made landfall near New Orleans, killing more than 1,800 people http://t.co/ts8NqVz"

The ***Geo-Tagger*** aims at anchoring a tweet to one or multiple geographical locations, i.e., performs *spatial grounding* of the reported event. Therefore, it potentially needs to incorporate the four different location types encountered in tweets, i.e., *User's Location Profile* (e.g., the user's home town specified in her profile), *User's Current Location* (if the tweet has been sent from a mobile device), *Locations in Text* (any location mentioned in a text, which could also be, e.g., 'London Press') and *Focused Locations* (locations mentioned in text that are indeed the locations of mentioned events) (Ikawa et al. 2013), and finally decide upon which location(s) the tweet should be mapped to. In the tweet above, the ***Geo-Tagger*** should detect "near New Orleans" as focused location, and thus map the hurricane event to the area around New Orleans.

All textually specified locations furthermore require *toponym resolution* or *geo-coding*, i.e., the mapping of the location name to the actual geographic positions (comprising latitude and longitude). For this task, publicly available ontologies like the GeoNames ontology and geo-coding service[18] can be employed. However, both *geo-* and *non-geo-ambiguity* (i.e., common words need to be distinguished from proper names, e.g., *Reading* may refer to a verb or a city in the UK), as well as the *ambiguity of location names* (e.g., Sydney may refer to a city in Australia or Canada) (Leidner 2007), again require additional contextual information. Therefore, feedback-loops from the subsequent processing levels are used in order to provide additional context, i.e., situation-adaptive perception: For instance, if a specific

[16]http://dbpedia.org.

[17]https://github.com/jendarybak/GATE-DBpedia_Spotlight.

[18]http://www.geonames.org.

tweet in question is missing location mentions, but the *User's Location Profile* is set to New York, USA, and a situation has been assessed that a severe flooding event is happening in New York, which matches that tweet's content, the tweet may be consequently tagged with the location New York.

Finally, after inferring the messages' content, the system needs to determine what types of information are actually relevant for monitoring the crisis situation at hand. Whereas previously proposed approaches for informativeness filtering predominantly suggest machine-learning based methods operating on vectors assembled from the messages' text (e.g., Imran et al. 2014b; Sakaki et al. 2010), we seek to consider the *semantics* of the tweet (in order to resolve synonyms, complex relations, and negations). Thus, based upon NERC performed by the ***Semantic Annotator*** in the previous step, these semantic annotations are employed in the subsequent ***Informativeness Filtering***. Based upon this semantic expansion by annotating entities with respect to a domain ontology, the system can attempt at classifying the tweets according to the following basic informativeness categories, following the informativeness classification proposed in Olteanu et al. (2015):

- off-topic: Tweets which are not related to a crisis, but retrieved since matching the ***Adapter***'s keywords, which are used in a different semantic context, such as:

> "It's like a hurricane with wind of sorrow and rains of helplessness that flood your very existence"

> "Re watching Miami Hurricane games, getting pumped for the season! #TheU"

- on-topic, irrelevant: Tweets that are related to the crisis, but apparently do not contain on-the-ground or situational update information. These mostly correspond to emotionally focused tweets commenting on the crisis, which, however, may be of value for characterizing the evolutionary phase and severity of a crisis, such as:

> "RT @JonnyBones: My thoughts and prayers are with everyone in Hawaii #HurricaneIselle"

- on-topic, relevant: This category comprises tweets which contain crisis-related information, such as:

> "Worst flooding I've ever seen on #LongIsland & I've lived here since childhood & been thru several hurricane including #SuperStormSandy"

> "Remains of Hurricane Bertha hitting hard this morning. Torrential rain all morning and it feels a bit chilly #fb #prayforoursouls #london"

> "Crazy flooding in our Queen's Park street from ex-hurricane Bertha… never seen flooding this bad. (cc @liamdutton) http://t.co/ox6T0PzBdu"

> "A shelter has opened up at Aunty Sally's Luau Hale for those displaced by Iselle. NO Pets allowed at this shelter. #prayersforPuna"

To assign a semantically ***Annotated Tweet*** to one of these categories, the ***Quality Assessor*** aims at determining its *information content*: It computes dedicated quality metrics which aggregate over the encountered entity annotations, i.e., annotations of ontologically relevant information, such as mentions of infrastructure elements (e.g., bridges, buildings), crisis preparation, mitigation and recovery actions, and natural phenomena (e.g., flooding, hurricane). Thus, the mentioning of a location or infrastructure entity may increase a tweet's information content value, whereas emotional expressions (e.g., the phrases "pray for", emoticons such as ";)"), for example, may decrease this value. This approach has been motivated by findings based on a manual inspection of large crisis data sets from Twitter reported in Vieweg et al. (2010), where it has been concluded that tweets that comprise *situational update information*, and have been posted with the aim to broadcast these information, contain higher information content, and are more often marked with location information, than commentary tweets, for instance:

> "New York experiencing flooding worse than Hurricane Sandy. All parkways and L I Expressway Closed. All South Shore Towns flooded"

Therefore, we propose a dedicated ***Quality Assessor*** component, which comprises a ***Quality Metric Calculator*** that can be configured towards various ***Metric Calculation Strategies***. Whereas our initial experiments on this knowledge-based ***Informativeness Filtering*** yielded promising results with respect to the identification of crisis-situation relevant tweets, we currently experiment with incorporating suitable ***Quality Metrics*** and their thresholds for category assignment. For instance, the co-occurrence of temporal information, location information, crisis types, and infrastructure or rescue action entities in a tweet could be attributed with larger weights. Furthermore, more specific location information, such as *Pahoa* and *Puna*, should be attributed with larger weights than coarse-grained information such as *Hawai'i*.

How to Infer Real-World Events from Social Media Data? Message Aggregation

Crowd-sensing essentially bases on the assumption that large-scale real-world events can be reconstructed by aligning the observations from multiple human sensors [as, for example, described in Sakaki et al. (2010)]. Therefore, JDL L1 *Object* or *Event Assessment* bases on grouping similar messages, in order to infer the underlying real-world incident mentioned therein. This furthermore also follows the finding that social media information is mainly of use on an aggregate level (Dugdale et al. 2012). However, this consequently also raises the challenge of how to ensure that valuable but sparse situational update information will not be dominated by general news and comments (which we will address in the subsequent processing steps).

In the *crowd*^{SA} architecture, after semantics-based ***Quality*** and ***Informativeness Filtering***, only the ***Relevant Annotated Tweets*** are retained for the further processing steps. Whereas for grouping related observations, most crowd-sensing approaches perform clustering based on text vectors, for instance, by employing the cosine-similarity between the tweets' word vectors (as in Rogstadius et al. 2013), we intend to investigate on how we can incorporate features derived from semantic annotations in the clustering procedure. Therefore, *crowd*^{SA}'s ***Message Aggregation*** component consists of a ***Feature Generator***, allowing for the specification of custom features, and a general ***Cluster Framework***, allowing to employ different cluster algorithms. Whereas the ***Cluster Framework*** simply returns sets of tweets that have been determined to be similar with respect to the defined features, i.e., should presumably comprise tweets discussing the same underlying real-world event, the ***Keyword Extractor*** aims at identifying common terms and keywords from this tweet set, resulting in a so-called ***Story*** (following the terminology proposed in Rogstadius et al. (2013)), i.e., a set of tweets presumably discussing the same topic, and keywords or entities that can be considered as descriptive of this tweet cluster.

How to Extract Object-Level Information from Social Media Data? Object Extraction

9.4.14 Event Detection and Tracking

In order to finally infer the underlying *real-world* event that is presumably reported within this tweet cluster, suitable *spatio-temporal-thematic* descriptors need to be extracted from this ***Story*** (Nagarajan et al. 2009), i.e., the system needs to determine *what* has happened *where* and *when*. Whereas the ***Message Pre-Processing*** phase aimed at extracting these kinds of information for each tweet separately, the ***Event Detection*** phase operates on aggregates of tweets to assemble a complete spatio-temporal-thematic coverage of the inferred real-world ***Events***, which is performed by ***Topic Fencing***, a ***Geo-Fencing*** and a ***Temporal Fencing*** component, respectively. From the ***Stories*** shown in Fig. 9.5, for instance, the following real-world events can be inferred: Tweets mentioning the damage after Hurricane Iselle passed the Hawai'ian islands, tweets referring to the event of the tropical storm Bertha crossing UK, and tweets addressing the flooding of Long Island.

In order to be interpretable for the ***Situation State Assessor***, these ***Event***-level descriptions need to be mapped to ***Object***s of the ***Domain Ontology***. Thus, the ***Object State Generator*** aims at mapping events to the ***Domain Ontology***, and instantiating corresponding ***Object States***. Finally, the ***Object Evolution Detector*** needs to infer whether the detected ***Object State*** corresponds to a new, i.e., previously unseen object, or denotes the evolution of an existing object, finally yielding the inferred ***CrowdObjects***.

It is worth to mention that a single ***Story*** may describe multiple ***Sub-Events***, characterized by specific subclusters on these subjects within the ***Story***, and thus trigger the instantiation of multiple ***Objects***. The stories depicted in Fig. 9.5, for instance, trigger the instantiation of the ***Objects*** *Power Outage* and *Fallen Trees* due to an ***Event*** *Hurricane* (hurricane Iselle in Hawai'i), and the ***Objects*** *Torrential Rain* and *Flooding* due to another ***Event*** *Hurricane* (hurricane Bertha in UK). Ultimately, the correlation of evolving ***Events*** and ***Objects***, as triggered by crises and discussed in ***Stories***, could potentially provide a means for *learning SETs* from the crowd, i.e., crowd-based knowledge acquisition.

Whereas the aggregation to ***Stories*** increases the confidence in the reported events, the SAW system would lose valuable but sparse information (e.g., posted by on-the-ground observers) if it solely operated on aggregated content, which will be likely dominated by general news and comments, and thus could only establish rather coarse-grain, event-level information. Therefore, we propose an *aggregation-segregation-based* approach, by introducing an additional ***Event Miner*** component into the $crowd^{SA}$ architecture: After confirming the general ***Event***-context based on ***Message Aggregation*** ("net-fishing" related tweets), or a feedback-loop from ***Situation Assessment*** or ***Situation Prediction***, the ***Event Miner*** specifically seeks to retrieve single tweets which comprise such highly relevant situational information ("line fishing" specific tweets), which should be identifiable based on their high information content. These may individually trigger the instantiation of the corresponding ***Crowd Objects***, such as the ***Object*** "DownedTrees" with location Kapoho, which has been only mentioned in a single tweet in our data set, out of 82 tweets reporting on hurricane damage in Kapoho:

> "Family trapped in home from large downed trees needs chainsaw help! 14-4490 Government Beach Road in Kapoho, HI @HICivilDefense #Iselle"

9.4.15 *Object Integration and Situation Assessment: Merging Crowd-Sensed Data with Other Information Sources*

In order to obtain a coherent perspective on the monitored environment, being the prerequisite for situation assessment, the object observations retrieved from crowd-sensed data (which we term *crowd objects*) need to be fused with observations obtained from the authority sensors (i.e., *authority objects*), cf. Fig. 9.5. This integration step thus requires appropriate merging and conflict resolution strategies in case the different sources report adversarial information. Finally, ***Crowd Objects*** may be fused with ***Authority Objects***, i.e., data obtained from authority sources, and ultimately serve as input for situation assessment and projection.

Closing the Situational Feedback Loop: Reconfiguring Crowd-Sensing Based on Established Situational Context

Based upon the crowd-sensing and processing of tweets summarized in Fig. 9.5, an example situation that can be derived from the *HawaiiHurricanes* data set is shown in Fig. 9.7, which depicts the assessment and evolution of a crisis situation instance capturing and summarizing the chain of events triggered by Hurricane Iselle's landfall on the island of Hawai'i. This has been derived as follows: After querying the Twitter API with crisis-related keywords (such as *hurricane* and *flooding*), and crisis-specific keywords (such as the hurricanes' names *Iselle* and *Julio*), the *crowd*^{SA} system has retrieved a broad set of tweets (cf. Fig. 9.5). After message pre-processing and aggregation, one resulting ***Story*** represents a cluster of tweets referring to damage types on the East side of the island of Hawai'i (notably around the area of Pahoa), as the clustering of semantically expanded tweets yields a cluster of tweets referring to power outages and toppled trees. Consequently, this triggers the instantiation of the corresponding ***Crowd Objects***, and subsequently, ***Crisis Objects***. These match the SST "PowerOutageInCity", and thus instantiate a situation state on the power outage situation in Pahoa, representing an update of the overall situation *SituationHurricaneIselle* (cf. Fig. 9.7). Consequently, based on this assessed situation, the now established situational context on the power outage situation in Pahoa can be reused in order to gather additional information on this

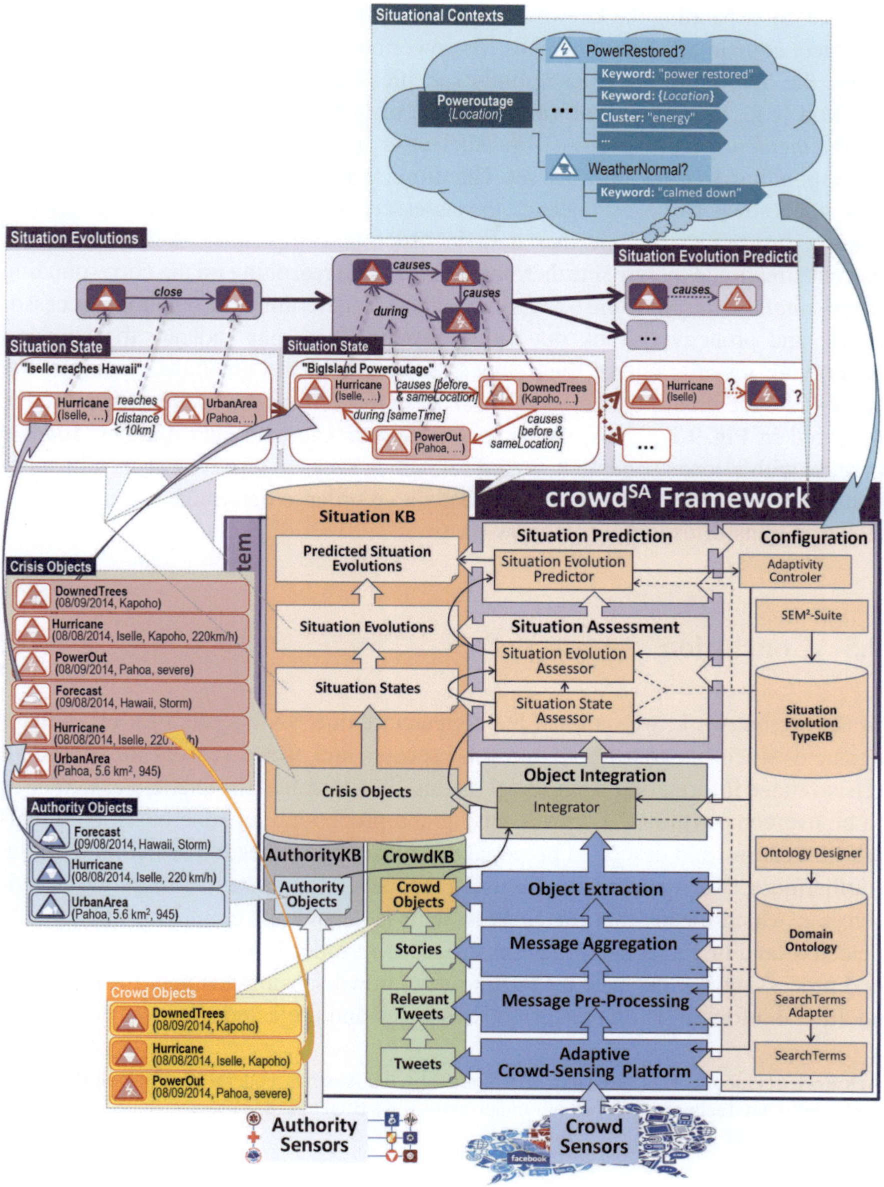

Fig. 9.7 Deriving a *situational context* from the currently assessed (or projected) situation states, which adapts the crowd-sensing configuration

specific situation instance: The ***Sensor Manager***, acting as the central interface for realizing both, JDL L4 and L5 functionality, *situation-adaptively* adjusts the crowd-sensing configuration by including this instance level information in the ***Adapter***'s keyword set. Thus, it adds the keywords "Pahoa", "power outage" and the like, by expanding the keyword set based on the crisis ontology (which may, for instance, contain a list of synonyms for power outages, and related damage types). Thus, the ***Sensor Manager*** essentially specifies a *situation profile* comprising the location (e.g., Pahoa) and information types of interest [expanding the idea of the *initial incident profile* proposed in Abel et al. (2012)], which is translated to a corresponding ***Adapter*** keyword set. Therefore, whereas this information previously was not known to be of interest, the assessed situation thus drives the crowd-sensing process to gain further data in order to complement and enhance the current situational picture. Furthermore, based upon reasoning on the corresponding *HurricaneDisaster* SET, the SAW system also can determine the potential successor states, and proactively look out for information that may indicate the situation is evolving towards one of these successor states. In the illustrated example, the successor state would be the SST *Power Restored* (cf. the *HurricaneDisaster* SET specified in Fig. 9.2). Thus, the ***Sensor Manager*** can also start another ***Adapter*** process with a corresponding keyword set, which seeks to retrieve tweets indicating that power supply has been restored in Pahoa, in order to determine when the crisis situation transforms into its recovery phase.

9.5 Conclusion

In the present work, we outlined the architecture of a SAW system for crisis management which integrates both *authority* sensors and *crowd* sensors for extracting crisis-related information from social media. We illustrated the challenges that need to be overcome within the crowd-sensing and processing pipeline based on a real-world scenario, and described potential technical approaches for overcoming these. Furthermore, we specifically discussed concepts for *situation-adaptive* crowd-sensing, which means that the SAW system is capable of employing the already detected (and/or projected) situation instances in order to adapt its crowd-sensing and processing, for instance by starting new crowd-sensing adapters for retrieving additional information on the currently assessed and projected situation.

Acknowledgements This work has been funded by the Austrian Federal Ministry of Transport, Innovation and Technology (BMVIT) under grant FFG BRIDGE 838526 and under grant ÖAD WTZ AR10/2015.

References

Abel F, Hauff C, Houben GJ, Stronkman R, Tao K (2012) Semantics + Filtering + Search = Twitcident. Exploring information in social web streams. In: Proceedings of the 23rd ACM conference on hypertext and social media, HT '12. ACM, New York, pp 285–294. http://doi.acm.org/10.1145/2309996.2310043

Baumgartner N, Gottesheim W, Mitsch S, Retschitzegger W, Schwinger W (2010) BeAware!—situation awareness, the ontology-driven way. Int J Data Knowl Eng 69(11):1181–1193

Baumgartner N, Mitsch S, Müller A, Retschitzegger W, Salfinger A, Schwinger W (2014) A tour of BeAware! – a situation awareness framework for control centers. Inform Fusion 20:155–173

Bontcheva K, Derczynski L, Funk A, Greenwood MA, Maynard D, Aswani N (2013) TwitIE: an open-source information extraction pipeline for microblog text. In: Proceedings of the international conference on recent advances in natural language processing. Association for Computational Linguistics Stroudsburg, PA, USA

Cameron MA, Power R, Robinson B, Yin J (2012) Emergency situation awareness from twitter for crisis management. In: Proceedings of the 21st international conference companion on world wide web. ACM, WWW '12 companion, pp 695–698. http://doi.acm.org/10.1145/2187980.2188183

Corvey WJ, Vieweg S, Rood T, Palmer M (2010) Twitter in mass emergency: what NLP techniques can contribute. In: Proceedings of the NAACL HLT 2010 workshop on computational linguistics in a world of social media, WSA '10. Association for Computational Linguistics, pp 23–24. http://dl.acm.org/citation.cfm?id=1860667.1860679

Cunningham H, Maynard D, Bontcheva K, Tablan V (2002) GATE: a framework and graphical development environment for robust NLP tools and applications. In: Proceedings of the 40th anniversary meeting of the Association for Computational Linguistics (ACL'02)

Dashti S, Palen L, Heris M, Anderson KM, Anderson S, Anderson TJ (2014) Supporting disaster reconnaissance with social media data: a design-oriented case study of the 2013 Colorado floods. In: Proceedings of the 11th international conference on information systems for crisis response and management (ISCRAM 2014)

Dugdale J, Van de Walle, B., Koeppinghoff, C (2012) Social media and SMS in the Haiti earthquake. In: Proceedings of the 21st international conference companion on world wide web. WWW '12 companion. ACM, New York, pp 713–714

Edlund J, Grönkvist M, Lingvall A, Sviestins E (2006) Rule-based situation assessment for sea surveillance. In: Belur V. Dasarathy (eds) SPIE Proceedings Vol. 6242: Multisensor, Multisource Information Fusion: Architectures, Algorithms, and Applications

Endsley MR (1995) Toward a theory of situation awareness in dynamic systems. Hum Factors: J Hum Factors Ergon Soc 37(1):32–64

Fuchs G, Andrienko N, Andrienko G, Bothe S, Stange H (2013) Tracing the German centennial flood in the stream of tweets: first lessons learned. In: Proceedings of the second ACM SIGSPATIAL international workshop on crowdsourced and volunteered geographic information, GEOCROWD '13. ACM, New York, pp 31–38. http://doi.acm.org/10.1145/2534732.2534741

Hong L, Convertino G, Chi E (2011) Language matters in twitter: a large scale study. In: Proceedings of the fifth international AAAI conference on weblogs and social media

Ikawa Y, Vukovic M, Rogstadius J, Murakami A (2013) Location-based Insights from the Social Web. In: Proceedings of the 22nd international conference on world wide web companion. International World Wide Web Conferences Steering Committee, WWW '13 Companion, pp 1013–1016

Imran M, Castillo C, Diaz F, Vieweg S (2015) Processing Social Media Messages in Mass Emergency: A Survey. In: ACM Computing Surveys 47(4), pp 67:1–67:38

Imran M, Castillo C, Lucas J, Meier P, Vieweg S (2014b) AIDR: artificial intelligence for disaster response. In: Proceedings of the companion publication of the 23rd international conference on world wide web companion. International World Wide Web Conferences Steering Committee, WWW companion '14, pp 159–162

Jakobson G, Parameswaran N, Buford J, Lewis L, Ray P (2006) Situation-aware multi-agent system for disaster relief operations management. In: Van der Walle B, Turoff M (eds) 3rd international conference on information systems for crisis response and management (ISCRAM)

Kenneth J, Landwehr PM, Carley KM (2014) An approach to selecting keywords to track on twitter during a disaster. In: Proceedings of the 11th international conference on information systems for crisis response and management (ISCRAM 2014)

Kogut P, Cranefield S, Hart L, Dutra M, Baclawski K, Kokar M, Smith J (2002) UML for ontology development. Knowl Eng Rev 17(1):61–64

Kokar M, Letkowski JJ, Dionne R, Matheus C (2008) Situation tracking: the concept and a scenario. In: IEEE military communications conference, 2008. MILCOM

Kumar S, Barbier G, Abbasi MA, Liu H (2011) TweetTracker: an analysis tool for humanitarian and disaster relief. In: Fifth international AAAI conference on weblogs and social media

Laxhammar R (2008) Anomaly detection for sea surveillance. In: 2008 11th international conference on information fusion, pp 1–8

Leidner JL (2007) Toponym resolution in text. Ph.D. dissertation

Llinas J (2010) A survey and analysis of frameworks and framework issues for information fusion applications. In: Hybrid artificial intelligence systems. Lecture notes in computer science, vol 6076. Springer, Berlin/Heidelberg, pp 14–23

Llinas J, Bowman C, Rogova G, Steinberg A, Waltz E, White F (2004) Revisiting the JDL data fusion model II. In: Svensson P, Schubert J (eds) Proceedings of the seventh international conference on information fusion (FUSION 2004), pp 1218–1230

MacEachren AM, Jaiswal A, Robinson AC, Pezanowski S, Savelyev A, Mitra P, Zhang X, Blanford J (2011) SensePlace2: GeoTwitter analytics support for situational awareness. In: 2011 IEEE conference on visual analytics science and technology (VAST), pp 181–190

Matheus C, Kokar M, Baclawski K (2003) A core ontology for situation awareness. In: Proceedings of the sixth international conference of information fusion, 2003, vol 1, pp 545–552

Meyer-Delius D, Plagemann C, Burgard W (2009) Probabilistic situation recognition for vehicular traffic scenarios. In: IEEE international conference on robotics and automation, 2009. ICRA '09. pp 459–464

Nagarajan M, Gomadam K, Sheth AP, Ranabahu A, Mutharaju R, Jadhav A (2009) Spatio-temporal-thematic analysis of citizen sensor data: challenges and experiences. In: Proceedings of the 10th international conference on web information systems engineering, WISE '09. Springer, Berlin, pp 539–553

Niklasson L, Riveiro M, Johansson F, Dahlbom A, Falkman G, Ziemke T, Brax C, Kronhamn T, Smedberg M, Warston H, Gustavsson P (2008) Extending the scope of situation analysis. In: 2008 11th international conference on information fusion

Olteanu A, Castillo C, Diaz F, Vieweg S (2014) CrisisLex: a lexicon for collecting and filtering microblogged communications in crises. In: Proceedings of ICWSM

Olteanu A, Vieweg S, Castillo C (2015) What to expect when the unexpected happens: social media communications across crises. In: Proceedings of the 18th ACM conference on computer supported cooperative work & social computing, CSCW '15. ACM, New York, pp 994–1009

Ozdikis O, Senkul P, Oguztuzun H (2012) Semantic expansion of tweet contents for enhanced event detection in twitter. In: Proceedings of the 2012 international conference on advances in social networks analysis and mining (ASONAM 2012). IEEE Computer Society, Silver Spring, MD, pp 20–24. http://dx.doi.org/10.1109/ASONAM.2012.14

Pröll B, Retschitzegger W, Schwinger W, Kapsammer E, Mitsch S, Baumgartner N, Rossi G, Czech G, Högl J (2013) crowdSA - crowdsourced situation awareness for crisis management. In: Proceedings of social media and semantic technologies in emergency response (SMERST)

Purohit H, Sheth AP (2013). Twitris v3: From Citizen Sensing to Analysis, Coordination and Action. In: 7th Intl AAAI Conference on Weblogs and Social Media, pp 746–747

Purohit H, Hampton A, Bhatt S, Shalin VL, Sheth AP, Flach JM (2014) Identifying seekers and suppliers in social media communities to support crisis coordination. Comput Supported Coop Work 23(4-6):513–545

Rhodes BJ, Bomberger NA, Freyman TM, Kreamer W, Kirschner L, L'Italien AC, Mungovan W, Stauffer C, Stolzar L, Waxman AM, Seibert M (2007) SeeCoast: persistent surveillance and automated scene understanding for ports and coastal areas, pp 65781M–65781M–12

Rogstadius J, Vukovic M, Teixeira CA, Kostako V, Karapanos E, Laredo JA, (2013) CrisisTracker: Crowdsourced social media curation for disaster awareness. In: IBM Journal of Research and Development, 57(5), pp.4:1–4:13, http://dx.doi.org/10.1147/JRD.2013.2260692

Sakaki T, Okazaki M, Matsuo Y (2010) Earthquake shakes twitter users: real-time event detection by social sensors. In: Proceedings of the 19th international conference on world wide web, WWW '10. ACM, New York, pp 851–860

Sakaki T, Toriumi F, Matsuo Y (2011) Tweet trend analysis in an emergency situation. In: Proceedings of the special workshop on internet and disasters, SWID '11. ACM, New York, pp 3:1–3:8

Salfinger A, Neidhart D, Retschitzegger W, Schwinger W, Mitsch S (2014a) SEM^2 suite — towards a tool suite for supporting knowledge management in situation awareness systems. In: 15th IEEE international conference on information reuse and integration (IRI 2014)

Salfinger A, Retschitzegger W, Schwinger W (2014b) Staying aware in an evolving world — specifying and tracking evolving situations. In: 2014 IEEE international inter-disciplinary conference on cognitive methods in situation awareness and decision support (CogSIMA). IEEE, New York, pp 172–178

Salfinger A, Girtelschmid S, Pröll B, Retschitzegger W, Schwinger W (2015) Crowd-sensing meets situation awareness - a research roadmap for crisis management. In: Proceedings of the 48th annual Hawaii international conference on system sciences (HICSS-48)

Scott PD, Rogova GL (2004) Crisis management in a data fusion synthetic task environment. In: 7th international conference on information fusion, pp 330–337

Shaw F, Burgess J, Crawford K, Bruns A (2013) Sharing news, making sense, saying thanks: patterns of talk on twitter during the Queensland floods. Aust J Commun 40(1) pp 23–40

Sheth A (2009) Citizen sensing, social signals, and enriching human experience. IEEE Internet Comput 13(4):87–92

Smid H, Mast P, Tromp M, Winterboer A, Evers V (2011) Canary in a coal mine: monitoring air quality and detecting environmental incidents by harvesting twitter. In: CHI '11 extended abstracts on human factors in computing systems, CHI EA '11. ACM, New York, pp 1855–1860

Ulicny B, Moskal J, Kokar MM (2013) Situational awareness from social media. In: Proceedings of the eighth conference on semantic technologies for intelligence, defense, and security, Fairfax VA, USA, November 12–15, 2013. CEUR-WS.org, CEUR workshop proceedings, vol 1097, pp 87–93

Vieweg S, Hughes AL, Starbird K, Palen L (2010) Microblogging during two natural hazards events: what twitter may contribute to situational awareness. In: Proceedings of the SIGCHI conference on human factors in computing systems, CHI '10. ACM, New York, pp 1079–1088

Walle B, Turoff M (2008) Decision support for emergency situations. Inform Syst e-Bus Manag 6(3):295–316

Yin J, Lampert A, Cameron M, Robinson B, Power R (2012) Using social media to enhance emergency situation awareness. IEEE Intell Syst 27(6):52–59

Chapter 10
Data Fusion Across Traditional and Social Media

Werner Bailer, Gert Kienast, Georg Thallinger, and Gerhard Backfried

Abstract Crises and disasters are covered continuously and without interruption by today's media, especially social media. There is not a single significant occurrence within the flow of events which they do not document. Consequently, the information contained in media—especially social media like Facebook and Twitter—provides an often neglected potential which should not be overlooked. Through fusion of sources, diverse, mixed, and complementary types of information can be tapped into and combined. The difficulty of this process is to view, channel, prepare, and exploit this inhomogeneous and enormous amount of information. Automatic monitoring of traditional as well as social media sources allows to deriving risk factors and risk indicators for crises and disaster events quickly. Intelligence derived from this process allows for earlier and swifter reaction to potential situations of crisis and interrelationships. Current publicly described technical and electronic infrastructure for national and international crisis and disaster management is not able to perform comprehensive analyses of all media channels automatically. The continuous developments in the areas of multimedia and social media demand the creation of adequate methods of processing. Relevant manifestations of events are to be identified automatically from documents from traditional (TV, radio, web) as well as social media and document clusters of the examined multimedia documents are to be presented to situational awareness experts. The focus of the Quelloffene IntegrierteMultimedia Analyse (QuOIMA) project is on the research on and development of algorithms and methods to achieve this goal. Automatic analysis of content in the multimedia and social media domain forms a fundamental innovation. From a technical, social studies, and scientific point of view, the targeted insights and findings of this project, form a fundamental contribution to security research, reaching far beyond the quality of existing systems. The integration of findings regarding situational awareness will provide more realistic risk assessment

W. Bailer (✉) • G. Kienast • G. Thallinger
JOANNEUM RESEARCH, Graz, Austria
e-mail: Werner.Bailer@joanneum.at; Gert.Kienast@joanneum.at; Georg.Thallinger@joanneum.at

G. Backfried
SAIL LABS Technology, Vienna, Austria
e-mail: Gerhard.Backfried@sail-labs.com

G. Rogova, P. Scott (eds.), *Fusion Methodologies in Crisis Management*,
DOI 10.1007/978-3-319-22527-2_10

increasing their possibilities to (re)act. End users extend their expertise and as a consequence the ability of the overall organizations to act.

10.1 The Transition from Traditional Media to Social Media

Until recently only traditional media, such as TV, radio, and web sources (blogs, online newspapers, etc.) have been used to gain up-to-date information about disasters and crises. Even today these curated media channels may still be an important source for first responders and concerned citizens.

But with the rise of social media the way people communicate has changed significantly. This does not only apply to people's everyday lives but also to extreme situations such as emergencies and disasters.

Today in many cases significant parts of the population do indeed use social media during crises situations (Johansson et al. 2012), especially micro-blogging systems with an open communication structure (e.g. Twitter), as they provide a new form to disseminate and receive crisis information (Nilsson et al. 2012; Bruns et al. 2012).

When moving from using only curated traditional media sources to including information from uncurated social media information a challenge for emergency responders is that they need to make sure to get the right amount of adequate and accurate information (Brugghemans et al. 2013). Even though a much wider scope of information can be accessed through social media, the gathered information may be too little, biased, inaccurate, late, unreliable, irrelevant, or conflicting (Muhren and Van de Walle 2010).

Social media platforms enable targeted two-way communication in crisis scenarios, allowing information to be collected *and* distributed through these channels. This type of communication is of great value not only because it allows first responders to have access to critical updates from individuals who eyewitness the situation on site but also enables them to send important disaster-related information to the persons who need it most.

Common requirements and assets of currently shared information are (near) real-time communication/information, personal involvement, reliable, critically challenged and questioned sources, as well as multi-media content. As technology is still evolving people will continue to have rising expectations of first responders and vice versa. Affected persons may soon indeed expect authorities to quickly respond to information shared via social media (Johansson et al. 2012). At the moment it can be safely assumed that the impact of social media on crisis communication is already significant and will continue to grow in significance in the years to come.

Originally social media have not been created with first responders in mind, but for the general public to connect and share their thoughts with family and friends; hence certain constraints remain when considering the scope of crisis communication. To offer citizens faster and more effective responses, national and international risk and communication experts and projects are starting to focus on new ways to systematically incorporate and integrate social media and the related technology into risk and crisis communication. Technological advances are accompanied by establishing of best practices which would enhance the goals of communication in crisis and disaster situations. From the point of view of a first responder, this concerns communication in both directions—gathering information from citizens and the media as well as communication of messages from first responders out to the world.

The public has thus turned into an important stakeholder in the community of crisis management, today everybody is a potential information source and message target. They share their observations, emotions, and opinions and communicate via social media platforms and thus form a crowd-sourced means for gathering and dissemination of data. This data needs to be combined, reconciled and contrasted with information provided by traditional sources to achieve a more complete picture of the situation at hand.

The combination of traditional media and social media enables gaining information which would otherwise not be available. On the one hand, information from traditional media (i.e., trusted sources) can be used to assess the trustworthiness of social media messages. For example, it might be possible to verify the location of a social media message by identifying that the image included in the traditional news and the image referenced by the social media message contain overlapping visual content (e.g., same background). On the other hand, social media messages are typically more up-to-date than traditional news and available also when news teams are not present at the location of a disaster. Thus they may give valuable insight to emergency responder organizations (e.g., identifying locations where help is needed or early identification of emerging disasters).

But social media turn out to be useful not only for the gathering of information. Outbound information activities include active dissemination of information, and targeted communication with effected persons immediately affected by the disaster or mere observers as well as corrective measures such as clarifying rumors or correcting misinformation (Mendoza et al. 2010).

10.2 Introducing the QuOIMA Project

The usage of open information sources (especially social media) in crisis and disaster management allows generation of knowledge which is necessary to perform a more realistic risk assessment. The QuOIMA project developed methods to mostly automatically analyze open-source media content—text, image, and video-from traditional and social media sources. To reduce the potential information overload media items are analysed and filtered through a clustering process.

This will enable essential users to detect trends and structures in a crisis or disaster situation much faster and more effectively. In addition to using only information available within their organizations, as it is common today, using new media channels can improve the process of situation assessment.

10.2.1 Motivation for Clustering

The motivation for clustering media items which are documenting a disaster or crisis situation is manifold.

First and foremost it is necessary to *reduce the amount of information* that needs to be screened by human operators, a time-consuming and expensive task. Given the sheer number of media documenting a disaster on traditional and social media channels and the time-criticalness, it is virtually impossible even for a large team of operators to keep up with the constant inflow of information.

Removal of duplicates (or near-duplicates): When it comes to user-generated content very often the same media item is uploaded multiple times to social media platforms. Two images may show identical content, but use different file formats and resolutions or one may be a cropped version of another. Videos may be of different aspect ratios, have additional text overlays, or show up as picture-in-picture behind a news presenter. In all of these cases the files are technically different, but no additional information is added by having multiple versions.

As shown in Fig. 10.1 it is very likely that different users capture images or videos at the *same location* (or same event). These media items will show the same scene but from a different viewing angle, zoom or at different points in time. Having all media taken at a certain location at hand allows operators to judge developments over time more easily. Media showing the same location will include reposts of prior disasters at the same location (e.g., the flood at the same river some years ago) or manipulated media (e.g., with inserted objects). Putting them in direct relation enables users to assess the credibility of the media based on other metadata (e.g., time stamps) or comparing the content.

The development of *new types of user interfaces* will be made possible by combining the various clustering steps described above. By providing a reduced set of media grouped by different criteria they can support emergency responders in gaining a faster and more accurate assessment of an unfolding situation (Bailer et al. 2010).

The research work within the QuOIMA project addresses clustering media items in the visual, audio and textual domains. There is a two-step process in place where clustering is initially performed independently for each of the modalities. Audio documents are first transcribed, converted into text, and enriched (e.g., by tagging of named entities) before being considered for clustering. This is followed up by a fusion mechanism where the results of both modalities are combined yielding a final clustering. The remainder of this chapter describes the steps for achieving such a clustered cross-media content set.

Fig. 10.1 Visualization of an example set of cross-media data from a local flooding consisting of photos and videos collected from blogs, TV, and social media platforms. The content can be clustered based on visual similarity with clusters corresponding to various locations. Depicted images and videos CC-BY Gipfelrast and Martin Podhovnik

10.3 Automatic Content Analysis and Clustering

10.3.1 Text Mining

Textual data can be ingested into the QuOIMA framework in a variety of ways. So-called text-feeders are employed to gather textual data from a wide range of sources, such as web pages, news-feeds, press-agencies, blogs, word documents, e-mails, or social media such as Twitter or Facebook. Furthermore, comments posted on image and video sharing sites can be integrated. The texts come in a wide range of formats, representations, and styles as well as in multiple languages. Specialized text-feeders extract textual information from specific sources and pass their results on to subsequent components for cleaning and tokenization. Further processing steps include the tagging of named entities, topic detection, translation, and sentiment analysis. Finally, enriched documents are passed on to the backend-server for storage and later information retrieval purposes. All text-feeders use public access tools (APIs) only.

A specialized text-feeder, operating much like a web-crawler, is used to gather and process data from internet sources such as web pages and news-feeds. Every web page which is crawled is also turned into a PDF file which is stored alongside the processed text. This ensures the persistence of information over time which may be crucial for correct interpretation of content as well as for providing evidence for decisions. Web content often is removed from sites after a certain period of time—e.g., due to legal restrictions—and also altered over time. In both instances, preserving content for later justification and explanation of resulting decisions is crucial. Boiler-plate detection similar to the one described in Kohlschütter et al. (2010) is applied to web content to extract only those portions of a web page which carry the actual content.

Another text-feeder gathers textual data from e-mail accounts and attached documents. The vast majority of blogs are available via feeds and are processed via the Web-Collector. Likewise, platforms providing e-mail access for specific events (such as Ushahidi[1]) can be integrated into the QuOIMA framework in this way.

Specific text-feeders for press agencies of significance in the Austrian and Central European region such as Reuters, AFP, or MENA are supplied with the system. These feeders rely on the availability of such news services (which have to be acquired and configured separately) and ingest news in terms of NewsML-G2 (IPTC 2014) documents.

Special feeders for Twitter and Facebook use the publicly accessible APIs to retrieve content from these social media platforms. All links mentioned in any of the textual or social media documents are preserved and can be clicked on and followed in the Media Mining Client. The subsequent processing of content is described below.

All text-based technologies perform their processing either on data provided by the text-normalization components or on the output of automatic speech recognition (ASR). As such, these technologies have to be able to deal robustly with a variety of formats and styles: ASR output does not contain proper punctuations, and may include recognition errors and segments spanning partial phrases up to several sentences. Furthermore, as ASR aims at transcribing as truthfully as possible the incoming signal, hesitations, false-starts, and non-speech events like coughing or breathing noises might be transcribed. However; the individual words contained in a transcript are (usually) proper words (full forms) in the given language. On the other end of the spectrum, data from social media sources likely include quite specific jargon, abbreviations and acronyms, many spelling mistakes, and occasional sections of foreign language and foreign character sets (e.g., Arabic social media texts are often spelled in Latinised Arabic, using Western characters and digits to mimic Arabic letters). Downstream processing has to be robust enough to cope with such differences in text and language. Balancing robustness and speed as opposed to accuracy is extremely important in such diverse settings.

[1]http://www.ushahidi.com

Text processing includes the pre-processing, cleaning, normalization, and tokenization steps. Language-specific processing of text (e.g., special handling of numbers, compound words, abbreviations, acronyms) textual segmentation and normalization of spellings are all carried out by these components.

Named entity detection (NED) of persons, organizations, locations, and disaster types as well as numbers is performed on the output of the text normalization components. (further types of NEDs can easily be added to the NED system). The NED system is based on patterns as well as statistical models defined over words and word-features and is run in multiple stages (Bikel et al. 1997).

The topic detection (TD) component first classifies sections of text according to a specific hierarchy of topics. Coherent stories are found by grouping similar adjacent segments. Subsequently, the already classified sections are compared to each other and similar, adjacent sections are merged (story segmentation). The learning models used for TD and story segmentation are based on support vector machines (SVM) with linear kernels (Boser et al. 1992).

Sentiment analysis is performed using a lexicon-based method, for which the discriminative features are predefined lists of words and patterns including a sentiment lexicon with a positive or negative score associated with each entry. Additional features include boosters (intensifiers), negations, idioms and phrases, emoticons, slang expressions, and abbreviations. The emotion emphasis expressed in social media by repeated letters, capitalization and exclamation marks is likewise detected. The sentiment-analysis component employs the same basic methodology as SentiStrength (Thelwall et al. 2010), but extends it by (1) the application to the domains of general news and natural disasters, (2) improving the support of multiple languages (German, Russian, and to some extent English), (3) handling of data from both traditional as well as social media, (4) processing of short texts as well as full articles, and (5) a four-class classification scheme (Shalunts et al. 2014). Unlike the majority of sentiment analysis approaches in the literature, it solves a dual-classification task by classifying a text into one of the following four classes: positive, negative, mixed (both positive and negative), or neutral (neither positive nor negative).

10.3.2 Visual Analysis

The aim of image and video analysis is to determine links between images and time segments of videos based on visual similarity. This similarity may be global (i.e., the content and composition of the entire images are similar) or local (i.e., only parts are matching). The degree of similarity may also vary, and lies in a continuum from actual duplicates (e.g., two tweets using the same image, repeated news story on TV), over near-duplicates (e.g., two videos of a press conference taken from different positions) to partial matches (e.g., different action taking place at the same location, picture-in-picture in TV news).

The aim of visual analysis of media content (both traditional and social media) is to provide complimentary information to textual analysis for filtering and clustering the information about a crisis situation. For the related problems of topic tracking approach in broadcast news, Hsu and Chang (2006) have shown that visual near-duplicate detection can provide a 25 % improvement of the tracking performance over using only textual information. Duplicates are quite common in social media, and thus identifying them helps reducing the number of images and video that need to be analysed. Near-duplicates received independently from different sources provide information for the assessment of credibility of the data. In addition, the information about who reposted which media provides insight into the relation of the active users, even if the media are distinct copies and the identity is not clear from the metadata. In connection with the textual information, the reuse of others' media items may provide information about the originality and credibility of the information provided by a user. Partial visual similarity can be used to cluster images taken at the same or nearby location, either at the same or at different times. Analysing the clustered images together thus enables better assessment of the situation as well as its evolution over time.

The initial step is to generate links between images and/or video segments based on their visual similarity. For images it is straight forward to determine the similarity between them. For videos, matching segments may be located anywhere in the video; thus matching means sliding the timeline of one video against the timeline of the other and determine the similarity at different positions. Determining the links consists of three steps:

1. Extraction of descriptors, i.e., compact information describing the visual properties of the image/video
2. Matching of the descriptors per image/frame pair
3. Determining matching segments of video based on frame results

The implementation of these steps used in the QuOIMA system is based on extraction and matching of image areas around salient key points, using the SIFT (Scale Invariant Feature Transform) algorithm (Lowe 2004). The SIFT algorithm has become very popular due to its powerful performance and is still used as a basic tool in the area of object recognition, near-duplicate detection, and other various related tasks. Although now almost 14 years old, SIFT's average performance on object recognition is still state of the art. Related algorithms designed to replace some parts of SIFT, e.g., different key point extraction methods (Hessian, Hessian-Affine, etc.), usage of other descriptors (Mikolajczyk and Schmid 2005; Bay et al. 2008; Dalal and Triggs 2005), or focusing rather on reduction of computational complexity without or with only a small loss of recognition performance, can demonstrate limited advantages, but SIFT's recognition performance is still competitive.

Both extraction and matching of the descriptors have been implemented on graphics processors (GPUs) using NVIDIA CUDA[2] in order to speed up processing. Only one field of the input video is used in order to avoid possible side effects of interlaced content. Descriptors are extracted from every fifth frame of the video. The SIFT implementation used here follows the original algorithm, but nearest-neighbor search of key points and their descriptors is implemented differently. Instead of using Best Bin First (BBF) for nearest-neighbor search, brute force matching using the GPU is performed, which is still faster and easier to implement than a CPU based nearest-neighbor search based on *kd*-trees recommended in Lowe (2004). Matching verification is done by using a homography between the matching points in the two images. This helps removing coincidental matches of small areas scattered across the image. The process of key point matching results in a number of key points that support a meaningful homography from one frame to the other. Reliable matches require 25–30 matching descriptors for at least some of the frames of a video. This number is used as a score and input to a two-dimensional matrix, where each axis denotes the timeline of a video segment. Subsequent matches are grouped into matching video segments, being robust against short time issues (e.g., motion blur). We create a similarity matrix between images and/or video segments based on the number of matching key points between them.

In order to reduce the computational complexity of visual matching, methods such as the bag-of-words model (Fei-Fei and Perona 2005), or more compact representations such as Fisher vectors (Perronnin and Dance 2007) or Vectors of Locally Aggregated Tensors (VLAT) (Picard and Gosselin 2011) can be used. A similar approach is the recent MPEG standard specifying compact descriptors (Duan et al. 2013) for still image search. However, while these approaches are much faster than matching the raw SIFT descriptors, they only provide comparable performance for duplicate and near-duplicate matches, but perform significantly worse for partial matches. Also, none of the current approaches makes adequate use of the temporal redundancy of video, this is still ongoing research.

Depending on the application, all values in the similarity matrix below a certain value can be set to zero, thus eliminating weak links between segments. For applications with few distinct visual environments this threshold can be rather low, while in applications where visual properties are shared by a high number of images and videos (e.g., floods along a river), the threshold must be chosen higher in order to keep similarities based on salient visual content. First, actual duplicates and possibly also near-duplicates with very high similarity scores (e.g., identical video clips only different by text overlay) are grouped and filtered. For these near-duplicates with high similarities, also the homography needs to be checked to be close to identity. After filtering those actual duplicates, unsupervised classification, i.e., clustering, of the images and video segments can be performed. Any clustering method, such as *k*-means or single-linkage clustering (Duda et al. 2000) can be used. As with many clustering problems, it may be hard to determine the optimal

[2]http://www.nvidia.com/object/cuda_home_new.html

number of clusters from the data. A possible approach in practice is to produce several clusterings with different parameters (e.g., number of clusters) and provide them as input to the fusion with textual and speech information (see below).

10.3.3 Audio and Speech Analysis

Audio processing includes the segmentation and classification of the incoming audio-stream, the identification of speakers (SID), and/or gender of the speaker and automatic speech recognition (ASR). After having been converted to the appropriate format by the feeding process, the audio signal is processed and segmented into homogeneous sections containing speech and non-speech. A first decoding pass, employing a set of sound-classes is used to segment and classify the incoming signal. The detection of silence, music and background noises is applied at this stage to allow for early pre-filtering of segments according to their audio content. Automatic speech recognition is only applied to those segments which are deemed appropriate for speech processing. The Sail Labs speech recognition engine itself is designed for large-vocabulary, speaker-independent, multi-lingual, real-time decoding of continuous speech. Decoding in real-time—or faster for those sources available offline—has been given particular attention to deal with large amounts of data and the real-time information needs of clients. Recognition is performed in a multi-pass manner, each phase employing more elaborate and finer grained models refining intermediate results, until the final recognition result is produced (Kubala et al. 1997). Subsequently, text-normalization as well as language-dependent processing (e.g., handling compound words for German as described in Hecht et al. 2002) is applied to yield the final decoding result in an XML format. As decoding proceeds, intermediate transcription results are made available via an API call to allow for subsequent processing with minimal latency. The engine per se is language independent and can be run with a variety of models created for different choices of language and bandwidth. Currently, speech recognition is supported in 14 languages with ongoing extensions to further languages. The models supplied with ASR are aimed at general broadcast-news, typically comprising terminology from sectors such as politics, economy, or sports. However, as vocabularies change from day to day as the news shift their attention to new persons, topics, locations or products, clearly models need to be updated and adjustable to the particular situation at hand. On the one hand, this is possible via a set of tools allowing users to tailor the models to their particular needs. This can be achieved, for example, by adding words pertaining to a particular event such as the name of a storm (Typhoon *Hayan*) or to a locations (such as *Maidan* in Kiev)—words and names which previously had not been part of the recognizer's vocabulary but became important terms within a matter of days. On the other hand, different news sources, such as RSS feeds, can be used to guide the vocabulary adaptation process on a daily or event-triggered basis. This way, new and important terminology mentioned in trusted news sources can be integrated into the speech recognition engine's vocabulary on an automatic basis.

A combination of frequency, recency, and variety of newly occurring terms is used in the process to select adequate candidates for vocabulary inclusion. Pronunciations for new terms are generated automatically, but can also be supplied manually along with manually selected further terms. Currently, this automatic update scheme is implemented to create refreshed and updated models on a daily basis, thus reflecting the most recent events and terms.

10.3.4 Fusion of Results

All processing on text, audio and video data is carried out in separate components, which can readily be plugged-in into the overall system. The results of the individual visual components are collected and output in XML format. This XML is subsequently merged (based on time tags) with the XML output produced by audio and textual processing (late fusion). The combined XML is then uploaded to the server and made available for search and retrieval. The different modalities, metadata, text, and visual information provide different views on the data. However, users are in most cases interested to be provided with a view of the data that is based on using the contributions from these different modalities in order to reduce the deviations between them. There are two basic approaches for fusing information from the different modalities. Early fusion operates on the level of similarity calculation and yields a joint similarity score from the different modalities, on which further steps, such as clustering are based. Late fusion performs the entire processing independently for each of the modalities, and applies the fusion only on the results. Late fusion is more flexible, as the user can control fusion parameters to quickly adjust the way the results are fused, without repeating large parts of the processing chain. For supervised semantic classification of concepts in video, it has been shown that late fusion of visual and text features outperforms early fusion (Snoek et al. 2005). Late fusion has also been successfully applied for linking similar video segments to search results using visual and textual information (Bailer et al. 2014). This work also shows that approaches using named entities (e.g., locations) are quite robust against imprecise temporal segmentations of the video, as they might occur when matching segments are found that only cover parts of a story.

Late fusion of the clustering results from the different modalities leads to a problem also known as ensemble clustering. The aim is to find an optimal clustering based on clustering results obtained from different methods, different parameterizations of different similarity measures (such as the different modalities in our case). An overview of methods can be found in Vega-Pons and Ruiz-Shulcloper (2011), organized into median partition, and object co-occurrence-based methods. The latter include relabeling/voting, methods based on co-association matrices and graph based methods. In the following, we discuss some methods that are relevant for our application.

A voting-based method for fusing multiple runs of non-deterministic clustering methods is proposed in Frossyniotis et al. (2002). Similar to hierarchical clustering, this greedy approach merges the most similar clusters obtained from the multiple clustering results used as input until the target number of clusters is reached. Bruno and Marchand-Maillet (2009) start building probabilities of joint cluster memberships based on the initial clustering results. The global clustering is obtained using probabilistic latent clustering, maximizing the posterior probability for each document with respect to the latent variables. The values of the latent variables are estimated using either the Expectation-Maximization algorithm or Non-negative Matrix Factorization.

Fred and Jain (2005) use Evidence Accumulation Clustering (EAC) for combining multiple clusterings. The method has no constraints on the number of clusters or cluster structures. It uses a co-association matrix containing the counts in how many of the input clusterings two elements occur in the same cluster. The authors apply single- or average-linkage clustering on the co-association matrix. When using single-linkage clustering, they use a matrix of p nearest neighbors for each data element in order to avoid handling a large n^2 similarity matrix.

Among the graph-based methods, Gao et al. (2005) propose a co-clustering scheme for visual and textual features, modelled as a tripartite graph. The optimal clustering is obtained from jointly optimising cutting the bipartite graphs representing image and text features. Rege et al. (2008) also use a representation as a tripartite graph, but use a variant of isoperimetric co-clustering for obtaining the two globally optimal bipartite graph segmentations. Recent work proposes a family of efficient algorithms for combining clustering results (Mimaroglu and Erdil 2013), working at a cluster rather than object level to reduce execution times. The approach is a variant of the graph-based COMUSA algorithm (Mimaroglu and Erdil 2011), working on cluster level. Clusters are considered nodes of a graph and edge weights between clusters based on co-association.

Related methods deal with clustering/combining retrieval results, requiring a user request and a list of ranked elements with respect to this request. In cases where a user selects a media item, these methods can also be relevant in order to provide on the fly relevant clusterings, taking the user interaction into account. An early approach (Voorhees et al. 1995) addresses the related problem of collection fusion, i.e., building a ranked result list from individual result lists, with typically different numbers of relevant documents and different similarity scores. The proposed approach is to learn model parameters from past queries and the optimal combination of their individual results. However, this requires training on comparable data, which is not always feasible in a crisis management context. Costantini and Nicolussi (2011) propose a cluster fusion strategy building a graph adjacency matrix from joint cluster memberships of images in the input clustering. The method is in the context of image retrieval, assuming a ranking of images. Starting from the first image, breadth first search in the graph is used to obtain the final clustering.

10.4 Conclusions

In this chapter we have shown, that there is currently a change in the process of gathering and disseminating information by emergency responders in crisis scenarios. Social media channels have become an important news-feed providing text, images, and video directly from and to victims or witnesses of critical situations. However, emergency responders are not always capable of dealing with the sheer speed and amount of information, which hence needs to be reduced in a pre-filtering step.

The QuOIMA project described here proposed a clustering approach which removes near-duplicates from the feeds, reducing the amount of information to be presented to an operator. This clustering is done separately for text and for audiovisual media items. The extracted information is then fused to achieve a final clustering, applying spectral clustering (Von Luxburg 2007) to the co-association matrix of the input clusterings.

The clustering does not only allow the removal of duplicates, but also enables the grouping of media items by location and will enable new types of service and user interface for emergency responders.

Acknowledgements The research work leading to these results is partially funded by the Austrian Federal Ministry for Transportation, Innovation and Technology[3] (BMVIT) within the KIRAS[4] program.

References

Bailer W, Weiss W, Kienast G, Thallinger G, Haas W (2010) A video browsing tool for content management in post-production. Int J Digit Multimed Broadcast 2010, 856761. doi: 10.1155/2010/856761

Bailer W, Lokaj M, Stiegler H (2014) Context in video search: is close-by good enough when using linking? In: Proceedings of ACM International Conference on Multimedia Retrieval, Glasgow, UK, pp 209–217

Bay H, Ess A, Tuytelaars T, Van Gool L (2008) Speeded-up robust features (SURF). Comput Vis Image Underst 110(3):346–359. doi:10.1016/j.cviu.2007.09.014

Bikel D, Miller S, Schwartz R, Weischedel R (1997) Nymble: a high-performance learning name-finder, ANLC'97 Proceedings of the fifth conference on applied natural language processing, Stroudsburg, PA

Boser BE, Guyon I, Vapnik V (1992) A training algorithm for optimal margin classifiers. In: Proceedings of the 5th Annual Workshop on Computational learning theory. ACM Press, New York, NY, pp 144–152

Brugghemans B, Van de Walle B, Milis K (2013) Impact of the distribution and enrichment of information on the coordination of a human-made fast-burning crisis. In: Proceedings of the

[3]http://www.bmvit.gv.at/

[4]http://kiras.at/

10th International Conference on Information Systems for Crisis Responses and Management (ISCRAM 2013), Baden-Baden, Germany, pp 89–93
Bruno E, Marchand-Maillet S (2009) Multiview clustering: a late fusion approach using latent models. In: Proceedings of the 32nd international ACM SIGIR conference on research and development in information retrieval (SIGIR'09), Boston, MA
Bruns A, Burgess J, Crawford K, Shaw F (2012) #qldfloods and @QPSMedia: crisis communication on Twitter in the 2011 South East Queensland Floods. ARC Centre of Excellence for Creative Industries and Innovation, Brisbane, QLD
Costantini L, Nicolussi R (2011) Image clustering fusion technique based on BFS. In: Proceedings of the 20th ACM international conference on Information and knowledge management (CIKM'11), Glasgow, UK
Dalal N, Triggs B (2005) Histograms of oriented gradients for human detection. In: Proceedings of the IEEE Computer Society Conference on Computer vision and pattern recognition (CVPR'05), vol 1, San Diego, CA, pp 886–893
Duan L-Y, Gao F, Chen J, Lin J, Huang T (2013) Compact descriptors for mobile visual search and MPEG CDVS standardization. In: IEEE International Symposium on Circuits and Systems (ISCAS), Beijing, China, 19–23 May 2013, pp 885–888
Duda RO, Hart PE, Stork DG (2000) Pattern classification. Wiley-Interscience, New York, NY
Fei-Fei L, Perona P (2005) A Bayesian hierarchical model for learning natural scene categories. In: 2005 IEEE Computer Society Conference on Computer Vision and Pattern Recognition (CVPR'05), vol 2, San Diego, CA, pp 524
Fred ALN, Jain AK (2005) Combining multiple clusterings using evidence accumulation. IEEE Trans Pattern Anal Mach Intell 27(6):835–850
Frossyniotis DS, Pertselakis M, Stafylopatis A (2002) A multi-clustering fusion algorithm. In: Proceedings of the Second Hellenic Conference on AI: methods and applications of artificial intelligence (SETN'02), Thessaloniki, Greece, April 11–12, 2002
Gao B, Liu T-Y, Qin T, Zheng X, Cheng Q-S, Ma W-Y (2005) Web image clustering by consistent utilization of visual features and surrounding texts. In: Proceedings of ACM Multimedia, Singapore, November 6–11, 2005, pp 112–121
Hecht R, Riedler J, Backfried G, German broadcast news transcription. In: Proceedings of the International Conference on Spoken language processing, ICSLP 2002, Denver, CO, USA, September 2002
Hsu WH, Chang S-F (2006) Topic tracking across broadcast news videos with visual duplicates and semantic concepts. In: IEEE International Conference on Image Processing, October 8–11, 2006, pp 141–144
IPTC. International Press Telecommunications Council. Available from: http://www.iptc.org/site/News_Exchange_Formats/NewsML-G2/. Accessed on 24 June, 2014
Johansson F, Brynielsson J, Narganes Quijano M (2012) Estimating citizen alertness in crises using social media monitoring and analysis, EISIC 2012, Odense, Denmark
Kohlschütter C, Frankhauser P, Nejdl W (2010) Boilerplate detection using shallow text features. In: Proceedings of the third ACM international conference on Web search and data mining (WSDM 2010), New York, NY
Kubala F, Jin H, Nguyen L, Schwartz R, Matsoukas S (1997) Broadcast News Transcription. In: Proc of the ICASSP'97, vol 1, pp 203–206
Lowe DG (2004) Distinctive image features from scale-invariant keypoints. Int J Comput Vis 60(2):91–110
Mendoza M, Poblete B, Castillo C (2010) Twitter under crisis: can we trust what we RT? In: SOMA 2010, Washington, DC
Mikolajczyk K, Schmid C (2005) A performance evaluation of local descriptors. IEEE Trans Pattern Anal Mach Intell 27(10):1615–1630, http://lear.inrialpes.fr/pubs/2005/MS05
Mimaroglu S, Erdil E (2011) Combining multiple clusterings using similarity graph. Pattern Recogn 44(3):694–703
Mimaroglu S, Erdil E (2013) An efficient and scalable family of algorithms for combining clusterings. Eng Appl Artif Intel 26(10):2525–2539

Muhren WJ, Van de Walle B (2010) Sense-making and information management in emergency response. Bull Am Soc Inf Sci Technol 36(5):30–33

Nilsson J et al. (2012) Making use of new Media for Pan-European crisis communication, ISCRAM 2012, Vancouver, Canada

Perronnin F, Dance C (2007) Fisher kernels on visual vocabularies for image categorization. In: Proc IEEE Conference on Computer vision and pattern recognition, Minneapolis, MN, pp 1–8

Picard D, Gosselin P-H (2011) Improving image similarity with vectors of locally aggregated tensors. In: IEEE International Conference on Image Processing, Brussels, Belgium, pp 669–672

Rege M, Dong M, Hua J (2008) Graph theoretical framework for simultaneously integrating visual and textual features for efficient web image clustering. In: Proceedings of the 17th international conference on World Wide Web, Beijing, China, pp 317–326

Shalunts G., Backfried G, Prinz K, Sentiment Analysis of German Social Media for Natural Disasters, , 11th International Conference of Information Systems for Crisis Response and Management, ISCRAM, 2014, University Park, PA

Snoek CGM, Worring M, Smeulders AWM (2005) Early versus late fusion in semantic video analysis. In: Proceedings of the 13th annual ACM international conference on Multimedia (MULTIMEDIA'05), Singapore, November 06–12, 2005

Thelwall M, Buckley K, Paltoglou G, Cai D, Kappas A (2010) Sentiment strength detection in short informal text. J Am Soc Inf Sci Technol 61(12):2544–2558

Vega-Pons S, Ruiz-Shulcloper J (2011) A Survey of Clustering Ensemble Algorithms. Int J Pattern Recogn Artif Intell 25(3):337–372

Von Luxburg U (2007) A tutorial on spectral clustering. Stat Comput 17(4):395–416

Voorhees EM, Gupta NK, Johnson-Laird B (1995) Learning collection fusion strategies. In: Proceedings of the 18th annual international ACM SIGIR conference on Research and development in information retrieval (SIGIR'95), Seattle, WA

Part IV
Reasoning About Situations and Threats

Chapter 11
Empowering the Next-Generation Analyst

David Hall, Guoray Cai, and Jake Graham

Abstract Situation analysis for activities such as crisis management, military situation awareness, homeland security, or environmental monitoring is both enabled and challenged by access to enormous data sets. The advent of new sensing capabilities, advanced computing and tools available via cloud services, intelligent interconnections to mobile devices, and global interconnectivity with ever-increasing bandwidths provide unprecedented access to data and to computing. In addition, emerging digital natives freely share data and collaboration. Thus, on one hand situation analysts have great opportunities to access unprecedented amounts of information from sensors, human observers and online sources to assist in understanding an evolving situation. On the other hand, this access to huge data sources and computing can create a type of intelligence attention-deficit disorder, in which analysts are overwhelmed by the urgent, but lack the ability to focus on important data. This chapter provides a summary of this dilemma, describes a new analysis paradigm that links data-driven and hypothesis driven approaches, introduces a new prototype analyst workbench, and discusses an educational approach to empower the next generation of analyst.

11.1 Introduction

The modern information analyst is faced with simultaneous opportunities and challenges. Whether the application involves military situation assessment, crisis management, environmental monitoring, business logistics, or many other areas, emerging trends in information technology and its use provide ready access to enormous amounts of data and computing capabilities. Key trends, previously noted by Hall and Llinas (2014) include (1) *new sensing capabilities*—new nano-scale and micro-scale sensors provide near-ubiquitous embedded and distributed sensing of people, places, and things; (2) *advanced computing capabilities*—advances in Moore's law have enabled enormous computing capabilities in mobile

D. Hall (✉) • G. Cai • J. Graham
College of Information Sciences and Technology, The Pennsylvania State University, University Park, PA 16802, USA
e-mail: Dhall@ist.psu.edu; Cai@ist.psu.edu; Jgraham@ist.psu.edu

G. Rogova, P. Scott (eds.), *Fusion Methodologies in Crisis Management*,
DOI 10.1007/978-3-319-22527-2_11

devices; (3) *intelligent interconnections*—as intelligent devices become nearly universal in smart phones and embedded in nearly every entity, and as connectivity becomes ubiquitous with increasing bandwidth, the "intelligence" of the infrastructure also increases via web services, emerging standards for universal sharing of data, and command of remote devices; and (4) *global interconnectivity and increasing bandwidth*—universal wireless communications with increasing bandwidth allows distribution of sensor data such as imagery and video while also providing access to cloud computing capabilities to access sophisticated modeling and data analysis tools. Two additional trends are human centric: (1) *digital natives*—the continued emergence of the comfort and capability that human digital natives exhibit while freely collecting and reporting information, sharing information, and collaborating in dynamic teams, and (2) *human-data interaction*—increasingly sophisticated computer/data access provides guided interaction and focus of attention for improved access to and analysis of potentially huge data sets.

These trends allow individual analysts to access huge sources of data including human observations and remote sensors, access sophisticated computing models and tools via cloud-based computing resources such as Heroku (https://www.heroku.com), collaborate with other analysts in a kind of crowd-sourcing of analysis for pattern recognition and model development, and even "outsource" analysis tasks using sites such as Mturk (https://www.mturk.com/mturk/welcome) and CloudFlower (www.clowdflower.com). However, these resources can provide so much data and little knowledge that users become overwhelmed and ineffective, focusing on the urgent, rather than on the important (McNeese and Vidulich 2002). Cai et al. (2006) found that the information needs of crisis managers in real-time crisis response situations have a unique set of characteristics (immediacy, relevance, and sharing) that impose stringent requirement for information design. Hall et al. (2006b) introduced this problem and argued for the need to link the data-driven inference process (driven by sensors and source collection) with human hypothesis-driven reasoning. More recently, Cai et al. (2014) described a visual analytic framework to support hypothesis-driven analysis, and Rimland et al. (2013) presented some automated tools to assist in focusing the attention of analysts.

This chapter describes the evolution of that analysis framework and a visual analytics toolkit for investigative intelligence. Some comments are also provided on emerging tools. In addition, we briefly describe some thoughts on the education of the next generation of analysts.

11.2 Conceptual Frameworks for Analysis

An extensive legacy is available describing functional and cognitive models related to information fusion, situation awareness, and crisis management. Examples of models include the Joint Directors of Laboratories (JDL) data fusion process model used extensively in the data fusion community (see for example (Hall and McMullen 2004), the Boyd Control Observe-Orient-Decide-Act (OODA) Loop (Boyd 1987),

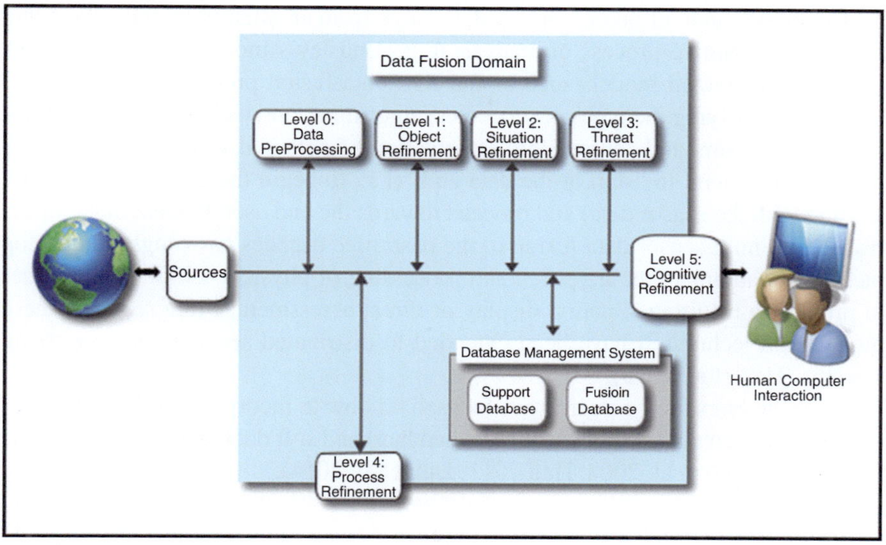

Fig. 11.1 Top level of the JDL data fusion processing model (Hall and McMullen (2004))

Dasarthy's model (Dasarathy 1994), the Omnibus Model (Bedworth and O'Brien 2000), Endsley's model for Situation Awareness (Endsley 1995), and a number of others). An excellent survey and review of many of these models is provided by Pek Hui Foo and Gee Wah Ng (2013), and also by Hall et al. (2006a).

The most commonly referenced model in military situation assessment, threat assessment, support for disaster relief and related domains is the JDL data fusion process model. The top level of the JDL data fusion process model is shown in Fig. 11.1. The model shows sources of information on the left hand side (e.g., sensor data, human observations), and a data fusion process represented by the blue box, with a link to end-users on the right hand side. Within the data fusion processing box are shown six "levels" of processing: (1) Level 0—Data Preprocessing/Sub-Object Data Assessment—estimation and prediction of signal/object observable states at the data level, (2) Level 1—Object refinement—estimation and prediction of the location, characteristics/attributes and "identity" of an entity or object, (3) Level 2—Situation Refinement—estimation and prediction of the relations among entities to understand an evolving situation, (4) Level 3—Threat refinement/impact assessment—estimation and prediction of the effects on situations of planned or estimated/predicted actions by the participants (or the environment), (5) Level 4—Process refinement—adaptive data acquisition and processing to improve the dynamic performance of the data fusion process, and (6) Level 5—Cognitive refinement—processes and interfaces to provide an effective link between the cognitive and decision-making processes of a user or users and the fusion process. It should be noted that the original JDL model only included levels 1, 2, 3, and 4. Level 0 processing was subsequently recommended by Steinberg et al. (1998), while the concept of Level 5 processing was introduced independently by Blasch and Plano (2002), and by Hall, Hall and Tate (2000).

The broad range of processing models range from an attempt to understand and model the cognitive process of analysis (e.g., Endsley's model, the Boyd OODA Loop), to functional models such as the JDL data fusion process model which is aimed at identifying the types of processes, algorithms, and techniques to transform data from sensors and sources into information of potential value to users. Many of the models tend to "start at the data end" (i.e., to begin the analysis process by starting with the sensor data) and proceed towards the end user. Indeed, much of the research in multisensor data fusion in the past three decades has sought to develop methods to automate the ingestion and processing of physical sensor data in order to produce a resulting situation display or threat assessment. More recently, these models and techniques have been extended to distributed sensing and processing systems (Hall et al. 2013).

In recent years, new research has focused on how to incorporate physical sensor data with human observations (the so-called hard and soft data fusion problem) (see for example Hall et al. 2008; Hall 2011; Hall et al. 2014).

11.3 The Role of the Human in Information Fusion Systems

While extensive research has focused on ingesting ever new types of data (including video, high-fidelity imagery, 3-D sources such as Lidar, and human observations) into an information fusion process and increasing automation of the data processing, research on the role of the human as participant in the fusion process has lagged. Hall and Jordan (2010) argued that humans should play a central role in information fusion for situation awareness and decision making. They suggested that humans can play three main roles in the fusion process: (1) as human observers, (2) as participants in pattern recognition and model (hypothesis) development, and (3) as collaborative team-based decision makers. This is illustrated in Fig. 11.2.

Advances in information technology and the tendencies of digital natives increase both the opportunities and demand for these three roles. For example, the proliferation of mobile computing, communications and sensing devices (e.g., on smart phones) enable human observers to provide inputs regarding an emerging situation or crisis event, and increasingly sophisticated human computer interactions provide increased opportunities for humans to interact with an ongoing automated reasoning process (e.g., via advanced search engines, advisory software agents, complex event processing and other methods). Finally, social media, virtual world technologies, and emerging "hive-mind" collaboration websites such as mTurk allow "crowdsourcing" of all types of services. A directory of over 2000 such websites is provided by the crowdsourcing directory at www.crowdsourcing.org/directory. As an example, the emergence of "citizen science" encourages amateurs to become involved in scientific endeavors by collecting data, helping to analyze patterns in data, and even the development of new models (see for example: https://www.zooniverse.org/).

Our primary focus in this chapter involves the role of humans as analysts participating in the formulation of hypotheses, analysis of data, and "sense-making"

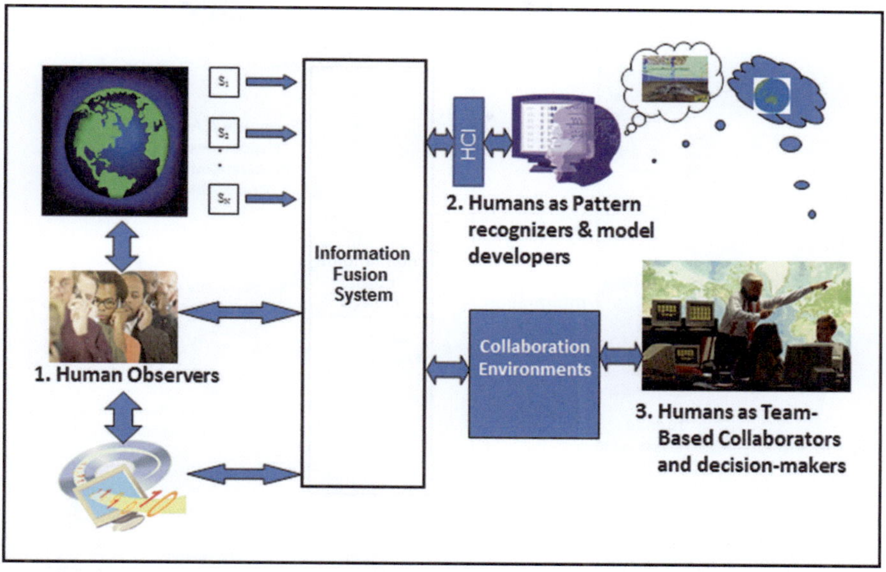

Fig. 11.2 Three roles of humans in information fusion

(role # 2 in Fig. 11.2). It should be noted that an enormous amount of research has been conducted on human-computer interaction, especially focused on data visualization. There is an extensive amount of free software, commercial software, academic tools, and other sources for such visualization tools. However, despite this research and the ready availability of tools, there are remaining challenges.

Consider Fig. 11.3 below (Hall 2002). The figure is similar to Fig. 11.2, but expands the roles and functions performed by the human analyst. In particular, we see two interacting cycles. The bottom cycle illustrates the concept of reaction to incoming or new data from the data fusion system. As data and reports are received, functions to be performed include; (1) receiving and processing the observations and reports (e.g., formatting, transformations, database updates, etc.), (2) meta-data tagging and transformations (e.g., generation of semantic meta-data to augment signals and images), (3) data correlation and fusion with existing data, (4) hierarchical composition and linking of data to other information for higher level inferences (e.g., using graph matching techniques, logical templates, complex event processing, intelligent agents, Bayesian inferencing methods), and finally, (5) data discovery and anomaly detection—use of a wide range of tools to discover new relationships, identify anomalous conditions, and perform a data for aging process. Each of these functions can include both computer automation as well as human in the loop processing. For example, meta-data tagging could be performed by machine learning algorithms in which an algorithm presents sematic “labels” to describe the content and context of image data. A human analyst can verify or refute these annotations and perhaps provide additional labels.

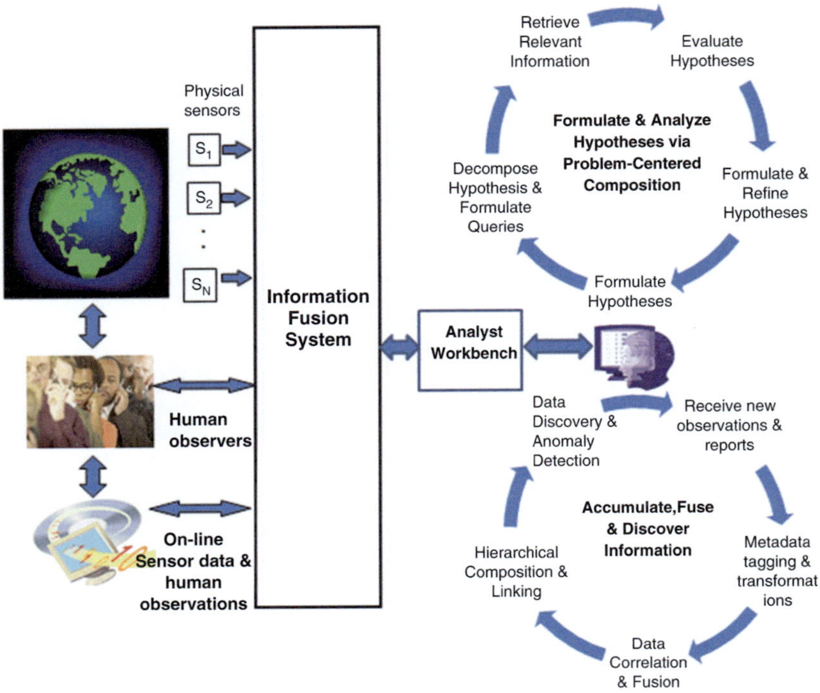

Fig. 11.3 Expanded view of human analysis/inferencing processes

The top cycle in Fig. 11.3 illustrates the analyst's activities related to formulating hypotheses to interpret or explain the data, development of stories or themes related to events, activities, trends, etc. Examples of functions include (1) the formulation of hypotheses, (2) composition of the hypotheses into data queries to determine if key supporting evidence might be available to support or refute the hypotheses, (3) retrieval of relevant information (and perhaps tasking of sensors or resources to obtain the requisite information), (4) evaluation of the validity of the hypotheses by judging the accuracy, extent, and relevance of supporting and refuting information, (5) reformulation and refinement of the hypotheses. Ultimately the top part of the cycle in Fig. 11.3 is aimed at "sense-making" or "storification" (development of narrative stories or "parables" to help understand an evolving situation).

While numerous visual toolkits support the data acquisition, fusion and discovery inference cycle, few are available to support the formulation, analysis and composition of hypotheses and stories to an evolving understanding of a situation. Moreover, limited work has been done of tools to integrate the data discovery and the hypothesis development inference cycles. The next section of this chapter describes an initial attempt by researchers at Penn State University to develop such an analyst workbench, with further comments on understanding the sense-making cognitive process.

11.4 Visual Analytic Framework for Risk Analysis

Crisis managers must make their decisions based on awareness and assessment of crisis situation. Both the understanding of risk situation and the assessment of pending threats require making sense of a large amount of data to form judgment. It is important to understand and support how risk analysts perform analytical reasoning with data. In our work, we use the science of analytical reasoning (Thomas and Cook 2005a, b) to build a conceptual model of how risk analysts reason with multiple sensor data sources. Figure 11.4 shows our conceptual model of data fusion driven by the analytical reasoning process of intelligence analysis. This model highlights the centrality of human analytical reasoning in the process of fusing data for making judgment.

This conceptualization emphasizes the roles of human analytical reasoning supported by highly interactive visual data representations in risk analysis. It goes beyond traditional focus of automated data fusion methods, and allows human analysts to be on the driver seat when navigating and synthesizing data. It recognizes that analytical reasoning with data is generally not a linear process. During an analytic session, the analyst engages in multiple, iterative dialogues with the information available. This process is known as *analytical discourse* (Thomas and Cook 2005b; Cai 2007). Information is consulted and extracted to (1) refine and elaborate the issues and questions, (2) gather data for identifying and evaluating evidence, and (3) using evidences and assumptions to evolve knowledge and insights. To make assessments about a situation, analysts perceive, understand, and reason about complex and dynamic data in order to collect evidences for reaching conclusions or judgment, and they often do so under significant time pressure and with incomplete and conflicting information. When forming their judgment on pending threats, analysts often consider multiple competing hypotheses and evaluate these alternative explanations in light of evidences and assumptions.

Analytical work itself is inherently difficult, due to the cognitive difficulty of critical thinking and sense-making. This difficulty is exacerbated by heavy

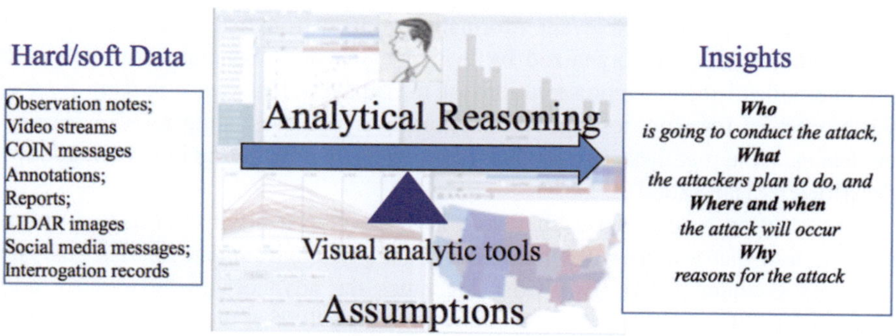

Fig. 11.4 Data fusion driven by analytical reasoning

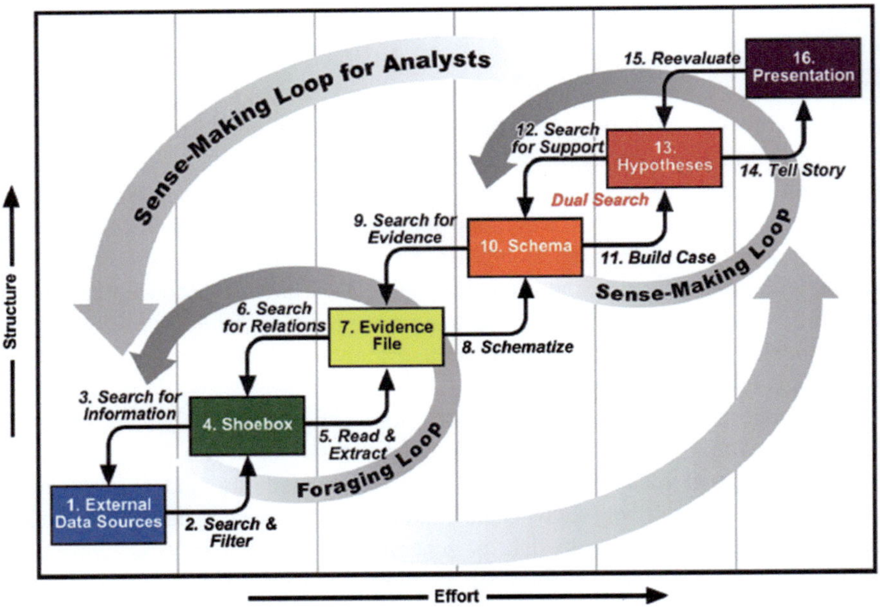

Fig. 11.5 A theory of sense-making (after Pirolli and Card (2005))

workload, time pressure, high stakes, and high uncertainty. Analysts must deal with data that are dynamic, incomplete, often deceptive, and evolving. Problem solving with such data involves breaking data into elements, examining the patterns to reveal evidence, accumulating and relating evidences, and assembling and harmonizing many different insights from different observations. This is consistent with the theory of sense-making developed by Pirolli and Card (2005) (See Fig. 11.5). Boxes in the figure represent data and arrows represent processes. It characterizes analytical work with data as the process that contains two sub-processes: data foraging loop and sense-making loop. In the *data foraging loop*, an analyst filters messages and actively searches for information and collects it in an information store (called a shoebox). Relevant information nuggets from this store are assembled into evidence files and organized by evidence schemas. Schemas correspond to the internalized mental representations of the analysts. The *sense-making loop* is the process of forming, testing, and supporting hypotheses using relevant data as evidences. Note that these two loops are similar to the interacting inferencing loops described by Hall (2002).

Our solution to the above challenge is to develop a highly interactive visual analytic tool that provides the flexible data analytic environment for risk analysts (Cai and Graham 2014). This work leverages the advances in the science and technologies of visual analytics (Thomas and Cook 2005a) that marry techniques from information visualization with techniques from computational transformation and analysis of data. In particular, we have designed and implemented a

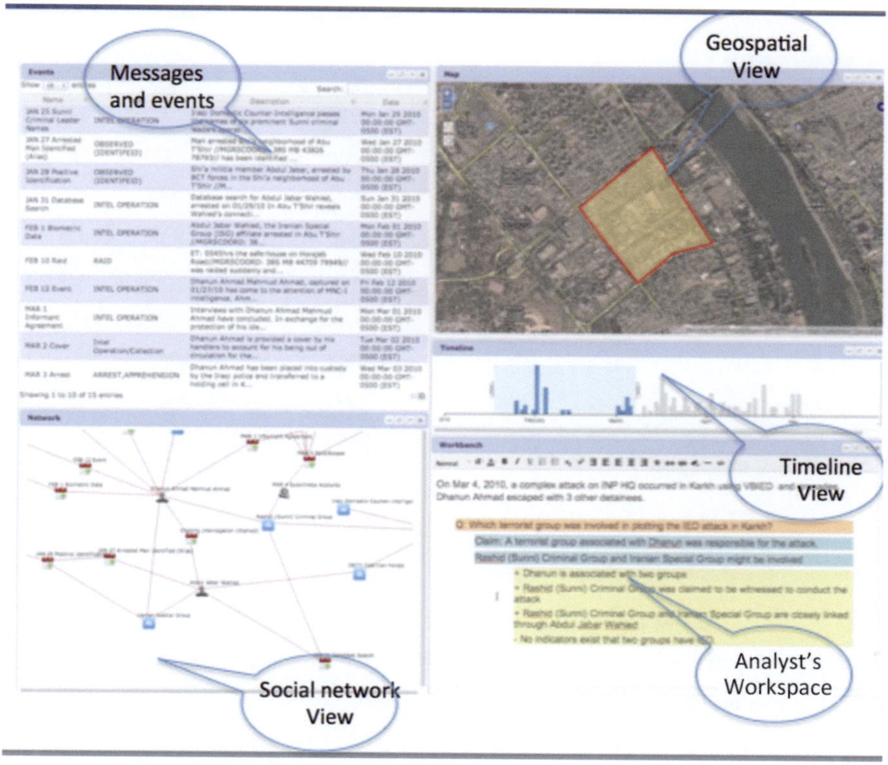

Fig. 11.6 THREATSight—a visual analytic environment for risk analysis

test-bed called THREATSight where human-analytical reasoning and data fusion capabilities are woven together using visual analytic methods. THREATSight (see Fig. 11.6) consists of two coupled components: a *visual dashboard* and an *analyst's workbench*. The visual dashboard is a flexible information exploration environment based the technology of coordinated multiple views (Carrico and Guimarares 1998) and is designed to support the "information foraging loop" of Fig. 11.5. The analyst's workbench (corresponding to the "sense-making loop" of Fig. 11.5) is an interactive workspace where the analyst constructs analysis by weaving visual representation of data into evidences in support of hypotheses and judgment.

Our design of visual analytic dashboard considers several models of multiple view systems (North and Shneiderman 2000; Roberts Jr 2007). It supports three different dimensions visual data exploration: *selection* of views, *presentation* of views, and *interaction* among views (Wang et al. 2000). Coordinated multi-views have been widely used for visual analytic support to exploratory data analysis. Here we use a specific design of multiple views to support exploration of events, messages, criminal groups, Improved Explosive Devices (IED), and their relationships through space, time, and domain knowledge schemas. Among many choices of visualization forms (Pike et al. 2009; Bostock et al. 2011) we have chosen four

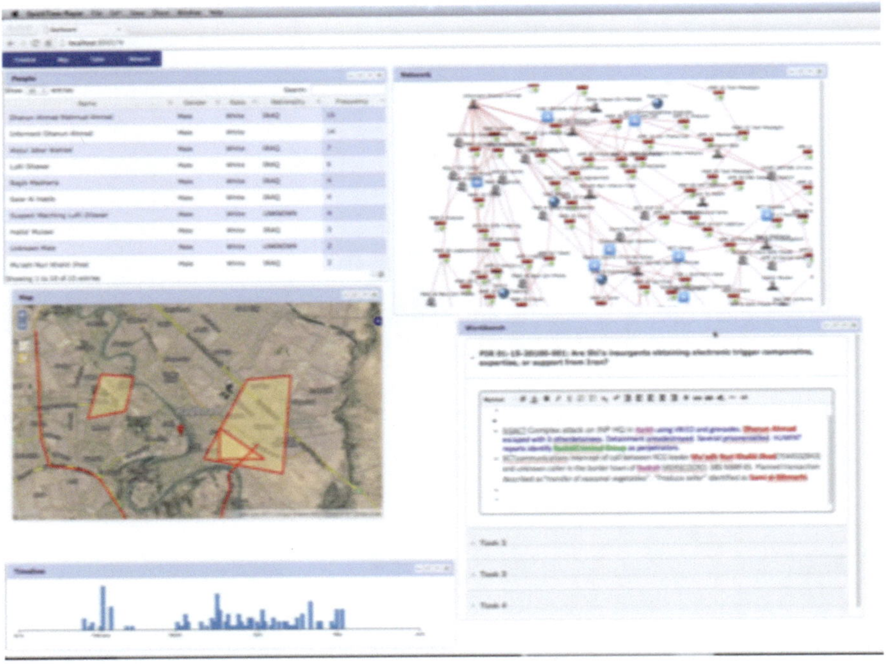

Fig. 11.7 Flexible layout and evidence collector functions

basic representations: tables, timeline, map, and network. Table views are used to visualize and interact with messages and events. A map view is used to discover associations through spatial relationships. Timelines allow effective query for filter and zero into specific time range. A network view is constructed from the relational data model that reflects the semantic relationships of the underlying data, and it allows exploration of associations among entities and objects. By making live linking across all views, different visualizations are coordinated to ease the analyst effort to find corresponding items depicted in different views.

One important design decision is to give the analysts the maximum flexibility in controlling their reasoning and interactions with data. For example, Fig. 11.7 shows that the system allows any number of windows to be opened, resized, re-positioned, and closed, so that they can organize their workspace. Coordination across views are also controllable by the analysts so that the analysts can use multiple views to provide "overview + detail," "filter/subset," and "comparison" analysis.

The true power of this system originates from the joint use of the workspace and dashboard functions, where the evaluation of evidence, the development of inferences and the assessment of competing hypotheses in the workspace creates needs for gathering additional evidences about situations from the dashboard. These analytic needs are translated into new requests to the data fusion subsystems to decide what aspects of the situation are to be further characterized and analyzed.

11.5 Training the Next Generation of Analysts

In addition to the development of analytical frameworks and new tools, we are also concerned with the education of future analysts. At the Pennsylvania State University College of Information Sciences and Technology (IST), a new undergraduate Bachelor of Science program in Security and Risk Analysis (SRA) provides a special track in intelligence analysis and modeling. This track is focused on issues such as the role of information and intelligence, deception and counter deception, visual analytics, decision making and risk analysis, and background information in economics, statistics, data mining, and knowledge management. The program is available both in residence at the Pennsylvania State University, University Park campus, as well as on-line via Penn State's World campus program (see http://www.worldcampus.psu.edu/degrees-and-certificates/security-and-risk-analysis-bachelors/courses).

Key aspects of this curriculum include:

(a) *Coursework*—extensive coursework in areas such as information science and technology, threat of terrorism and crime, information security, decision theory and analytics, risk management and assessment, analytic statistics, analytical writing, and cyber security
(b) *Problem-based learning*—"hands on" problem-based learning using advanced scenarios related to crime, terrorism and crisis events
(c) *Real-world experiences*—all students must take at least one internship with a corporate or government organization
(d) *Capstone course*—a full semester, team-based project in which students must integrate their knowledge to address a real-world problem
(e) *Integrated skills*—integration of key skills such as critical thinking, clear analytical writing, individual and team-based problem solving, and ability to present results to a critical audience

One aspect of the Penn State SRA training involves the use of analytic decision game (ADG) methods to assist in training and teaching the next generation of analysts (Graham and Hall 2012). In particular, the SRA senior level capstone course utilizes sophisticated data analysis and decision-making exercises for student teams. For example, a scenario was developed entitled, *Piracy on the High Seas* (Graham 2011). This task requires students to develop a comprehensive study of the then current state of Somali Piracy. The *High Seas* scenario placed students in the Intelligence Section (C-2) of a Combined Task Force (CTF). Data supporting this scenario has the look and feel of intelligence messages that could be received in a CTF command post. The scenario deals with more than just piracy however, causing students to sort through the ambiguities of media and international political commentary to understand what groups pose the real danger, all while the threat of piracy moves from the Gulf of Aden to tourist areas closer to home. Other scenarios address cyber futures, crime and terrorism in Washington District of Columbia (Graham 2010), financial scenarios related to cyber threats to the US financial sector, and many others.

Graham has noted that the use of simulations in the undergraduate classroom is challenging (Graham and Hall 2012). Exercise design, scenario, and data creation and exercise facilitation all require significant effort. Care must be taken to ensure that the data should sound and feel authentic and that it should be presented in a realistic fashion to engage and hold the attention of the students. Never the less, the development and use of such scenarios and exercises significantly prepare students for real-world applications and environments. We routinely obtain feedback from our internship and postgraduation employers concerning how well prepared our students are. It is also helpful that they gain "hands-on" experience with tools such as Analyst's Notebook and other geographical information system tools such as the previously described toolkit.

Finally, it is interesting to note that the students have self-organized to create opportunities for addressing challenging real-world problems. A student-run *Red-Cell Analytics Club* and associated *Red-Cell Analytics Laboratory* (https://rcal.ist.psu.edu/) have provided opportunities for students to conduct support activities for challenges such as (1) participating in the DARPA Red Balloon challenge (finish 10th out of nearly 9000 competing teams) (Tang et al. 2011), (2) assisting in improving the cyber and physical security of the Penn State large football stadium, (3) performing analyses of current crises and events, (4) working with agencies such as the Pennsylvania State Police to understand the tendencies of the digital natives use of social media, and (5) addressing challenge problems via "hackathons," and individual study assignments. Such activities provide enrichment opportunities for our undergraduate students as well as provide opportunities for faculty members to better understand new modes of thinking and interaction.

11.6 Summary

Rapid changes in information technology and the use of such technology by digital natives provide both opportunities and challenges in situation analysis for domains such as crisis management, military support of humanitarian relief, environmental monitoring, and global neighborhood watch for homeland security. While enormous new resources are available via global distribution of sensors and communications devices, access to advanced computing resources and sophisticated software via cloud computing, and reporting via participatory observers, a gap exists in assisting analysts in data understanding and development and assessment of hypotheses. We believe that work is required to (1) develop new understandings and paradigms for simultaneous data foraging and sense-making, (2) develop new tools and analyst workbench environments for enabling analyst analysis and understanding, and (3) new methods to educate the next generation of analysts. We have provided here a glimpse of our efforts at the Pennsylvania State University College of Information Sciences and Technology to address all of these areas.

References

Bedworth M, O'Brien J (2000) The Omnibus model: a new model of data fusion? IEEE Aero Elect Syst Mag 15(4):1903–1910

Blasch E, Plano S (2002) DFIG Level 5 user refinement issues supporting situational assessment reasoning. In: Proceedings of SPIE, vol 4729. SPIE, Wyndham, PA, pp 270–279

Bostock M, Ogievetsky V, Heer J (2011) Data-driven documents. IEEE Trans Vis Comput Graph 17(12):2301–2309

Boyd J (1987) A discourse on winning and losing. USAF, Maxwell AFB, Montgomery, AL

Cai G (2007) Formalizing analytical discourse in visual analytics. In: VAST-IEEE Symposium on Visual analytics and technology, Sacramento, CA, pp 217–218.

Cai G, Graham J (2014) Semantic data fusion through visually-enabled analytical reasoning. In: Proceedings of Fusion 2014: The Seventeenth International Conference on Information Fusion, International Society of Information Fusion, Salamanca, Spain

Cai G, Sharma R, Maceachren A, Brewer I (2006) Human-GIS interaction issues in crisis response. Int J Risk Assess Manag 6(4/5/6):388–407

Cai G, Hall DL, Llinas J, Gross G (2014) A visual framework for data fusion in investigative intelligence. In: SPIE DSS Conference on Next-Generation Analyst II. SPIE, Baltimore, MD

Carrico L, Guimarares N (1998) Integrated multi-views. J Vis Lang Comput 9(3):287–297

Dasarathy B (1994) Decision fusion. IEEE Computer Society Press, Washington, DC

Endsley M (1995) Towards a theory of situation awareness in dynamic systems. Hum Factors 37:32–64

Foo P, Ng G (2013) High-level information fusion: an overview. J Adv Inf Fusion 8(1):33–72

Graham JL (2010) Analytic decision game # 2: diabolical deeds in the district: instructor course guide for SRA 231: decision theory and analysis. The Pennsylvania State University, College of Information Sciences and Technology, University Park, PA

Graham JL (2011) Analytic decision game # 3: piracy on the high seas, instructor course guide for SRA 440W: security and risk analysis capstone course. The Pennsylvania State University College of Information Sciences and Technology, University Park, PA

Graham JL, Hall DL (2012) The use of Analytic Decision Game (ADG) methods for test and evaluation of hard and soft data fusion systems and education of a new generation of data fusion analysts. In: MSS National Symposium on Sensor and Data Fusion. GTRI, Washington, DC

Hall DL (2002) Semantic information fusion: breaking the bits-barrier. The Pennsylvania State University, College of Information Sciences and Technology, University Park, PA

Hall DL (2011) Challenges in hard and soft fusion: worth the effort? In: Proceedings of the SPIE Defense, Security and Sensing Symposium, Orlando, FL

Hall DL, Jordan JM (2010) Human-centered information fusion. Artech House, Inc., Boca Raton, FL

Hall DL, Llinas J (2014) The emergence of the millennium analyst: How advances in information technology and use enable asymmetric ISR. In: Proceedings of the MSS National Symposium on Sensor Data Fusion. Springfield, VA

Hall DL, McMullen SA (2004) Mathematical techniques in multisensor data fusion. Artech House, Inc., Boca Raton, FL

Hall DL, Hellar BD, Llinas J, McNeese M (2006) Assessing the JDL model: a survey and analysis of decision and cognitive process models and comparison with the JDL model. In: Proceedings of the National Symposium on Sensor and Data Fusion, McLean, VA

Hall DL, Hellar BH, McNeese MD (2006b) Rethinking the data overload problem: closing the gap between situation assessment and decision making. In: MSS National Symposium on Sensor and Data Fusion. MITRE Corporation, Washington, DC

Hall DL, McNeese M, Llinas J, Mullen T (2008) A framework for hard/soft fusion. In: Proceedings of the 11th International Conference on Information Fusion, Cologne, Germany

Hall DL, Chong C-Y, Llinas J, Liggins M II (2013) Distributed data fusion for network-centric operations. CRC Press Taylor & Francis Group, Boca Raton, FL

Hall DL, Rimland J, Shafer S (2014) A hitchhiker's guide to distributed hard and soft information fusion infrastructure development. In: Proceedings of the International Conference on Information Fusion, Salamanca, Spain

Hall MJ, Hall SA, Tate T (2000) Removing the HCI bottleneck: How the human computer interface (HCI) affects the performance of data fusion systems. In: Proceedings of the 2000 MSS National Symposium on Sensor and Data Fusion, San Diego, CA

McNeese M, Vidulich MA (2002) Cognitive systems engineering in military aviation environments: Avoiding Cogminutia Fragmentosa. CSERIAC Press, Wright Patterson AFB, OH

North C, Shneiderman B (2000) Snap-together visualization: can users construct and operate coordinated visualizations? Int J Hum Comput Stud 53(5):715

Pike W, Bruce J, Baddeley B, Best D, Franklin L, May R et al (2009) The scalable reasoning system: lightweight visualization for distributed analytics. Inf Vis 8(1):71–84

Pirolli P, Card S (2005) The sensemaking process and leverage points for analyst technology as identified through cognitive task analysis. In: Proceedings of the 2005 International Conference on Intelligence Analysis, McLean, VA

Rimland J, McNeese MM, Hall DL (2013) Conserving analyst attention units: use of multi-agent software and CEP methods to assist information analysis. In: SPIE DSS Conference on Next-Generation Analyst. SPIE, Baltimore, MD

Roberts Jr JC (2007) State of the art: coordinated and multiple views in exploratory visualizations. In: CMV'07 Fifth International Conference on Coordinated and Multiple Views in Exploratory Visualization, Zurich, Switzerland, pp 61–71

Steinberg AN, White F, Bowman CL (1998) Revisions to the JDL model. In: Proceedings of the Joint NATO/IRIS Conference, Quebec City, QC, Canada

Tang J, Cebrian M, Giacobe N, Kim H-W, Kim T, Wickert D (2011) Reflecting on the DARPA red balloon challenge. Trans ACM 54(4):78–85

Thomas JJ, Cook KA (2005a) Illuminating the path: the research and development agenda for visual analytics. IEEE Computer Society, Washington, DC

Thomas JJ, Cook KA (2005) The science of analytical reasoning. In: Illuminating the path: the research and development agenda for visual analytics. National Visualization and Analytics Center

Wang BM, Woodruff A, Kuchinsky A (2000) Guidelines for using multiple views in information visualization. In: Proceedings of the Working Conference on Advanced Visual Interfaces. ACM, New York, NY, pp 110–119

Chapter 12
Abductive Inferencing for Integrating Information from Human and Robotic Sources

John R. Josephson

Abstract Abductive inference (best-explanation reasoning) is a useful conceptual framework for analyzing and implementing the inferencing needed to integrate information from human and robotic sources. Inferencing proceeds from reports, to explanations for these reports, given in terms of hypothesized real-world entities and the processes by which the entities lead to the reports. Reports from humans and robotic sources are subject to different kinds of corruption, so they require different treatment as sources of evidence. The best explanation for a certain report might be that it presents a reliable statement that results from a chain of causality from the events reported, to their effects on human or robotic senses, and from there through transduction, processing, and reporting. Confidence in this explanation will be undercut by evidence supporting a rival explanation, such as one involving error or intended deception.

12.1 Multisource Fusion Analyzed as Abduction

12.1.1 Shared Ontology

What ties together human and robotic sources is that they all participate in, and report about, the same world external to those sources. Accordingly, successful fusion of information from disparate sources requires information to be interpreted into representations intended to refer to that external world. This requires points of contact in the ontologies that are used to interpret information from the different kinds sources so that it becomes possible to identify the entities that are used to interpret reports from the different kinds of sources as referring to the same entities in a world external to the sources. Sources of either type might report the presence of an entity, of a certain type and characteristics, at a certain place and time. Human individuals, vehicles, terrain features, and human structures are physical entities

J.R. Josephson (✉)
Computer Science and Engineering Department, The Ohio State University, Columbus, OH, USA
e-mail: jj@cse.ohio-state.edu

G. Rogova, P. Scott (eds.), *Fusion Methodologies in Crisis Management*,
DOI 10.1007/978-3-319-22527-2_12

that can be expected to be common referents, along with certain physical events such as explosions and radio transmissions. Besides agreement on types of entities, successful fusion requires agreement on (or inter-translatability of) reference frames and units of measurement for time and space.

12.1.2 Abductive Inference

Abductive inference, also called "abduction" (Peirce 1955), "abductive reasoning" (Walton 2004), "inference to the best explanation" (Harman 1965; Lipton 2004), and "retroduction" (Hanson 1958), is a useful conceptual framework for analyzing the information processing that takes reports as input, and produces, as output, estimates of objective states of affairs. The key idea is that objective states (including entities, events, relations) are critical elements of the causes for the reports, and so reports may be *explained by* hypotheses about those objective states.

For example, suppose an explosion causes acoustic disturbances, which then propagate, and impact upon a sensor, where the signal is transduced, processed by computations, and causes a report of an acoustic event to be sent to a fusion system. Suppose the same explosion causes several sensors to report. Then the information from these reports can be fused by inferring the best explanation for the reports. The occurrence of the reports, and the reported timing, intensity, and other event characteristics, is explained by hypothesizing an objective event, an explosion, taking place at a certain time and location, with certain characteristics, that accounts for the reports, because of the chain of causality connecting the explosion with the reports. The hypothesis is supported by evidence that it is the best explanation for the reports. Alternative hypotheses include explosive events with different locations and times of occurrence, multiple explosive events, non-explosive events that cause acoustic disturbances such as collisions, reporting errors with various causes such as hypersensitivity, processing errors, communication errors, and miscalibrations, and also intentional deception by tampering with sensors or injecting false signals. One hypothesis is a better explanation than another in light of such considerations as explanatory power, internal consistency, consistency with background information, plausibility, and simplicity.

Similarly, suppose an explosion causes acoustic disturbances, which are propagated to human observers, and cause human reports to be sent, although in this case the time estimates in the reports will be much less precise. Again, the occurrence and contents of the reports may be explained by hypothesizing one or more explosive events as causes of the reports. As before, alternative hypotheses include explosive events with different locations and times of occurrence, multiple explosive events, and non-explosive events that cause acoustic disturbances. However, for human reports, alternative hypotheses include deliberate deception by the source, as well as honest reports that are based on poor evidence or perceptual mistakes. Abstractly, the logic is the same for both robotic and human sources, although the details will be different, especially with regard to alternative hypotheses positing various kinds of error or deception. In both cases, the evidential basis for a fused estimate, its logical justification, will take the form of an abductive argument that this estimate is the best explanation for the occurrence and contents of the reports, in contrast with rivals.

Likewise, fusion that takes both human and robotic reports of acoustic events as inputs can be usefully analyzed in terms of abductive inferencing. The fused estimate will be supported by its ability to account for the reports, in contrast with the ability of rival hypotheses to account for them. A hypothesis about the occurrence of an explosion at a certain time and location might offer to explain both human and robotic reports. Thus we see that abductive inferencing is a useful way to describe the evidential basis for the conclusions, or outputs, of fusion processing, although such inferencing might be entirely implicit in the actual processing.

12.1.3 Human, Human-Interpreted, and Robotic Sources

As illustrated in Fig. 12.1, the story is a bit more complicated than simply human verses robotic sources. Information from many kinds of non-biological sensors is commonly processed by humans before it becomes available for fusion with other sources. For example, humans interpret imagery from video cameras and write reports, satellite imagery is interpreted by human analysts, and so on. Moreover, even reports based on human senses may rely on physics-based augmentations of human senses, such as infrared viewers, sound amplification, binoculars, or even eyeglasses. This complicates the story about potential sources of corruption, although it does not fundamentally change the analysis of the evidential basis for the outputs of fusion processing. It still makes sense to view the justification for fusion outputs as best explanations for the occurrence and contents of fusion inputs.

Figure 12.1 also illustrates the relationship between the causal processes connecting events with reports, the explanations of reports in terms of events and causal processes, inferencing, and the evidential support for outputs of fusion processing.

12.1.4 Abductive Arguments

Charles Sanders Peirce (1839–1914) contended that there occurs in science and in everyday life a distinctive pattern of reasoning wherein explanatory hypotheses are formed and accepted, calling that kind of reasoning "abduction" (Peirce 1955). Abductive arguments are ubiquitous at or near the surface of typical arguments offered in science, situation analysis, and ordinary life, and may be considered part of commonsense logic.

An abductive argument has a pattern approximately as follows (Josephson and Josephson 1996):

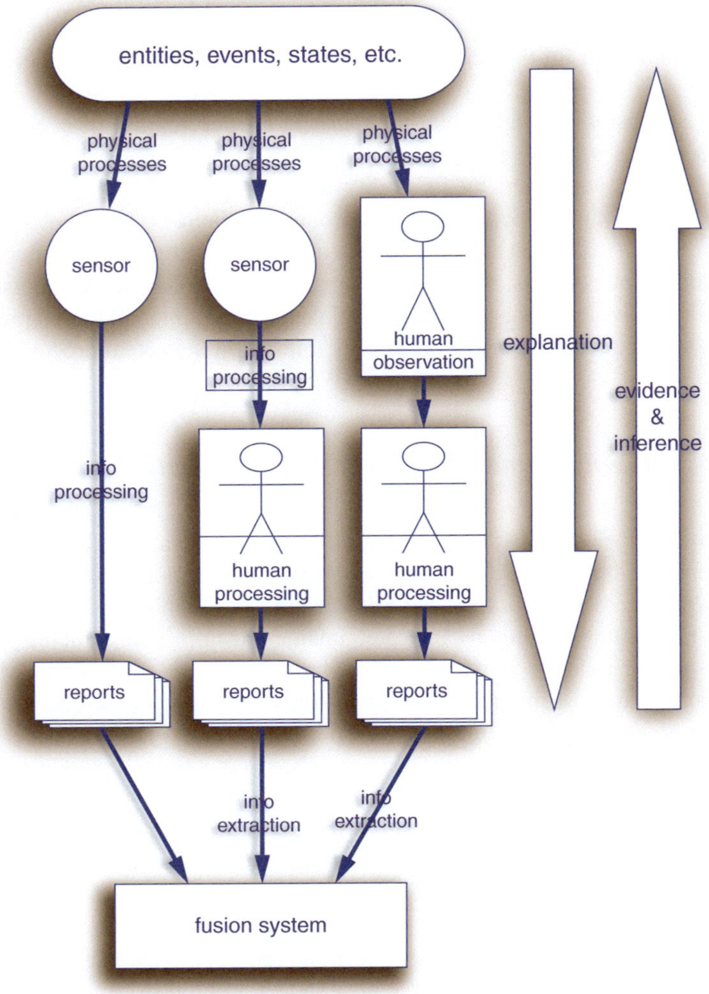

Fig. 12.1 Causal processes, explanations, and evidence

D is a collection of data (facts, observations, givens).
Hypothesis *H* explains *D* (would, if true, explain *D*).
No other hypothesis explains *D* as well as *H* does.

Therefore, *H* is *probably* correct.

Note that the conclusion is justified, not simply as a possible explanation, but as the *best explanation* in contrast with alternatives. The strength of the conclusion *H*, the force of the "probably" in the conclusion statement, reasonably depends on the following considerations:

- How decisively the leading hypothesis surpasses the alternatives
- How good this hypothesis is by itself, independently of considering the alternatives
- How thorough was the search for alternative explanations
- Confidence in the accuracy of the data (but "bad data" or "noise" can be considered to be an alternative explanation)

Besides confidence in its correctness, willingness to accept a conclusion also reasonably depends on pragmatic considerations, including:

- How strong is the need is to come to a conclusion at all, especially considering the possibility of gathering further evidence before deciding
- The costs of being wrong and the benefits of being right

Failure of predictions counts as evidence against a hypothesis and so tends to improve the chances of other hypotheses emerging as best. Failure of predictions may improve the margin of decisiveness by which the best explanation surpasses the failing alternatives. Thus, abductive inferences are capable of turning negative evidence against some hypotheses into positive evidence for alternative hypotheses.

Besides the failure of predictions, several other factors may enter into the evaluation of a hypothesis, either in isolation or in contrast with rivals. These factors include internal consistency, plausibility (including precedentedness and consistency with background knowledge), likelihood (prior, after local-match evaluation, and all things considered), simplicity, explanatory power, and specificity.

It is important to distinguish abduction, considered statically, as a pattern of argumentation or justification, from the dynamic inferential processes whereby explanatory hypotheses are generated and evaluated. It is also important to recognize that best explanations are almost always composite hypotheses or "theories" comprised of parts. Coherent "scenes" and "situations" are composite hypotheses. Abductive arguments provide justifications for the conclusions or outputs of "abductive processes," which in general comprise generation and evaluation of explanatory hypotheses, and decisions about which hypotheses to accept.

12.2 Explanatory Hypotheses and Error Models

12.2.1 Source Credibility

It can be readily seen that estimations of source credibility rely on abductive inferences. The simplest explanation for why fact *F* is reported by some source *S* is commonly that *F* is known by *S*, and that *S* truthfully reports his or her knowledge of *F*. To say that *F* is *known* by *S*, implies more than simply that *S* believes *F*, and that *F* happens to be true. It also implies that *S* has good reasons to believe *F*, i.e., that *S* is *justified* in believing *F*, according to the classical analysis of knowledge from Plato according to which knowledge is justified, true belief (Plato). Alternative explanations are that the source does not actually believe *F*, or believes *F* but is not justified in believing *F* because he or she lacks good evidence for *F*.

Let us use the term *canonical explanation* to refer to the explanation, for an item of the form "fact *F* is reported by *S*," that *F* is known by *S*, and that *S* truthfully reports his or her knowledge of *F*. According to the Platonic account of knowledge, acceptance of the canonical explanation commits one to accept the proposition that *F* is true (and thus to accept *F*).

Harman (1965) gave an account of law court testimony, which argues that when we infer that a witness is telling the truth, we are using best-explanation reasoning. According to Harman our inference goes as follows:

(a) We infer that he says what he does because he believes it.
(b) We infer that he believes what he does because he actually did witness the situation which he describes.

Our confidence in the testimony is based on our conclusions about the most plausible explanation for that testimony. Our confidence fails if we come to think that there is some other plausible explanation for his testimony—for example, that he stands to gain from our believing him.

It is clear that Harman's account generalizes to cover robotic as well as human sources of "testimony." It is reasonable to believe the facts attested to only if their truth is part of, or implied by, the best explanation for the occurrence and contents of the reports. To the degree that the evidence supports alternative explanations involving error or deception, confidence that these are indeed facts is reduced or removed.

12.2.2 Veracity, Competence, Access, and Objectivity

Schum and Morris (2007) have recently given a detailed analysis of proper estimates of the credibility of human sources of intelligence. They take pains to distinguish between the competence and the credibility of sources in regard to specific allegations. Overall *credibility* should be assessed by considering both the *veracity* (truthfulness), and the *competence* of a source, where competence includes both *observational sensitivity* and *objectivity*. We may generalize "observational sensitivity" to include all considerations regarding the *access* of sources to good information about the subjects reported, including whether reports are based on direct observation or hearsay, and whether observing conditions were adequate for reliable observation. We may generalize "objectivity" to include all considerations regarding distortion of interpretations, including whether a source is biased by loyalties, ideology, or prejudice, and whether a source has poor memory, perhaps as a result of Alzheimer's disease, or from hindsight influenced by others.

Combining the analysis of Schum and Morris with that of Harman, we consider whether a source really believes the facts attested to, and consider alternative hypotheses such as that the source may have colluded with others on the story, intends to provide misinformation, or is perhaps willing to say anything they think we want to hear. As alternative hypotheses garner evidential support, the hypothesis of truthful reporting is undercut. Support for these alternative hypotheses might come from the demeanor of the source, evidence of prior misconduct, or evidence of collusion, among other things.

Even if the source believes the facts attested to, the source may not be justified in believing the facts, given the evidence that was available to the source. The source may have lacked sufficient access, or have lacked objectivity in processing the available evidence. Here again, as alternative hypotheses find evidential support, the hypothesis of competence is undercut. Support for these alternative hypotheses might come from evidence that the source relied on hearsay, was not wearing their eyeglasses at the time, or was biased by cultural prejudice.

Considerations of observational sensitivity (i.e., access), objectivity, and veracity also apply to humans in the causal chain between physics-based sensors and reports. Evidence of bias, for example, supports alternative explanations for why the contents of a report are as they are, and undercuts confidence in the facts attested to by the report.

The credibility of reports from robotic sources can be analyzed analogously to those from human sources. Robotic reports might be unreliable because of failures of observational sensitivity or objectivity. Perhaps the events reported were a great distance away from a sensor, or weather conditions were poor, so that observational sensitivity was compromised. A sensor might exhibit a classification bias because of poor design or biased training. The analog of untruthfulness for robotic sensors is false reporting due to tampering, injection of false reports, or sensor construction compromised by negligent disregard for reliability.

12.3 Enhancement by Higher Level Fusion

As illustrated by Figs. 12.2 and 12.3, low-level fusion of information from "hard" sources (robotic sources or robotic sensors with competent human-based processing) and "soft" sources (humans) can be enhanced by fusion at higher levels. "Low-level" fusion refers here to processing that integrates information to arrive at estimates regarding localized physical objects and events (JDL level-1). "Higher level" fusion here refers to processing that integrates information to arrive at estimates of an agent's plans or intentions (JDL level-3). (JDL level-2 is not considered here in the interest of brevity.)

As shown in Fig. 12.2, hypotheses about localized physical objects and events might be supported by reports from hard and soft sources. In turn, hypotheses about plans and intentions might be supported by evidence from localized physical objects and events, but they might also be supported directly by reports from soft sources.

Returning to the acoustic example of Sect. 12.1.2, we imagine a system in which the occurrence of explosions of various characteristics, at various times and places, are explained by using hypotheses about the intentions and plans of various parties,

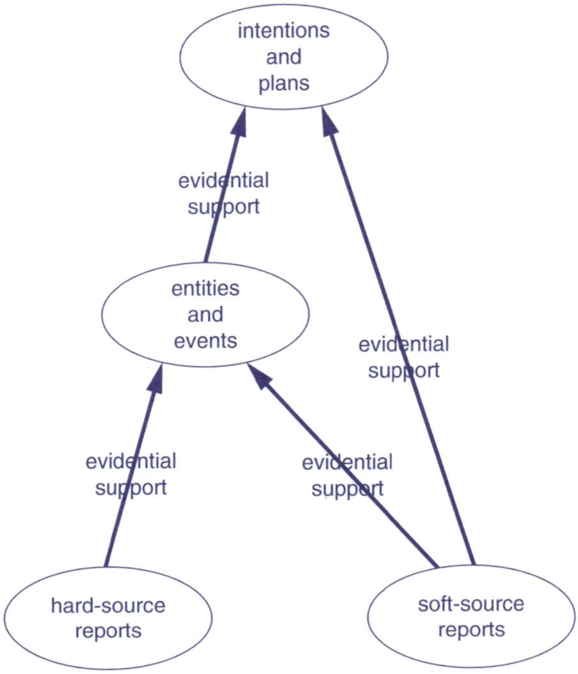

Fig. 12.2 Support for lower and higher level hypotheses

including friendly forces. Some explosions might be small-arms fire, some IED attacks, some friendly fire, and some celebrations.

In general, higher level hypotheses may provide predictions by which they can be tested against the stream of events. Failed predictions provide evidence against a hypothesis, in proportion to the strength of the predictions. Predictions lead immediately to questions about the occurrence of the predicted events. Thus a hypothesis that an insurgent group is intentionally targeting markets frequented by foreigners, leads to the prediction that the pattern is likely to continue for at least a short while, perhaps until the plan achieves its political objectives. Predicting that the pattern will continue may lead to specific predictions about the likelihoods of specific types of explosive events, in certain specific places, over certain specific time periods. These specific predictions lead immediately to questions regarding the fulfillment of the predictions, which may guide the questioning of human sources, or the allocation of sensors.

Predictions from confident higher level hypotheses might be actionable for operations, but they might also aid low-level interpretation by suggesting hypotheses (car bomb again), and setting up quick interpretations (car bomb!), since predicted events can usually be best explained by the hypotheses that gave rise to them.

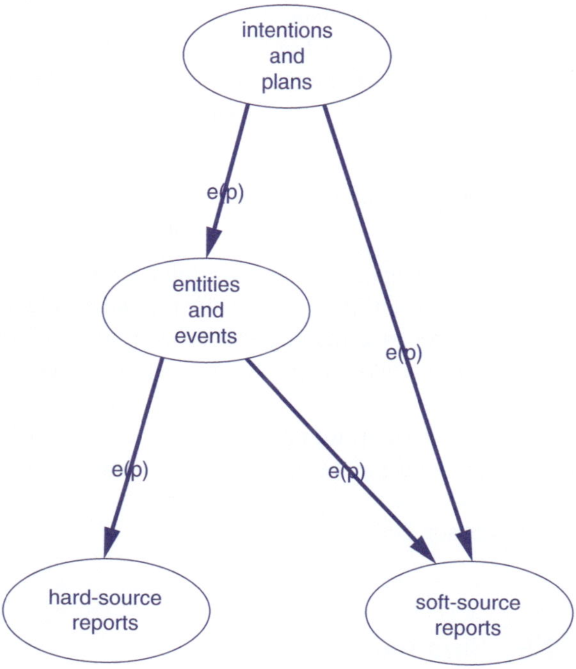

Fig. 12.3 Explanations and predictions

The generic information needs of higher-level processing include information needed for:

- Scoring hypotheses (e.g., by matching against predictions)
- Discriminating among rival hypotheses, analogous to "differential diagnosis" in medicine

These information needs may lead to questions about lower-level phenomena, which can be used to guide the collection of further information. Providing guidance to collection is a kind of adaptation of the fusion system, and thus may be considered to be a kind of "level-4 fusion."

Confident higher level hypotheses may lead to predictions that cast doubt on lower-level conclusions. For example, an explosion might be classified in a way that does not fit the prevailing pattern, so the classification might be doubted, which could lead to reconsideration and revision. Alternatively, the confidence of the classification might be downgraded so that no other higher level hypotheses rely too strongly on it.

Thus we see that higher level processing can enhance low-level processing in several ways, including by providing low-level processing with control, suggestions, and evidence.

Figure 12.3 schematizes how the evidential support for hypotheses at the two levels results from explanatory relations, that enable abductive inferencing, and predictive relations that may provide disconfirming evidence for hypotheses at their tails, and sometimes provide "circumstantial evidence" for hypotheses at their heads. "e (p)" in Fig. 12.3 stands for an "explanation" relation, possibly accompanied by some strength of "prediction" relation.

12.4 Implementation

Processing strategies for abductive inferencing can be investigated scientifically by implementing such strategies in software and testing their performance. Researchers in AI have been investigating abductive inferencing for many years (e.g., Josephson and Josephson 1996; Peng and Reggia 1990; Hobbs et al. 1993), using a variety of formalisms.

Prototype systems for military information fusion based on abductive inferencing have recently been developed that have shown some degree of success, for level-1 fusion as well as for higher levels (e.g., Pantaleev and Josephson 2006; Bharathan and Josephson 2006).

12.5 Summary

We have seen how abductive inferencing provides an analysis of the evidential basis for the outputs of fusion processing. Fusion outputs are logically justified as best explanations, in contrast with rivals, for the occurrence and contents of fusion inputs. This analysis applies equally well to fusion from robotic and human sources, as well as to fusion that makes use of all three of human, robotic, and human-interpreted hard sources.

The hypotheses that are used to explain reports from any of the sources refer to external events and states, and, at least implicitly, refer to possible causal processes connecting with the reports. Explanations climb chains of hypotheses from effects to causes.

Source credibility can be assessed analogously for human and robotic sources. Source assessment uses abductive arguments that consider rival hypotheses regarding the access, objectivity and veracity of the sources. The canonical explanation, that what is reported is known and reported correctly, is undercut by any evidence for rival hypotheses, such as those that challenge the claim to objectivity.

Higher level processing can be used to enhance low-level processing in several ways, including by providing low-level processing with control, hypothesis suggestions, and circumstantial evidence.

Besides the analysis of evidence and credibility in fusion processing, abductive inference may provide useful computational models.

Acknowledgements This research was supported through participation in the Advanced Decision Architectures Collaborative Technology Alliance sponsored by the US Army Research Laboratory under Cooperative Agreement DAAD19-01-2-0009. The views and conclusions contained in this document are those of the author and should not be interpreted as representing the official policies, either expressed or implied, of the Army Research Laboratory, Defense Department or the US Government.

References

Bharathan V, Josephson J (2006) An abductive framework for level one information fusion. In: Fusion 2006: The 9th International Conference on Information Fusion, Florence, Italy

Hanson NR (1958) Patterns of discovery. Cambridge University Press, London

Harman G (1965) The inference to the best explanation. Philos Rev 74:88–95

Hobbs JR, Stickel M, Appelt D, Martin P (1993) Interpretation as abduction. Artif Intell J 63:69–142

Josephson JR, Josephson SG (eds) (1996) Abductive inference: computation, Philosophy, Technology. Cambridge University Press, New York, NY

Lipton P (2004) Inference to the best explanation, 2nd edn. Routledge, New York, NY

Pantaleev A, Josephson J (2006) Higher-level fusion for military operations based on abductive inference: proof of principle. In: Conference on Multisensor, multisource information fusion: architectures, algorithms, and applications, part of The International Society for Optical Engineering (SPIE) Defense and Security Symposium, Orlando, FL

Peirce CS (1955) Abduction and induction. In: Buchler J (ed) Philosophical writings of Peirce. Dover, New York, NY

Peng Y, Reggia JA (1990) Abductive inference models for diagnostic problem solving. Springer, New York, NY

Plato. Theaetetus.

Schum D, Morris J (2007) Assessing the competence and credibility of human sources of intelligence evidence: contributions from law and probability. Law Probabil Risk 6:247–274

Walton D (2004) Abductive reasoning. University of Alabama Press, Tuscaloosa, AL

Chapter 13
High-Level Fusion for Crisis Response Planning

Kathryn B. Laskey, Henrique C. Marques, and Paulo C.G. da Costa

Abstract Each year, natural and anthropogenic crises disrupt the lives of millions of people. Local, national, and international crisis response systems struggle to cope with urgent needs during and immediately after a crisis. The challenges multiply as population grows, density of urban areas increases, and coastal areas become more vulnerable to rising sea levels. A typical crisis scenario requires coordinating many diverse players, including local, national and international military, other governmental, and non-governmental organizations. Often, no single entity is in charge of the response, making coordination even more difficult. There is an urgent need for better ways to allocate resources, maintain situation awareness, and reallocate resources as the situation changes. Information fusion is vital to effective resource allocation and situation awareness. Some of the greatest inefficiencies stem from the inability to exchange information between systems designed for different purposes and operating under different ownership. Information integration and fusion are too often entirely manual. Greater automation could support more timely, better coordinated responses in situations where time is of the essence. The greatest need is not for low-level fusion of sensor reports to classify and track individual objects, but for high-level fusion to characterize complex situations and support planning of effective responses. This paper describes challenges of high-level fusion for crisis management, proposes a technical framework for addressing high-level fusion, and discusses how effectively addressing HLF challenges can improve efficiency of crisis response. We illustrate our ideas with a case study involving a humanitarian relief operation in a flooding scenario.

K.B. Laskey (✉) • P.C.G. da Costa
Department of Systems Engineering and Operations Research and C4I Center, George Mason University, Fairfax, VA 22030, USA
e-mail: klaskey@gmu.edu; pcosta@gmu.edu

H.C. Marques
Aeronautics Institute of Technology - ITA, São José dos Campos, SP 12228-800, Brazil
e-mail: hmarques@ita.br

G. Rogova, P. Scott (eds.), *Fusion Methodologies in Crisis Management*,
DOI 10.1007/978-3-319-22527-2_13

13.1 Introduction

Each year, tens of millions people are displaced by natural hazard events, violent conflict, or a combination of the two. Population growth, urban density, and climate change contribute to an increasing need for humanitarian assistance. Despite increases in aid resources provided by the international community, the unmet need continues to grow (Yonetani 2014). Rapid and effective response to crisis is vital to reducing loss of life and negative impacts on health and well-being. Unfortunately, crisis response by local and international communities has too often been characterized by inadequate coordination, poor information sharing, and failures of leadership (Humphries 2013). Suggestions have been made for addressing the shortfalls (e.g., Lewis and Lander 2011; Bajoria 2011; Martin et al. 2014; Seybolt 2000), and initiatives by the international community aim to improve responses to future crises. The Inter-Agency Standing Committee (IASC) (Inter-Agency Standing Committee 2013) was established by United Nations Resolution 46/182 in 1991 as a forum for coordination, policy development, and decision-making on humanitarian assistance. In 2005, the Humanitarian Reform process was established to improve effectiveness, timeliness, and predictability of humanitarian response (Street 2009). A key tenet of the reform process was improving coordination among humanitarian assistance partners. As a mechanism to improve coordination, the IASC established the Cluster Approach (Inter-Agency Standing Committee 2006). A cluster is a group of organizations within a given sector (e.g., health, water, shelter) of humanitarian response. In the cluster approach, the organizations in a cluster work collaboratively in a spirit of partnership to address issues in their sector. Effective collaboration among cluster organizations requires shared situation awareness, for which information fusion is an essential enabler.

Crisis planning is an ongoing activity that begins well in advance of a crisis with a continuous cycle of preparedness activities. As one example, Fig. 13.1 shows the National Preparedness Cycle (Federal Emergency Management Agency 2014a), which identifies the cycle of preparedness activities conducted by the US Federal Emergency Management Agency as part of its overall National Preparedness System. Similar preparedness plans have been developed at the international, national, and local levels by jurisdictions around the world. Preparedness activities include planning, organizing, equipping, training, exercising, evaluating, and taking corrective action, with the objective of being ready to provide agile, coordinated, effective response in the event of a crisis.

Once a crisis has occurred, a crisis-specific response must be conducted. This requires mobilizing crisis response assets, often at multiple levels and operating under different jurisdictions, to develop, plan, and execute a coordinated response. A generic sequence of crisis response activities is shown in Fig. 13.2. This cycle can be mapped to Boyd's OODA loop (Saucedo 2013): Observe (assess situation), Orient (identify/evaluate response options), Decide (develop response plan), and Act (execute plan). The sequencing is not rigid: activities may overlap and earlier activities may be revisited as required by the crisis situation.

Fig. 13.1 United States National preparedness cycle

Fig. 13.2 Crisis response activities

The first step is to assess the situation as it evolves in order to support a timely and effective response. Endsley (1995) defines situation awareness informally as "knowing what's going on," and formally as "the perception of the elements in the environment within a volume of time and space, the comprehension of their meaning, and the projection of their status in the near future" (Endsley 1995, p. 36). Situation awareness requires gathering and combining information from available sources to form a more complete picture than could be gained from each individual source alone. Organizations responsible for crisis management often establish fusion centers, where information is gathered and fused into a common operating picture that can be shared by responders. Crisis mapping tools such as FEMA's Situation Awareness Viewer for Emergency and Recovery System (SAVER^2) (Federal Emergency Management Agency 2014b) or the open source Ushahidi platform (Ushahidi 2008), provide a means to support situation understanding by visualizing the fused information.

The next steps are to identify and evaluate options for responding to the crisis. Response options include issuing warnings and instructions to the public, evacuating affected individuals, establishing public shelters, disseminating food, blankets and other necessities, search and rescue, medical response, law enforcement or military action, hazardous substance control and mitigation, and actions to ensure health and safety of responders. Assessment of the current and projected situation supports identification of needs and determination of response capabilities that are projected to be available. Once options are evaluated, they are assessed for feasibility and predicted effects. Based on this assessment, a set of responses is selected for implementation.

After responses are selected for implementation, a concrete plan is developed for implementing the selected set of responses. Planning involves determining which capabilities will be activated, the timing and staging of responses, and how the overall response will be coordinated. The specifics of the planning process depend on the type and nature of the crisis. People and resources may be put on alert. Resources may be pre-positioned. Emergency response or incident management teams may be deployed and activated.

The final step is plan execution. The famous von Moltke statement that no plan survives first contact (Hughes 1993) applies equally well to both military action and non-military crisis response. Newly arriving information can invalidate a plan and necessitate replanning. It is essential to account for uncertainty, to develop plans that are resilient to deviations from assumptions, to build in the capability to replan as the response operation evolves, and to provide a means to coordinate among responding organizations.

Figure 13.2 shows the planning and execution cycle of a single responder. Crises of any complexity require coordinated actions by many responders. Depending on the scale of the crisis, responders may span international, federal, state, tribal and local agencies, as well as non-governmental, private sector, and community organizations. Each of these organizations has distinct roles, responsibilities, and information sources. A more accurate depiction of the response process must explicitly provide for coordination among these players, as shown in Fig. 13.3.

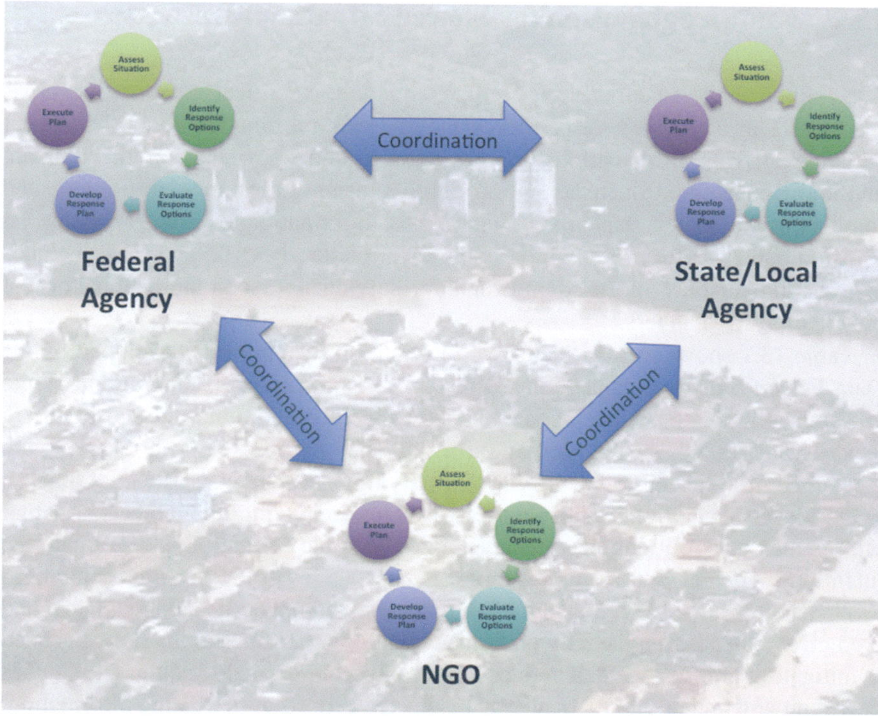

Fig. 13.3 Response must be coordinated among responding organizations

Especially in a fluid environment requiring coordination among diverse responders, automated support is critical to effective crisis response. Decision support tools can provide up-to-date geo-located displays of population needs and response asset availability in order to facilitate allocation of assets to needs. In addition, candidate asset allocations can be generated, evaluated for predicted effectiveness, and presented to responders to facilitate rapid response. These products can be updated as information becomes available, supporting rapid replacing as the situation changes. Existing tools can support some of this functionality, but many aspects still must be performed manually, leading to errors and inefficiencies.

The remainder of this chapter presents a framework for semi-automated situation assessment in support of planning for crisis situations. The chapter is organized as follows. Section 13.2 discusses information fusion for shared situation awareness. Section 13.3 describes the overall effects-based planning process and the role of situation assessment within the process. Section 13.4 illustrates the approach on a case study in a humanitarian relief operation for a flooding scenario. The paper ends with a conclusion and suggestions for further work.

13.2 Knowledge Fusion for Shared Situation Awareness

A comprehensive understanding of the situation is critical to effective crisis response. During a crisis, responders must integrate information about the current situation arriving from various sources, background knowledge of the domain, and knowledge about doctrine and procedures, using these inputs to produce and implement a plan for responding to the crisis. Information is brought together at a fusion center to form a common operational picture (COP). The COP includes a geospatial display of fused information, serving as a common repository of information for planners.

Table 13.1 shows a classification of information fusion levels developed by the Data Fusion Subpanel of the Joint Directors of Laboratories (Franklin and White 1987). The model has since been extended to include additional levels (Steinberg et al. 1999), but these four levels suffice for our purposes. Levels 0 and 1 are concerned with detecting, identifying, classifying, and tracking individual objects. Higher levels deal with establishing relationships among multiple objects, together with contextual information, to characterize a complex situation and evaluate the intentions of actors. Automation is more advanced at lower levels. At present, high-level fusion products, such as situation displays, automated decision support, and predictive analysis, rely heavily on human cognition. Greater automation is needed to mitigate cognitive overload, reduce errors, and achieve timely results. The COP of the future will update automatically by fusing new reports with existing information; identify and highlight critical missing information and information needs; discover conflicts and inconsistencies with semi-automated capability to repair errors; and provide explanations to users of inferences about the situation.

Syntactic translation alone is insufficient to support high-level fusion. Human reasoning relies on an implicit understanding of domain entities and the relationships among them. Automating this kind of reasoning requires formalizing this tacit knowledge so it can be retrieved and used when needed. Ontologies are the current paradigm for explicit, formal specification of domain knowledge. Most languages

Table 13.1 Joint Directors of Laboratories classification of fusion levels

Data fusion level	Association process	Estimation process	Product
L3 impact assessment	Evaluation (situation to actor's goals)	Game-theoretic interaction	Estimated situation utility
L2 situation assessment	Relationship (entity-to-entity)	Relation	Estimated situation state
L1 object assessment	Assignment (observation-to-entity)	Attributive state	Estimated entity state
L0 signal assessment	Assignment (observation-to-feature)	Detection	Estimated signal state

for expressing ontologies, such as the World Wide Web Consortium (W3C) Recommendation OWL (W3C 2009, 2004), do not have a standardized means to represent uncertainty or support uncertainty management. Clearly, reasoning under uncertainty is essential to effective crisis management. Inputs from low-level fusion systems come with uncertainty, as do the domain relationships used by analysts to draw inferences about a complex situation.

Semantically rich uncertainty representation and reasoning has been an active area of recent research (e.g., Costa 2005; Ding 2005; Predoiu and Stuckenschmidt 2008; Carvalho 2011). The PR-OWL Probabilistic Ontology Language (Costa 2005) and its updated version, PR-OWL 2 (Carvalho 2011), extends OWL with new classes, subclasses and properties that support uncertainty representation and reasoning. PR-OWL is written in OWL (W3C 2004) and provides a consistent framework for knowledge representation and reasoning in domains with uncertainty.

The mathematical basis for PR-OWL is multi-entity Bayesian networks (MEBN), which integrates first-order logic with Bayesian probability. MEBN provides a semantically rich language and reasoning formalism for representing and reasoning about a collection of interrelated entities, their attributes, and the relationships among them (Laskey 2008). Knowledge about the domain is represented as a collection of repeatable patterns, expressed as parameterized fragments of graphical probability models called MEBN fragments, or MFrags. An MFrag represents probabilistic relationships among uncertain attributes of and relationships among domain entities. MFrags are templates that can be instantiated and combined to form a graphical model encoding a joint probability distribution involving many random variables. A set of MFrags that collectively satisfies constraints ensuring a unique joint probability distribution is called a MEBN Theory (MTheory).

An MFrag can have three different types of nodes, as depicted in the example MFrag of Fig. 13.4. Resident nodes (yellow ovals in the figure) are random variables that form the core subject of an MFrag. Input nodes (gray trapezoids in the figure) are essentially "pointers" referencing resident nodes in other MFrags, providing a mechanism for connecting resident nodes between MFrags at instantiation time. Finally, context nodes (green pentagons in the figure) represent conditions that must be satisfied for the probability distribution of an MFrag to be valid. By allowing uncertainty on context nodes, MEBN can represent several types of sophisticated uncertainty patterns, such as relational uncertainty or existence uncertainty.

An MFrag can be seen as a "chunk" of domain knowledge encapsulating a pattern that can be instantiated as many times as needed to represent a specific situation. A Bayesian network (BN) formed by instantiating and combining MFrags needed to respond to a query is called a situation-specific Bayesian network, or SSBN. MEBN provides a composable modeling framework, which can be used to represent and reason about complex crisis situations.

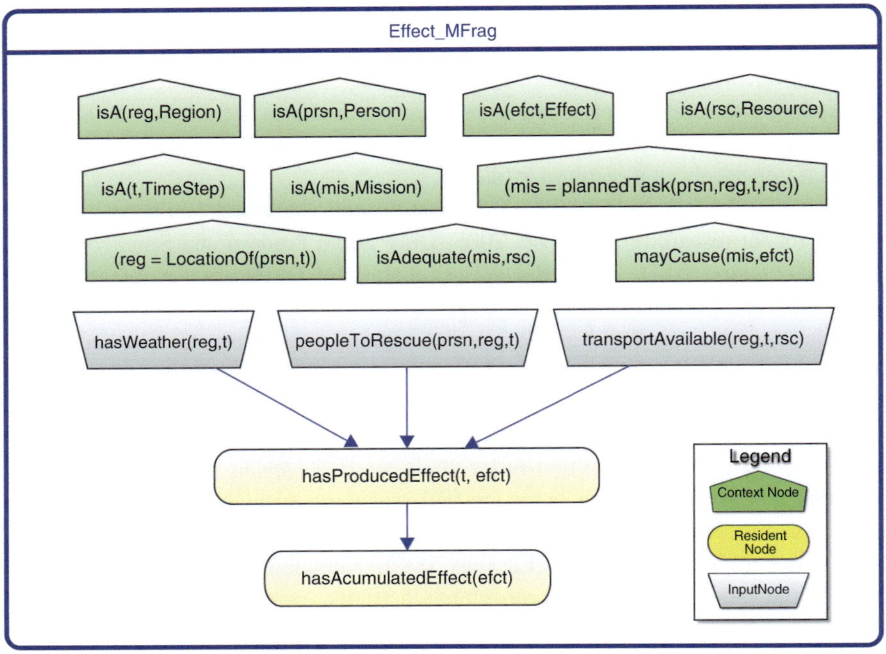

Fig. 13.4 Example of MFrag exemplifying effect modeling

13.3 Effects-Based Planning for Crisis Management

This section places information fusion into the context of planning a crisis response operation. We adopt the point of view of a military organization in the role of defense support of civil operations. We adopt the language of effects-based operations, a framework for planning and conducting combined civil-military operations aimed to achieve a given effect (Smith 2002). Effects-based planning aims to achieve objectives defined in terms of human behavior described according to multiple dimensions and at multiple levels. Achievement of these objectives is evaluated according to measures of effectiveness defined to capture the desired effects. Effects can occur simultaneously at multiple levels of a military operation, as well as in political, civil, and economic arenas. Effects can be both physical and psychological in nature, and can accrue over time. Effects may be interrelated, and can cascade into successions of indirect effects. Thus, it is necessary to consider uncertainties and to consider the likelihood of occurrence of different effects under the different options being considered.

Different types and levels of responder organizations (e.g., federal, state, local, NGO) have different objectives and require different information granularity. Therefore, planning for each organization has a different focus and addresses different problems. As an example, the operational level works with higher-level tasks, and therefore does not need to consider plan details such as the exact entity that will produce a desired effect, but only how the effect will achieve or interfere with the desired end state. Effects modeling must determine which tasks have to be executed to support the effect realization at each hierarchical level. Therefore, effects modeling can make planning more efficient by developing detailed plans at lower levels only for those tasks that are likely to lead to the desired effects. As another example, an NGO charged with providing medical care will have concerns different from a military organization charged with rescue and transport. Effects modeling can help to determine information exchange requirements between rescuers and the emergency medical centers to which injured persons are transported.

13.3.1 Planning Process Flow

Figure 13.5, adapted from Marques (2012), models the flow of documents, processes, decisions, and data to support the effects-based planning process. The figure is based on a naturalistic decision-making model adapted from the Integrated Model of Real-World Decision Making (Rasmussen 1993). The model is appropriate for scenarios involving time constraints and incomplete information.

The process starts with the receipt of orders/requests (1) that convey the command intent (2), which will be used as parameter by the Inference Engine (3). The orders/requests will also be added to the knowledge base, which is structured through the Task Probabilistic Ontology (i). Next, the analyst invokes the inference engine (4) to fuse reports from external sources (iii) and to pose a series of queries against the ontology (ii) to produce a scored task list (5). The Task Selection step uses the Scored Task List as input (6) for the analyst to decide (7) whether a new list is needed (8) or the current one is good enough for initiating the process of planning context generation (9). If the latter, then the resulting context is used to form the planning problem (10), which is submitted to a planning system (11). The generated plan (12) is then translated into the simulation tool's language (13). The simulation tool receives the input (14) and executes the simulation, generating output (15) to be viewed by the analyst (16). The analyst uses the simulation output to revise and refine the task list (7). This process continues until the analyst determines that a satisfactory plan has been developed. Figure 13.5 is color-coded as follows: orange denotes documents and process inputs/outputs; green denotes processes developed during our research that are either manual or implemented in a general programming language; red denotes processes based on existing tools; white denotes decisions; and blue denotes data from the simulation output.

13.3.2 Orders and Requests

The process starts by incorporating documents that describe the intent and the orders to be translated into actions (arrow 1 in Fig. 13.5). The documents that describe the orders and requests are established by doctrine and/or policy. Planners receive information on the desired effects, measures of performance (MOP), phases of the

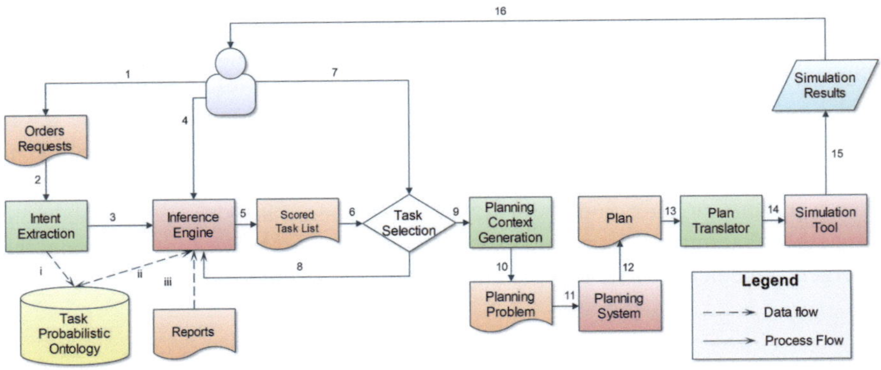

Fig. 13.5 Effects-based planning process flow

operation, activities, and a scenario description. Those documents are utilized in the process of intent extraction and then translated and included into the knowledge base (arrow 2 and dashed arrow i in Fig. 13.5).

13.3.3 Intent Extraction

Intent extraction is the process of translating orders and requests into the desired effects to be achieved by the planner. This involves analyzing the mission, the command intent, and the scenario where the operations will be conducted. We use a hierarchical planning approach (Ghallab et al. 2004; Yang 1997) because the hierarchical decomposition can be structured to correspond to task models of human performers, leading to plans that are acceptable to users (Goldman 2004). Orders are translated to the Battle Management Language (BML) format (Schade and Hieb 2007; Simulation Interoperability Standards Organization 2012) and the intent is extracted to map between the intent and the desired effect to be achieved through a set of tasks. The mapping occurs through the elicitation of a prioritized target list and other relevant information. At this stage of research, this process remains manual. For future work, we are devising an automated extraction implementation using the same ontologies currently used to support the task inference. The output of intent extraction is a prioritized list of effects to be achieved, represented according to the Task Probabilistic Ontology.

13.3.4 Task Probabilistic Ontology

The Task Probabilistic Ontology comprises a set of interconnected ontologies that together represent knowledge needed for information fusion and response planning in complex crisis scenarios (Marques et al. 2011). Figure 13.6 shows the six included ontologies. The *domain ontology* represents concepts and relations from the problem domain. This ontology is specific to the given crisis type. For our case study, it represents concepts relating to air rescue of citizens in the flooding scenario of our case study. The planning framework can be extended to different types of operation in different kinds of crisis scenario by extending the domain ontology. The *Battle Management Language (BML) ontology* represents concepts needed to represent shared intent. BML is an international standard defining an unambiguous language to command and control forces and equipment conducting military operations (Carey et al. 2001; Schade and Hieb 2006). Its aim is to support shared situational awareness and a shared common operational picture by providing a machine-interpretable representation of command intent to be used for human, robotic, and simulated forces. The *planner ontology* represents concepts and relationships needed to support automated planning. Figure 13.6 shows only tasks related to hierarchical planning; other planning concepts could be included if a different kind of planner were to be used. The *course of action (COA) ontology* represents concepts and relationships needed to model a course of action for our hierarchical planning system, which represents courses of action in terms of phases, activities, tasks, and effects. The *scenario ontology* represents aspects of the scenario needed for representing the situation and planning a response. The *PR-OWL ontology*, described in Sect. 13.2 above, contains concepts and relationships needed for representing uncertainty in semantically rich domains.

The Task Probabilistic Ontology provides support for information fusion and automated planning under uncertainty. As currently implemented, it includes knowledge sufficient for planning for the case study described in Sect. 13.4. Its modular nature provides a flexible framework for extending to new scenarios, new domains, and new planning methods. This can be achieved by integrating new concepts into the scenario ontology, the domain ontology, and the planning ontology, respectively. The Task Probabilistic Ontology is specified in OWL 2.0 (W3C 2009), a W3C recommendation. Figure 13.6 shows concepts from the ontology. The ontology also includes relationships among these concepts. Some examples of relationships include: *mayCause* represents uncertain cause and effect relationships between *Action_Type* and *Effect_Type*; *performs* represents a relationship in which an *Aircraft_Type* performs an *Action_Type*; and *hasActionType* represents that an *Action_Type* is associated with a *Mission_Type*.

13.3.5 Inference Engine, Reports, and Scored Task List

The inference engine is responsible for fusing reports from multiple sources to form a comprehensive picture of the situation, addressing relevant uncertainties, to support planning. This process is the locus of high-level information fusion in support of effects-based planning.

Fig. 13.6 Partial semantic structure of the Task Probabilistic Ontology

The analyst invokes the inference engine by a sequence of queries (dashed arrow (ii) in Fig. 13.5) to generate a list of tasks capable of achieving the effects identified during intent extraction (Sect. 13.3.3) for a given phase of the operation. This is achieved by posing a set of connected queries, resulting in a task list, where each task specifies Target Type, Resource Type, Configuration, Action Type, Effect Type, Desired Effect, Object Type and Object of Interest. These queries are described in the Description Logics Query Language (DL Query) which is implemented in the Protégé Tab using the Manchester syntax (Horridge and Patel-Schneider 2012; Stanford University 2009). If no resource capable of generating the Desired Effect is available, the result will be an empty list. The task list declares the possible and authorized tasks capable of producing a specific effect.

After generating the list of tasks, the probabilistic part of the ontology is invoked to fuse reports and assess the likelihood of success for the phase being queried. Reports come from multiple sources and provide current information on the situation. Report types will depend on the mission, the scenario, and the objects declared as selected for that specific phase. The allowable report types are represented in the probabilistic ontology, and are connected to the hypotheses of interest to planners and relevant to the likelihood of success. The probabilistic ontology includes a priori knowledge stored in local probability distributions for each RV in the model. In response to a query, the reasoner applies Bayesian inference to calculate updated marginal distributions for RVs of interest based on prior information and the available data at the time of the query (e.g., from sensors, reports, etc.). In other words, during a campaign, as new information accrues, this belief updating process is used to calculate the posterior probabilities. The result of belief updating represents the best knowledge possible to support new planned actions given the available information.

The task list generated by the DL reasoner is tagged with the posterior probability that its accumulated effect will achieve the desired effect for the phase being queried. The task list with success probability is sent to the analyst, who is responsible for determining whether the planned tasks are sufficient to achieve the desired effect.

13.3.6 Task Selection

The task list reasoner employs a heuristic algorithm to return a subset of the tasks identified by the inference engine as potentially capable of achieving the desired effect. The algorithm attempts to find a subset that contains a small number of tasks whose accumulated effect meets a threshold on the probability of mission success. The analyst examines the scored task list generated by the inference engine and the simulation output (see arrows (7) and (16) in Fig. 13.5). Based on this examination, the analyst can return to the inference engine to query for additional tasks (arrow (8) in Fig. 13.5), or send the current list of tasks to the planner and simulator. Upon completion of task selection and approval by the analyst, a list with the tasks that are most likely to produce the desired effect will be sent to the Planning Context Generation process (see arrow (9) in Fig. 13.5).

13.3.7 Planning Context Generation

Planning context generation is the process of establishing the problem context that will be used by the planning system. It is composed of two activities: planning problem definition and parsing. Planning problem definition declares the initial state, the initial task network, the set of operators, and the set of methods that will be sent to the planning system. The process has to establish the initial state based on the objects of interest, the location of the resources, and the actions to be developed. After task selection, each object and resource position will be retrieved in order to define the initial state, as well as the axioms that will provide the planning system with the necessary information. Parsing transforms the planning problem into the language used by the planning system.

Planning context generation is currently a manual process that occurs after the inference procedure identifies the actions that will generate the tasks submitted in the task list (see (9) and (10) in Fig. 13.5). Future work will include partial or full automation of this process.

13.3.8 Planning Problem

The planning problem document contains the initial state specification and the planning problem to be solved by the planning system. The present work uses the Simple Task Network formalism and the ADL language (Pednault 1989).

The document specification declares the initial state with all objects and relations necessary for the operators and methods also declared in the document. The planning problem will then be input to the planning system as indicated by arrow 11 in Fig. 13.5.

13.3.9 Planning System, Plan and Plan Translator

The planning system is responsible for generating a plan if one can be found. The ontology describes the planning system methods and operators, according to the problem-solving method the planning system utilizes. The planning system searches for possible plans (arrow 12 in Fig. 13.5), and receives only the necessary methods and operators, as well as the minimum set of objects to generate the initial state, thus reducing the search space. This is one of the benefits of using the previous task as input to inference.

The plan is sent to the translator (arrow 13 in Fig. 13.5), which will parse it using the simulation tool. Parsing depends on the simulator language, requiring a translator for each simulator. Finally, the plan is sent to the simulation tool (arrow 14 in Fig. 13.5). If a general planning language can be utilized in this process, such as the Planning Domain Definition Language (PDDL) or one of its derivatives (McDermott et al. 1998; Fox and Long 2003; Edelkamp and Hoffmann 2004; Gerevini and Long 2005; Younes and Littman 2004), the effort can be reduced.

13.3.10 Simulation Tool

The simulation tool of choice may be different depending on the scenario purpose and the analyst decision. That is, the same scenario can be simulated using diverse simulation tools. The purpose of the simulation is to support analyses of the probability of success of each planned task and to allow the analyst to understand phenomena about each mission of interest. The simulation produces results (arrow 15 in Fig. 13.5), which are sent to the analyst (arrow 16 in Fig. 13.5). On the basis of these results, the analyst can adjust the task selection (arrow 7 in Fig. 13.5), fine tuning the task list.

13.4 Case Study: Humanitarian Relief in a Flooding Scenario

The methodology is illustrated using a case study originally reported in (Marques 2012) involving defense support of a humanitarian relief operation. The case study is concerned with planning under uncertainty in a flood response scenario. The objective of the case study is to provide a set of requests to be planned in order to achieve the shared intent of rescuing people affected by a flooding event.

13.4.1 Description

The scenario unfolds in the city of Itajaí in Santa Catarina, Brazil. The location is prone to seasonal flooding, triggering rescue efforts nearly every winter. Major floods can be complex and unpredictable events. Water levels and flows in flood-prone areas can go from normal to extreme in a very short time frame. Timing and severity of flooding is often very difficult to predict. While some flood-prone areas have dedicated early-warning systems, citizens, and businesses in most areas must rely on reports from national and local meteorological services to assess the potential for flooding. The difficulty of prediction, coupled with inadequate warning systems and rapid water rise, can leave large numbers of people stranded and necessitate major rescue efforts. For example, in 2008, heavy rains in the state of Santa Catarina generated floods and landslides affecting eight million people. Severe and rapid flooding, with the river rising to 10 m (32.8 ft) above normal level, left many residents stranded. A massive relief operation was required, involving many different organizations and about 12,000 collaborators.

For this case study, we considered a flooding event with water levels of 1, 2, and 5 m above normal, with small groups of people needing to be evacuated. Figure 13.7 shows the Itajaí region with the victims' positions (green boxes below the river). The figure also shows the Navegantes airport in the north-eastern area of the picture next to the blue icons. The white boxes in the figure indicate locations A1–A10, and the icons indicate the objects' position.

The objective is to generate a plan to rescue all stranded people in the region. This can be cast as a resource allocation problem, in which rescue assets are assigned to locations. Following the process of Fig. 13.5, the inference system performs information fusion to support situation awareness and provide updated information on aspects of the situation. The results of inference can be displayed in the common operational picture to support responder situation awareness, as well as being provided to a planning system that generates a plan to be verified through simulation.

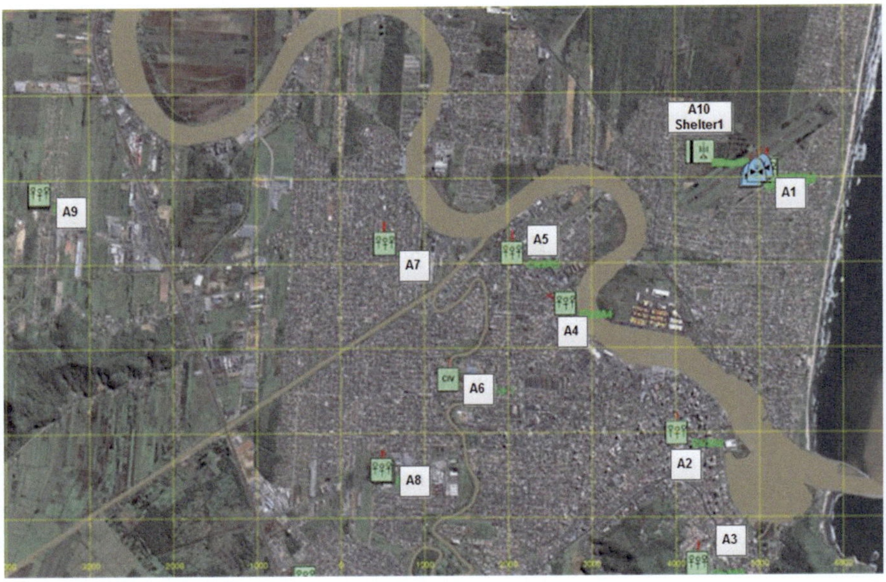

Fig. 13.7 The Itajaí scenario

Our simplified scenario considers only person transport/rescue missions. Thus, the orders/requests were either to rescue civilians from one of the locations shown in Fig. 13.7 and transport them to a shelter, or to transport responders from one location to another. The resources available for planning consisted of two configurations of the UH-60 helicopter, one with capacity ten persons and the other with capacity seven persons. It is relatively straightforward to extend the scenario and model to incorporate other missions such as transporting of medicine, clothes, water, etc., and other resources such as boats or ground vehicles. Scaling up to a much larger number of locations is also relatively straightforward, but will introduce tractability concerns. The present case study was intended as a proof of concept, and did not attempt to address scalability to larger scenarios.

13.4.2 Information Sources

Information fusion provides decision-makers with an understanding of key aspects of the situation to guide the crisis response. In our flooding scenario, key uncertainties to be addressed through information fusion include:

1. *Flooding*. Understanding the locations and extent of current flooding, as well as predictions of future flooding, is crucial to effective response. Information about current flooding can be obtained through weather reports, satellite imagery, 911 calls, tweets and other social media reports, and reports from responders on the

ground. Responder reports can come through the responder developing the plan, or through reports from other responder organizations. Here, interoperability is key: automated fusion is supported by common formats and semantics of reports. Flood predictions can come from models that take account of topography, frozen or saturated ground conditions, location and condition of defenses such as dams and levees, current river and/or tide levels, and tide stage, as well as predictions of future precipitation and storm surge.

2. *Rescue Targets*. The planners need to know the locations of stranded individuals, the number of individuals at each location, and their status (e.g., healthy with normal mobility, healthy with impaired mobility, mildly injured, severely injured, deceased). Relevant information sources include 911 calls, tweets, and reconnaissance reports. Report processing must deconflict duplicate reports that occur when people send rescue requests through multiple channels.
3. *Weather Conditions*. Severe weather conditions can hinder rescue efforts by reducing trafficability and increasing the danger to responders. Severe weather can also contribute to worsening of flood conditions. It is essential for response planners to have accurate reports on current and projected future weather conditions. Information on current conditions can come from weather radar, reports on the ground from responders, 911 calls and tweets from victims, and weather sensors in the field. Future weather conditions are informed by weather models from national and local weather agencies. Models exist at different resolution and granularity. Weather forecasts usually integrate the results of multiple weather models to arrive at a combined forecast.
4. *Resource Availability*. When multiple organizations are contributing to the crisis response, there is typically uncertainty about the availability of response capabilities under the control of other organizations. There is typically also uncertainty about assets under one's own control, due to hazardous conditions and unpredictable response times. Timely communication of plans and status reports is essential to prevent duplication of effort (e.g., sending multiple rescue assets to the same location) and to ensure the most comprehensive coverage possible. Interoperable reporting systems are essential to effective response coordination.

13.4.3 Crisis Planning Probabilistic Ontology

Ontologies represent domain knowledge and information needed to maintain situational awareness and to support planning. This requires formal representation of time, space, actions, effects, resources, and uncertainty over a dynamic future. Our case study made use of a probabilistic ontology expressed in PR-OWL to represent and reason with uncertainties inherent in the flood response problem.

Uncertainties associated with our crisis management case study are represented in the crisis planning probabilistic ontology. This MTheory was adapted from the effects-based military planning MTheory developed by Marques (2012), which was designed to support planning of military operations. For this work, we extracted portions of the military planning MTheory relevant to crisis management (e.g., we excluded portions dealing with use of weapons against an enemy). We then extended the MTheory to include domain knowledge and uncertainties specific to the humanitarian relief domain, such as availability of rescue transport resources and injury status of persons to be rescued. The MTheory was kept simple for illustrative purposes, but could be extended to a more comprehensive representation of domain uncertainties.

Figure 13.8 shows the graphical structure of the MTheory. Node types are depicted using the convention described in Sect. 13.2: green pentagons for context nodes, yellow ovals for resident nodes, and gray trapezoids for input nodes. The diagram was produced in the UnBBayes-MEBN probabilistic ontology tool (Costa et al. 2008).

The MTheory contains nine types of entity: *Persons* needing to be rescued; *Regions* where a person to be rescued may be located; *Resources* that can be assigned to rescue a stranded person; *Missions* of which our case study considers only *Rescue*; *Phases* of a mission; *COAs* or courses of action; *Effects* intended to result from a mission; *Reports* that provide evidence about some aspect of the situation; and *TimeSteps* that occur during execution of a mission. Instances of these entity types can be substituted for the parameters of the MFrags, thus allowing multiple instances of each random variable.

The crisis planing MTheory contains seven MFrags:

1. The *Effect MFrag* is the core MFrag specifying the factors that produce individual and accumulated effects. The *hasProducedEffect* random variable represents whether or not a task has produced the intended effect. The probability is higher when the weather is good and the people to be rescued are not injured or killed; and the effect can happen only if the assigned transport is available. The random variable *hasAccumulatedEffect* accumulates all the planned effects for the mission to determine the overall mission success.
2. The *Situation MFrag* defines uncertain random variables characterizing the situation, such as transport availability, the injury status of the people to be rescued, the locations of persons, and the weather in the region.
3. The *Phase MFrag* describes the probability of success of a phase of a course of action. It accumulates the probability of all the effects associated with the mission and the phase.
4. The *COA Context MFrag* represents factors associated with the course of action, such as the planned tasks; whether or not an effect is associated with a phase and a mission; whether or not a mission can cause an effect.
5. The *COA MFrag* describes the probability of success of an entire course of action, which may include multiple phases and missions.
6. The *Report MFrag* describes reports about the situation. For simplicity, we have included only one type of report (weather), although additional report types could easily be added.

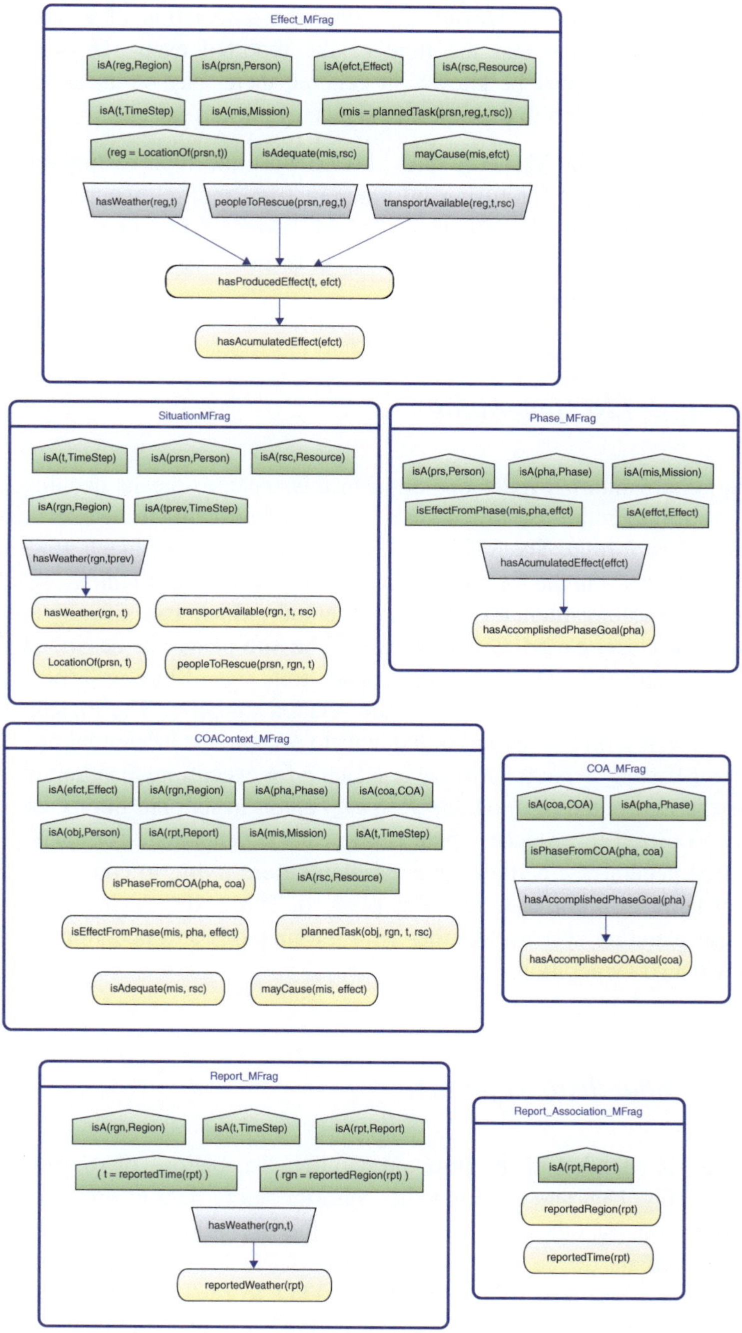

Fig. 13.8 The crisis planning probabilistic ontology

7. The *Report Association MFrag* represents the association of a report to the region reported upon and the time of the reported event. In our scenario, these associations may be known; in general, there may be uncertainty in the association of a report to the reported object, event, or time.

The crisis planning MTheory provides support for representing uncertainty about the situation and for fusing information from multiple sources to provide comprehensive situational awareness. The MTheory of Fig. 13.8 has been kept simple for purposes of this illustrative case study, but is comprehensive enough to illustrate the basic principles.

13.4.4 Task List Generation

Planning a humanitarian relief operation in this scenario means deciding how to respond to rescue requests, which are described as missions to be executed in the planner method format. The STN planner language (ADL, Pednault 1989) is used to describe the tasks in the format *Rescue_Person (person, destination)*. In addition, the planner ontology includes the class Method, which has an individual (i.e., instance of a class) called *Rescue_Person*. The sub-plans are described through the ontology and the individual *Air_Rescue* from class Mission_Type is related to the Method *Air_Rescue_Person* by the relation *executes*. The logical reasoner uses these classes, properties and individuals to execute the inferential process.

After a Mission_Type *Air_Rescue* is selected for execution, the planning system receives the associated Method *Air_Rescue_Person* in its plan library as the sub-plan to be utilized to produce a plan if one can be generated. The STN planning system first generates a partially ordered plan before trying to concatenate all the actions if more than one task has to be planned. Then, all tasks to be planned are sent to the planning system, together with the methods and operators needed for Planing Context Generation.

13.4.5 Probabilistic Inference

Uncertainty is encoded in the prior probability distributions of the probabilistic ontology of Fig. 13.8 and used together with incoming data to produce posterior probabilities on hypotheses of interest. In the scenario there are three small groups of civilians to be rescued, labeled as CIV1A1, CIV2A1, and CIV3A1. The resources allocated are Helicopter_FAB1111 to rescue CIV1A1 at time T0, Helicopter_FAB3333 to rescue CIV2A1 at time T1, and Helicopter_FAB2222 to rescue CIV3A1 at time T1. Figures 13.9, 13.10 and 13.11 show the output of the probabilistic reasoner (Costa et al. 2008) when evaluating the probability distribution for the accumulated effect of a rescue mission in Sector ALFA 1.

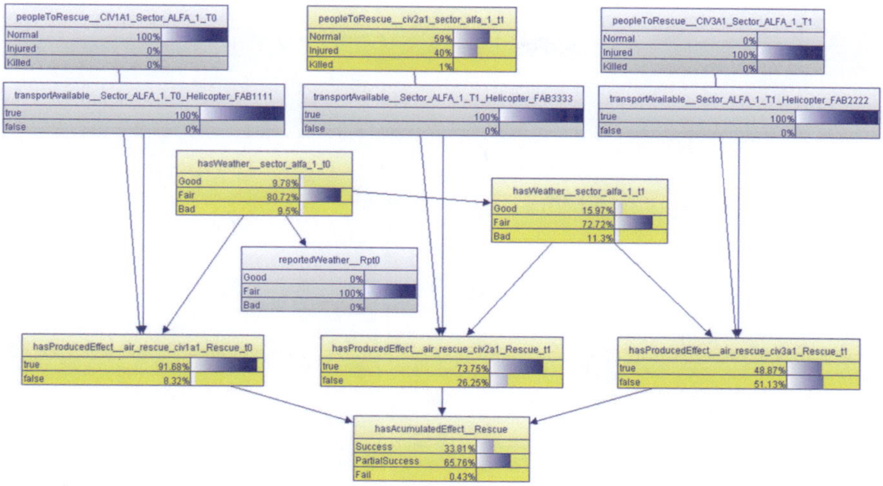

Fig. 13.9 The accumulated effect of a rescue mission with no evidence about group *CIV2A1*

The figures present three different situations, all for the same basic humanitarian relief scenario, but with different combinations of evidence items. Three possible outcomes are considered: *Success*, meaning all missions succeed; *PartialSuccess*, meaning at least one of the missions succeeds; and *Fail*, meaning none of the missions succeeds.

In the first situation (Fig. 13.9), weather reports indicate the weather is fair in Sector ALFA 1 at time T0 and the CIV2A1 group status is unknown. The weather at time T0 influences the probability distribution for the weather at time T1, even if we don't know anything about the weather forecast at that time, giving the model the ability to reason about the probability of success in Sector ALFA 1 at time T1. The accumulated probability of success of all rescue missions is nearly 34 %; the probability of all three missions failing is less than 0.5 %.

In the second situation (Fig. 13.10), weather reports indicate the weather is fair in Sector ALFA 1 at time T0 and the CIV2A1 group status is killed. The accumulated probability of success of all rescue missions has dropped to about 23 %, and the probability of all three missions failing has increased to about 1 %. The reasoner considers that the presence of killed people decreases the probability of success of a rescue mission, as stated in the prior probability tables. The belief update gives us the opportunity to consider new information, enhancing the analyst's ability to understand the new picture.

In the third situation (Fig. 13.11), weather reports indicate the weather is fair in Sector ALFA 1 at time T0, the weather is bad in Sector ALFA1 at time T1, and the CIV2A1 group status is killed. The accumulated probability of success of all rescue missions now is only about 19 %, and there is now a 1.3 % probability that all three missions fail. The reasoner considers that bad weather decreases the probability of success of a rescue mission, as stated in the prior probability tables. As new evidence

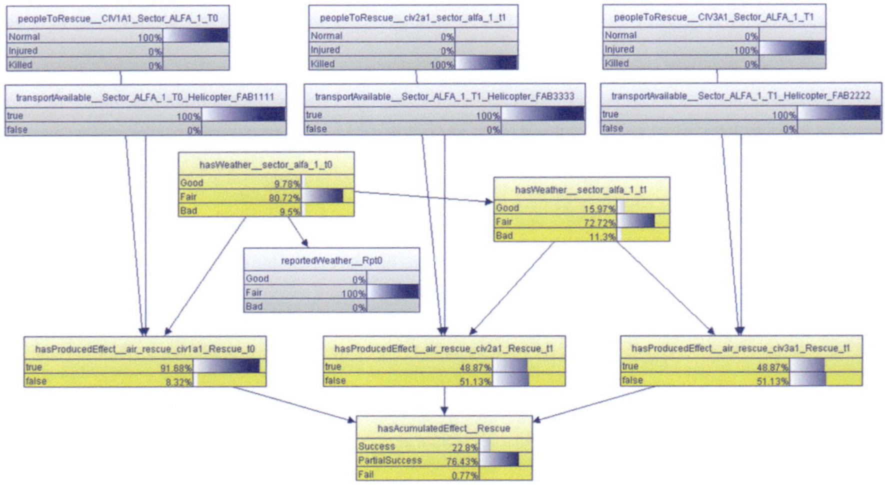

Fig. 13.10 The scenario probabilities given the evidence *CIV2A1? = killed* and *Report T0 A1? = fair*

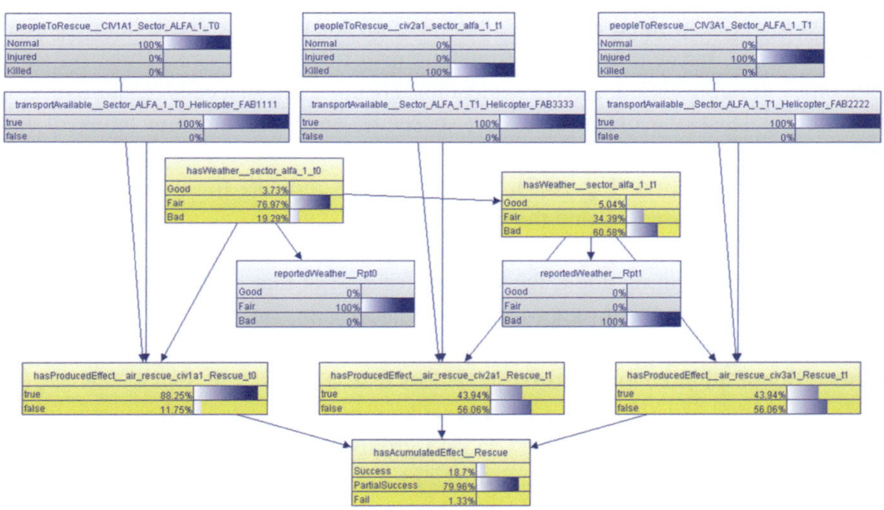

Fig. 13.11 The scenario probabilities given the evidence *CIV2A1? = killed* and *Report T0 A1? = fair* and *Report T1 A1? = bad*

accrues, the model adjusts the probability of success, supporting automation in task selection to support planning.

While the probability of at least partial success in all these scenarios is quite high, the probability that all missions succeed is low. The probability of success depends on scenario variables such as the weather and the availability of resources, as well as the defined success criteria. If the analyst considers the likelihood of success to be too low, he can consider missions that assign additional resources to the regions of interest. Of course, this must be balanced against making the resources unavailable for other missions.

13.4.6 Planning Context Generation

Once the tasks are defined, the inference engine can provide the information needed to support planning. Through a series of queries, it is possible to obtain the classes related to the domain of interest and also the object properties establishing the initial state. The end state is declared as a set of orders/requests, the same ones that initiated the inference process.

13.4.7 Planning Implementation

The planning system utilized was SHOP2 (Nau et al. 2003) in an Ubuntu distribution (Ubuntu 2015) running in VMWare Player (VMWare 2014). The implementation was written in steel bank common lisp (SBCL) (Newman 1999), which was convenient since SHOP2 is based on Lisp and SBCL is open source. SHOP2 is a partial order planner and the output is not optimal, but it has important features such as the ability to provide a visualization of the sequence of actions that will lead to the desired end state.

Figure 13.12 presents a small portion of the planner ontology, which works in conjunction with the domain ontology for this specific case study. Figure 13.13 pro-

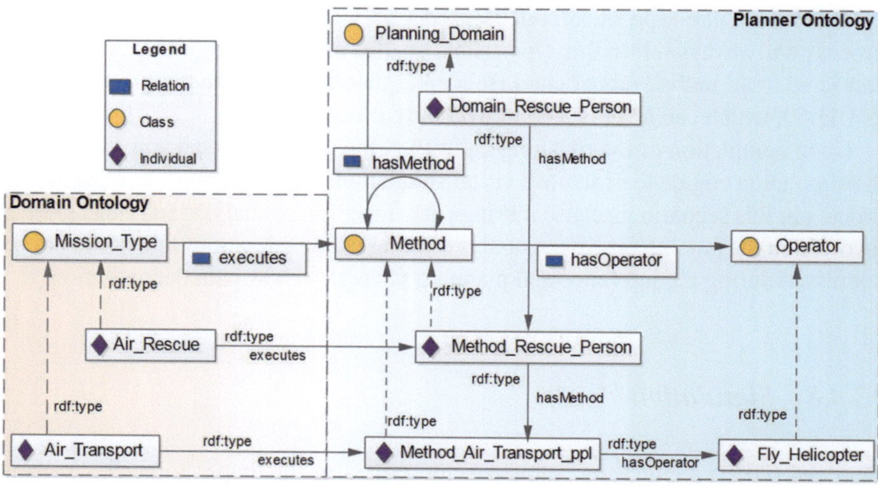

Fig. 13.12 Partial ontology showing the relations between the classes Mission_Type and method

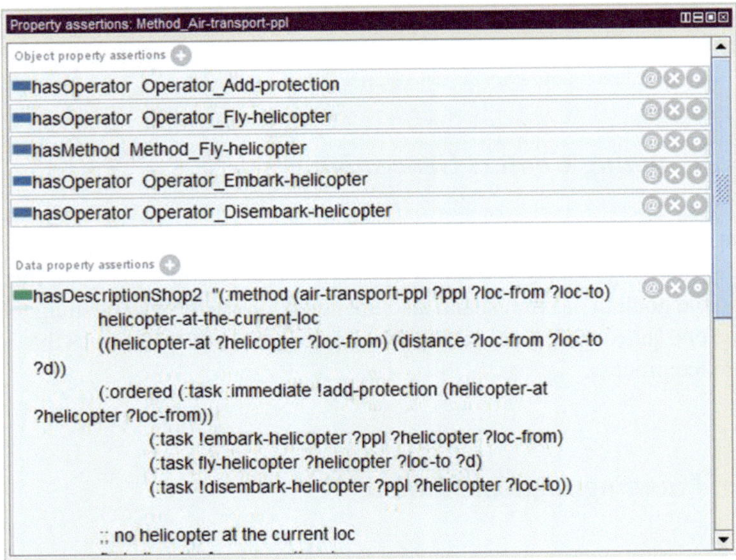

Fig. 13.13 Data and object property assertions for method *Method_Air_Transport_ppl*

vides the data and object property assertions of Method *Method_Air_Transport_ppl* with its description, in ADL language, that will be included in the planning problem file to be submitted to the planning system. The relation *executes* (blue rectangle depicted on the right of the domain ontology square) is used for mapping the Mission_Type to its associated planning method in STN formalism, which represents the mission to be executed in the real world. Reasoning is performed using the Protégé tool (Stanford University 2012) reasoner.

Since only one type of aircraft is used (the UH-60 helicopter), the inference process will always return this type. Thus, queries to the ontology will be basically restricted to all individuals of class Helicopter that have the type parameter defined as UH-60, which can be allocated individually if necessary.

Upon completion of the planning algorithm, the result is parsed and received by the simulation engine to generate a visual simulation. This is an important part of the process in this scenario, because it brings new insights for analysts and other people involved in the planing process, usually resulting in considerations that could not be addressed during the inference and planning phases, such as path deconfliction.

13.4.8 Simulation

Marques (2012) applied the framework described in this paper to generate plans and simulate their execution on the flood relief scenario. The MTheory used by Marques was designed for general military operations and was modified to address

Fig. 13.14 The simulation execution

the humanitarian relief domain. The MTheory of Fig. 13.8 further extends the humanitarian relief portion of this military operations ontology to address the flood relief scenario.

The simulation was conducted using VT MÄK VR-Forces 4.0.3 (VT MÄK 2012) with the High-Level Architecture 1.3 protocol for distributed simulation running in Windows 7. VR-Forces' simulation begins by loading a scenario file with the initial state. All entities have no plan at this stage. After parsing, all entity plans are injected and the simulation starts. The parsing implementation was written in C^{++} using the VR-Forces' API and translates the hierarchical plan output by SHOP2 planner into a conditional plan for each entity represented in the scenario as requested by VR-Forces engine.

The simulation was executed ten times with different departure times for the helicopters. In each run, all people where rescued in less than 20 min with only four helicopters in use for this small scenario. Figure 13.14 shows four different moments during the simulation. The first is the helicopter taking off from Navegantes airport. The second shows a helicopter arriving at location A6, where three civilians have to be rescued and four policemen have to be transported. Moment three shows two helicopters arriving at A10 (Shelter1), and the fourth moment shows civilians disembarking from the helicopter.

The generated plan is not optimal in terms of distance flown and helicopter allocation, but the result is considered reasonable, taking an acceptable amount of time to rescue all individuals. In this case, no conflicting actions were identified, which means that no adjustment to the defined plans was required.

13.5 Conclusion

This research established a knowledge representation and reasoning model to support automation in effects-based operations planning at the tactical level. At that level, the commander's intent has to be translated into specific tasks that have a probability of achieving the desired effect. Multi-source information fusion supports situation awareness and informs the automated planner. The inference model can reason about various aspects of interest, such as command intent, aspects of the scenario, the planning domain, and a probabilistic description of the problem structure.

The knowledge representation made use of a probabilistic ontology written in OWL (W3C 2004) and extended by the PR-OWL language (Costa 2005; Carvalho 2011), which we used to describe our Task Ontology. The traditional part of the ontology is responsible for task identification, while the probabilistic part is responsible for supporting inference about the probability of plan success. The probabilistic part was modeled in MEBN logic (Laskey 2008) using a COA MTheory developed for the research.

Depending on the scenario at given specific time, the planned task may produce different probabilities of achieving the desired effect. By accumulating the expected effects, the inference algorithm generates a task list that will be the base for a planning problem description. Then, only the planned tasks that will most likely produce the desired effect will be sent to the planning system, reducing the search space to find a solution to the planning problem.

The planning output generates a sequence of actions that can be translated into missions for a planning cell or to a simulator language in order to generate simulations and identify issues that are not currently possible to be inferred such as path deconfliction. Using simulation and task inference, the task list may be refined until a satisfactory result is achieved.

The approach was conceptually validated through a case study in humanitarian relief during a flooding event, in which a probabilistic ontology was able to identify correctly the tasks capable of producing a set of desired effects and support a planner in reasoning about the likelihood of plan success.

The present work fills a gap that existed in the automation of tactical planning for EBO scenarios. It is necessary to make inferences under uncertainty and severe time constraints that incorporate command intent and consider the many options available to decision-makers. Command and control scenarios are becoming increasingly complex. There may be an overwhelming number of actions that could reach a desired effect. Simply adding more analysts to address the exploding complexity is neither feasible nor effective. In such challenging situations, reducing the search space before using a planning system can significantly reduce the time to find a solution and increase the possibility to find a feasible one that is applicable to the problem.

The Task Ontology developed for this work can be extended in order to support other scenarios that have a similar planning process, such as operations in support of epidemics, earthquakes, or terrorist incidents. For example, in an epidemic scenario, the intent would be to focus on treating patients while ensuring geographical containment of the disease. Possible actions would be to transport medical teams, to establish a quarantine area, and to identify the root of the endemic.

The approach described here is the first to utilize probabilistic ontologies to support effects-based planning. It demonstrated the ability to use the results of higher-level fusion to inform a planner, and demonstrated a semi-automated process of interleaved planning and simulation, to allow planners to define a plan with a sufficiently high probability of success.

Acknowledgements The authors gratefully acknowledge VT MÄK for providing the VR-Forces simulation and visualization tools for research use, and thank Shou Matsumoto for UnBBayes support and software engineering.

References

Bajoria J (2011) Improving UN responses to humanitarian crises. UN Chron XLVIII(4). http://unchronicle.un.org/issue/7-billion-people-1-united-nationshandhands/

Carey S, Kleiner M, Hieb M, Brown R (2001) Standardizing battle management language—a vital move towards the army transformation. In: IEEE fall simulation interoperability workshop, Orlando

Carvalho R (2011) Probabilistic ontology: representation and modeling methodology. Ph.D. dissertation, Systems Engineering and Operations Research, George Mason University

Costa PC (2005) Bayesian semantics for the semantic web. Ph.D. dissertation, Information Technology, George Mason University

Costa PC, Ladeira M, Carvalho RN, Laskey KB, Santos LL, Matsumoto S (2008) A first-order Bayesian tool for probabilistic ontologies. In: FLAIRS conference, pp 631–636

Ding Z (2005) BayesOWL: a probabilistic framework for semantic web. Ph.D. dissertation, Science of Computation, University of Maryland

Edelkamp S, Hoffmann J (2004) PDDL2.2: the language for the classical part of IPC-4. Technical Report No. 195, Albert-Ludwigs-UniversitätFreiburg, Institut für Informatik. https://courses.cs.washington.edu/courses/cse473/06sp/pddl2.2.pdf

Endsley MR (1995) Toward a theory of situation awareness in dynamic systems. Hum Factors J Hum Factors Ergon Soc 37:32–64(33). doi:10.1518/001872095779049543. http://www.ingentaconnect.com/content/hfes/hf/1995/00000037/00000001/art00004

Federal Emergency Management Agency (2014) National preparedness cycle. https://www.fema.gov/national-preparedness-cycle. Accessed 30 Dec 2014

Federal Emergency Management Agency (2014) Private section fact sheet. http://www.fema.gov/pdf/privatesector/saver_factsheet.pdf. Accessed 30 Dec 2014

Fox M, Long D (2003) PDDL2.1: An extension to PDDL for expressing temporal planning domains. J Artif Intell Res 20(1):61–124

Franklin E, White J (1987) Data fusion lexicon. Technical report, Joint Directors of Laboratories, Technical Panel for C3, Data Fusion Sub-Panel, Naval Ocean Systems Center, San Diego

Gerevini A, Long D (2005) Plan constraints and preferences in PDDL3: the language of the fifth international planning competition. Technical Report, Department of Electronics for Automation, University of Brescia, Italy

Ghallab M, Nau D, Traverso P (2004) Automated planning theory and practice. Elsevier/Morgan Kaufmann, Amsterdam

Goldman RP (2004) Ontologies and planners a statement of interest. In: AAAI spring symposium. Technical report, vol 4, pp 103–105

Horridge M, Patel-Schneider P (2012) OWL 2 web ontology language Manchester syntax, 2nd edn. http://www.w3.org/2007/OWL/wiki/ManchesterSyntax. Accessed 05 Jan 2015
Hughes DJ (ed) (1993) Moltke on the art of war: selected writings. Presidio Press, New York
Humphries V (2013) Improving humanitarian coordination: common challenges and lessons learned from the cluster approach. J Humanit Assist. http://sites.tufts.edu/jha/archives/1976
Inter-Agency Standing Committee (2006) IASC guidance note on using the cluster approach to strengthen humanitarian response. Technical report. http://www.refworld.org/docid/460a8ccc2.html. Accessed 28 Dec 2014
Inter-Agency Standing Committee (2013) Official site. http://www.humanitarianinfo.org/iasc/. Accessed 28 Dec 2014
Laskey KB (2008) MEBN: A language for first-order Bayesian knowledge bases. Artif Intell 172(2–3):140–178
Lewis G, Lander B (2011) Only as strong as our weakest link: can the humanitarian system be collectively accountable to affected populations? Humanit Exch Mag 52. http://www.odihpn.org/humanitarian-exchangemagazine/issue-52/only-as-strong-as-our-weakest-link-can-the-humanitarian-system-be-collectivelyaccountable-to-affected-populations
Marques HC (2012) inference model with probabilistic ontologies to support automation in effects-based operations planning. Ph.D. thesis, Instituto Tecnológico de Aeronáutica, São José dos Campos
Marques HC, de Oliveira JMP, da Costa PCG (2011) Representing COA with probabilistic ontologies. In: Proceedings of the international command and control research and technology symposium
Martin SF, Weerasinghe S, Taylor A (eds) (2014) Humanitarian crises and migration: causes, consequences and responses. Routledge, New York
McDermott D, Ghallab M, Howe A, Knoblock C, Ram A, Veloso M, Weld D, Wilkins D (1998) PDDL—the planning domain definition language, version 1.2. Technical report, TR-98-003, Yale Center for Computational Vision and Control. http://homepages.inf.ed.ac.uk/mfourman/tools/propplan/pddl.pdf. Accessed 03 Jan 2015
Nau D, Ilghami O, Kuter U, Murdock JW, Wu D, Yaman F (2003) SHOP2: an HTN planning system. J Artif Intell Res 20:379–404
Newman W (1999) Steel bank common Lisp (SBCL). http://www.sbcl.org/. Accessed 2015-01-03
Pednault EPD (1989) ADL: exploring the middle ground between STRIPS and the situation calculus. In: Proceedings of the international conference on principles of knowledge representation and reasoning. Morgan Kaufmann Publishing, San Francisco, pp 324–332
Predoiu L, Stuckenschmidt H (2008) Probabilistic extensions of semantic web languages: a survey. In: Ma Z, Wang H (eds) The semantic web for knowledge and data management: technologies and practices. Idea Group, Hershey
Rasmussen J (1993) Deciding and doing: decision-making in natural contexts. In: Klein G et al (ed) Decision-making in action: models and methods. Ablex, Norwood
Saucedo AA (2013). Encyclopedia of military science. In OODA (Observe, Orient, Decide, and Act), vols 1–4. SAGE Publications Inc., Thousand Oaks, pp 1016–1019. Retrieved from http://knowledge.sagepub.com/view/encyclopedia-of-military-science/n335.xml?rskey=Pcof8j&row=2
Schade U, Hieb MR (2006) Formalizing battle management language: a grammar for specifying orders. In: IEEE spring simulation interoperability workshop
Schade U, Hieb MR (2007) Improving planning and replanning: using a formal grammar to automate processing of command and control information for decision support. Int C2 J 1(2):69–90
Seybolt TB (2000) Systemic network analysis of refugee relief: a business-like approach. Text. http://reliefweb.int/report/world/systemic-network-analysis-refugee-relief-business-approach
Simulation Interoperability Standards Organization (2012) Official site. http://www.sisostds.org/Home.aspx. Accessed 03 Jan 2015
Smith EA (2002) Effects based operations: applying network centric warfare in peace, crisis and war. CCRP, Alexandria

Stanford University (2009) Description logic query language: Dl query. http://protegewiki.stanford.edu/wiki/DLQueryTab. Accessed 02 Jan 2015

Stanford University (2012) Protégé project. http://protege.stanford.edu. Accessed 02 Jan 2015

Steinberg AN, Bowman CL, White FE (1999) Revisions to the jdl data fusion model. doi:10.1117/12.341367. http://dx.doi.org/10.1117/12.341367

Street AM (2009) Humanitarian reform a progress report. Humanit Exch Mag 45. http://www.odihpn.org/humanitarian-exchange-magazine/issue-45/humanitarian-reform-a-progress-report

Ubuntu (2015) Ubuntu OS. Available at: http://www.ubuntu.com/. Accessed 03 Jan 2015

Ushahidi (2008) Official site. www.ushahidi.com. Accessed 27 Dec 2014

VMWare (2014) Official site. http://www.vmware.com. Accessed 03 Jan 2015

VT MÄK (2012) VR-Forces. http://www.mak.com/. Accessed 03 Jan 2015

W3C (2004) OWL web ontology language. http://www.w3.org/TR/owl-features. Accessed 01 Jan 2015

W3C (2009) OWL 2 web ontology language document overview. http://www.w3.org/TR/owl2-overview/. Accessed 01 Jan 2015

Yang Q (1997) Intelligent planning: a decomposition and abstraction based approach. Springer, London

Yonetani M (2014) Global estimates 2014: people displaced by disasters. Technical report, Norwegian Refugee Council. http://reliefweb.int/sites/reliefweb.int/files/resources/201409-global-estimates.pdf. Accessed 27 Dec 2014

Younes H, Littman M (2004) PPDDL: the probabilistic planning domain definition language. http://www.cs.cmu.edu/. Accessed 03 Jan 2015

Chapter 14
Network Methods and Plan Recognition for Fusion in Crisis Management

Lauro Snidaro and Ingrid Visentini

Abstract Building and updating a situational picture of the scenario under consideration is the goal of the Situation Assessment (SA) Information Fusion (IF) process. The scenario generally involves multiple entities and actors where possibly only a few under direct control of the decision maker. SA aims at explaining the observed events (mainly) by establishing the entities and actors involved, inferring their goals, understanding the relations existing (whether permanently or temporarily) between them, the surrounding environment, and past and present events. It is therefore apparent how the SA process inherently hinges on understanding and reasoning about relations. SA is a necessary preparatory step to the following phase of Impact Assessment (IA) where the decision maker is interested in estimating the evolution of the situation and the possible outcomes, dangers and threats. SA and IA processes are particularly complex and critical for large-scale scenarios with nearly chaotic dynamics such as those affected by natural or man-made disasters. This chapter will discuss recent developments in information fusion methods for representing and reasoning about relational information and knowledge for event detection in the context of crisis management. In particular, network methods will be analysed as a means for representing and reasoning about relational knowledge with the purpose of detecting complex events or discovering the causes of observed evidence.

14.1 Introduction

Crisis or disaster response is probably one of the most complex and difficult scenarios where data/information fusion techniques can be called upon for SA. Modern emergency response centres (ERCs) (e.g. American Red Cross Digital Operation center) try to compose a picture of the dramatically evolving situation in order to task appropriate resources to the areas of distress. These nodal points of information aggregation and decision making can rely on reports from units

L. Snidaro (✉) • I. Visentini
Department of Mathematics and Computer Science, University of Udine, via delle Scienze, 206, Udine, Italy
e-mail: lauro.snidaro@uniud.it; ingrid.visentini@gmail.com

G. Rogova, P. Scott (eds.), *Fusion Methodologies in Crisis Management*,
DOI 10.1007/978-3-319-22527-2_14

Table 14.1 Typical data available after a disastrous event

Data	Source	Sink	Type	Format
Phone calls	Citizens	ERCs	Soft	Audio
Reports	Deployed units	ERCs	Soft	Audio/textual
News	News agencies	All	Soft	Textual/audio-video
Satellite/aerial imagery	Satellites/aerial vehicles	ERCs	Hard	Images/video
Social messages	Citizens	All	Soft	Textual/images

deployed on the field (e.g. ambulance operators, police forces, fire brigades, civil protection), direct requests of help and spontaneous accounts from citizens, and more recently information derived from social media and networks.

The bulk of data generated during a crisis, whether it be as a consequence of natural or man-made disaster, is created and needs to be processed and consumed in a very limited amount of time. The amount of data coming into play goes rapidly beyond human-manageable size. Even though the operations of most crisis management centres are still performed manually by operators (even sifting thousands of social network messages per day by hand), there is a strong potential for introducing automatic systems leveraging information fusion methods and techniques to assist decision makers.

ERCs have to deal with very heterogeneous data including both hard (device generated) and soft (human generated) types. Table 14.1 shows typical data available after a disastrous event. As it can be seen, there is a potentially enormous number of data sources (e.g. citizens) while the data consumers are limited. Some information is conveyed directly to ERCs in unsolicited way (e.g. phone calls by citizens), some is broadcast by news and press agencies, some can be the result of orders given to deployed emergency units to report back on some specific event or condition, and some can be found browsing the continuous flow of messages on social media. Typical characteristics are (extended from Hristidis et al. 2010):

- Large number of producers and consumers of information
- Time sensitivity of the exchanged information
- Various levels of trustworthiness of the information sources
- Lack of common terminology
- Combination of static (e.g. maps) and streaming (e.g. damage reports) data
- Heterogeneous formats, ranging from free text to XML and relational tables (multimedia data is out of the scope of this paper)

In the seminal work (Llinas 2002), Llinas describes a multi-year research project directed at designing IF-based automatic system for crisis management. Rogova et al. report on the progress of the activities of the project in a series of papers touching different research topics: in Scott and Rogova (2004) the characteristics of a disaster-stricken domain and of the data emerging from it are presented; in Little and Rogova (2005) an ontology describing the elements needed for building a situational picture of catastrophic events is commented; the work then approaches how to reason about disaster data in order to support SA (Rogova et al. 2006), in particular for discovering unobserved events; the series also comprises the works (Steinberg and Rogova 2008; Rogova 2009) where the importance and role of contextual information in SA for crisis management are discussed.

Of the above mentioned series of papers, Rogova et al. (2006) is key to inspire and motivate our work here. We definitely concur that situation and impact assessment techniques have to deliver consistent current and predicted situational pictures in order to assist decision makers. As a matter of fact, building and updating a situational picture of the scenario under consideration is the goal of the SA Information Fusion (IF) process. The scenario generally involves multiple entities and actors where possibly only a few under the direct control of the decision maker. SA aims at explaining the observed events (mainly) by establishing the entities and actors involved, inferring their goals, understanding the relations existing (whether permanently or temporarily) between them, the surrounding environment, and past and present events. It is therefore apparent how the SA process inherently hinges on understanding and reasoning about relations. SA is a necessary preparatory step to the following phase of Impact Assessment (IA) where the decision maker is interested in estimating the evolution of the situation and the possible outcomes, dangers and threats. SA and IA processes are particularly complex and critical for large-scale scenarios with nearly chaotic dynamics such as those affected by natural or man-made disasters.

This chapter presents the main elements and issues to be considered for developing a fusion system that could help the decision maker for crisis management. From the discussion above, it is clear how a complex scenario could be understood by considering the many relations that interweave entities, actors, objects and how they evolve in time, possibly predicting the future state of things. Representing and evolving this complex network of real-world relationships would be clearly infeasible to do in the continuous time domain since observations from information sources are typically provided at discrete time intervals.[1] Therefore, it is reasonable to think that inferential processes could be performed considering as input the observation of discrete events that can be viewed as the sampling in time of some problem variables. This is why events are here considered as the building blocks for building the situational picture of the scenario under observation. Sequences of specific events (e.g. a sequence of effects, a cause-effect chain) could have particular meaning when considered together as they might hint that a bigger phenomenon is taking place. As already mentioned, one of the objectives of an SA system, especially when assessing future impacts, is to predict the next developments of the current situation. That is, if a known dangerous occurrence is known to be the result of a specific chain of other events, then the system should be able to detect it before

[1] Although, this typically depends on the specific process and variables under consideration. For example, target tracking (filtering) can be performed in fine-grained time intervals since sensor observations can be frequent enough so to have a "continuous-like" response of the output. For higher level fusion processes, observations are typically much more coarse grained.

its completion. This is a functionality that a fusion system for SA should clearly possess. Interestingly enough, there is a Computer Science community entirely dedicated to what is known as *plan recognition* (Sukthankar et al. 2014). As it will be exposed later, there is an interesting match between the goals, objectives and methods studied by this community and some of those of interest to the fusion community, especially for what regards high-level fusion (Llinas et al. 2004). This, to the best of our knowledge, has not yet been clearly highlighted.

Network methods is something that can be found in both communities as a powerful means to represent complex relationships existing between uncertain entity data and reason about events. Specifically, focus will be given to the representation and exploitation of structured relational knowledge in order to capture the complex relationships existing between entities, actors, objects in the observed scenario and the events in which they take part. The most common network methods will therefore here discussed, especially in terms of the type of reasoning they can support. Particular attention will be given to abductive inference to discover the causes of observed effects. As it will be discussed later, this aspect is of fundamental importance in a crisis management scenario.

The chapter is organized as follows: Sect. 14.2 provides a listing of the typical requirements for SA fusion systems to be used in crisis management scenarios; Sect. 14.3 explains the main functions of the SA process; Sect. 14.4 defines what is here considered by "event" and "anomaly"; Sect. 14.5 provides an overview of the field of plan recognition and shows the links to the themes here discussed; Sect. 14.6 describes how network structures can be used to represent and maintain relational information; Sect. 14.7 discusses the different types of reasoning mechanisms and their relevance for crisis management fusion systems, in particular network methods are discussed as the main instruments for reasoning about relational knowledge; Sect. 14.8 proposes a system architecture that accommodates the aspects highlighted in this chapter describing the interactions between the different processing blocks and functions, and Sect. 14.9 concludes the chapter.

14.2 Requirements for Crisis Management

Summarizing and further developing from Rogova et al. (2006), this section provides a concise listing of objectives, requirements and constraints that would need to be addressed when developing an SA fusion system for crisis management.

Starting with the objectives, the following three are the top-level goals that the system should reach in order to be able to assist the decision makers in reacting to a possibly chaotic situation in a disaster-stricken scenario.

Objectives

- Generation of a situational picture from collected data/information
- Dynamic generation of hypotheses about the possible evolution of the observed situation

- Dynamic generation of hypotheses about possible causes of the observed situation

While the first one is common to all SA systems and actually defines what an SA is supposed to do, the second aims at projecting the current situation in the future reasoning about possible (dangerous) impacts. This objective is typical of systems that process a continuous flow of data/information and have to suggest to the decision makers possible outcomes of the current state of things (e.g. automatic surveillance systems Snidaro et al. 2011). The third one is instead characteristic of systems that have to react to an unexpected situation that was not being "tracked" or that was not predictable in any way. In this case, the system receives a number of observations regarding the new state of things (typically a wide-spread disruption condition) and has to reason back to possible causes that might have triggered the reported effects in order to create and understand the current situational picture and allow to react accordingly.

From the above list of objectives, a number of requirements naturally follow that specify functionalities of interest to the decision-makers:

Requirements

- Evaluation of plausibility of hypotheses
- Evaluation of costs and risks associated with hypotheses
- Causal inference is needed
- Knowledge base and beliefs revision

The third objective described above, in order to be helpful for the system operator, calls for an evaluation of the different hypotheses formulated to explain the possible causes of the observed facts. This evaluation would allow a ranking to quickly pin-point the most plausible hypotheses given the evidence provided by the information sources. In addition to this, also a ranking of the hypotheses depending on other factors rather than plausibility can be critical in a crisis domain. For example, a ranking of the most disruptive future developments of the current situation in terms of human lives and damage to structures. Therefore, options with theoretically low plausibility to become true, and thus with a low position in the plausibility ranking, could still be highlighted to the decision maker (Rogova et al. 2006). The second and third objectives require the system to be able to perform deductive and abductive causal reasoning in order to predict future outcomes of the current situation and to abduce the most probable causes of the observed effects, respectively. Given the typical inherent uncertainty of all reports from information sources and observations from sensors, the system should be able to dynamically revise collected knowledge in light of new evidence. This calls for the ability to manage possible conflicting observation and perform belief revision accordingly. One last but fundamental aspect of knowledge base (KB) update and revision in systems that need to perform abductive reasoning (like in this case to meet the third objective) is the following: particular attention has to be paid when reasoning about possible causes and incorporating them into a new set of beliefs. As it will be discussed later, abduced knowledge is inherently *weak* since multiple causes could be available to choose from to explain the observed effects and thus the most likely abduced premise is in any case shrouded by uncertainty. To make things worse, if new facts are deduced from abduced knowledge, this could rapidly drive the KB in a highly hypothetical state possibly making it totally unreliable.

In addition to the requirements above, a number of critical constraints given by the specific characteristics of a post-disaster crisis scenario have to be taken in consideration, these include:

Constraints

- Rapidly changing wide scenario
- Bursty huge amounts of data and information
- Data/information typically heterogeneous and mostly unstructured
- Uncertainty of observations
- Severe time and resource constraints
- High cost for wrong decisions

A post-disaster scenario is typically of metropolitan scale, where the areas under severe distress might even be very limited but with important repercussions to the rest of the city. The scenario is also typically abruptly changing (e.g. unforeseen explosions, floods, injuries, etc.), thus generating huge amounts of reports in a bursty fashion after sudden changes. As described in the Introduction, information sources are usually very heterogeneous and most of the reports are spontaneous and unstructured (e.g. phone calls to emergency centres, Twitter messages, etc.) and thus difficult to consider and process as a whole. In addition to this, citizens-generated reports are generally imprecise and/or unreliable (e.g. phone calls by people under panic) since not provided by trained personnel. These might have to be combined with sensor observations and reports from emergency units thus making the case for the fusion of huge amounts of hard/soft (Hall et al. 2008; Ferrin et al. 2008) data and information with different degrees of uncertainty and reliability. As already stressed, the system is called to work under severe time and resource constraints thus for example, limiting the possibility to perform huge batch processing on the information collected. Also, controllable information sources (i.e. emergency response units deployed in the area) are also very limited (with respect to the size of the scenario), thus redirecting information sources to acquire additional data needs to be done under the contrasting requirements of timeliness and cost (in the sense of limitedness of resources). This brings us to the last constrained mentioned in the list: in this setting, every decision has a cost but wrong ones are tragically expensive (e.g. human lives, wide-spread damage to structures, etc.).

14.3 Situation and Intent Assessment

As depicted in Fig. 14.1, if objects, entities and actors and their relationships constitute what is present in the observed scenario (ontological knowledge), events constitute the alphabet by which it is possible to recount what is happening in

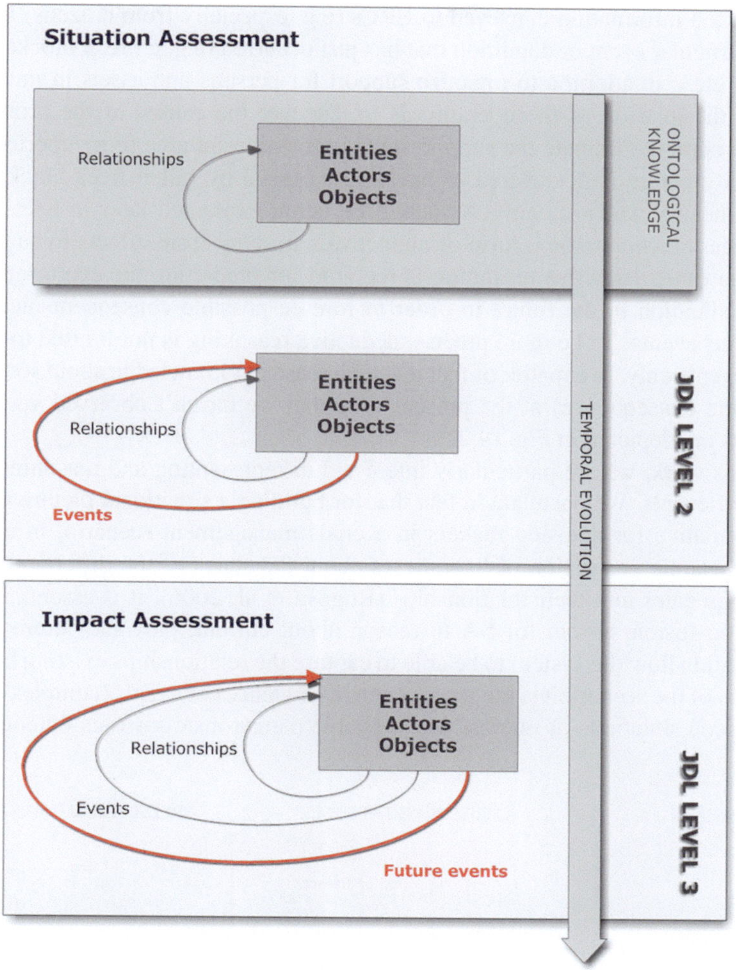

Fig. 14.1 The role of events in situation and impact assessment processes and corresponding JDL levels

the scene. The combination of all these elements allow to build a situational picture describing the current condition according to available information. According to the terminology of the JDL fusion model (Llinas et al. 2004), this SA process belongs to Level 2 which allows to present to the decision maker the current state of things: what is in the scenario, what is happening and why. Reasoning mechanisms and a KB of known events and their possible causes allow to predict future events and better estimate the current condition in light of possible future impacts (IA, JDL level 3).

Reasoning about possible causes of observed effects and events is particularly important for crisis management as already pointed out in Rogova et al. (2006).

Most of the information conveyed to ERCs (e.g. especially from citizens) is about some particular event or condition that has just occurred (e.g. a fire, a blocked road, a flood, etc.); in addition to organize support for persons and assets in immediate danger, the mission of the operators is to discover the causes of the problem in order to better coordinate the support units in the relevant area (e.g. reported flood in the city centre is discovered to have been caused by fallen trees blocking the flow of nearby water stream). As it will be better explained later in Sect. 14.7.1, this process involves some form of abductive reasoning from effects to causes. At the same time, deductive reasoning is required for predicting the evolution of the current situation in the future in order to foresee possible consequent incumbent dangerous events. To be more precise, deductive reasoning is not limited to predict future events only, as a matter of fact it can increase our knowledge about something (i.e. some consequence) at the present time that we haven't observed yet. These concepts are depicted in Fig. 14.2.

In this work, we are particularly interested in representing and reasoning about chains of events. We postulate in fact that for building a situational picture that can be informative for decision makers in a crisis management scenario, in addition to the dynamic generation of hypotheses about the states of the different entities, their aggregates and their relationships (Rogova et al. 2006), it is essential for an automatic fusion system for SA to reason about current, past and future events. This would allow the system to be able to capture the relationships existing between elements of the scenario via the recognition of sequences of events (complex events) that encode situations of interest with possible dangerous/disastrous outcomes. To

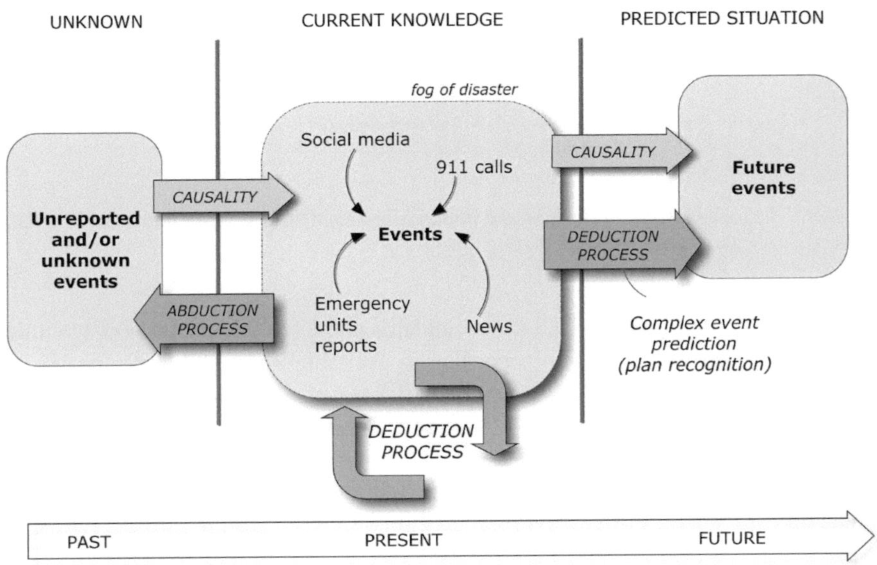

Fig. 14.2 Past, current and future event analysis for SA and IA

this end, we survey a few main lines of research existing in the literature for representing and recognizing complex events and their relationships, performing event prediction in light of observed actions, and reason about their (possibly unobserved) causes.

14.4 Events and Anomalies

Events and anomalies can be considered fundamental building blocks for developing an informed situational picture of the environment. In this section, we provide the necessary definitions of these concepts and relate them to the JDL fusion model (Llinas et al. 2004) in light of the typical functionality and requirements of an SA system. In the following, the term *level* will be used as per JDL terminology.

While there are many papers in the literature that deal with events and provide various definitions (Snidaro et al. 2007), we here break down the main concepts in light of the typical functionality and requirements of an SA system, following the approach and terminology given in Snidaro et al. (2015c).

For our purposes, an **event** is a "significant occurrence or happening". It can be subdivided in *simple*, when we consider the variation of a quantity or state, or *complex*, which is a combination of atomic or complex activities (Snidaro et al. 2007) which is here reported for the sake of completeness. Figure 14.3 gives an intuitive representation of the idea.

A **simple event** is any significant variation of input data, at any level, discernible by the system. Also called *atomic* in the literature, we here use the term *simple* to avoid confusion with *ground atoms* defined in Sect. 14.7.3. They can be directly observable or not, and can be either instantaneous or last for an arbitrarily long period of time. As the name implies, this is the most basic type of occurrence and cannot be further decomposed into simpler constituting events.

More in general, variations of input signals (Level 0), of a target's state (e.g. speed, direction, etc. that can be included in Level 1), of a target's relation with other entities (Level 2) are all examples of simple events.

Complex events are a combination of two or more component events (simple or complex) that can be arbitrarily combined through logical operators ($\wedge$, $\vee$, $\neg$) to

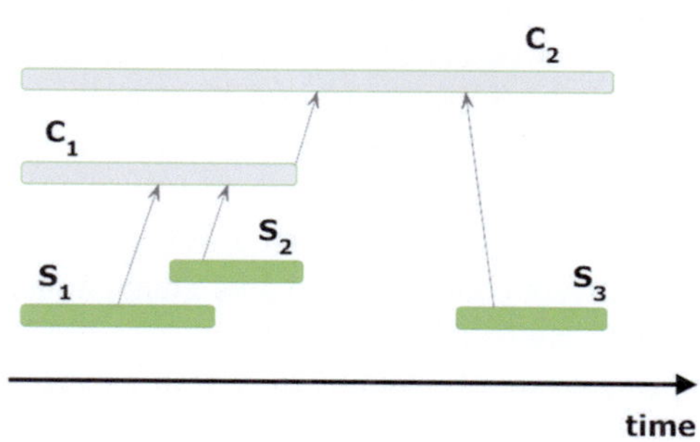

Fig. 14.3 Example of the detection of a complex event by observing the occurrence of its components. C_2 is composed by complex event C_1 and simple event S_3. All the events in this example are non-instantaneous as each of them spans a certain interval of time

encode articulated expert and domain knowledge. Complex events can be either triggered by a specific time-ordered sequence of component events, or be just an unordered collection of them. In addition, a complex event can be composed by a heterogeneous combination of events generated by data at different levels. Generally, complex events span a certain interval of time. They could have a fixed time-frame, that is constituting events have to occur within a given time window, or not as they wait indefinitely for all the component events to happen.

An **anomaly** can be considered a critical event to which the system is generally called to react to. Usually, a *threshold* establishes if input data can be considered unexpected or anomalous, thus raising an exception. Thresholds, provided by domain experts or learned automatically by the system from data, are therefore used to immediately spot an anomalous condition. However, anomalies provide no notion whatsoever on the meaning of the exceptional input. Following this definition, an anomalous event is an occurrence of some type that deviates from expected values or behaviour. In an SA system, the KB is consulted to infer a possible conclusion from the anomalous condition.

Events and anomalies can be defined by **explicit or implicit models** in the system. In the former case the model encodes, usually exploiting expert and contextual knowledge (Snidaro et al. 2015a), the complete description of what an (anomalous) event is. On the contrary, implicit modelling means that samples of activities are unsupervisedly learned by the system in order to detect a deviation from common patterns.

14.5 Plan Recognition

This section discusses how concepts known in the plan recognition literature can be applied to an IF system for SA in a crisis management scenario (Sect. 14.5.2). In the following, a brief overview of plan recognition is given first, along with a survey of the few available literature in the fusion domain (Sect. 14.5.1).

Plan recognition is the process of inferring the top-level plans of an agent from its observed actions (Raghavan et al. 2014). This generally assumes that the agent is pursuing one or more top-levels goals and that its actions are directed to reach those goals. If this is true, then the objectives of the agent, unknown to the observer, could be predicted by reasoning about the observed actions. The latter are in turn perceivable to the observer in form of events. Therefore, formulating hypotheses regarding the agent's plans means reasoning about sequences of events. It is interesting to see how this process is typically abductive (see Sect. 14.7) since the agent's actions are caused by its goals, only the actions are observable, and multiple hypotheses could be formulated to explain the observations.

A recent collection of works giving a picture of the state of the research on this subject can be found in Sukthankar et al. (2014). Here, only the main concepts are briefly presented along with exposing the connections to relevant fusion concepts.

Plan recognition can have many different applications and the works in the literature can use different approaches depending on a few assumptions on the agent to be observed (Avrahami-Zilberbrand and Kaminka 2014):

- *Key-hole recognition*: no interaction occurs between the agent and the observer. The agent does not know, or does not care, of being observed and the observer only perceives the agent's actions;
- *Intended recognition*: typical applications are interactive systems, where the user is actively communicating her/his intentions in order to receive relevant services (e.g. navigation systems);
- *Adversarial recognition*: the agent is considered an adversary and the observer is generally interested in thwarting the agent's plans as in games or surveillance. On the other hand, the agent might actively put into practice strategies to confuse or hinder the recognition process carried out by the observer.

The terminology, and even maybe the approaches, might appear inappropriate when dealing with non-self-conscious agents, such as natural events. However, the term "adversary", common in the plan recognition literature, can be more generally intended as something or someone acting (even not deliberately as in the case of natural causes) against the natural order of things, or the "normality" as understood by the observer. For example, in the case of natural "adversarial" forces, knowledge of physical rules and contextual information might provide leads to inferring possible causes of observed disastrous effects.

14.5.1 Plan Recognition in Fusion

As already anticipated in the Introduction, there is an interesting correspondence between the goals of the plan recognition community and those of the high-level fusion one. As highlighted by Llinas et al. (2004), most fusion processes, even high-level ones, have been up to now of *deductive* type, where behaviour models (whether anomalous or not) are generally used to predict future outcomes from current observations in a *deductive* fashion. However, in many scenarios such as in the case of crises originated by human deliberate acts (e.g. terrorist attacks) or catastrophic natural causes it is not reasonable to assume the availability of a significant library of models that would allow to keep track of the situation. Instead, an *abductive* approach seems to be most appropriate in order to discover the causes of the observed effects (Llinas et al. 2004). And this is what is most compelling about the approaches addressed by many works in the plan recognition literature: the effort is primarily directed to the understanding of an underlying cause or causes to the chain of events being observed.

To the best of our knowledge, and as already mentioned, plan recognition has not received much attention in the Information Fusion community and very few works can be found. For example, it has been considered for a C4ISR[2] system in DiBona et al. (2006). The authors present an architecture in which planning, monitoring, and information fusion modules cooperate in a closed-loop. This is probably the first paper to expose a possible coupling between plan recognition and information fusion, and provides a break-down of the coupling by specifying the processing steps required in a C4ISR system. However, planning is intended to be functional for the fusion system by directing the fusion process to better adapt to the current plans of the decision maker. That is, plan understanding is used to decode the operational and information needs of the commander so to properly cue sources to gather the relevant context needed to drive the fusion process. This type of adaptation of the fusion process is considered to be belonging to Level 4 of the JDL model dedicated to fusion process refinement and resource management (Llinas et al. 2004). This type of plan-driven fusion is also considered here, as discussed in Sect. 14.5.2, but more focus is given to applying plan understanding techniques for SA.

Another work that can be found in the fusion literature and that is relevant to the topics discussed here is Allouche and Boukhtouta (2010), which proposes a general framework for distributed plan modelling and monitoring. Specifically, the paper focuses on the issue of coordinating plans and actions in a multi-agent domain. The proposed solution is fusion mechanism to fuse the plans of a set of agents and generate a global coordinated plan.

It is interesting to notice that plans and goals are tightly connected with the concept of *internal context* discussed in Snidaro et al. (2015b).

14.5.2 Plan Recognition and SA for Crisis Management

The second and third objectives of an SA fusion system for crisis management stated in Sect. 14.2 can be addressed by plan recognition techniques. However, there are more aspects related to *planning* and *plan recognition* that might come into play in a complex fusion system for SA. Here we want to clarify some concepts and show their applicability to our discussion:

1. *User's plan recognition*. This process exploits plan recognition to decode and understand the requirements of the user as a plan to achieve certain goals. As discussed in Sect. 14.5.1, this process is described in DiBona et al. (2006). This process would be followed by the planning step (next point) to properly cue sources to gather the relevant context and information needed to accomplish the user's goals.
2. *Planning*. Planning is a deeply studied field of AI devoted to finding action sequences that achieve the agent's goals (Russell and Norvig 2010). In a SA

[2]Command, Control, Communications, Computers, Intelligence, Surveillance and Reconnaissance.

fusion system for crisis management, this is a Level 4 Resource Management process where the system, starting from the user's plan extracted by plan recognition (previous point), is self-adjusting and reconfiguring (possibly re-deploying, cuing and selecting available sensors, adjusting filters on information inputs, setting algorithms' parameters, etc.) in order to maximize the probability to obtain the information needed to satisfy the needs of the decision maker.[3]

3. *Complex event (plan) recognition.* This process is the actual detection of the patterns directly specified in the user's requirements (or derived from them). The user might be interested in detecting specific (maybe possibly dangerous) events that have not occurred yet.[4] We here highlight the connection—not fully exposed, to the best of our knowledge—between what is known as *complex event recognition* in the surveillance domain (Vishwakarma and Agrawal 2013) and the approaches taken in the plan recognition literature (Sukthankar et al. 2014). Specifically, the issues in representing complex relational knowledge, and in detecting complex sequences of events from noisy observations, is a major focus of the plan recognition literature. Recent advancements make use of hybrid techniques that incorporate with promising results both logic and probability. We here make the case that recognizing (known) patterns of events as they occur is not different from recognizing the steps of an unfolding plan. Network approaches, ideal for relational knowledge and that support reasoning about observed events, are discussed in the following sections.
4. *Abductive plan recognition.* As expressed in Sect. 14.5, plan recognition is typically an abductive process when the plan needs to be inferred backwards from the observed effects to the possible causes. This type of reasoning is fundamentally different from the typically forward-chaining deductive inference process described in the previous point (see Sect. 14.7). Therefore, this is a key element for crisis management SA fusion systems where the causes need to be discovered from the disastrous effects observed, as stated in Sect. 14.2.

14.6 Networks and Graphs

As already discussed in the previous sections, understanding and discovering relations existing between the entities in the observed scenario is a fundamental step of SA. This means that what is observed is typically a network of elements that

---

[3]One of the interests of the planning algorithms community is providing the most effective (in some sense) sequence of *actions* that would satisfy the user's goals. This is particularly relevant for software agents or robotic systems where the software can perform actions that can modify the environment (software environment or the real word as in the case of robots). Here, the SA fusion system is mostly intended at situation monitoring and assessment, with limited direct capacity of modifying the real word (e.g. relocating information sources).

[4]This is typical in surveillance and crisis management where the system is processing events as they occur. However, detection of patterns could be performed even over past events, that is mining offline data in search for relevant patterns in a forensic fashion.

are linked together in some way, or for which a number of relations connecting them can be established. In SA this is done to come to an estimate of their state and to an explanation of their behaviour. The study of networks as collection of connected parts spans many disciplines (e.g. computer networks, human societies, predatory interactions between animal species, etc.) and has developed a solid theoretical background (Newman 2010).

A graph is the key model used to represent a network as a collection of points joined together in pairs by lines. Points are referred to as *vertices* or *nodes* and the lines are called *edges* (Newman 2010). Since its initial development by Euler in 1736 to solve the problem of the Seven Bridges of Königsberg, this elegant representation has given raise to the powerful graph theory (Biggs et al. 1976). Although many variations of the graph model exist, of notable interest for the scope of this paper are attributed relational graphs (ARGs) which have found application as a means for intelligence analysis, given their intuitive structure which allows to easily represent people, organizations, objects or events as nodes and relationships like interaction, ownership or trust as edges (Coffman et al. 2004; Biermann et al. 2013). In addition to conveniently represent data in an intuitive way for humans, they also allow two main strands of computation to be performed that can significantly ease the work of the intelligence analyst: graph matching and social network analysis. These have found considerable fortune in the information fusion community since they facilitate representing relational knowledge, reasoning over the relationships between actors, and detecting anomalous patterns of activity. Specifically, subgraph isomorphism algorithms are being used to detect known patterns which typically constitutes threat models (Template Graph) in a much larger graph or set of graphs representing the actual instances of observed entities and their relationships (Data Graph). Since of course it is very unlikely that real exact matches are going to occur due to noise or missing information on the part of the Data Graph and because of imprecise modelling of the template, the problem generally resolves to computing subgraph inexact matching. Being its complexity NP-complete, research is generally directed to finding suitable approximations or heuristics that can solve the problem in a reasonable amount of time (Sambhoos et al. 2008).

14.7 Reasoning

The main goal of a reasoning engine or probabilistic inference system is to associate a posterior probability distribution with a set of queries (Russell and Norvig 2010), given observed evidence. The incorporation of abductive/inductive and deductive inferencing processes is a vital element in an automatic fusion system, and it represents a fundamental step for situational awareness. How this involvement can be obtained, on both theoretical and applicative levels, is a crucial point, and is subject of ongoing research (Llinas et al. 2004).

The reasoner is usually fired by low-level observations provided by sensors, covering in this way the majority of abnormal situations in the domain; however, it is interesting to notice how anomalous behaviours do not always follow standard trends or well-known patterns, especially if related solely to vessel movements, but sometimes they take the form of seemingly unrelated activities on a larger scale (Munson 2009).

14.7.1 Reasoning Directions

Deduction

As clearly expressed in Rogova et al. (2006), the process of building a situational picture comprises the dynamic generation of hypotheses about the states of the environment. This involves reasoning about situational items, their aggregations, the relationships between them, their behaviour and goals, and the events in which they are involved. This calls for both deductive and abductive reasoning methods. Deductive reasoning allows to derive new knowledge when the antecedent is known to be true (Fig. 14.4 left) by, say, direct observation. Knowing that a certain event C has occurred and knowing that the event has a causal relationship with effect E, even without observing E we might reasonably conclude that E has verified as well.[5] Of course C could be the antecedent of other rules as well and those would also fire deducing the truth values of the involved consequents. In turn, the consequents could appear in the antecedents of other rules thus possibly (if all conditions are met) fire other rules in a cascading fashion. Whenever a rule is fired new knowledge is created (something new is known to be true) and this allows to possibly continue reasoning if there are rules involving this new knowledge in the antecedents. This type of reasoning is called *forward chaining* and is typical of experts systems that systematically employ *modus ponens* to deduce new knowledge and make reasoning proceed.

Abduction

Abduction follows instead the opposite direction (Fig. 14.4 right), that is looking for causes that can explain the observed events or effects. This reasoning direction is much less direct than the previous one. While in deduction, *modusponens* allows

[5]The relationship just described is assumed to be of causal type where the cause C is known to produce the effect E. $C \Rightarrow E$ is then a rule expressing a causal relationship between the antecedent and the consequent. This is a typical condition when the KB is built from experts' knowledge that usually express rules in terms of causation giving to the rules much more strength and semantics than it would happen if the rules were to be learned automatically (in this latter case only correlation can be learned).

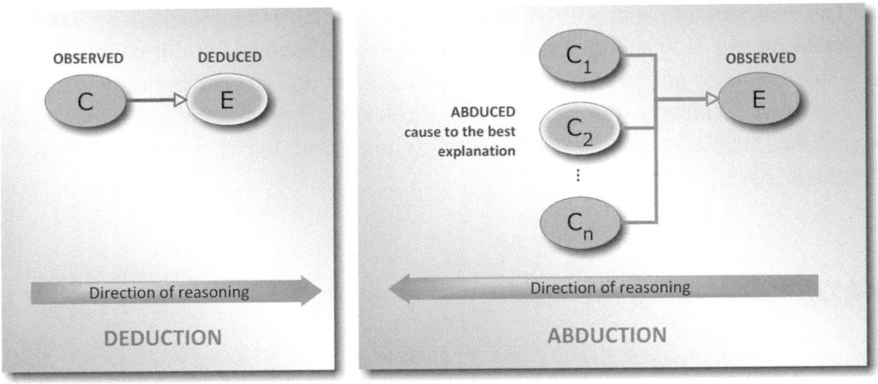

Fig. 14.4 Directions of logical inference. Deduction (*left*) concludes E after having observed C and knowing $C \Rightarrow E$. Abduction looks for the antecedent that best explains the observed event E

to directly derive the truth of the consequent from the truth of the antecedent, here no direct result is readily available. First of all, say E is the observed event then E could appear as the consequent of multiple rules in the KB, thus opening the investigation on a number of possible antecedents. Secondly, the truth of the consequent E says nothing about the truth value of the possible antecedent C_i as the clause $C_i \Rightarrow E$ is satisfied whatever the truth value of C_i.

Therefore, besides a "Peircean" use of the term *abduction*, basically seen as "the third" mode of inference (along with deduction and induction) and further studied by Hanson (1958) who investigated the logic of discovery in the 1950s, another way of using it considers it a synonym for the inference to the best explanation model (IBE). Such a model, which has a great diffusion in the technological domains, from multimedia automated analysis (see Möller and Neumann 2008) to the fusion domain itself (see Bharathan and Josephson 2006; Josephson 2008) is based on the studies of Gilbert Harman's, who formulated the idea in the 1960s (Harman 1965). Referring again to Fig. 14.4(right), IBE means finding the set of values for all the random variables involved such that a certain C_* maximizes the probability of being the antecedent to have triggered E, thus concluding that C_* is the best explanation for E.

A survey of some key concepts around abductive inference, including different types of abduction, relevant to the IF domain can be found in Ferrin et al. (2010).

14.7.2 Bayesian Networks

Bayesian Networks (BNs) are a popular instrument in classification problems with uncertain knowledge. A BN is a directed acyclic graph whose nodes correspond to random variables that can be discrete or continuous. The nodes can be connected by directed edges indicating conditional dependencies among the random variables

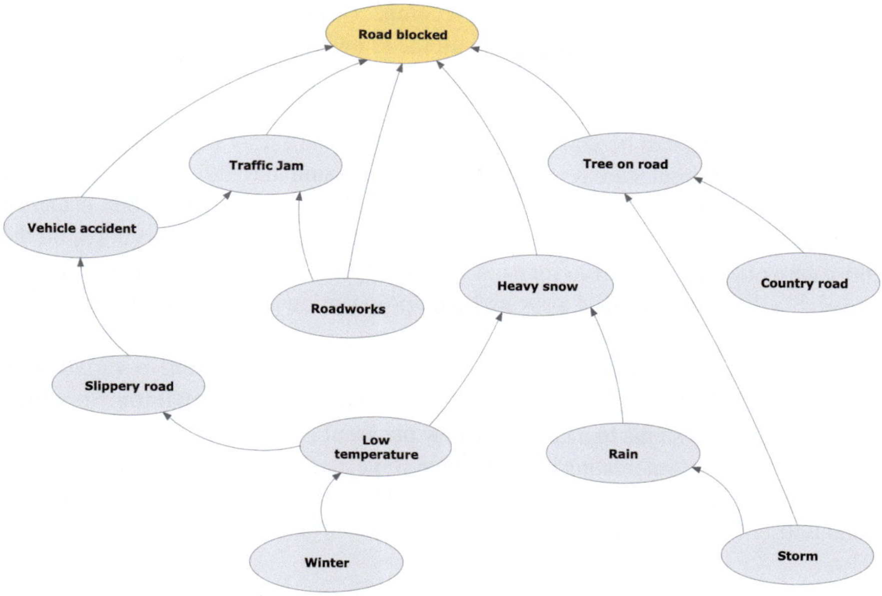

Fig. 14.5 Example of a directed graph to encode the possible cause-effect chains that can explain the observation of a "Road blocked" complex event

associated with the nodes. In particular, an edge e from node X to node Y indicates that X is a parent of Y. That is, e encodes the fact that (the random variable associated to) X has a direct influence on (the variable associated with) Y. This allows to intuitively encode causal and influence relationships, thus greatly facilitating the knowledge engineering process where domain expert knowledge needs to be formally represented. The last element that is needed to define a BN is the conditional probability distribution $P(X_i|Parents(X_i))$ (where $Parents(X_i)$ are the random variables directly parent of X_i), for every node X_i, so to express probabilistically the influence of the parents on the child node. Figure 14.5 shows an example of (simple) events that can explain the report of a blocked road. The arrows indicate causal relationships.

The explicit encoding of independence and conditional independence relationships allows to greatly reduce the number of probabilities that need to be specified to represent the full joint distribution $\mathbf{P}(X_1, \ldots, X_n)$ where n is the number of nodes (variables) in the network. That is, the joint probability of a realization $x_1, \ldots, x_n$ of the n variables X_i can be written as:

$$P(x_1, \ldots, x_n) = \prod_{i=1}^{n} P(x_i|\text{parents}(X_i)), \tag{14.1}$$

where $\text{parents}(X_i)$ are the values of the variables in $\text{Parents}(X_i)$ that appear in $x_1, \ldots, x_n$.

A BN allows therefore to express in a compact way the full joint distribution of the random variables X_i, thus providing all the knowledge needed to answer queries on them. For example, finding the impact of observing $X_j = x_j$ on another variable X_i can be done by considering the conditional probability distribution $P(X_i|x_j)$ where each entry $P(x_i|x_j)$ can be computed as $P(x_i|x_j) = P(x_i, x_j)/P(x_j)$. Entries $P(x_i, x_j)$ of the marginal joint can be found by marginalizing the full joint distribution $\mathbf{P}(X_1, \ldots, X_n)$ over all variables except X_i and X_j. This is generally called *evidence propagation* or *belief updating* and to see perturbation effects of observing x_j on the entire network it should be run for each variable X_i. This is typically computationally intractable, however message-passing belief propagation algorithms exist that allow to considerably reduce the effort (Pearl 1988).

Belief propagation allows to sequentially update current beliefs in variables of the network by including the evidence received. This is fundamental for supporting the ability of SA systems to incorporate new pieces of evidence as they are collected. However, the most interesting capability of Bayesian Networks for our discussion here is *belief revision* or abductive inference. This allows to find the most probable instantiation of all unobserved variables, given the observed data, that is

$$x_U^* = \arg\max_{x_U} P(x_U|x_O), \tag{14.2}$$

where x_U^* is the *most probable explanation* (MPE), X_U is the set of unobserved variables, x_O is the observed evidence, and x_U is a possible configuration of states of variables X_U. This type of abduction is referred to as probabilistic abduction as it provides a probabilistic mechanism that can be used effectively to provide the best explanation of the observed effects in terms of the unobserved causes. The MPE cannot in general be obtained by belief propagation as (14.2) is an optimization problem that requires dedicated algorithms (Pearl 1988).

Bayesian networks provide an attractive framework as they can effectively handle uncertainty, support both deductive and abductive inference, and can be trained both learning the structure of network and the parameters. Despite being so largely used for probabilistic representations of uncertain knowledge, Bayesian networks have strong limitations, including the fact that they allow reasoning about the same fixed number of attributes, as their nature is essentially propositional: the set of random variables is fixed and finite, and each has a limited domain (Russell and Norvig 2010). As result, their application to complex problems is often impeded, as they require to define in advance with confidence how many entities will be involved, and what type of relationships intercur among them. Therefore, they cannot support the complexity of the relational knowledge that would be required in a crisis management scenario (see Sects. 14.1 and 14.2) that can be better expressed via first-order logic. Even in Hidden Markov Models, which have been used in case of temporally and spatially distributed observations for event recognition (Li and Porikli 2004; de Vries et al. 2008), the number and type of states must be specified in advance. This last condition largely impacts on performances when scaling up to a larger size scenario.

Many approaches have been proposed to link the probabilistic reasoning under uncertainty provided by Bayesian Networks and first-order logic. One of the first attempts can be found in Poole (1993) where the proposed framework supports probabilistic Horn abduction. The interest in combining the expressiveness of first-order logic and the uncertainty management of probabilistic approaches has given raise lately to the field of statical relational learning (SRL) (Getoor and Taskar 2007) which aims at fostering the unification of the two fields. Markov Logic Networks are one of the most recent and interesting developments in SRL and are briefly discussed in the next section.

14.7.3 Markov Logic Networks

We here provide essential background notions of Markov Logic Networks, but the reader is advised to refer to Richardson and Domingos (2006) for further details. MLNs are a powerful tool for combining logical and probabilistic reasoning. While a KB of logic formulas is satisfiable only by those worlds (truth values of atomic formulas) in which it is true, an MLN relaxes this hard constraint by associating a probability value with the worlds that do not fully satisfy the KB. Therefore, the fewer formulas a given world violates the more probable it is.

An MLN is then a set L of pairs (F_i, w_i) where F_i is an FOL formula and w_i its corresponding real-valued weight. The set of all formulas F_i in L constitutes the KB while the weight w_i associated with each F_i reflects how strongly the constraint imposed by the formula is to be respected. This impacts directly the probability assignment: worlds which satisfy a high weight formula are going to be much more probable than those that do not.

A Markov Logic Network L together with a finite set of constants C defines a Markov network $M_{L,C}$ that models the joint distribution of the set of random (binary) variables $X = (X_1, X_2, \ldots, X_n) \in \mathscr{X}$. Each variable of X is a ground atom (predicate whose arguments contain no variables) and $\mathscr{X}$ is the set of all possible *worlds*, that is the set of all possible truth value assignments of n binary variables. The network is built as follows:

- $M_{L,C}$ contains one (binary) node for each possible ground atom given L and C
- An edge between two nodes indicates that the corresponding ground atoms appear together in at least one grounding of one formula in L. Ground atoms belonging to the same formula are connected to each other thus forming cliques.
- A feature f_i is associated for each possible grounding of a formula F_i in L. Each f_i assumes value 1 if the corresponding ground formula is true and 0 otherwise.

The probability distribution over X taking values $x \in \mathscr{X}$ specified by $M_{L,C}$ is given by:

$$P(X = x) = \frac{1}{Z} \exp\left(\sum_{i=1}^{|L|} w_i n_i(x)\right), \tag{14.3}$$

where $|L|$ indicates the cardinality of L, thus counting the number of formulas of the KB, and $n_i(x)$ is the number of true groundings of F_i in the world x.

$$Z = \sum_{x' \in \mathscr{X}} \exp\left(\sum_{i=1}^{|L|} w_i n_i(x')\right) \tag{14.4}$$

is a normalizing factor often call *partition function*.

Given the joint distribution function in (14.3), it is possible to calculate the probability that a given formula F_i holds given the Markov Network $M_{L,C}$ as follows:

$$\begin{aligned} P(F_i|M_{L,C}) &= \sum_{x \in \mathscr{X}_{F_i}} P(X = x|M_{L,C}) \\ &= \frac{1}{Z} \exp\left(\sum_{x \in \mathscr{X}_{F_i}} w_i n_i(x)\right), \end{aligned} \tag{14.5}$$

where $\mathscr{X}_{F_i}$ is the set of worlds where F_i holds.

While (14.3) provides the probability of configuration x of truth values for the ground atoms in the Markov Network, Eq. (14.5) can be used instead to evaluate the probability that a formula F_i (e.g. a predicate representing an event) holds given $M_{L,C}$ where C is composed by observed entities and other constants provided by contextual knowledge. This gives a glimpse of the power of the framework, an arbitrary formula that can be grounded in $M_{L,C}$ can be queried to get the probability of being true. Thus not only the formulas in L but also any logical combination of them that can be grounded in the Markov network can be queried. This is extremely important for an SA system where the operator might want to evaluate the truth degree of a new (complex) event or condition as the combination of existing evidence in the KB.The framework can also provide the probability that a formula F_2 holds given that formula F_1 does or provide an answer to whether the KB entails a given formula.

According to the definitions given in Sect. 14.4, an MLN provides an explicit way of encoding knowledge. However, both rule weights and the rules themselves can be learned from data. These capabilities make MLNs a powerful tool that combines the benefits of both implicit and explicit modelling.

It should be noticed that MLNs despite many advantages over existing similar SRL approaches, according to their original formulation (Richardson and Domingos 2006), present also several shortcomings. One that is often highlighted is the non-trivial way of assigning weights to the formulas (if derived from expert knowledge and not learned directly from data). Another one, which is directly connected to one of the main themes of this work, is the inability to perform abductive inference. That is, only deductive inference is supported as exploited in Snidaro et al. (2015c) for detecting anomalous patterns in vessel traffic. However, several solutions have been proposed recently to address this problem such as the Pairwise Constraint model, the Hidden Cause model, and the Abductive Model Construction. A good overview and discussion of these extensions for supporting abductive inferences for plan recognition can be found in Raghavan et al. (2014).

14.7.4 Addressing the Requirements

The previous sections have illustrated the basics of graphs and networks to represent relational knowledge and the reasoning algorithms that exploit network structures. In particular, the discussion has led to MLNs which provide a powerful framework based on FOL with probabilistic support for uncertainty. MLNs can also be used to encode other formalisms such as BNs and HMMs, thus constituting a more general tool. We here discuss how MLNs address the objectives and requirements for a fusion system for crisis management as stated in Sect. 14.2.

Objectives

- Generation of a situational picture from collected data/information: the ground Markov network generated upon instantiation of variables in the KB according to observed evidence constitutes a tool to answer arbitrary queries involving the ground atoms. That is, a reasoning framework is available to answer queries on observed data returning a probability value associated with the truth value of the query expression;
- Dynamic generation of hypotheses about possible evolution of the observed situation: typical FOL forward-chain reasoning of MLNs allows deductive (predictive) reasoning as in Snidaro et al. (2015c);
- Dynamic generation of hypotheses about possible causes of the observed situation: the Pairwise Constraint model, the Hidden Cause model, and the Abductive Model Construction (Raghavan et al. 2014) are mechanisms developed to add abductive reasoning capabilities to MLNs thus allowing to reason from effects to causes.

Requirements

- Evaluation of plausibility of hypotheses: as already mentioned above, each query performed on the ground network is answered by the framework with a probability value for its plausibility according to the current state of the variables;
- Evaluation of costs and risks associated with hypotheses: this part is more related to decision making and would need to be implemented as an additional processing step to be paired with SA;
- Causal inference is needed: as per the third objective, causal reasoning can be supported through abductive inference. However, as already discussed,

abductive inference simply provides the most likely antecedent p of the observed consequent q. If the KB encodes causal rules, then the antecedent is also a cause of the observed effect. Of course, in general the logical implication $p \Rightarrow q$ does not specify a causal relationship between p and q;

- KB and beliefs revision: as explained in Sect. 14.2, this requirement enforces that some form of belief revision be present for the KB. Particularly if explanations produced by abductive inference are to be added as new facts in the KB. Since abduction does not provide guarantees on the truth of the antecedent, deducing e given the rule $p \Rightarrow e$ and the abduced premise p should be performed only with the availability of some roll-back algorithm in case p is later found not to hold. Algorithms for multi-hypotheses management and KB revision have been largely studied in AI. For example, assumption-based truth maintenance systems (ATMS) (de Kleer 1986) mechanisms could be paired with abductive MLNs to fully support abductive reasoning and belief revision.

The constraints expressed in Sect. 14.2 can be condensed into the following which relate to the capabilities of the reasoning system:

Constraints

- Data/information typically heterogeneous and mostly unstructured: the reasoner takes as input pre-processed data thus significant processing is required to extract symbolic information from raw-data. The vast number of possible sources and the heterogeneity of the data generated in a crisis scenario calls for some form of structured data already generated at sensor/platform level. Structured languages able to couch both soft data and hard data such as Controlled English and Battle Management Language (BML) (Biermann et al. 2014) should be considered as pre-processing step;
- Uncertainty of observations: this is one of the strongest points in favour of a logic-probabilistic relational reasoning model such as MLN. While retaining the expressiveness of FOL, MLN allows and handles contradicting observations and configurations of the values of the variables (ground atoms) that violate the satisfiability of the KB. That is, the weights assigned to the KB formulas (see Sect. 14.7.3) specify how strictly the rules have to be respected. Worlds (configurations of ground atoms truth values) that don't fully satisfy the KB are simply given less probability but not completely ruled out as they would be in FOL.
- Severe time and resource constraints: while exact inference in MLNs is #P-complete, which would make the approach completely useless if for small size problems, in practice efficient approximate reasoning algorithms (e.g. Gibbs sampling) coupled with clever heuristics (e.g. typed variables, partial grounding, lazy inferences, etc.) that allow to attain results in reasonable time (although approximate).

14.8 Architecture

An interesting and well-written presentation of a recent architecture framework for high-level fusion can be found in Ao et al. (2014). The paper takes inspiration from the STDF fusion model (Lambert 2006) and proposes three stages for the inference process: Abduction (explanation), Deduction (Prediction) and Induction (Generalization). The work provides an interesting account of recent probabilistic, logic and machine learning approaches, discussing how the already mentioned three stages can be applied to every fusion level taking into account both uncertainty management and learning. The paper concludes that recent graphical models approaches possess all required features to implement all three inference stages. While agreeing on this conclusion, we here depart from the model proposed in Ao et al. (2014) particularly for what regards the sequence of inference steps.

This section discusses some of the fundamental process models, functions and algorithms that would need to be involved in a fusion system for crisis management. Figure 14.6 sketches an architecture comprising classic elements in a fusion process. The diagram is rather informal and is geared more to summarizing and visualizing the elements addressed in the previous sections, rather than presenting a formal architecture model. In particular, what follows is seen from a high-level fusion perspective, focusing on reasoning algorithms for SA and prediction:

1. [Sources]: Input data is intended to be comprised of detected events;

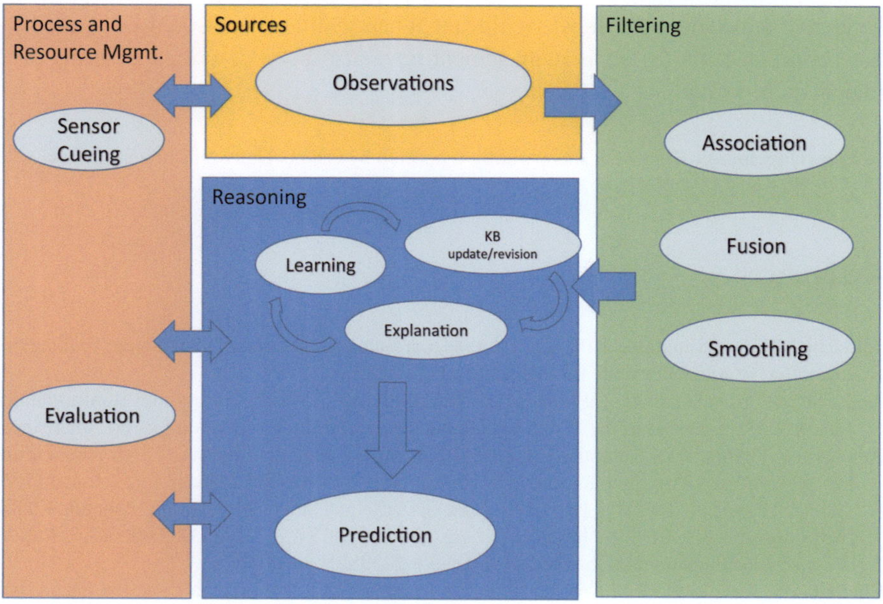

Fig. 14.6 Process modules and functions

2. [Filtering]: Here the filtering process is intended to prepare the data by associating observations with known variables, fusing multiple measurements, and smoothing the data for removing repetitions. Probabilistic temporal models (HMM or DBN, both expressible as MLNs) are typically used for complex-event recognition.
3. [Reasoning]: The reasoner is in charge of assessing the current state by including the new observations, performing inference on the KB to assess the current state, abducing possible causes, revising the beliefs if necessary in case of contradictions, and performing online learning. The process should also be able to formulate hypotheses about future states;
4. [Process and resource management]: This process evaluation step should continuously monitor the overall process in light of the current state and the goals expressed by the decision maker. This module should be able to control assets (e.g. relocate sensors) in order to better address the goals.

14.9 Conclusions

This chapter has highlighted several key elements that should be taken into consideration while developing a fusion system for crisis management. Specifically, it threads together concepts from SA, event detection, plan recognition, and data fusion, showing and exposing previously unknown commonalities and interconnections. Specific attention has been dedicated to available approaches for supporting abductive reasoning. The most significant functionalities that would be needed in SA fusion systems for crisis management have also been described in terms of a logical architecture.

Acknowledgements This work was partially supported by ONRG Grant N62909-14-1-N061.

References

Allouche MK, Boukhtouta A (2010) Multi-agent coordination by temporal plan fusion: application to combat search and rescue. Inf Fusion 11(3):220–232

Ao Z, Scholz J, Oxenham M (2014) A scientific inquiry fusion theory for high-level information fusion. In: Proceedings of the 17th International Conference on Information Fusion, Salamanca

Avrahami-Zilberbrand D, Kaminka GA (2014) Chapter 4 - keyhole adversarial plan recognition for recognition of suspicious and anomalous behavior. In: Sukthankar G, Goldman RP, Geib C, Pynadath DV, Bui HH (eds) Plan, activity, and intent recognition. Morgan Kaufmann, Boston, pp 87–119. http://dx.doi.org/10.1016/B978-0-12-398532-3.00004-X. http://www.sciencedirect.com/science/article/pii/B978012398532300004X

Bharathan V, Josephson JR (2006) An abductive framework for level one information fusion. In: Proceedings of the 9th International Conference on Information Fusion, Florence

Biermann J, Horling P, Snidaro L (2013) Experiences and challenges in automated support for intelligence in asymmetric operations. J Adv Inf Fusion 8(2):101–118

Biermann J, Nimier V, Garcia J, Rein K, Krenc K, Snidaro L (2014) Multi-level fusion of hard and soft information. In: Proceedings of the 17th International Conference on Information Fusion. ISIF, Salamanca

Biggs N, Lloyd EK, Wilson RJ (1976) Graph theory 1736–1936. Oxford University Press, Oxford

Coffman T, Greenblatt S, Marcus S (2004) Graph-based technologies for intelligence analysis. Commun ACM 47(3):45–47

de Kleer J (1986) An assumption-based {TMS}. Artif Intell 28(2):127–162. http://dx.doi.org/10.1016/0004-3702(86)90080-9. http://www.sciencedirect.com/science/article/pii/0004370286900809

de Vries G, Malaisé V, van Someren M, Adriaans P, Schreiber G (2008) Semi-automatic ontology extension in the maritime domain. In: Proceedings of twentieth Belgian-Dutch Conference on Artificial Intelligence (BNAIC 2008)

DiBona P, Belov N, Pawlowski A (2006) Plan-driven fusion: shaping the situation awareness process using empirical plan data. In: Proceedings of the 9th International Conference on Information Fusion. ISIF, Florence

Ferrin G, Snidaro L, Canazza S, Foresti G (2008) Soft data issues in fusion of video surveillance. In: Proceedings of the Eleventh International Conference on Information Fusion, Cologne, pp 1882–1889

Ferrin G, Snidaro L, Foresti GL (2010) Revisiting the role of abductive inference in fusion domain. In: 13th Conference on Information Fusion. IEEE, New York

Getoor L, Taskar B (eds) (2007) Statistical relational learning. Adaptive computation and machine learning. MIT, Cambridge

Hall D, McNeese M, Llinas J, Mullen T (2008) A framework for dynamic hard/soft fusion. In: Proceedings of the 11th International Conference on Information Fusion, Cologne

Hanson NR (1958) Patterns of discovery. Cambridge University Press, Cambridge

Harman GH (1965) The inference to the best explanation. Philos Rev 74(1):88–95

Hristidis V, Chen SC, Li T, Luis S, Deng Y (2010) Survey of data management and analysis in disaster situations. J Syst Softw 83:1701–1714

Josephson JR (2008) Abductive inferencing for integrating information from human and robotic sources. In: Proceedings of the 11th International Conference on Information Fusion, Cologne, pp 109–114

Lambert DA (2006) A unification of sensor and higher-level fusion. In: Proceedings of the 9th international conference on Information Fusion, Florence

Li X, Porikli F (2004) A hidden Markov model framework for traffic event detection using video features. In: International Conference on Image Processing, 2004 (ICIP '04), vol 5, pp 2901–2904

Little EG, Rogova GL (2005) Ontology meta-model for building a situational picture of catastrophic events. In: Proceedings of the Eighth international conference on Information Fusion. International Society of Information Fusion, Philadelphia

Llinas J (2002) Fusion for natural and man-made disasters. In: Proceedings of the fifth International Conference on Information Fusion. International Society of Information Fusion, Annapolis

Llinas J, Bowman CL, Rogova GL, Steinberg AN, Waltz EL, White FE (2004) Revisiting the JDL data fusion model II. In: Proceedings of the Seventh International Conference on Information Fusion, Stockholm, vol II, pp 1218–1230

Möller R, Neumann B (2008) Ontology-based reasoning techniques for multimedia interpretation and retrieval. In: Semantic multimedia and ontologies: theory and applications. Springer, New York

Munson LM (2009) Looking for anomalies all in the wrong places. In: USNI Proceedings Magazines 135/7/1,277

Newman M (2010) Networks: an introduction. Oxford University Press, Oxford

Pearl J (1988) Probabilistic reasoning in intelligent systems: networks of plausible inference. Morgan Kaufmann, San Francisco

Poole D (1993) Probabilistic horn abduction and Bayesian networks. Artif Intell 64(1):81–129

Raghavan S, Singla P, Mooney RJ (2014) Chapter 3 - plan recognition using statistical-relational models. In: Sukthankar G, Goldman RP, Geib C, Pynadath DV, Bui HH (eds) Plan, activity, and intent recognition. Morgan Kaufmann, Boston, pp 57–85. http://dx.doi.org/10.1016/B978-0-12-398532-3.00003-8. http://www.sciencedirect.com/science/article/pii/B9780123985323000038

Richardson M, Domingos P (2006) Markov logic networks. Mach Learn 62:107–136

Rogova GL (2009) Context-awareness in crisis management. In: Proceedings of the Military Communications Conference. IEEE, New York, pp 1–7

Rogova GL, Scott PD, Lollett C, Mudiyanur R (2006) Reasoning about situations in the early post-disaster response environment. In: 9th International Conference on Information Fusion

Russell SJ, Norvig P (2010) Artificial intelligence - a modern approach, 3rd edn., vol I-XVIII. Pearson Education, Upper Saddle River

Sambhoos K, Llinas J, Little E (2008) Graphical methods for real-time fusion and estimation with soft message data. In: 11th International Conference on Information Fusion. IEEE, New York, pp 1–8

Scott P, Rogova G (2004) Crisis management in a data fusion synthetic task environment. In: Proceedings of FUSION 2004

Snidaro L, Belluz M, Foresti G (2007) Domain knowledge for surveillance applications. In: Proceedings of the Tenth International Conference on Information Fusion, Quebec

Snidaro L, Visentini I, Foresti G (2011) Data fusion in modern surveillance. Studies in computational intelligence, vol 336, chap. 1. Springer, Berlin, pp 1–21

Snidaro L, García J, Corchado JM (2015a) Context-based information fusion. Inf Fusion 21:82–84. doi:10.1016/j.inffus.2014.02.001. Guest Editorial

Snidaro L, García J, Llinas J (2015b) Context-based information fusion: a survey and discussion. Inf Fusion 25:16–31. doi:10.1016/j.inffus.2015.01.002

Snidaro L, Visentini I, Bryan K (2015c) Fusing uncertain knowledge and evidence for maritime situational awareness via Markov logic networks. Inf Fusion 21:159–172. doi:10.1016/j.inffus.2013.03.004

Steinberg A, Rogova G (2008) Situation and context in data fusion and natural language understanding. In: Proceedings of the Eleventh International Conference on Information Fusion, Cologne

Sukthankar G, Goldman RP, Geib C, Pynadath DV, Bui HH (eds) (2014) Plan, activity, and intent recognition. Morgan Kaufmann, Boston

Vishwakarma S, Agrawal A (2013) A survey on activity recognition and behavior understanding in video surveillance. Vis Comput 29(10):983–1009. doi:10.1007/s00371-012-0752-6. http://dx.doi.org/10.1007/s00371-012-0752-6

Chapter 15
A Model for Threat Assessment

Alan N. Steinberg

Abstract A concept for characterizing, predicting and recognizing threat situations is developed. The goal is to establish a systematic approach to automation of some of these functions. Approaches are discussed to address the fundamental problems of (a) sparse and ambiguous indicators of potential or actualized threat activity buried in massive background data; and (b) uncertainty in threat capabilities, intent and opportunities.

15.1 What Is a Threat?

We speak of threats in a variety of ways in informal language. A threat can be:

- A *threatening act*, e.g., a verbal threat: "Putin said, 'I will occupy Crimea'";
- A *threatened action or situation* (a pending situation or event): "Putin is threatening to control Ukraine." A threatened situation is presumed to be an *unwelcome* situation: we might say "a threat of rain," but probably not "a threat of sunshine";
- A *threatening situation* (an extant situation that increases the likelihood of future unwelcome situations): "Russian troops in Crimea are a threat to world peace";
- A *threatening agent*: "Putin is a threat to world peace."

Threat Assessment involves assessing situations to determine whether undesirable events are occurring or will occur. Undesirability presumes an entity—and usually a *first-person* entity, the threat assessor—to whom an event is undesirable.

There is particular concern when threats are intentional, i.e., when an agent acts with the intent of causing outcomes that are undesirable to another agent. In such cases we speak of *intentional threatening acts and situations or of threatened acts, events* and *situations.* We can follow ordinary usage and use "threat event or situation" as equivalent to "threatened event or situation" (Steinberg 2005, 2014).

A.N. Steinberg (✉)
Independent Consultant, Lake Ridge, VA, USA
e-mail: alaneilsteinberg@gmail.com

G. Rogova, P. Scott (eds.), *Fusion Methodologies in Crisis Management*,
DOI 10.1007/978-3-319-22527-2_15

Various researchers have concentrated on one or another aspect of threat assessment in defining "level 3" data fusion:

- *Temporal aspect*: predicting future states (Salerno 2007; Endsley 2000); projecting states (whether past, present, or future) (Steinberg and Bowman 2009);
- *Conditional/causality aspect*: estimating and predicting scenarios and outcomes that have not (yet) occurred or are not known to have occurred; *vice* observed situations (Lambert 2006);
- *Utility aspect*: estimating cost of situations (usually of future or contingent situations) (Salerno 2007; White 1988).

Fusion levels and their involvement in threat assessment are discussed in Steinberg (2015a, b) and in Sect. 15.2.

Generally speaking, threat assessment includes the following functions:

- *Threat Event Prediction*: estimating the characteristics and likelihoods of pending bad situations or events ("attacks" in case of intentional threats): who, what, where, when, why, how;
- *Indications and Warning*: estimating that a threat event is occurring or is imminent;
- *Threat Event Assessment* (*Attack Assessment*, for intentional threats): determining responsible parties (e.g., country, organization, individuals) and roles; threat capability (e.g., weapon and delivery system characteristics; intended target(s) of attack); intended outcomes (physical, political, economic, psychological states); tactics, plans, and procedures (goal and plan decomposition) (Steinberg 2005);
- *Outcome Assessment*: Estimation and prediction of event outcome states and their cost/utility to the responsible parties, to affected parties, or to the threat assessor him/herself. These can include both intended and unintended consequences.

Of course, "tactics, plans, and procedures" and "intended targets" only apply literally for intentional threats. The remainder of these functions apply to unintentional threats as well: to natural disasters or inadvertent consequences of human action.

15.2 Representing and Reasoning About Threats

Threat assessment intrinsically involves reasoning about interactions and other relationships among entities, about situations, their progression over time and outcomes. Reasoning about relationships and situations, situation dynamics and outcomes has been considered the province of "level 2" and "level 3" data fusion.

A data fusion process has the role of estimating entity states of interest within a problem domain. It may operate at one or more of the "data fusion levels";

generating corresponding estimates of (level 0) patterns, (level 1) individuals, (level 2) relationships/situations, (level 3) scenarios/outcomes, or (level 4) the state of the inferencing system itself—e.g., of the threat assessor.[1]

Level 1 data fusion is concerned with estimation of states of entities considered as individuals. That is to say, it is concerned with *attributive* states: with values of 1-place state variables such as target location, type, or attributive parameters.

In contrast, fusion levels 2 and 3 are concerned with estimation of entities considered as aggregates: respectively, as relationships situations and courses of action or scenarios (Steinberg 2009, 2015a, b; Steinberg and Rogova 2015). In level 2 and level 3 fusion, both attributive and *relational* states are pertinent: values of *n*-place state variables, $n \geq 1$. Level 2 and 3 inferences have direct analogy to those at level 1: situation recognition is a problem akin to target recognition. Situation/scenario tracking is akin to target tracking (Lambert 2006; Steinberg 2009).

Level 2 fusion—Situation Assessment, whether implemented by people, an automatic process or some combination thereof—requires the capability to make inferences of the following types:

- Inferring relationships among entities on the basis of their estimated attributes;
- Inferring the presence and the states of entities on the basis of relationship in which they participate;
- Inferring relationships on the basis of other relationships (e.g., coordinated actions);
- Recognizing and characterizing extant situations.

Level 3 fusion is distinguished from level 2 in the way information used in estimation relates to the estimated states. The distinction is between states that are observed and states that are not, but rather are inferred based on indirect evidence (Steinberg 2015a, b).

Types of level 3 fusion inferences include:

- *Conditional Event/Situation Prediction*: "If *x* were to follow this course of action, what would be the outcome?" This involves a reactive environment (one that responds differentially to actions) and often involves one or more responsive agents; often assumed to be capable of intentional activity. In Military

[1]We argue in Steinberg (2015a) and Steinberg (2015b) that it makes more sense to distinguish inference problems on the basis of type of entity state variables rather than by type of entity: a given entity can be addressed at more than one level. For example, a vehicle can be the "target" of a level 1 fusion process if level 1 states (e.g., its location, velocity, size, weight, type, identity, or activity) are being estimated. The same vehicle can be the "target" of a level 2 process if it is considered as a complex or a *structure*, such that level 2 states (e.g., the relationships among its components or its subassemblies) are being estimated. It could also be the subject of a level 3 process if it is considered as a *dynamic process*, such that level 3 states (e.g., its course of action and outcome states) are being estimated. It could even be the subject of a level 4 process if it is considered as the system performing the estimation and level 4 states (e.g., the operating conditions and performance relative to users' objectives) are being estimated.

Threat Assessment, the reactive element consists of "our" forces, concerned with intentional actions and reactions of hostile agents;
- *Forensic Projection*: "What past scenario caused the present evidence?"
- *Conditional/Counterfactual Event/Situation Prediction*: "If x had followed this course of action, what would have been the outcome?" Inferencing includes estimation of (conditional/counterfactual) outcome and cost, which may be performed using Bayesian cost analysis. Once again, the focus is usually on outcomes of "our" alternative courses of action.

In the first two of these cases, inferencing is of the temporal evolution of a situation—i.e., of a scenario—some of which has not been observed. In the third case, inference is on the basis of assumptions, possibly counterfactual assumptions.

In general, inference of unobserved states involves estimation of evolving situations: of courses of actions and their interactions and outcomes, i.e., of *scenarios* (Lambert 2006).

Clearly, threat assessment can have elements of level 1 fusion (e.g., agents' capabilities); of level 2 fusion (e.g., relationships with various assets of concern that might present opportunities for attack or desires to attack); and of level 3 fusion (e.g., prediction of evolving interactions, to include attacks).

Given the above definition, threats are modeled in terms of potential and actualized relationships between threatening entities (which may be people or human agencies) and threatened entities, or targets (which often include people or their possessions).

Characterizing situations and scenarios is generally a matter of assessing the states of constituents and their interrelationships. We can model a relationship as a graph having nodes for attributes/relations and attributed/related entities. A situation is represented as a network of such relationships. A *concrete* situation is one consisting of concrete relationships: ones in which all variables are anchored (i.e., have instantiated values in the assumed universe of discourse).

As we will want to reason about relations and relationships, it will be convenient to treat these as entities, i.e., as values of sets of random variables. Figure 15.1 is a graphical representation of a situation, in this case a situation of an aircraft being tracked by a surface-to-air missile (SAM) system. The situation is comprised of the entities involved and their attributes, relations, and relationships.

We can depict a level 2 hypothesis as a belief network after the pattern of Fig. 15.1. This figure is in the form of a factor graph, in which variables are represented as circles and functions on these variables are represented as squares (Kschischang et al. 2001). Examples of such functions are causal conditions or conditional probabilities, but they can represent any relationship among variables. Nodes in the belief network represent estimated values of state variables at any fusion level (corresponding to an individual, instantiated attribute or relation, or an abstract attribute or relation). Each such node in the belief network combines the effects of evidence from its immediate neighbors and distributes its own evidence to them (requiring provisions that information is not circulated back to an originating node).

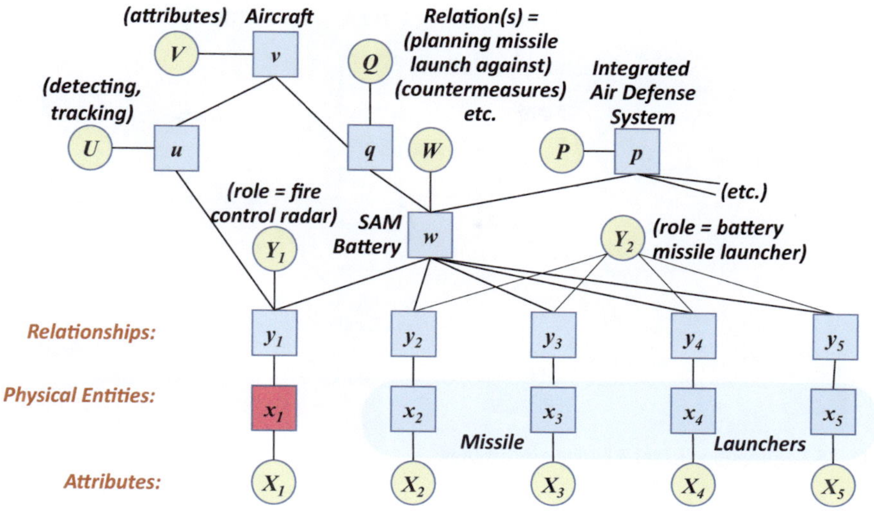

Fig. 15.1 Factor graph representation of example situation

In our application, functions can include instantiations: individual entities, relationships and situations, etc. In this example, nodes x_i are individual physical elements of the missile SAM battery—radar and launchers—(or, more correctly, the corresponding instantiated attributive variables); the X_i are the abstract variables themselves. Similarly, the y_i are relationships—the relationships of fire control radar or of missile launcher to a SAM battery—being instantiations of the relational variables Y_i, etc.

A scenario can be modeled as a dynamic network and a level 3 hypothesis as a dynamic belief network. A hypothesis of a threat scenario includes

- Inference of threatening situations: of the conditions for one entity x to (adversely) affect entities y;
- Prediction of threat actions and outcomes: interactions whereby entities adversely affect others.

Threat situations, scenarios, and events are inferred on the basis of the attributes and relationships of the entities involved. Figure 15.2 is an example of a level 3 hypothesis, simplified to show only instantiation entities and only at five particular times. As discussed in Sect. 15.3, pre-incident indicators concern the capability and opportunity ("Cap" and "Opp" respectively in the figure) to cause various events. In the case of intentional threats, indicators of intent "(Int" in the figure) are also pertinent. In the illustrated example, the concern is about such acts as they affect a given asset (e.g., a person or facility).[2]

[2]McMichael, Jarrad, and colleagues at CSIRO (Commonwealth Scientific and Industrial Research Organisation) developed methods for efficient storage, search, and manipulation of such dynamic

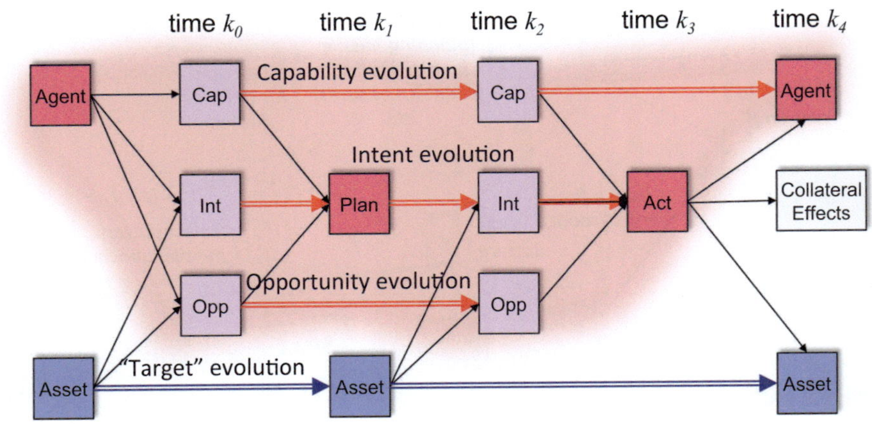

Fig. 15.2 Simplified Level 3 hypothesis example

15.3 Threat Prediction

15.3.1 Elements of Prediction

In the best-selling book *The Gift of Fear* (de Becker 1997), de Becker describes factors that have been bound to be useful in assessing predictions, slightly paraphrased in Table 15.1:

Elements 1, 3, 4, 5, 7, and 10 concern factors of the class of outcomes under consideration; while elements 2, 4, 6, 8, 9, and 11 concern the knowledge, skills, biases, and priorities of the person making the prediction. Element 4 concerns both as it relates to the contextual dependencies of the outcome and of the predictor's knowledge of the pertinent contextual information.

As de Becker notes, pre-incident indicators ("PINs") are generally the most valuable of these predictive elements. Although indicators of events differ with the type of event, a useful general model was proposed by Little and Rogova (2006) and adapted by this author (Steinberg 2005, 2009, 2014), as well as by Garcia et al. (2012). Indicators of threat situations relate to the capability, opportunity, and—where purposeful entities are involved—intent of entities to carry out such actions against various "targets."

Capability, Opportunity, and Intent factors in predicting purposeful action relate directly to the military's "Ready, Willing, and Able" and to the "Means, Motive, and Opportunity" of criminal law. Roughly, these address the following questions:

---

belief networks, employing what they call grammatical methods for representing and reasoning about situations and scenarios (McMichael and Jarrad 2005).

Table 15.1 The elements of prediction [(de Becker 1997), pp. 114ff]

1. Measurability of outcome	How measurable is the outcome you seek to predict? Will it be clear if it happens or does not happen?
2. Vantage	Is the person making the prediction in a position to observe the pre-incident indicators and context?
3. Imminence	Are you predicting an outcome that might occur soon, as opposed to some remote time in the future?
4. Context	Is the context of the situation clear to the person making the prediction?
5. Pre-incident indicators (PINs)	Are there detectable pre-incident indicators that will reliably occur before the outcome being predicted?
6. Experience	Does the person making the prediction have experience with the specific topic involved?
7. Comparability	Can you study or consider outcomes that are comparable?
8. Objectivity	Is the person making the prediction objective enough to believe that either outcome is possible?
9. Investment	How much is the person making the prediction invested in the outcome?
10. Replicability	Is it practical to test the issue being predicted by trying it elsewhere?
11. Knowledge	Does the person making the prediction have knowledge about the topic?

- *Capability*: "What do they have and what can they do with it?"
- *Intent*: "What do they want to do and how determined are they to do it?"
- *Opportunity*: "How ready are they to do it?"

Capability, intent, and opportunity can be seen as collectively *sufficient* conditions for an event to occur. Also, capability and opportunity are individually *necessary* conditions for an agent to cause an event; intent is not, as unintended events can occur (either as inadvertent consequences of intentional acts or because no purposeful actor is involved).

Figure 15.3 shows the structure of a model that combines these factors in generating and evaluating possible threat events. The Threat Assessment process described in Sect. 15.7 and illustrated in Sect. 15.3.2 uses such a model to (a) generate, evaluate, and select hypotheses concerning entities' capability, intent, and opportunity to carry out an attack and (b) identify indicators, warnings, and characterizations of attacks that occur.

An entity's capability to carry out a specific type of threat activity depends on its capability: its ability to control physical assets and (as appropriate to the

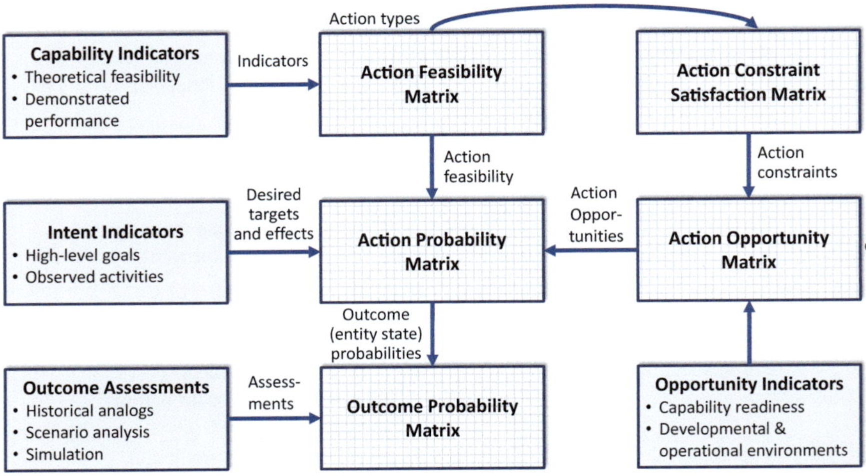

Fig. 15.3 Elements of a threat situation hypothesis (Steinberg 2005)

type of threat entity) the skills and knowledge to use them to perform the action. The hypothesis generation process searches for indicators of such capability and generates a list of feasible threat actors and activities.

An entity's intent can be described in terms of

(a) Desired outcomes: desired states of "target" individuals, relationships, situations, and scenarios and in the threat agent itself ("What will this attack do for ME?");
(b) Planned courses of action to achieve these outcomes.

Assuming a putative rational agent, planning would involve a utility function on outcomes with Bayesian cost.

A postulated type of threat activity constrains a threat entity's opportunity to carry out an attack against particular targets, e.g., to deploy or deliver a type of weapon as required to effect the desired attributive or relational state changes in these targets. Other constraints are determined by the target's accessibility and vulnerability, and by the threat entity's assessment of its capabilities and opportunities to perform the proposed course of action and of the outcome.

Capability assessment makes use of physical, informational, and cognitive models as available. Intent assessment uses indicators of investment of resources in a postulated capability and opportunity, as well as models of agents' goals and methods. Opportunity assessment uses models of agents, of potential targets, and of operating environments: of agents' physical deployment, that of targets and their susceptibility and vulnerability to various actions and of operating environment as needed to evaluate access and action outcomes.[3]

15.3.2 An Example: T3 (Technology, Technique, Tactic) Forecasting

Let us examine the application of Threat Assessment using Capability/Opportunity/Intent methods. The particular concern is the predictions of disruptive developments of technologies, techniques, and tactics (T3). Such assessment is important in Scientific and Technology Intelligence for National Security: what technologies, techniques, or tactics are being or could be developed that might be used militarily or otherwise to harm our national interests? Similar assessments are also useful in business: what products or methods are competitors developing that could put us at a competitive disadvantage?

In these applications, the COI-associated questions become:

- *Capability*: "What technology, technique, or tactic do they have? What can they do with it?"
- *Opportunity*: "How ready are they to use it?"
- *Intent*: "What technology, technique, or tactic do they want? How determined are they to get it? What do they want to do with it?"

Each of these dimensions used in T3 forecasting regarding entities of concern (e.g., those involved in developing or acquiring threatening capabilities) is approached from two complementary directions (Fig. 15.4):

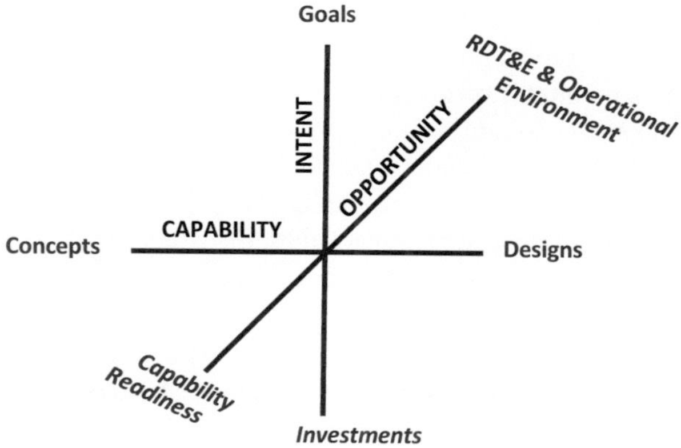

Fig. 15.4 Dimensions of threat assessment

[3]Means of threat assessment in the absence of complete, high-confidence models of such factors are addressed in Sect. 15.6.

Capability assessment involves both

- Concept-driven assessment: concerning the applicability and effectiveness of technologies, techniques, and tactics (T3) in various mission application; mapping from problems to solutions; and
- Design-driven assessment: determining the impact of specific observed or imputed designs.

Opportunity assessment involves both

- Internal opportunity assessment: identifying indicators of capability maturity (state-of-the-art); and
- External opportunity assessment: determining the availability of an enabling developmental and operating environment, e.g., target (or market) access and vulnerability; materials and systems; management command, control; required data; personnel, training, and readiness.

Intent assessment involves both

- Top-down, goal-driven assessment: analyzing high-level goals of subject agents, identifying technologies/techniques/tactics (T3) of interest; and
- Bottom-up, investment-driven assessment ("How much effort is being put into it?"): accounting for observed activities, such as reported system and component designs, their performance/maturity/application; and magnitude, maturity, and persistence of pursuit of T3 relevant to mission capability; to include investment in pertinent infrastructure.

A process for performing and integrating these factor assessments is shown in Fig. 15.5 and described below. Throughout this process, analysts are asked to assign probabilities and influence weights to various factors in evaluating the capability, opportunity, and intent of subject agents (in this case, scientific, engineering, or operational organizations) to develop and employ novel technologies, techniques, or tactics.

Such probabilities and weights will be derived from predictive models developed either from historical statistics or analytic models as available.[4] In many instances, subjective weightings and probability ranges provided by subject-matter experts or a Dempster–Shafer plausibility and belief calculus will be useful to capture uncertainly in specific probability assignments. If attainability and impact decisions and resulting response decisions (e.g., decisions to invest in pertinent research and development) are found to be highly sensitive to uncertainty in probability weighting assignment, detailed process-flow modeling tools can be employed to refine assessments. In many cases, analytic results have been found to be remarkably insensitive to errors in subjective probabilities and weights (Kahneman 2011).

[4]Section 15.6 discusses categories of inference problems in which predictive statistical models of entities and activities of concern are not available.

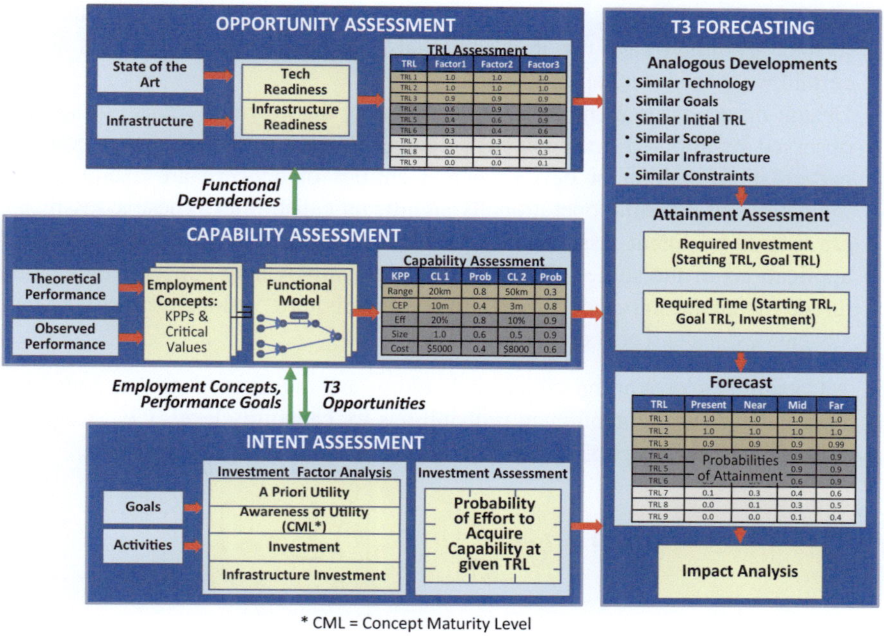

Fig. 15.5 Technology, technique, and tactic (T3) forecasting process

At the very least, the analyst-entered values permit the assumptions used in analysis to be explicitly articulated, facilitating sensitivity analyses of the conclusions as well as comparative analysis of alternative hypotheses.

Capability Assessment

Capability Assessment involves evaluating the probabilities that specific performance values are achievable by the use of a given technology, technique, or tactic (T3). Measures of performance for a given T3 area are expressed as Critical Technology Parameters (CTPs). For example, CTPs for a ballistic missile might include range, payload capacity, cost, reliability, robustness, mobility, and life cycle cost.

Capability Assessment is based on models of potential technology employments and can include both concept-driven and design-driven assessments:

- Concept-driven assessment involves maintaining cognizance of theoretical and T3 developments worldwide to ascertain the expected applicability and effectiveness of such developments in operational applications. Such assessments reduce the likelihood of "technology surprise." In concept-driven assessment, the analyst postulates potential employment concepts for a new technology, technique, or tactic. For example, the analyst might speculate that a new nanomaterial technology might have use in biomaterials, various consumer products and in certain weapon system applications.

- Design-driven assessment involves ascertaining the potential impact of specific observed or imputed developments. Technical information sources specific to particular acquisition or developmental entities of concern are examined for indications of potential operationally significant capabilities. These assessments are filtered by Intent Assessments to indicate potential system and operational applications of the subject T3. In such cases, employment concepts may come directly from the source documentation or may require "red-team" postulation of system designs and concepts of operation.

A Capability score for the subject technology, technique, or tactic T is computed as a normalized sum of attainment probabilities for capability levels CL_n, weighted by priorities for such capability levels to various employment concepts EC_m:

$$Cap(T) = \frac{1}{MN}\sum_{n=1}^{N}\left[p\left(CL_n \middle| T\right)\sum_{m=1}^{M} W_{m,n}\right].$$

Opportunity Assessment

Opportunity Assessment concerns the availability of conditions necessary for implementing and using of subject technologies, techniques, or tactics. These enabling conditions include the maturity of the subject T3 itself, expressed as a Technology Readiness Level (TRL).[5] Enabling conditions also include factors within operational environments and pertinent research, development, test, and evaluation (RDT&E) environments. Such factors may include the availability of critical components, producibility, affordability, and related risks.

Thus, Opportunity Assessment involves an assessment of current or baseline TRLs, both for endogenous factors in T3 performance (e.g., material purity, device scale), as well as for supporting factors (materials availability, producibility, test and evaluation facilities, personnel skills, etc.).

An Opportunity score for a given technology/technique or tactic T is defined as the expected TRL expressed as the average expected TRL of enabling factors F_m weighted by their importance to T:

[5]Technology Readiness Levels as defined by the US Department of Defense are (TRL1) basic principles observed and reported; (TRL2) technology concept and/or application formulated; (TRL3) analytical and experimental critical function and or characteristic proof-of-concept; (TRL4) component and/or breadboard validation in laboratory environment; (TRL5) component and/or breadboard validation in relevant environment; (TRL6) system/subsystem model or prototype demonstration in relevant environment; (TRL7) system prototype demonstration in operational environment; (TRL8) actual system completed and qualified; and (TRL9) actual system proven in successful mission operations (paraphrased from Nolte et al. (2003)).

$$Opp(T) = \widehat{TRL}(T) = \frac{1}{M_T}\sum_{m=1}^{M} W_T(F_m)\frac{1}{N_{TRL}}\sum_{n=1}^{N_{TRL}} np\left(TRL\left(F_m\right) = n\right);$$

where M_T is the number of pertinent factors and $N_{TRL} = 9$ the number of TRLs.

Intent Assessment

Intent Assessment involves both (a) a top-down assessment of the goals of the entity of concern, and likely tactics, plans, and procedures for achieving those goals (goal decomposition); and (b) a bottom-up assessment of perceived investments of money and effort in factors that affect the development of capabilities.

Top-down, goal-driven assessment of the given entity is performed by analyzing the presumed strategic goals of an entity of concern. Plausible courses of action and development/employment concepts are postulated.[6] These are used to estimate the cost-effectiveness of the subject technology (etc.) in various employment concepts.

Such top-down reasoning may provide indications that it would be very much in the interest of the given entity to develop a certain technology. For example, a country that has suffered repeated air attacks by a neighboring country can be expected to seek improved air defense capabilities.

In such cases, employment concepts are defined that trigger Capability and Opportunity Assessments as described above. Relevant indicators of such intent include indications of interest in an operational capability (e.g., "over-the-horizon air surveillance radar") together with indications that the given entity considers the specific technology to be an effective and achievable solution (expressed as "Concept Maturity Levels," as defined below).

Bottom-up, activity driven Investment Assessments involve evaluating the likelihood that reported or predicted design, development, test, acquisition, etc., activities are being undertaken to achieve increased TRLs for given capability levels. Investment Assessment involves evaluating factors that affect the pace of TRL progression for particular capability levels, i.e., level of investment and effort in the development of the above opportunity factors. Relevant indicators of investment include

- Indications of unusual interest in a given technology, technique, or tactic via information-gathering activities (conferences, seminars, etc.) and expressions of interest (cultivation of expertise, requests for information, etc.);
- Indications (in technical publications, conference presentations, advertisements, trade publications, etc.) concerning the magnitude, maturity, financial and political support for activities in pursuit of specific technologies, techniques, or tactics; and
- Investments in the relevant infrastructure. These can include existing and projected system environments (sensors, communications, processing, controls,

[6] Such inferences are examples of Category 3 reasoning as discussed in Sect. 15.7.

databases, operator training, etc.). Key to ascertaining the relevance of such activities is the funding and organizational relationships of T3 developments.

In order to want to use a technology, technique, or tactic for some purpose, an agent must appreciate the utility of such employment. History would undoubtedly have been much different had Hannibal been aware of certain uses of saltpeter, sulfur, and charcoal. We express such awareness as an agent's Concept Maturity Level (CML).[7] The utility of a technology, technique, or tactic as perceived by an agent α is determined as a function of the respective CML together with the cost-effectiveness of the given T3 across employment concepts as perceived by α. A weighted-sum using analyst-modifiable weights is calculated across CML, cost-effectiveness and investment factors. This defines an Investment score:

$$\omega_\alpha(T) = \sum_{EC} \left[\omega_\alpha\,(EC)\,CostEff_\alpha\,(T, EC)\,\frac{1}{N_{CML}} \sum_{n=1}^{N_{CML}} p\,(CML\,(T, E, \alpha) = n) \right];$$

where $N_{\mathrm{CML}} = 7$ is the number of CMLs. This Investment score will be to determine a probability of sufficient investment as used described in the next section.

15.4 Attainment, Technology Convergence, and Impact Assessment

Capability, opportunity, and intent factors are combined to project future TRL milestones for given Capability Levels. This process combines opportunity and intent factors to estimate the effects of the assessed level of investment (in the broadest sense) on the TRL progression of significant Capability Levels.

A knowledge base of historical T3 developments should be maintained to facilitate projection to the development being assessed. These are supplemented by knowledge acquired by subject-matter experts.

The probability that entity α will develop a subject technology from TRL m to TRL n within a certain number of years can be estimated from experience with similar activities:

$$p\left(TRL_m(T)\middle|A_i\right) = \sum_j \frac{\$A_i}{A_j} Sim\,(A_i, A_j)\,Success\left(A_j, m, n\right);$$

[7] JPL defined Concept Maturity Levels (CMLs) as (CML1) Cocktail Napkin (i.e., rudimentary understanding of the concept); (CML2) Initial Feasibility; (CML3) Trade Space; (CML4) Point Design; (CML5) Concept Baseline; (CML6) Initial Design; and (CML7) Integrated Baseline (Wessen et al. 2010).

where Success(A_j, m, n) is the historical degree of success in moving from TRL m to n by means of activity A_j and Sim(A_i, A_j) is a metric between 0 and 1 indicating the degree of similarity between the technology development activities. Assuming independence among factors, Sim(A_i, A_j) is a weighted summation over such factors as technology types, developers, theoretical foundations, materials, processes, initial TRL, goal TRL, infrastructure/available resources, development time, and sociopolitical constraints (e.g., permissibility of live subject testing).

By projecting from analogous historical developments, determination is made of the probability that a particular TRLs are achievable in near-, mid- and far timeframes for the subject technology, technique, or tactic, given the estimated level of investment.

Level 3 hypotheses at system or mission levels extend the single-technology/technique/tactic forecasts to forecasts of investment and opportunities for future capabilities afforded by the integration of multiple T3 items. T3 convergence points are occasions when there are opportunities for integration of multiple emerging technologies, techniques, and tactics, for significant new system-level or mission-level capabilities. T3 convergence points are forecast on the basis of capability development models; of a set of structural equations that represent the causal dependencies of various factors in attaining a particular capability. These structural models can be documented as level 3 hypotheses after the pattern of Fig. 15.2; indicating causal dependencies, relationship qualifications and process state transitions.

As necessary, the impact of new developments is assessed by modeling and simulation of the postulated T3 insertion into an operational environment: involving simulated or actual collection, analysis and exploitation systems.

Short of these expensive and often infeasible methods, impacts are assessed analytically using simplified performance models of systems with the new development, using the predicted capability level in terms of values of Critical Technology Parameters (CTPs).

15.5 Why Is Threat Assessment Hard?

The relative difficulty of Scenario Assessment, and particularly of Threat Assessment, can largely be attributed to the following three factors:

- *Weak spatiotemporal constraints*: Evidence relevant to a level-1 estimation problem (e.g., target recognition or tracking) can be assumed to be contained within a small spatiotemporal volume, generally limited by kinematic or thermodynamic constraints. In contrast, many situation and threat assessment problems can involve evidence that is widespread in space and time, with no easily defined constraints;
- *Weak ontological constraints*: The types of evidence relevant to threat assessment problems can be very diverse and can contribute to inferences in

unexpected ways. This is why much of intelligence analysis—like detective work—is opportunistic, *ad hoc*, and difficult to codify in a systematic methodology. Rather, the methodology in threat assessment is second-order: not to discover instantiations of pre-scripted threat scenarios, but (1) to discover patterns of activity that are consistent with unanticipated threat scenarios and (2) to nominate searches for data that could either confirm or refute such possibilities;
- *Weakly modeled causality*: Threat assessment often involves inference of human intent and human behavior. Such inference is basic not only to predicting future events (e.g., attack indications and warning) but also in understanding current or past activity. Needless to say, our models of human intent, planning and execution are far less complete and far more fragile than the physical models used in target recognition or target tracking (Steinberg 2009).

These are the reasons that the problem of recognizing and predicting terrorist attacks is much different from that of recognizing or predicting the disposition and activities of conventional battlefield equipment.

Often the key indicators of potential, imminent or current threat situations are in the relationships among people and equipment that are not in themselves distinguishable from common, non-threatening entities.

15.6 The Human Threat

de Becker (1997, pp. 110ff) identifies four considerations in a person's decision to undertake violent acts. We paraphrase these as:

- *Perceived Justification*: does the person feel justified in using violence?
- *Perceived Alternatives*: does the person believer that he has available alternatives to violence that will move him toward the outcome he wants?
- *Perceived Consequences*: How does the person value the likelihood and cost/utility of consequence associated with the contemplated act?
- Perceived Ability: *does the person believe he can successfully perform the contemplated* act?

These are all factors that condition the agent's intent to undertake a contemplated action. They are his perceptions—internal estimates—of his own capability (ability), opportunities (alternatives), goals and priorities (justification) and of outcomes (consequences). To assess such an agent's likelihood of undertaking a given action, we would need a way to evaluate the mental processes and mental states that determine such perceptions.

Unlike physical state inference problems (target recognition and tracking); inferring mental states involves problems of:

- *Observability*: psychological states are not directly observed but must be inferred from physical indicators, often on the basis of inferred physical and informational states;

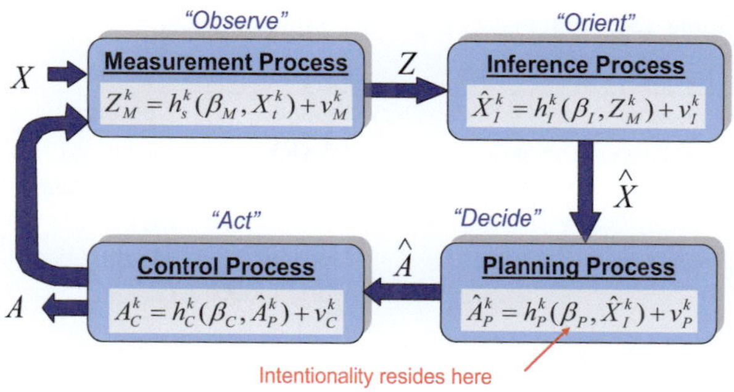

Fig. 15.6 Model of responsive agent behavior (Steinberg 2005)

- *Complexity*: the causal factors that determine psychological states are numerous, diverse, and interrelated in complex ways;
- *Model Uncertainty*: these causal factors are not well understood, in comparison with the physical models that allow us to recognize and predict target types and kinematics (Rogova and Steinberg 2015; Rogova and Nimier 2004).

Nonetheless, it is well to remember that, although human behavior is complex and poorly modeled, with factors that are difficult or impossible to observe, it is nonetheless not random.

Boyd's "OODA loop" (Boyd 1976) is an admittedly crude model of responsive human behavior; ignoring the may subtle feedback processes involved. Yet it may serve as a simple analytic tool to characterize the effects of sensing, inference, planning, and control on agents' behaviors. Figure 15.6 shows a formal interpretation of the OODA loop, decomposing a responsive agent's behavior in terms of process models corresponding to the four OODA elements (Steinberg 2005, 2014).

In the present discussion, we are concerned with these processes as they affect agents' behavior (a) as indications of allegiance, activity, and intent, and (b) as an means to qualify information received (directly or indirectly) from such agents. Observable activity—to include systematic and random errors—will be functions of the source's acquisition and inference of information and his planning and control of what and how to respond. Table 15.2 maps the Capability/Opportunity/Intent factors into the four functions of this responsive agent model.

Interactions among purposeful agents can be represented as networks of interacting OODA-loops, each of which represents the processes of a responsive agent. Network participants can "communicate" with one another intentionally or inadvertently by many means: by conversations, documents and activities that may be observed by intended or unintended recipient agents.

As illustrated in Fig. 15.7, each such agent "observes" (in the broadest sense of being stimulated in one way or another) the results of various agents' actions; he "orients" himself by drawing inferences from received stimuli and other available information; he "decides" how to respond to the perceived changing world; and he "acts" according to such decisions, given his ability and opportunity to act.

We can think of the ensemble of people who interact with one another in any way—intentionally or unintentionally, whether by radio or by gunfire—as an open

Table 15.2 COI factors in general response model

Process		Output	Factors		
Function	OODA corollary	Product	Opportunity	Capability	Intent
Measurement	"Observe"	Measurements (Z_k)	World and system state (X_k)	Sensor model	Sensor controls
Inference	"Orient"	Updated world model $\left(\widehat{X}_k\right)$	Measurements (Z_k)	Inference model, prior world model $\left(\widehat{X}_{k-1}\right)$	Information needsInference controls
Planning	"Decide"	Planned course of action ($\hat{A}_k$)	Updated world model $\left(\widehat{X}_k\right)$	Planning model	Goals/priorities, planning controls
Control	"Act"	Course of action (A_k)	World and system state (X_k)	Control model	Controls

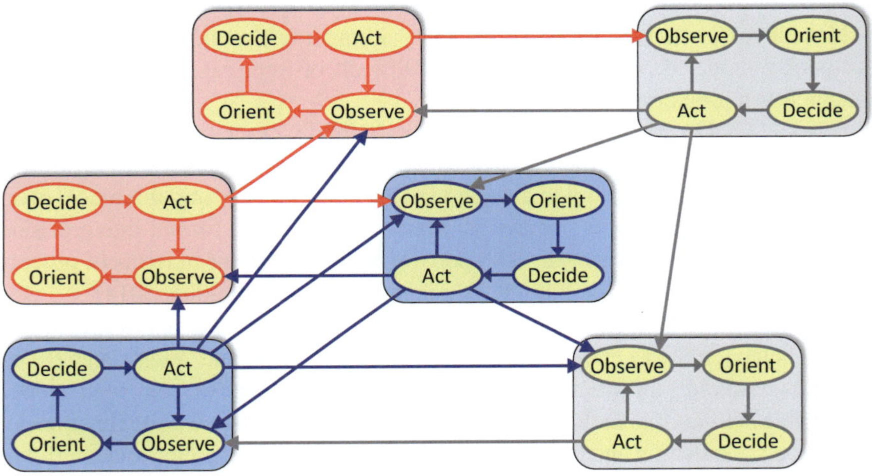

Fig. 15.7 Network of interacting agents (Steinberg 2005)

causal network. Clearly, understanding and predicting the behavior of any one of these interacting agents requires an ability to characterize the network of such interactions and the responsive behavior in each agent.[8]

[8]Interpersonal interactions and influences add to complexity and model issues. There are well-known issues of emergent group behavior that are arguably not predictable by constructive modeling.

15.7 Categories of Inference Problems

One would like to be able to exploit all aspects of the problem in estimating and predicting activity states of concern: objects and activities of interest as well as background objects and activities ("normal" states and behaviors). Unfortunately—as in many applications infused with unpredictable human behavior—we cannot rely on any or all of these to discriminate many kinds of threat agents and their activities.

Table 15.3 lists various categories of inference problems, adapted from a taxonomy developed by Waltz (2003). These categories employ different inference methods because of the differing availability of predictive models. Because higher-level categories are successively less able to rely on predictive target models, they rely successively more on clues or constraints provided by context.

15.7.1 Category 0: Model-Based Recognition

Category 0 includes the problems assumed in traditional data fusion systems, in which state estimation is addressed as a conventional model-based recognition problem. It is assumed that predictive models of the observable features and behaviors of entities of interest (at any fusion level) are sufficiently well-known that they can be recognized by comparing observations with predictions of such observations by deduction from models of entities and measurement conditions.

Often, however, target characteristics and behaviors are situation-dependent. In such cases it is important to use relevant situational information for context in selecting, conditioning or otherwise adapting target models. Contextual information in the form of known or estimated higher-order states—situations, courses of action, scenarios, etc.—can be used to provide expectations and resolve otherwise ambiguous low-level states. Among situational information that can readily be

Table 15.3 Approaches to diverse inference problems

Problem category	Approach
Cat 0: Model-based recognition	• Data-driven target model • Signature/behavior recognition
Cat 1: Anomaly-based detection	• Data-driven normalcy model • Anomaly detection/diagnosis
Cat 2: Hypothesis-based explanation	• In-operation composition of situation/scenario hypotheses to account for data • Predict and seek differential observables
Cat 3: Context-based feasibility	• Contextual indicators of concern • Feasibility inferred from contextual constraints • Hypothesis generation and testing (probe)

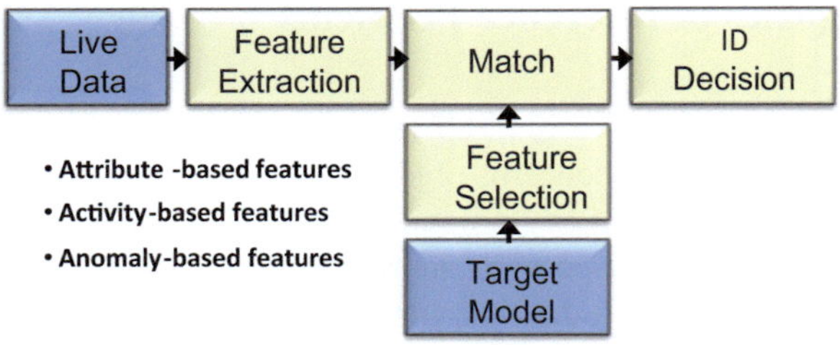

Fig. 15.8 Generic model-based recognition scheme

exploited are induced phenomenology in the measurements: shadows, wakes, contrails, plumes, etc. (Steinberg and Rogova 2015).

Figure 15.8 depicts the traditional recognition scheme whereby "targets" are recognized by comparison with reference data that represent expected target characteristics, whether in the form of historical data sets, templates of expected observables or functional models from which observables may be predicted (Steinberg 2013). Features extracted for comparison between observations and models may represent physical attributes (size, shape, materials, etc.) or activity patterns (motion, interactions, or other behaviors).

We can subdivide Category 0 to distinguish cases where target models are "given" from those in which target models are derived by statistical learning:

- *Category 0a*: in which the actual target state (at whatever state estimation level) is known absolutely, e.g., under controlled test conditions;
- *Category 0b*: in which definitive predictive models of targets and their behaviors are available explicitly from design documentation or are derivable analytically from first principles;
- *Category 0c*: in which predictive models are derived inductively from training data.

Unfortunately, model-based recognition is in many ways inadequate for the many applications in which the availability of required models cannot be assumed. Actors of concern in many threat scenarios are highly variable in their attributes and activities; often with highly nonlinear and dynamic statistical properties. The dependencies that would allow observable features and activities to be predicted are often complex and poorly understood, involving the caprices of human behavior.

15.7.2 Category 1: Anomaly-Based Detection

Category 1 concerns problems in which target signatures or behaviors might not be well characterized, but in which backgrounds can be characterized, allowing non-background activity to be detected by contrast. In other words, category 1 problems can be addressed using model-based recognition methods, but where recognition is primarily of background features and activities; counting on "target" entities and activities to be detectable as anomalies relative to recognized backgrounds.

Such methods are widely used in Automated Target Recognition when target models are either unavailable or suspected of being unreliable (Steinberg 2013). Anomaly-based methods have been adapted to such applications as Maritime Domain Awareness, attempting to detect aberrant human, social, cultural behaviors by contrast with the "patterns of life" that characterize normalcy. Recognition is based on models of classes of "normal" vessel activities: fishing, ferrying, pleasure cruising, etc. Contextual models are employed to account for variability in such activity patterns due to weather, regulations, hazards to navigation, etc. Deviations from such patterns—"anomalies"—are then examined as potential cases for concern (Steinberg 2013; Moore 2005).

Anomaly-based methods in such applications face problems in the enormous variability and complexity of "normal" targets and activities, i.e., in the great diversity of vessels and activities that have no threat association. Patterns of such "normal" activities as fishing, recreational boating, and distressed vessels—behaviors that pirates or terrorists may assume as deceptions—can be very difficult to model: highly variable, nonlinear, and ambiguous, and subject to numerous contextual factors. Anomalies may be very subtle, requiring deep understanding for modeling and very sophisticated detection processes. Failure to capture the range of such normal behaviors will result in false alarms and missed detection of threat activities (Steinberg 2013).

What is more, we would expect an intelligent adversary to do what he can to increase the ambiguity of his observable features or actions (Bennett and Waltz 2007). He will "pretend, portray, profess an intended lie" to prevent or dispel suspicion (Whaley 1969). He will work within the cycle of feasible responses, dropping cover only when there is no time for effective response. Pirate vessels will display the attributes, say, of small-scale fishing craft, while stealing close enough for a quick dash to accost an unsuspecting victim.

Figure 15.9 summarizes the general scheme of anomaly-based recognition methods and their deficiencies for threat assessment in a particular asymmetric threat domain: that of maritime counter-piracy.

15.7.3 Category 2: Hypothesis-Based Explanation

Category 2 inferencing employs abductive reasoning in place of or in addition to model-based deductive/inductive reasoning. Situations are understoood not simply by reference to stored predictive models, but by models adaptively composed to explain available data (Rogova and Nimier 2004). Such methods are appropriate when training data concerning targets and background are insufficient for modeling by statistical learning methods.

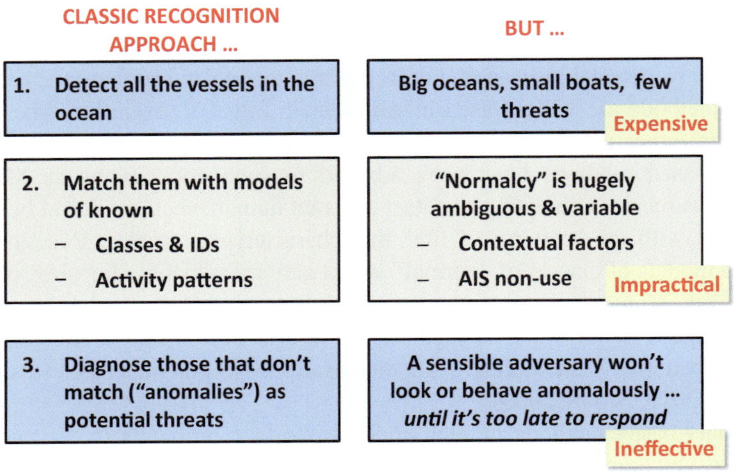

Fig. 15.9 Anomaly-based recognition and its problems (Waltz 2003)

An analyst (or possibly an automated process) constructs a hypothesized situation or scenario in an attempt to account for observed data. As in the classical scientific method, the hypothesis is evaluated to predict further observables that could either confirm, or refute the hypothesis. By acquiring such data as available, explanatory, predictive models of the observed situation/scenario are refined, selected, or rejected.

15.7.4 Category 3: Constraint-Based Plausibility

Category 3 involves inferencing from contextual information alone. It differs from category 1 in not requiring direct detection of targets or activities of concern.

Category 3 encompasses the most difficult of hidden target problems (again, using "targets" in the broadest sense) in which activities of interest may not be detectable or discriminable at all. Rather, contextual cues cause concern for general classes of activities: adversary capability developments, strategic planning, etc. A determination of feasibility can trigger a search for indicators of activities of concern (Snidaro et al. 2015). We can encounter problems of this type in geopolitical intelligence: what is being planned or being built behind closed doors? In these cases there may be no discernable signature associated with the activity of concern. But, as illustrated in Fig. 15.10, measurement, inference, and control capabilities and opportunities serve as constraints on actionable intent.

Various classes of threats are constrained in different ways. For example,

- The use of nuclear weapons is constrained primarily by capability (the required materials and assembly are difficult) and secondarily by intent (specifically, the

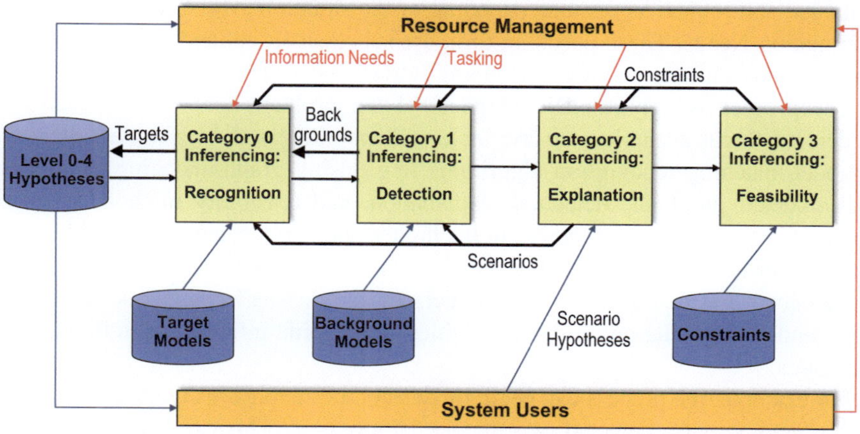

Fig. 15.10 Adaptive threat assessment across problem categories

intent factors cited by de Becker: perceived justification, perceived alternatives, and perceived consequences);

- Assassination of a political leader is constrained primarily by opportunity (leaders are well-protected) and, of course, by intent;
- Terrorist attacks against "soft" targets such as civilian populations are constrained primarily by intent, as targets of opportunity are generally plentiful and the required materials are often easily acquired.

The analytic process in category 3 problems is guided by the constraining factors. The potential for using nuclear weapons is first guided by searches for indicators of the required capability to design, develop and deliver such weapons. For agents having or likely to have such capabilities, intent is evaluated both (a) from the top down, in terms of the perceived justification, perceived alternatives and perceived consequences relative to agents' high-level goals, and (b) from the bottom up, as in Category 2, postulating the potential relationships of observed activities in postulated threat scenarios.

Because intent is the easiest of the three factors to obtain and the most difficult to detect, soft target attacks are far more difficult to prevent than the other examples.

15.8 An Architecture for Adaptive Threat Assessment

Because of the focus on human intentional threats and the difficulties listed in the previous section, we will want the threat assessment process to be adaptive and opportunistic.

Figure 15.10 depicts a notional system architecture for Threat Assessment. The process is adaptive, by which situation and event hypotheses are generated, evaluated and refined as the understanding of the situation evolves to meet current Information Needs. These needs, in turn, are successively refined as a function of current state estimates and users' evolving statements of need.

The process is opportunistic. Aspects of the problem for which predefined models of threat scenarios are available will be addressed by simple Category 0 model-matching recognition algorithms (e.g., using graph-matching). For less well-modeled problems, detection, explanation, and feasibility methods will be employed as warranted. Inferences in the higher-numbered categories provide context for constraining and resolving those in lower-numbered categories (Steinberg and Rogova 2015). Furthermore, products of the "easier" categories—resolved target and background entities and activities—can guide inferencing in the harder categories.

For example, the process can begin by detection of anomalies relative to normal background activity (detected by category 1 processes). This can trigger category 0 processing to attempt to recognize that activity. If the data matches no models, or matches ambiguously, category 2 processes could be triggered in an attempt to explain the anomalous data.

Alternatively, contextual factors alone may engender worries about the possibility of activities of concern; say, hidden developments of advanced weapons. If these contextual factors indicate that such activities are feasible (level 3 inferencing), a search for confirmatory or refuting indicators will be undertaken. Depending on the detectability of such indicators and the availability of predictive models, category 0 and category 1 methods might be employed to recognize the activity of concern or, at least, suspicious anomalies.

Accordingly, the process reasons across all the problem categories, integrating goal-driven and data-driven processes for each category.

15.8.1 Category 0 Inferencing

Goal-driven process: a system user posits an explicit set of Information Needs (INs)—or such needs are inferred automatically by decomposition of higher-level statements of needs—requesting information concerning specific entities or activities of interest (e.g., weapons assembly in cave number 253 in the region) or of classes of these (e.g., caves in the region of concern).

If a model exists for a target/activity/type of concern, data are assembled from knowledge bases and/or live sensors/sources and tested against the models as follows. Such data are also tested against models of other entities/activities of concern:

1. INs are parsed into capability, opportunity, and intent factors and these are decomposed into sets of endogenous ("problem") variables;
2. Category 3 processing is triggered to test for hypothesis feasibility (do the predicted values of endogenous variables fall within assumed constraints?);

3. Endogenous variables are mapped to sets of "observables," i.e., to variables that are feasible to evaluate—e.g., by sensor measurement or data mining—and which discriminate target states of interest;
4. The cost-effectiveness of candidate information acquisition plans is evaluated; if no cost-effective plan is found, the problem is referred to system users for possible pursuit as a Category 1, 2, or 3 problem; otherwise, information acquisition plans are selected and implemented;
5. Acquired information is evaluated for relevance and confidence and selected information is fused to refine estimates of problem variables[9];
6. If significant matches are found, a flag is raised on the user interface and Category 1 processing is initiated to test for alternative explanations of the data;
7. If information needs are not yet satisfied; the above process is iterated, given the refined state estimates.

Category 0 inferencing can also be triggered by anomalies detected in Category 1 or by indications of concern from Category 2 analyst-generated models or Category 3 constraint satisfaction.

Data-driven process: Data received from sensors or other live sources are compared with models of entities and activities of concern and processing proceeds per step 7 above.

If Category 0 processing indicates possible targets/activities of interest, Category 1 processing is triggered to examine other possible explanations of the data, allowing determination of probability assignments across alternatives.

15.8.2 Category 1 Inferencing

Goal-driven process: background/normalcy conditions are characterized in the region of interest;

Data-driven process: data received from sensors/sources or from Category 0 inferencing are tested for anomalies to be diagnosed by Category 0, 2, and/or 3 processes.

15.8.3 Category 2 Inferencing

Goal-driven process: a system user composes a hypothesized scenario (process model). This hypothesis is tested as described for Category 0 goal-driven processing;

[9]Methods for evaluating and compensating for source data quality are discussed in (Snidaro et al. 2015, Steinberg et al. 2014).

Data-driven process: user-generated scenario hypotheses are used to test hypotheses that were determined to be ambiguous by Category 0 or Category 1 inferencing, i.e., in cases where available validated models are either of insufficient resolution or of sufficient confidence to meet Information Needs.

15.8.4 Category 3 Inferencing

Goal-driven process: a system user tasks the system to determine the feasibility of generically defined activities (e.g., human habitation; movement of x kg of fissile material from location y to location z by time k).

Data-driven process: Category 0–2 inferencing triggers evaluation of constraint satisfaction.

15.8.5 Processing Functionality

Data Fusion functions infer the states of entities to meet current Information Needs:

- *Level 0–3 fusion*: inferring states of external entities: signals/features, individuals, relationships, situations, and scenarios (per Table 15.1); estimates of physical and perceptual states of entities are used to infer both actual and potential relationships among entities; and specifically to infer threat situations, in terms of one entity's capability, opportunity, and intent to affect (other) entities;
- *Level 4 fusion*: inferring the states of system resources: evaluating and diagnosing performance relative to current INs.

Resource management functions pertinent to threat assessment include

- *Performance management*: refining INs (direction by system users); planning and executing methods to meet information needs;
- *Data acquisition management*: sensor management and data mining to obtain reports of real-world entities, relationships, and events to meet current INs;
- *Process management*: selecting batches of data and processing modules to meet current INs;
- *Model management*: building and refining threat models. Advanced learning methods, as well as Category 2 compositional methods can be employed to develop and validate predictive models of threat activity.

User interfaces

- Facilitate the understanding of information sources and hypotheses at all fusion levels and models of resources, targets, backgrounds, and higher level constraints (e.g., generic physical, psychological, and sociological models);

- Facilitate the user in assigning factor weights and probabilities in developing capability, opportunity, and intent models, and employment concepts;
- Facilitate the visualization of manipulation of hypotheses for sensitivity analysis and analysis of alternatives;
- Provide indications of current and predicted threat events and situations for event prediction, indications and warning, threat characterization, attack assessment and consequence assessment.

References

Bennett M, Waltz E (2007) Counterdeception principles and applications for national security. Artech House, Boston, MA

Boyd JR (1976) Destruction and creation. U.S. Army Command and General Staff College, 3 Sept 1976

de Becker G (1997) The gift of fear (and other survival signals that protect us from violence). Dell Publishing, New York

Endsley MR (2000) Theoretical underpinnings of Situation Awareness: a critical review. In: Situation awareness analysis and measurement. Lawrence Erlbaum Associates Inc., Mahwah, NJ

Garcia J, Snidaro L, Visentini I (2012) Exploiting context as binding element for multi-level data fusion. Proc., Fifteenth international conference on information

Kahneman D (2011) Thinking, fast and slow. Farrar, Straus and Giroux, New York

Kschischang FR, Frey BJ, Loeliger HA (2001) Factor graph and the sum-product algorithm. IEEE Trans Inf Theory 47:498–519

Lambert DA (2006) A unification of sensor and higher-level fusion. Proc. 9th international conference on information fusion

Little EG, Rogova GL (2006) An ontological analysis of threat and vulnerability. Proc. Ninth international conference on information fusion, Florence

McMichael D, Jarrad G (2005). Grammatical methods for situation and threat analysis. Proc. Eighth international conference on information fusion, Philadelphia, June 2005

Moore KE (2005) Predictive Analysis for Naval Deployment Activities (PANDA), briefing to industry: PANDA Overview, 16 Sept 2005

Nolte WL, Kennedy BC, Dziegiel Jr RJ (2003) Technology readiness level calculator. NDIA Systems Engineering Conf., 20 Oct 2003

Rogova G, Nimier V (2004) Reliability in information fusion: literature survey. Proc. Seventh international conference on information fusion, Stockholm

Rogova GL, Steinberg AN (2015) Formalization of 'Context' for information fusion. Context Enhanced Information Fusion. Springer

Salerno JJ (2007) Where's level 2/3 fusion—a look back over the past 10 years. Proc. Tenth International Conference on Information Fusion, Quebec

Snidaro L, Garcia J, Llinas J (2015) Context-based information fusion: a survey and discussion. Inf Fusion 25:16–31

Steinberg AN (2005) An approach to threat assessment. Proc., Eighth international conference on information fusion, Philadelphia

Steinberg AN (2009) Foundations of situation and threat assessment. In: Liggins ME, Hall DL, Llinas J (eds) Chapter 18 of Handbook of multisensor data fusion. CRC Press, London

Steinberg AN (2013) Situation management for counter-piracy. In: Prediction and recognition of piracy efforts using collaborative human-centric information systems. Ios Press, Amsterdam

Steinberg AN (2014) Threat assessment with technical intelligence applications. Proc. CogSIMA, San Antonio, TX, Mar 2014
Steinberg AN (2015) Levels? Proc. 18th international conference on information fusion, Washington
Steinberg AN (2015) Situations and contexts. ISIF Perspectives on Information Fusion (Perspectives), Vol 1(1)
Steinberg AN, Bowman CL (2009) Revisions to the JDL data fusion model. In: Liggins ME, Hall DL, Llinas J (eds) Chapter 3 of Handbook of multisensor data fusion. CRC Press, London
Steinberg AN, Bowman CL, Blasch E, Morefield C, Morefield M, Haith G (2014) Adaptive context assessment and context management, Proc., Seventeenth International Conference on Information Fusion, Salamanca, Spain, 2014
Steinberg AN, Rogova GL (2015) System-level use of contextual information. In: Snidaro L, Garcia J, Llinas J, Blasch E (eds) Context-enhanced information fusion. Springer, Berlin
Waltz E (2003) Knowledge management in the intelligence enterprise. Artech House, Boston, MA
Wessen RR, Adler M, Leising CJ, Sherwood B (2010) Measuring the maturity of robotic planetary mission concepts II. SpaceOps 2010 Conference, Hosted by NASA Mars, Huntsville, Alabama, 25–30 Apr 2010
Whaley B (1969) Strategem: deception and surprise in war. Center for International Studies, MIT, Cambridge, MA
White FE (1988) A model for data fusion. Proc. First national symposium on sensor fusion, GACIAC, IIT Research Institute, Chicago

Chapter 16
Rule-Based Support for Situation Management

Patrícia Dockhorn Costa, João Paulo A. Almeida, Isaac S.A. Pereira, Marten van Sinderen, and Luís Ferreira Pires

Abstract The notion of situation enables designers, maintainers, and users to abstract from lower-level entities and properties and to focus on the higher-level patterns that emerge in time. Situation management concerns a number of tasks including situation specification, situation detection (which may involve composite situation pattern recognition), and situation's life cycle control. This chapter discusses how to approach situation management from a rule-based perspective. We present a rule-based situation management infrastructure to support the development of situation-aware applications and show its applicability to a scenario in the public health domain, concerning situations for detecting influenza epidemics.

16.1 Introduction

A central issue in reactive or proactive systems is the ability to bridge the gap between events that occur in the environment and the particular state-of-affairs of interest (*aka* situations) upon which the system is required to act (or react to). The field of human factors and ergonomics (HF&E) addresses this issue in a human-goal centric approach by means of a well-established concept called situation awareness.

In (1995), Endsley defines situation awareness as "the perception of the elements in the environment within a volume of time and space, the comprehension of their meaning, and the projection of their status in the near future." Although Endsley's definition has been proposed to support humans, we argue that it can be beneficially applied to support the development of situation-aware applications.

We argue that the situation awareness concept, as referred to by Endsley, should be exploited by reactive or proactive systems, such as context-aware applications. Context-awareness focuses on characterizing the user's environment (context) to promote effective interaction between applications and their users by autonomously

P. Dockhorn Costa (✉) • J.P.A. Almeida • I.S.A. Pereira
Federal University of Espírito Santo (UFES), Av. Fernando Ferrari, 514, Vitória, ES, Brazil
e-mail: pdcosta@inf.ufes.br; jpalmeida@ieee.org; pereira.zc@gmail.com

M. van Sinderen • L. Ferreira Pires
University of Twente, PO Box 217, 7500 AE Enschede, The Netherlands
e-mail: m.j.vansinderen@utwente.nl; l.ferreirapires@utwente.nl

G. Rogova, P. Scott (eds.), *Fusion Methodologies in Crisis Management*,
DOI 10.1007/978-3-319-22527-2_16

adapting application's behaviors according to the user's current situation. In the field of context-awareness, Dey, in (2001), was one of the first to make efforts in this direction by proposing the situation abstraction concept, which is an extension of the context concept and refers to a mean "to determine when relevant entities are in a particular state so they (applications) can take action." Therefore, context-aware applications in the sense of Dey (2001) could also be considered as situation-aware applications.

In order to leverage the benefits of the situation abstraction concept in the scope of context-aware application development, proper support is required at design-time (to specify situation types) and run-time (to detect and maintain information about situations). Various realization alternatives could be used to implement such support. A solution based on *rules* appears naturally given the nature of situation detection, in which the user's context is continuously monitored in order to check whether certain conditions are met. In a rule-based implementation, the designer defines rules which are repeatedly applied to a collection of *facts* in a *working memory*. In order to apply a rule based approach to situation awareness, the context conditions are represented as facts, and the rules describe the conditions under which certain situations exist.

As opposed to procedural solutions, rule-based solutions offer flexibility in that rules can be modified and added at application runtime with no need for code recompilation. Since situation specifications may change over time, and new situation specifications may be defined at application runtime, it is beneficial to use a mechanism that offers such flexibility. For these reasons, we explore in this book chapter rule-based solutions to support the development of situation-aware applications.

We start by examining the notion of situation carefully, which leads to general requirements for situation management. For our purposes, situation management encompasses support for situation type specification, deployment, situation detection (which may involve composite situation pattern recognition), and situation's life cycle control. We discuss how situation management requirements could be supported in traditional rule-based solutions. Despite the immediate benefits of such an approach, traditional rule-based solutions do not fully realize the potential of explicit support for situation awareness. In order to leverage these benefits, we propose a rule-based situation management infrastructure to support the development of situation-aware applications and show its applicability to a scenario in the public health domain.

The rule-based situation management infrastructure we propose includes SCENE (Pereira et al. 2013), a platform for situation management that builds up on the JBoss Drools engine (and its integrated Complex Event Processing platform) and enhances this engine's functionality to natively support rule-based situation-awareness. *SCENE* provides a means to specify situation types with a simple yet expressive rule pattern. Given a situation type specification, the platform makes situation detection and life cycle management nearly transparent for situation-aware application designers.

This chapter is further structured as follows: Sect. 16.2 introduces the concept of situations and discusses requirements for situation life cycle management; Sect. 16.3 presents a motivating application scenario in the public health domain, concerning situations for detecting influenza epidemics; Sect. 16.4 discusses the support for situation management in traditional rule-based systems; Sect. 16.5 presents the SCENE rule-based platform for situation management and illustrates its usage in the management of influenza epidemics; finally, Sect. 16.6 presents concluding remarks.

16.2 Requirements for Situation Management

This section discusses requirements for situation management. We first discuss the notion of situations and situation types, and then consider what kind of support would be needed in order to define situation types at application design time and detect situations at application runtime.

16.2.1 Situations and Situation Types

Situations are composite entities whose constituents are other entities, their properties and the relations in which they are involved (Costa 2007). Situations support us in conceptualizing certain "parts of reality that can be comprehended as a whole" (Rosemann and Recker 2006). Examples of situations include "John is working," "John has fever," "John has had an intermittent fever for the past 6 months," "John and Paul are outdoors, at a distance of less than 10 m from each other," "Bank account number 87346-0 is overdrawn while a suspicious transaction is ongoing"[1]

Situations are often reified (such as in Barwise (1989), Costa et al. (2006)), or ascribed an "object" status (Kokar et al. 2009), which enables one not only to identify situations in facts but also to refer to the properties of situations themselves. For example, we could refer to the duration of a particular situation or whether a situation is current or past, which would enable us to say that the situation "John has fever" occurred yesterday and lasted 2 h.

A situation type (Kokar et al. 2009) enables us to consider general characteristics of situations of a particular kind, capturing the criteria of unity and identity of situations of that kind. An example of situation type is "Patient has fever." This type is multiply instantiated in the cases in which instances of "Patient" (such as "John," "Paul") can be said to "have fever." Thus "John has fever" and "Paul has

[1]Technically, the sentences we use to exemplify situations are utterances of propositions which hold in the situations we consider; however, we avoid this distinction in the text for the sake of brevity.

fever" are examples of instances of "Patient has fever." These examples reveal the need to refer to entity types such as "Patient" as part of the description of a situation type. The same can be said for "has fever" which, in this case, is defined in terms of a property of entities which instantiate the entity type "Patient" (namely "body temperature").

Detecting situations (i.e., instantiations of a situation type) require detecting instances of the entity types involved in the situation whose properties satisfy constraints captured in the situation type. The situation is said to be active while those properties satisfy constraints captured in the situation type. A situation ceases to exist when those properties no longer satisfy the defined constraints. In this case, the situation is said to be a past situation. The point in time in which a particular situation instance is detected is called situation activation instant and the point in time in which the situation ceases to exist is called situation deactivation instant.

Figure 16.1 provides a graphical representation of the life cycle of three situations instantiating the same situation type. The vertical axis represents the possible states-of-affairs of the entities in the domain of interest. The horizontal axis represents the passing of time. For the sake of simplicity, suppose we are only concerned with a single property "temperature" of a single entity instance "John" of type "Patient," and we are interested in the situation type "John has fever." This situation type is

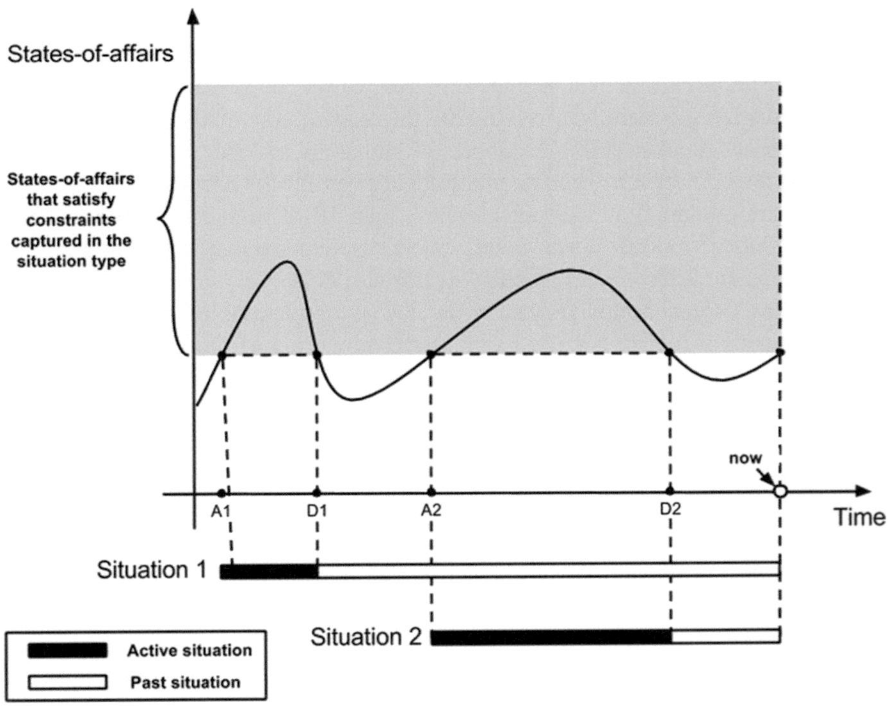

Fig. 16.1 Example of situation instances life cycle

characterized when John's temperature lies above a given threshold (gray area in Fig. 16.1).

These characteristics of situations lead us to the following basic requirements for our situation-based approach:

1. Situation types should be defined at design time, and situations instantiating these types should be detected at runtime;
2. Situation types should be defined with reference to entity types as well as constraints on entities' properties and relations;
3. Temporal properties of situations should be considered (such as initial time, and, for a past situation, final time and duration).

In addition to these requirements, we have also observed that the definition of complex situation types may be more manageable by defining these types in terms of a composition of simpler situation types. Thus, we also include the recursive composition of situation types in our approach. This allows us to consider different levels of situation assessment. The composition of situations into more complex situations may be driven by their temporal properties.

16.2.2 Situation Life Cycle Management

In addition to defining situation awareness, Endsley (1995) provides a theoretical framework for situation awareness, proposing three levels of situation awareness: (1) the perception level; (2) the comprehension level, and the (3) projection level. The first level is related to perceiving the status, attributes, and dynamics of relevant elements in the environment; the second level involves the synthesis of situation elements through the processes of pattern recognition, interpretation, and evaluation of what was perceived; and, the third level comprises the ability to project future actions of the elements in the environment. Although Endsley's reference model has been proposed to support humans, it can be beneficially applied to support the development of situation-aware applications. In this setting, situation-aware applications will support the users in some situation awareness tasks, e.g., automating situation detection from low-level sensors. We consider situation detection functionality to provide automated support for the so-called *situation assessment* phase, which comprises part of the perception level and the comprehension level (Fig. 16.2). Situation detection functionality receives information from the environment (or domain) and feeds high-level situation information to a situation observer (a situation-aware application).

As discussed by Kokar et al. (2009), "to make use of situation awareness [...] one must be able to recognize situations, [...] associate various properties with particular situations, and communicate descriptions of situations to others." The notion of situation enables designers, maintainers, and users to abstract from the lower-level entities and properties that stand in a particular situation and to focus on the higher-level patterns that emerge from lower-level entities in time.

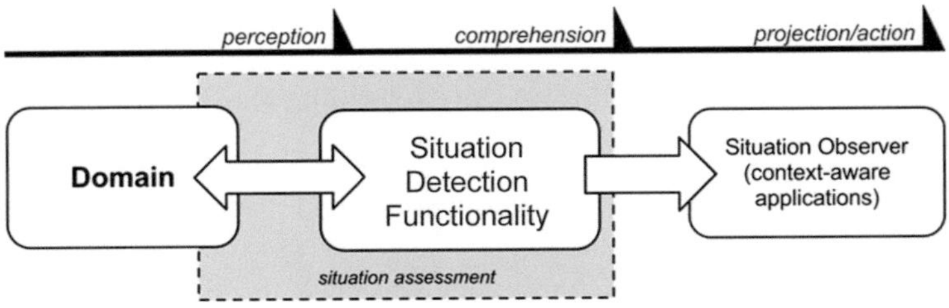

Fig. 16.2 Endsley's reference model and rule-based situation detection functionality

In order to leverage the benefits of the situation abstraction concept in the scope of context-aware application development, proper support is required at design-time (to specify situation types, as discussed in the previous section) and run-time (to detect and maintain information about situations). In other words, context-aware applications should provide support for *situation life cycle management*, which encompasses support for situation type specification, deployment, situation detection (which may involve composite situation pattern recognition), and situation's life cycle control.

16.3 Motivating Application Scenario: Influenza Surveillance

In order to illustrate the use of the notions of situations and situation types in a situation-aware application, we discuss in this section a healthcare application for the surveillance of seasonal influenza epidemics. This also serves as a running example in order to show how the situation management infrastructure is used in Sect. 16.5.

Seasonal influenza surveillance is particularly important as influenza infections still cause substantial morbidity and mortality every year. According to World Health Organization (2014), influenza annual epidemics are estimated to result in about three to five million cases of severe illness, and about 250,000–500,000 deaths, worldwide. Symptoms include a sudden onset of high fever, cough (usually dry), headache, muscle and joint pain, severe malaise (feeling unwell), sore throat and runny nose.

Seasonal influenza spreads easily and can sweep through schools, nursing homes, businesses, or towns. The main goal of influenza surveillance is to minimize the impact of the disease by providing useful information to public health authorities so they may better plan appropriate control and intervention measures, allocate health resources, and make case management recommendations (World Health Organization 2010). Public health programs could benefit from computational systems that help monitoring the population in influenza focus areas in order to assist decision-making. For example, such system could monitor the severity of annual epidemics in order to assist policy makers in making decisions about public interventions.

In that sense, we consider a situation-aware application scenario based on the influenza *routine sentinel surveillance system* described in World Health Organization (2010). Sentinel surveillance involves systematically collecting data on a routine basis from a limited number of surveillance sites. Data collected over time are use to "establish historical trends and baselines to provide a range of usual, expected values against which to compare outbreaks related to new viruses or unexpected events related to previously circulating viruses. Such historical data will allow rapid assessment of future pandemic severity and provide the necessary infrastructure to follow the impact of an event, such as an outbreak of a novel influenza virus, as it unfolds over time. The data will provide valuable information on the usual seasonality of influenza and the groups at risk for severe disease" (World Health Organization 2010).

The WHO global influenza surveillance standards define the surveillance case definitions for influenza-like illness (ILI) and severe acute respiratory infections (SARI), as follows (World Health Organization 2014):

- ILI case definition: an acute respiratory infection with measured fever of $\geq$38 °C and cough; with onset within the last 10 days.
- SARI case definition: an acute respiratory infection with history of fever or measured fever of $\geq$38 °C and cough; with onset within the last 10 days; and requires hospitalization.

Based on information provided by the sentinel sites, we define three situation types regarding patients' symptoms and diagnosis:

1. The *Fever Situation*, which is considered to exist whenever a patient's temperature is above 37 °C;
2. The *ILI Situation*, which is considered to exist for every patient presenting fever and cough with onset within the past 10 days;
3. The *SARI Situation*, which is considered to exist for every patient in ILI situation with recommendation for hospitalization.

   Considering the existence of historical data about the number of ILI and SARI cases, the WHO defines guidelines to calculate *alert thresholds* (World Health Organization 2010) that help identifying current influenza epidemics, in a given week. Therefore, if the number of reported SARI cases is above the alert threshold for that particular week, then an influenza epidemics alert for that region is registered.

   With respect to epidemics information, we define two situation types:
4. The *Epidemics Situation*, which is considered to exist for every region whose total number of reported SARI cases is above the alert threshold calculated for that region;
5. *The EpidemicSpread Situation*, which is considered to exist for every pair of regions within 100 km from each other, in which an epidemics alert is detected in one and spread to the other.

Note that ILI and SARI are situations that characterize individuals and depend on the simpler Fever situation. The Epidemics situation depends on a number of SARI situations in a particular region and in a particular time window. Further, EpidemicSpread depends on the patterns of occurrences of Epidemics both in time and space. These characteristics of these situations illustrate typical situation management requirements as discussed in Sect. 16.2: compositionality and the use of temporal constraints involved in the relation between situations.

16.4 Support for Situation Management in Traditional Rule-Based Systems

Rule-based systems address a need for capturing, representing, storing, distributing, reasoning about, and applying knowledge in a computational environment (Hayes-Roth 1985). In such systems, knowledge is expressed in terms of sets of rules, and can consist of various kinds of information, including: (1) inferences that follow from observations; (2) abstraction, generalization, categorization of given data, and (3) necessary and sufficient conditions for achieving some goal. Knowledge rules are repeatedly applied to a collection of facts that are considered to be part of the system's working memory. These two concepts are essential to a rule-based system:

1. Facts represent circumstances that describe a certain state-of-affairs in the real world; and,
2. Rules represent heuristics that define a set of actions to be executed in a given state-of-affairs.

Rules are similar to *if-then* statements of traditional programming, in which the *if* part is often called the left-hand side (LHS) or premises, and the *then* part is the right-hand side (RHS) or conclusions. The LHS consists of an expression, which can be a single expression or a combination of expressions (composite expression). For the rule to be applicable, i.e., to execute the RHS, or to derive the conclusions, the LHS should be *true*.

Considering the requirements for situation management as discussed in Sect. 16.2, proper support should be provided for situation type specification, situation runtime detection, and situations' temporal reasoning. In traditional rule-based systems, situation types would be typically defined in terms of fact templates, and situation instances would be represented as situation facts in the working memory. For example, the situation that "John has fever" could be represented by a fact PersonHasFever(John). Situation facts should exist (or be defined as "active") in the working memory while the predefined conditions for their existence hold.

Situation fact instantiation requires situation fact life cycle management, which consists of situation fact creation, activation, deactivation, and destruction. The activation of a situation fact occurs simultaneously to its creation. When the conditions for a particular situation fact begin to hold and that situation fact does

Table 16.1 Situation rules general templates

	Creation rule	Deactivation rule
LHS (as a *conjunction of Boolean expressions*)	(situation type conditions apply) (not (situation fact exists))	(negated situation type conditions apply) (situation fact exists)
=> RHS (*as a set actions*)	=> create (situation fact) [RaiseEvent()]	=> deactivate (situation fact) [RaiseEvent()]

not exist yet, the situation fact is created. The deactivation of a situation fact occurs when those conditions become false. Deactivated situation facts consist of historical records of situation occurrence, which may be used to detect situations that refer to past occurrences of a situation type.

In a traditional rule-based system, a pair of rules should exist in order to manage the life cycle of a situation fact: (1) a creation rule and (2) a deactivation rule (see Table 16.1). The condition part of a creation rule checks whether the situation type condition holds, and whether an instance of that particular situation is not currently active. If these conditions are met, a situation fact is created, and optionally, an event can be raised. Analogously, the condition part of a deactivation rule checks whether the situation type condition no longer holds, and whether a current situation fact active for this situation already exists. When these conditions are met, this situation instance is deactivated, and optionally, an event can be raised.

Note that it is the application designer that is responsible for manually specifying rules for situation detection and deactivation. Thus, instead of specifying directly the necessary and sufficient for a situation of a particular type to exist, the designer needs to translate these conditions into a condition for creation of a situation fact and a separate condition for deactivation of a situation fact.

With respect to temporal requirements, traditional rule-based systems have no native support for temporal reasoning. A time dimension (e.g., initial and final time attributes) would have to be included manually in situation facts in order to enable some temporal reasoning. In that sense, the initial time attribute would capture the moment a situation begins to hold, and the final time attribute, the moment a situation ceases to hold. All operations involving time (e.g., to detect Fever situations "within the last 10 days") would have to be coded manually with reference to these attributes.

16.5 SCENE: A Rule-Based Platform for Situation Management

In order to overcome the limitations of traditional rule-based approach, we propose SCENE, a platform for situation management that leverages on JBoss Drools engine (and its integrated Complex Event Processing platform) and enhances this engine's functionality to natively support rule-based situation-awareness. Situation management in SCENE encompasses support for situation type specification, deployment, situation detection (which may involve composite situation pattern recognition), and situation's life cycle control.

Our platform allows rule-based situation specification (and further situation life cycle management) by means of a simple rule pattern.

16.5.1 Drools

Drools is a rule-based system that runs an enhanced implementation of the Rete algorithm, which matches patterns by remembering past pattern matching tests. Only new or modified facts are tested against rules, which guarantees the efficiency of pattern matching.

Rules are stored in a component called the Production Memory and the facts that the Inference Engine matches against are kept in a component called the Working Memory. Facts are asserted into the Working Memory where they may then be modified or retracted. A system with a large number of rules and facts may result in many rules being true for the same fact assertion; these rules are said to be in conflict. The Agenda component manages the execution order of these conflicting rules using a Conflict Resolution strategy (Bali 2009).

Rules are defined in Drools by means of a domain-specific language called the Drools Rule Language (DRL). A DRL rule declaration comprises a condition and a consequence expression block, respectively referred to as Left Hand Side (LHS) and Right Hand Side (RHS). A rule specifies that when the particular set of conditions defined in the LHS occurs, the list of actions in the RHS should be executed. The LHS is composed of conditional elements which can be combined through logical operators, such as *and*, *or*, *not*, and *exists*; and set operators, such as *contains* and *member of*. A conditional element can be a pattern or a constraint. A pattern matches against a fact in the working memory (of the specified class type); constraints match against properties, and are defined as conditions within a pattern. The RHS allows the declaration of procedural code to be executed when the conditions defined in LHS are satisfied.

Declaring new fact types in Drools can be done with plain Java objects, or directly in the rule engine, by means of the DRL. For example, consider the DRL declaration of a new fact type Person (in Fig. 16.3), with two attributes name (of type String) and dateOfBirth (of type Date):

16.5.2 SCENE: User's View

Situation types are specified in SCENE by means of structural and behavioral aspects, which are realized by *Situation Classes* and *Situation Rules*, respectively.

```
declare Person
   name: String
   dateOfBirth: java.util.Date
end
```

Fig. 16.3 Example of a fact type declaration

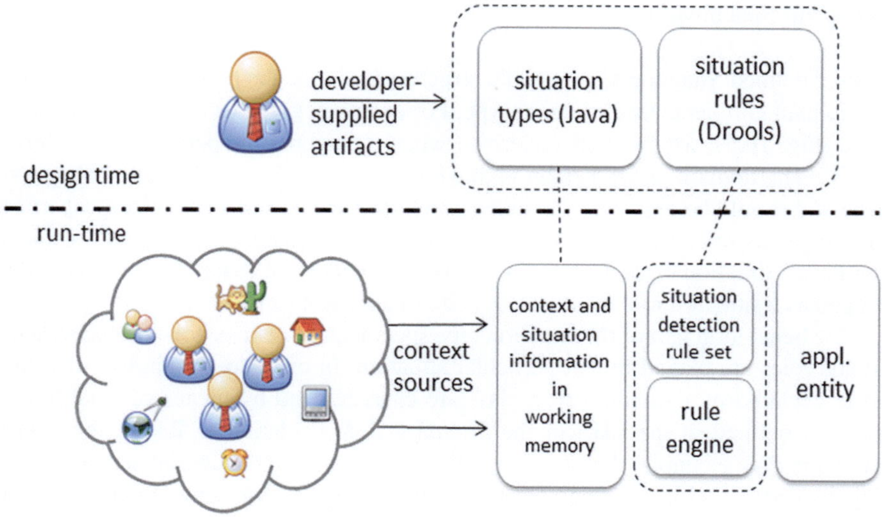

Fig. 16.4 SCENE's rule-based approach to situations

Every user-defined Situation Class specializes the predefined class *SituationType*, which is an abstract class that addresses the situation temporal properties and compositional characteristics.

Figure 16.4 depicts SCENE's rule-based approach, relating design time and run-time aspects. The upper part of the figure shows the design-time elements. At design-time situation types are specified by means of structural and behavioral aspects, which are realized by *Situation Types* and *Situation Rules*, respectively. The lower part of the figure reveals the run-time relations between the sources of context information and the rule-based implementation. Context sources provide context information, which is input as *facts* in the engine's working memory. The rule engine uses the situation detection rule set to detect situation instances from the facts in the working memory.

```
declare Fever extends SituationType
   febrile: Person @SituationRole   //defines the role of a participant
end
```

Fig. 16.5 The *Fever* situation type class declaration

Situation Specification

A user-defined Situation Class should structurally define the particular roles played by domain entities in that situation type. For example, consider the Fever Situation type (Fig. 16.5), which is characterized when a person's temperature rises above 38 °C. The domain entity Person is playing a role (febrile) in the Fever Situation type and should be explicitly defined as such. In our approach, situation properties are tagged as roles using the @SituationRole Java annotation. Figure 16.5 depicts the Fever Situation class declaration in DRL, in which the domain entity Person is tagged as a situation role by means of a @SituationRole annotation.

The behavioral part of the situation type specification defines how the abovementioned roles are played in that particular situation. In order to accomplish that, the roles declared in a situation type class are characterized by means of conditional patterns defined in the LHS of the *Situation Rule* declaration. Taking the Fever Situation type example, the Fever Situation Rule (i.e., the behavioral specification, depicted in Fig. 16.6) defines febrile as any person whose temperature exceeds 38 °C. The role febrile is specified as an LHS pattern identifier, which is a binding variable whose value is assigned for each person satisfying that particular condition. By means of these binding variables, we can handle matched facts as objects in the RHS of a rule. Therefore, LHS identifiers are used to handle situation participants, relating them to their respective situation role labels, which should have been previously declared in the situation type class. Note that identifiers names should match the property names tagged as situation roles in the Fever Situation class (as defined in Fig. 16.5). SCENE uses this information internally to allow proper situation type specification (and further situation life cycle control).

When a situation rule is fully matched (i.e., the conditions are satisfied), all the facts bound by LHS identifiers that refer to situation roles comprise the so-called *situation cast*. The situation cast is the set of all the entities that participate in the situation (including other situations in composite situation types).

The RHS of a situation rule invokes SCENE's procedural API through the *SituationHelper* module. The invocation of the *situationDetected* method starts the situation life cycle control (situation creation, activation, and deactivation), which is completely realized by the situation platform. When a situation is activated, a situation fact is inserted in the working memory representing that particular situation occurrence.

Situation Rules can also present particular metadata attributes, which are declared before the LHS block. The *@role* metadata is assigned to as situation

```
rule "Fever"
 @role(situation)
 @snapshot(on)
 @type(Fever)
   When
     $febrile: Person(temperature > 38)
   then
     SituationHelper.situationDetected(drools)
end
```

Fig. 16.6 The *Fever* situation rule

Table 16.2 Situation rule metadata annotation

Metadata annotation	Usage
@role	"*@role*role(situation)" informs the engine to handle the rule as a *situation rule.*
@snapshot	"@snapshot(on)" turns on snapshotting for the situation cast. This is mandatory for situation types that take part in complex situation compositions. The default value is "off".
@restore	establishes the participation state storage approach at situation deactivation. There are three options: @restore(first): stores the participants' state at the time of situation detection; @restore(stable): stores the most stable participants' state throughout the situation's life, or; @restore(last): stores the participants' state at the time of situation deactivation.

so that the engine can recognize the respective rule as a situation rule. The @type metadata specifies the situation type class that particular situation rule refers to. The other two metadata attributes are related to what we call situation's *snapshotting* setup. The situation *snapshotting* refers to the process of saving situation cast state snapshots throughout the situation's existence. Snapshotting allows composite situation types to constrain past situation occurrences based on situation cast states.

Consider, for example, that we may need to refer to John's temperature in a particular past occurrence of John's Fever Situation. Since John's temperature most probably changed throughout the active phase of that particular past situation occurrence, a decision should be made about the temperature value to be stored. Therefore, in addition to specifying the need to keep past situation occurrences, SCENE allows the specification of three strategies for participation state storage, namely *first*, *stable*, and *last*, which are specified in a rule by means of the *@restore* metadata. Table 16.2 explains in detail the metadata attributes currently supported.

With respect to the participation state storage strategies, consider, for example, a particular past occurrence of John's Fever Situation in which John's temperature (1) was 39 °C at situation activation, (2) has stabilized in 39.5 °C for the longest period of time during situation active state, and (3) was 37 °C at situation

```
1   rule "ILI"
2   @role(situation)
3   @type(ILI)
4       when
5           $fever: Fever($patient: febrile)
6           exists(ActivateSituationEvent(situation == $fever)
7                   over window:time(10d))
8           exists(CoughCase(patient == $patient)
9                   over window:time(10d))
10      then
11        SituationHelper.situationDetected(drools);
12  end
```

Fig. 16.7 The *ILI* situation rule

deactivation. Using the strategies first, stable, and last, the following temperature values would be restored, respectively: 39 °C, 39.5 °C, and 37 °C. The @restore metadata is optional; when omitted, the stable strategy is considered as default.

Sliding Time Window

Drools Fusion natively supports *sliding windows* aiming at scoping events of interest to a window that is constantly moving. The sliding time window, in particular, allows the definition of rules that only match events occurring in a particular time window (i.e., events that occurred in the last amount of certain time units).

The ILI situation rule specification (depicted in Fig. 16.7) uses the Drools sliding time window operator *window:time* to restrict the time occurrence of events. The LHS of the ILI situation rule defines that a person in an ILI situation should be febrile and having a *CoughCase*, with onset within the past 10 days. The rule's LHS constrains a Fever situation by means of a Fever situation type pattern (line 5), and attributes the matched situation instances to the *$fever* binding variable. The other patterns in this condition use the ***exists*** conditional element, which implements the first order logic's existential quantifier and checks whether objects exist in the working memory. In line 6, the *exists* condition element is used to check for the existence of an *ActivateSituationEvent* for that particular situation Fever (*situation == $fever*) within the last 10 days (*over window:time(10d)*). An *ActivateSituationEvent* is an initiator event generated by SCENE for every situation activation.

The CoughCase event type is a Drools event that has been defined with a time to live of 10 days. Therefore, a CoughCase event for a particular patient lasts for 10 days in the working memory. In line 8, the *exists* conditional element is used to

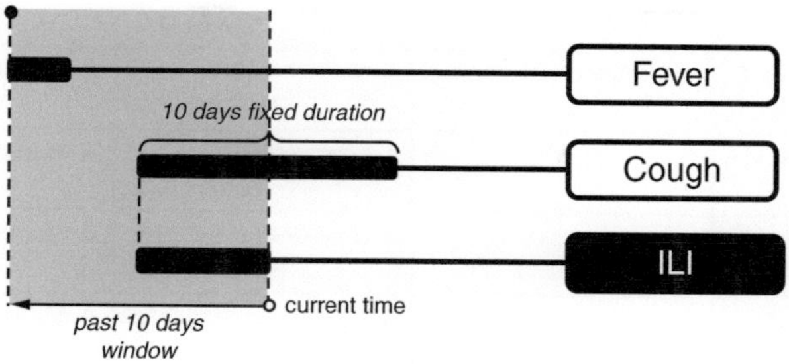

Fig. 16.8 Example timeline for ILI situation

check for the existence of at least one CoughCase event, for that particular patient (*patient* == *$patient*) within the last 10 days.

Figure 16.8 depicts an example timeline of an instance of ILI situation, in terms of an occurrence of situation Fever and an occurrence of the CoughCase event, for the same patient. Note that the ILI situation begins to exist simultaneously to the occurrence of event CoughCase, which has started within the past 10 days (from "current time"). When the time window from the creation of the *ActivateSituationFever* event to "current time" exceeds 10 days, the ILI situation ceases to exist.

Temporal Reasoning

Drools Fusion natively provides LHS operators to correlate events in a temporal perspective. All 13 Allen's operators (Allen 1983) are supported and also their logical complement (negation). For example, it is possible to define conditions in which an event happens *before* another one, or when both events *overlap* in time (among other possible event correlations). Nevertheless, events in Drools are always records of past occurrences; thus, differently from situations, there are no "active" (or current) events. This requires special treatment of temporal operations involving situations, as the final time of active situations is undetermined. We have thus enriched the situation reasoning engine, to allow the definition of constraints for situations using the temporal operators. This allows us to apply temporal operators to pairs of situations, to pairs of events (as supported natively in Drools) and to situation-event pairs.

Figure 16.9 shows all supported temporal operators. Time is represented in the horizontal direction and situations in black represent inactive situations (those that have ended). Their definitions rely on comparisons of the initial time and the converse final time of situations.

The SARI situation rule specification (depicted in Fig. 16.10) uses the *during* temporal relation to describe an ILI case with recommendation for hospitalization. The rule's LHS constrains an active ILI situation for a particular patient, capturing this patient in the binding variable *$patient*. Particular ILI situation instances

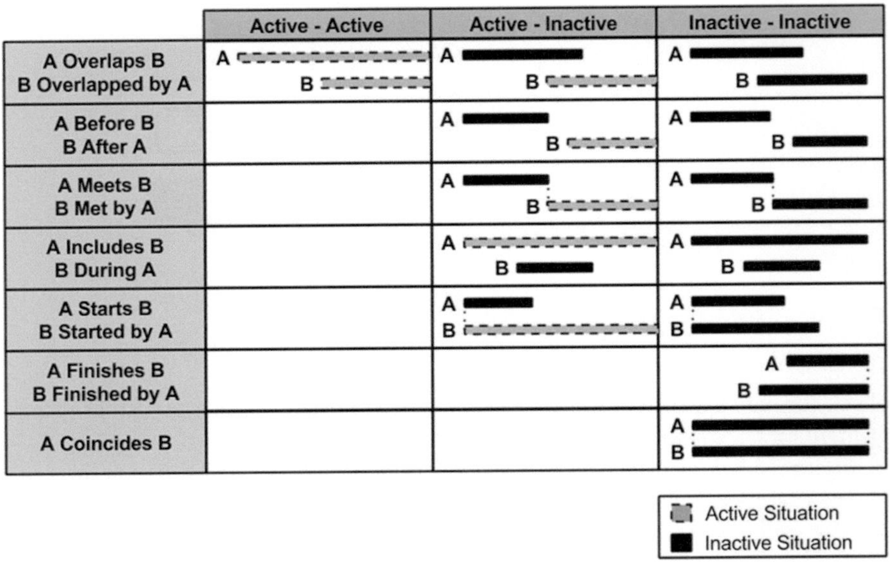

Fig. 16.9 Situation temporal relations

```
1      rule "SARI"
2      @role(situation)
3      @type(SARI)
4      when
5        $ili: ILI($patient: this.patient, active)
6        HospitalizationRequirement(this.patient==$patient,  $clinic:  clinic,
7    this during $ili)
8      then
9        SituationHelper.situationDetected(drools);
       end
```

Fig. 16.10 The *SARI* situation rule

are captured in the binding variable *$ili* (line 5). In addition, there should be a *HospitalizationRequirement* event thrown for that particular patient (line 6, *this.patient == $patient*) during the just captured ILI situation (line 6, *this during $ili*).

Figure 16.11 depicts an example timeline of an instance of SARI situation, in terms of an occurrence of ILI situation and an occurrence of a Hospitalization Requirement event, for the same patient. Note that the SARI situation begins to exist simultaneously to the occurrence of a Hospitalization Requirement event, which has started during an ILI situation occurrence, for that particular entity. When that

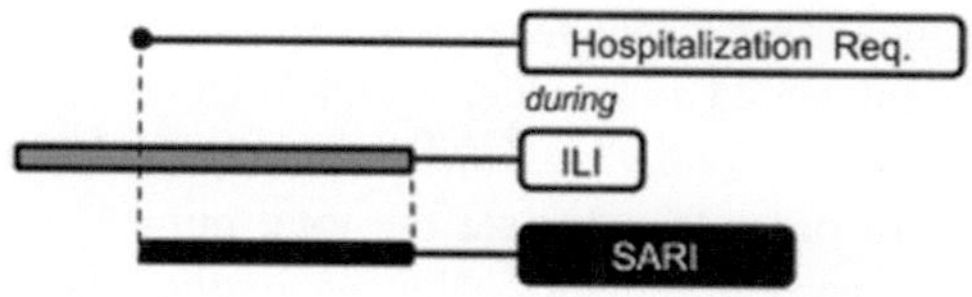

Fig. 16.11 Example timeline for SARI situation

```
1   rule "Epidemics"
2   @role(situation)
3   @type(Epidemics)
4       when
5           $week: Week() over window:length(1)
6           AverageEpidemicLevel( $region: region,
7                                   week.number == $week.number,
8                                   week.year == ($week.year - 1),
9                                   $caseCountThreshold: caseCountThreshold)
10          accumulate ($sari: SARI(clinic.region == $region,
11                              this during $week);
12                      $count: count($sari);
13                      $count > $caseCountThreshold)
14      then
15        SituationHelper.situationDetected(drools);
16  end
```

Fig. 16.12 The *Epidemics* situation rule

particular occurrence of the ILI situation ceases to exist, so does the SARI situation instance.

Figure 16.12 depicts the Epidemics Situation Rule, which aims at detecting possible influenza epidemics situations for a given region. An epidemics situation is considered to exist for every region whose total number of reported SARI cases is above the alert threshold calculated for that region. This situation specification assumes the existence of historical data in the working memory that captures the alert threshold of regional SARI cases for a given week and year. This information is captured by means of the *AverageEpidemicLevel* fact type. In addition, a weekly event is thrown in order to represent the weeks managed by the surveillance system, by means of the *Week* event type. This event type specifies attributes to capture the week's identification and year.

The rule's LHS first pattern (line 5) constraints the Week event to be the last occurrence, by means of the sliding length window (*window:length(1)*), that considers events based on their arrival. Therefore, the *$week* binding variable refers to an object that represents the last *Week* event occurrence. The pattern defined in line 6 constrains the *AverageEpidemicLevel* historical data to be the one of that particular week (*week.number == $week.number*), but of the previous year (*week.year == $week.year-1*). The *AverageEpidemicLevel* fact of a particular region for a given week/year contains the already consolidated alert threshold, which is calculated as specified by the WHO in (2010). The alert threshold value is kept in the *$caseCountThreshold* binding variable, in line 9.

In order to calculate the total number of SARI cases in the current week for that particular region, the *accumulate* Drools conditional element is used. The accumulate conditional element has three parts: one that defines a pattern, another that may use predefined functions, and a third that defines a constraint. In lines 10 and 11 a pattern is defined in the first part of the accumulate, aiming at capturing all the SARI cases, for all the clinics of a given region.

The *$sari* binding variable will have references to all SARI situations detected for the clinics in *$region*, during the *$week*. Line 12 specifies the second part of the accumulate, which uses the *count* predefined function, that counts the number of sari situations detected for that particular region, during that week. Line 13 compares the total number of SARI cases detected during this week with the alert threshold, captured in line 9. If the number of SARI cases is above the alert threshold, an Epidemics situation is detected for that particular region.

Figure 16.13 depicts the *EpidemicSpread* situation rule that detects cases in which pairs of neighbor regions (100 km from each other) are asserted with Epidemics situations. In order to calculate the distance between the regions, the rule uses a user-defined Boolean function (*near*) that compares the distance between two regions with a value, informed as parameter between brackets (*near [value]*). The pattern defined in line 5 constrains regions that are in epidemics situation, which are captured in the *$e1* binding variable. Lines 6, 7, and 8 also constraints regions that are in epidemics situations, as long as (1) they are different from the regions of the previous pattern (*$source != $target*); (2) both *Epidemics* situations overlap in time (*this overlapped by $e1*); and (3) these regions are within 100 km from each other (*near [100 km]*). If these conditions are met, the *EpidemicSpread* situation is detected for that particular pair of regions.

```
1   rule "EpidemicSpread"
2   @role(situation)
3   @type(EpidemicSpread)
4       when
5           $e1: Epidemics($source: region)
6           $e2: Epidemics($target: region, $source != $target,
7                          this overlappedby $e1,
8                          $source.location near["100km"] $target.location)
9   then
10          SituationHelper.situationDetected(drools);
    end
```

Fig. 16.13 The *EpidemicSpread* situation rule

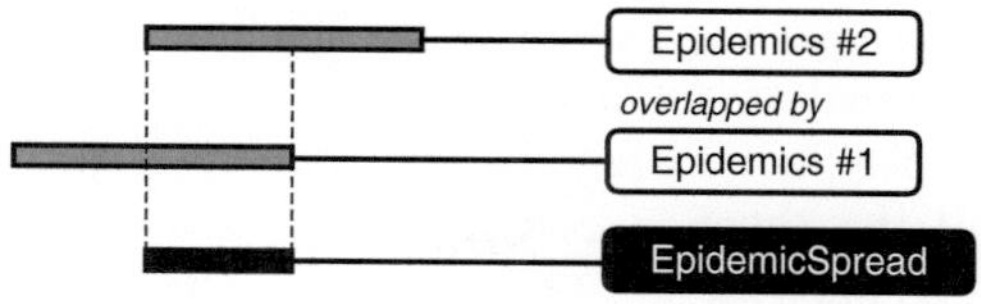

Fig. 16.14 EpidemicSpread timeline example

Figure 16.14 depicts an example timeline of an instance of EpidemicSpread situation, in terms of two occurrences of Epidemics situations, for neighbor regions (located within 100 km from each other). Note that the EpidemicSpread situation begins to exist simultaneously to the occurrence of an Epidemics situation for region number 2. When the occurrence of the Epidemics situation for region number 1 ceases to exist, so does the EpidemicSpread situation instance.

16.5.3 SCENE: Realization

Situation's Life Cycle Management

A situation life cycle consists of situation *detection*, *creation*, *activation*, and, possibly, but not necessarily, *deactivation*. As discussed in previous sections, situation detection occurs when the LHS of a situation rule is satisfied (for a particular situation cast). Note that conditions in the LHS of a situation rule hold true while the situation exists (John's temperature exceeds 37 °C during the existence of John's Fever Situation). However, although the situation rule may be executed several times while the conditions hold, only one *situation fact* should be created to represent that particular situation occurrence. In order to solve this issue, our approach separates situation detection from its creation.

The situation's life cycle management strategy benefits from a Drools feature called Truth Maintenance System (TMS). The TMS automatically ensures the logical integrity of facts that are inserted in the working memory in the RHS of a rule. A logical fact exists in the working memory while the conditions (in the LHS) of the rule that has inserted in the working memory remain true, and retracted from the working memory when conditions no longer hold. Thus, the solution we have used consists on a logically inserted fact produced by the firing of a situation rule to reflect the situation fact state (existence or nonexistence). This solution has enabled us to detect the activation and deactivation of a situation instance by means of a single rule specification, which otherwise, would require a pair of activation–deactivation rules.

Internally, the TMS maintains logical facts by verifying whether there is an equal object already in the working memory before inserting any object. This way, an object only becomes a logical fact in the working memory when it is unique; otherwise, it is discarded by the engine. Therefore, in our approach, a *situation logical object*, which we call *CurrentSituation object*, is created by the *SituationHelper* class for each time the *situationDetected* method is executed. When it is unique (in terms of its situation type and cast), a *CurrentSituation* object is inserted as a *CurrentSituation* fact in the working memory. When snapshotting capabilities are required for a particular situation, *SCENE* keeps serialized versions of the situation cast for each *CurrentSituation* objects creation (i.e., for each execution of the situation rule's RHS). As an example, consider John's Fever Situation. The first time the RHS of Fever situation rule (Fig. 16.6) is executed for a particular cast (which consists of John in this case), a *CurrentSituation* object is created and immediately, through logical insertion, becomes a *CurrentSituation* fact. Further executions of the situation rule's RHS for that particular cast (John) will only produce *CurrentSituation* objects which will be rejected as new facts by the TMS. In our example, serialized versions of *situation casts* referred to by these objects are kept since we have chosen to keep past occurrences of Fever Situation. The TMS automatically retracts John's Fever *CurrentSituation* fact when the LHS of the Situation Fever rule no longer holds for John.

Situation activation occurs simultaneously to its creation, and the deactivation occurs when the situation rule's condition no longer holds. Deactivated situation facts consist of historical records of situation occurrences, which may be used to detect situations that refer to past occurrences.

In order to handle situation activation and deactivation, our approach internally defines a predefined pair of rules, which are executed in terms of (existence or nonexistence of) *CurrentSituation* facts. The *Situation Activation* rule (Fig. 16.15) matches for every newly inserted *CurrentSituation* fact (i.e., *CurrentSituation* facts with *situation* attribute set to null). The RHS of the activation rule creates an instance fact of the Situation Type class and its properties are assigned by the corresponding entities involved in that particular *situation cast*. Considering our John's Fever Situation example, when the activation rule is executed, an instance of the Fever class (Fig. 16.5) is created and John is assigned to the attributed febrile. In addition, the *CurrentSituation* fact's attribute *situation* now refers to the newly created situation type instance. This way, the activation rule no longer matches for that particular *CurrentSituation* fact. Upon execution of the situation activation rule, *SCENE* also generates an *initiator* event (*ActivateSituationEvent*), which represents the activation timestamp for that situation (and is used for temporal reasoning).

The *Situation Deactivation* rule (defined in Fig. 16.16) matches for every *SituationType* fact yet active (attribute *active* is true) for which there's no corresponding *CurrentSituation*. The absence of the *CurrentSituation* is a consequence of the TMS logical retraction due to the no longer fulfillment of the situation rule's conditions by a particular *situation cast*. The RHS of the deactivation rule creates a *terminator* event (*DeactivateSituationEvent*) for that particular situation and also sets its transition to a *non active* state (attribute *active* of Situation Type class is set to false).

```
rule "SituationActivation"
   when   // newly inserted situation fact is detected
      $act: CurrentSituation(situation == null,
                             $type: type,
                             $castset: castset,
                             $timestamp: timestamp)
   then  // activate situation, creating situation instance
      SituationHelper.activateSituation(drools,
                                        $castset,
                                        $type,
                                        $timestamp));
end
```

Fig. 16.15 Situation activation rule

```
rule "SituationDeactivation"
      when  // situation fact is retracted by TMS
            $sit: SituationType(active==true)
            not(exists CurrentSituation(situation == $sit))
      then  // de-activate situation
            deactivateSituation(drools, (Object) $sit);
end
```

Fig. 16.16 Situation deactivation rule

Situation Profile Management

The Situation Profile Manager (SPM) is a module that stores profiles for each situation specification based on declared metadata information previously mentioned in Sect. 16.5.2. These profiles allow the situation engine to apply particular management strategies, such as the *cast snapshotting* and *participation state storage* strategies. The SPM assembles rules profiles by parsing the rule base at the execution of the session bootstrapping, capturing situation rule's metadata values. The SPM also maintains the rules profiles throughout the situation's life cycle execution.

Regarding the life cycle of *snapshot-enabled* situation facts, the *snapshotting* process takes place for every situation rule's LHS match, in which a serialized version of the assembled *situation cast* (tagged with a timestamp) is stored. When the situation ceases to hold, the chosen restoring strategy is carried on at situation deactivation (execution of the *deactivateSituation* helper method).

Temporal Evaluators

The Drools native API provides an extensible way to implement new LHS operators. This particular feature allowed us to implement proper evaluators to handle the situation temporal relations. Our approach applies the Allen's interval algebra over initiator and terminator situation events, which are created by the activation and the deactivation rules, respectively.

Given the dynamic nature of a situation occurrence, in which situations may be related to an initiator event only (active situation) or related to both initiator and terminator events (inactive situation), the situation temporal operators have to consider the absence of the terminator event.

In order to evaluate situation temporal relations, the situation operators' implementation extracts the events of interest from situations facts parameters and then evaluates the situation temporal relation by means of initiators and terminators events (using the temporal operators currently provided by Drools).

16.6 Concluding Remarks

We have explored in this chapter rule-based solutions to support the development of situation-aware applications. We argue that a solution based on *rules* appears naturally given the nature of situation detection, in which the user's context is continuously monitored in order to check whether certain conditions are met. Rule-based solutions offer flexibility with respect to the maintainability of the rules: rules can be modified and added at application runtime with no need for code recompilation. Since situation specifications may change over time, and new situation specifications may be defined at application runtime, it is beneficial to use a mechanism that offers such flexibility. We have discussed here how traditional rule-based platforms can be beneficially used to develop such systems and we also highlight the limitations of traditional rule-based solutions with respect to situation management. Situation type specification in traditional rule-based systems require two rules to specify a particular situation type: one to specify the situation detection condition and another to capture the situation deactivation condition. In addition, events are not natively supported by rule-based systems, which forces application developers to explicitly deal with time issues (e.g., initial and final time attributes and temporal relationships).

In order to overcome such limitations, we have proposed a rule-based situation management infrastructure to support the development of situation-aware applications. We have implemented a rule-based platform for situation management (coined SCENE) that leverages on JBoss Drools engine by adding functionality to natively support rule-based situation-awareness.[2] Situation specification requires a single rule pattern following the standard Drools rule's constraint dialect. Situations type

[2]The SCENE source code is available at https://github.com/pereirazc/SCENE

definitions can be composed of constraints over domain entities, and in addition can be composed of existing situation types. We have addressed the temporal aspects of applications, and included operators to relate situations based on their temporal aspects. The detection is rule-based, and is deployed on mature and efficient rule engine and complex event processing technology available off-the-shelf. The platform manages situations by implementing situations life cycle control, such as situation activation, state maintenance, and deactivation.

As part of our infrastructural efforts, we have extended SCENE in order to incorporate distribution facilities. The distribution support is based on the notion of shared public situations, which can be detected by an engine running in a particular node and shared with engines running in other nodes with a mirroring scheme implemented on top of a (publish-subscribe) situation distribution service. This enables us to support flexible distribution configurations, based on the detection of independent situations in parallel (possibly of the same situation type) and the allocation of different (related) situation types to different network nodes. In Raymundo et al. (2014) we present a performance assessment of the distributed infrastructure, showing that the overheads introduced by distributed communication are acceptable and that a distributed situation management scenario can outperform a centralized one.

In addition to providing infrastructural support for situation detection, we have also explored a graphical language (coined SML) for situation modeling. This work has been reported in Costa et al. (2012), in which we present model-driven transformations from SML models into situation rules to be executed on our rule-based situation platform.

We are currently working on extensions of SCENE to incorporate the notion of uncertainty in situation management. This means that situation rule developers will be able to explicitly account for variations in the quality of context information and of the situations inferred from context information.

Acknowledgements This research is partially funded by the Brazilian research agencies CNPq (311313/2014-0 and 485368/2013-7), CAPES/CNPq (402991/2012-5) and the Dutch national program COMMIT (project P7 SWELL).

References

Allen JF (1983) Maintaining knowledge about temporal intervals. Commun ACM 26:832–843

Bali M (2009) Drools JBoss rules 50 developer's guide. Packt Pub, Birmingham

Barwise J (1989) The situation in logic, vol 17, CSLI lecture notes. CSLI, Stanford

Costa PD (2007) Architectural support for context-aware applications: from context models to services platforms. Ph.D. Thesis, University of Twente

Costa PD, Guizzardi G, Almeida JPA, Ferreira Pires L, van Sinderen M (2006) Situations in conceptual modeling of context. Workshop on Vocabularies, ontologies, and rules for the enterprise (VORTE 2006) at IEEE EDOC 2006, IEEE Computer Society Press

Costa PD, Mielke IT, Pereira I, Almeida JPA (2012) A model-driven approach to situations: situation modeling and rule-based situation detection. Proc. 16th IEEE enterprise distributed object computing conference (EDOC 2012), IEEE Computer Society Press

Dey AK (2001) Understanding and using context. Pers Ubiquit Comput 5(1):4–7

Endsley MR (1995) Toward a theory of situation awareness in dynamic systems. Hum Factors Ergon Soc 37:32–64

Hayes-Roth F (1985) Rule-based systems. Commun ACM 28(9):921932

Kokar MM, Matheus CJ, Baclawski K (2009) Ontology-based situation awareness. Inf Fusion 10:83–98. doi:10.1016/j.inffus.2007.01.004

Pereira I, Costa PD, Almeida JPA (2013) A rule-based platform for situation management. IEEE International multi-disciplinary conference on cognitive methods in situation awareness and decision support (CogSIMA), IEEE Computer Society Press

Raymundo CR, Costa PD, Almeida JPA, Pereira I (2014) An infrastructure for distributed rule-based situation management. IEEE International multi-disciplinary conference on cognitive methods in situation awareness and decision support (CogSIMA), IEEE Computer Society Press

Rosemann M, Recker J (2006) Context-aware process design exploring the extrinsic drivers for process flexibility. Proc. 7th CAISE workshop on business process modelling, development, and support (BPMDS '06)

World Health Organization (2010) Global Epidemiological Surveillance Standards for Influenza [online]. http://www.who.int/influenza/resources/documents/WHO_Epidemiological_Influenza_Surveillance_Standards_2014.pdf?ua=1. Accessed 14 Aug 2014

World Health Organization (2014a) Influenza (Seasonal) [online]. http://www.who.int/mediacentre/factsheets/fs211/en/. Accessed 14 Aug 2014

World Health Organization (2014b) WHO surveillance case definitions for ILI and SARI [online]. http://www.who.int/influenza/surveillance_monitoring/ili_sari_surveillance_case_definition/en/. Accessed 14 Aug 2014

Chapter 17
From Argumentative Crisis to Critical Arguments: How to Argue in the Face of Danger

Laura Bonelli, Silvia Felletti, and Fabio Paglieri

Abstract Building on evidence from the field of risk perception and communication, two key roles of argumentation in crisis management are highlighted: (1) balancing trust construction and persuasive goals in crisis prevention and preparedness, and (2) ensuring time-efficient cross-examination of choice options in group decision making at a time of crisis. The implications for an information fusion approach to crisis management are discussed, suggesting a rich potential for future research.

17.1 Introduction

In everyday life, argumentation is the activity of exchanging and considering reasons for or against a certain conclusion, which could be either epistemic (what to believe) or practical (what to do), or both—as it is often the case, since we typically argue to establish what to believe *in order to* decide how to act. In philosophy, argumentation is sometimes defined narrowly, by focusing only on one of its many functions—typically, the resolution of a conflict of opinions via rational means (van Eemeren and Grootendorst 2004). Other authors, however, accept a broader understanding of the notion, acknowledging that argumentative processes can serve multiple purposes: not only conflict resolution but also deliberation, inquiry, persuasion, venting one's feelings, and so on (Walton 1998). Moreover, the formal study of argumentation, especially in Artificial Intelligence and knowledge engineering, tends to emphasize the abstract nature of argumentative relationships, and use the concept to operationalize complex networks of conflicts among multiple nodes (so called abstract argumentation structures; see Dung 1995). But even in

L. Bonelli (✉) • S. Felletti
Theory & Technology group (T3), Istituto di Scienze e Tecnologie della Cognizione, Consiglio Nazionale delle Ricerche, Rome & Department of Psychology, University of Rome "La Sapienza", Rome, Italy
e-mail: Laura.bonelli@istc.cnr.it; Silvia.felletti@stc.cnr.it

F. Paglieri
Goal-Oriented Agents Lab (GOAL), Istituto di Scienze e Tecnologie della Cognizione, Consiglio Nazionale delle Ricerche, Rome, Italy
e-mail: Fabio.paglieri@istc.cnr.it

G. Rogova, P. Scott (eds.), *Fusion Methodologies in Crisis Management*,
DOI 10.1007/978-3-319-22527-2_17

computer science there are many approaches that hinge instead on the semantics of arguments (for a survey of argumentation studies in AI, see Reed and Norman 2004; Besnard and Hunter 2008; Rahwan and Simari 2009): as a case in point, consider argumentation-based decision-support system (e.g., Ouerdane et al. 2011). The wealth of models and results achieved through the interdisciplinary study of argumentation can be applied to information fusion problems in a variety of ways, and it is likely to be especially relevant in the context of crisis management, as we shall discuss in this chapter. At a general level, we can identify two main areas of interest for argumentation models in the context of information fusion:

1. Argumentation-based aggregation of different sources of information, both at the lower level (e.g., using abstract argumentation to represent and evaluate large sets of interconnected sources/statements; Caminada and Pigozzi 2011) and at the higher level (e.g., using argument schemes to reason about the trustworthiness of sources/messages; Paglieri et al. 2014; Parsons et al. 2014);
2. The use of argumentation to ensure that information relevant for a crisis is used in a wisely and timely manner (*argumentation for decision making*).

In this chapter we will focus on the latter topic, building on a variety of theoretical approaches, and in particular on the decision-theoretic model of argumentation recently developed by Paglieri and collaborators (Paglieri 2009, 2013a, b, 2015; Paglieri and Castelfranchi 2010), and preliminarily corroborated by empirical studies on respondents from the US (Hample et al. 2012) and Romania (Cionea et al. 2011). The central tenet of the model is that the process of arguing requires several decisions by the arguer: looking at how these decisions are made illuminates not only the moves and strategies actually observed in real life dialogues (Cionea et al. 2011; Hample et al. 2012), but also provides a framework to analyze the rationality (or lack thereof) of such choices (Paglieri 2013b). The current process-based taxonomy of argumentative decisions is as follows (for further details, see Paglieri 2013a):

- *Engagement*: the decision to enter an argument or not, either by proposing one or by accepting to be drawn into one (Cohen 2005; Paglieri 2009).
- *Editing*: all decisions concerning what arguments to use (selection) and how to present them to the audience (presentation), in order to maximize their intended effects (Hample and Dallinger 1990, 1992; Hample 2005; Hample et al. 2009).
- *Timing*: the decision on when it is time to speak and when it is time to listen (not much studied at all, except in terms of highly stylized turn-taking; e.g., Walton and Krabbe 1995).
- *Interpretation*: if there are ambiguities in what the counterpart is saying, the decision on whether to criticize them, ask for more clarity or additional information, or solve them autonomously—and if so, favoring what interpretation, on what grounds, and to what ends? A well-studied case of argument interpretation concerns enthymemes, defined as incomplete arguments (Paglieri 2007; Paglieri and Woods 2011a, b).
- *Reaction*: decisions concerning whether to accept or challenge an argument, an objection, or a counter-argument raised by the counterpart (Gilbert 1997).

- *Termination*: the decision on when and how to end an ongoing argument, before it has reached its "natural" conclusion, e.g., agreement, victory of one of the parties, etc. (Hample et al. 1999; Paglieri 2013a).

The decision-theoretic perspective on argumentation immediately reveals the centrality of goals to make sense of argumentative decisions (Paglieri and Castelfranchi 2010). We need to know what motivates the arguers to engage in dialogical interaction to understand their moves (descriptive level), and we need to know what function argumentation is meant to serve in that context, if we hope to assess those moves against their adequate standard (normative level). This function in turn constrains what personal goals the arguers can *legitimately* pursue in that context, whether or not they coincide with their actual agenda. With respect to crisis management, two main argumentative functions stand out as absolutely crucial, and are roughly associated with different phases of a crisis:

1. Arguing to *build trust and/or reach a consensus* among the stakeholders (*prevention and preparedness*);
2. Arguing to make *good and time-effective decisions* (*intervention*).

These two functions of argument, and the different stages of crisis management they mostly pertain to, are certainly intertwined with each other. Nonetheless they are not to be confused or conflated, since argumentation has to satisfy very different constraints in each of these contexts. Thus in what follows we will discuss them separately: how to argue to build trust and shared values in crisis prevention will be the topic of Sect. 17.3, whereas argumentation in support of time-sensitive decision making at times of crisis will be analyzed in Sect. 17.4.

Before looking at the kind of argumentative choices required at each of these stages of crisis management, however, we have to gain a better understanding of the cognitive factors involved in our reactions to critical situations, especially regarding two key concepts: risk and trust. These affect both how a crisis is perceived and how relevant information about it should be communicated (Sect. 17.2), with important implications on how argumentation unfolds.

17.2 Risk Perception and Risk Communication

Technological advancement and industrialization have increased our living standards, but have also produced a growth in the risk potential carried by hazards such as chemical or nuclear technologies. We could not imagine our life without industrialized food, cars, or air conditioning, but we do not seem to tolerate well their associated risks. Instead of feeling more protected by modern safety systems and by the perspective of the longest life expectancy ever reached in history, people feel like facing more risk today than in the past (Slovic 1987, 1999). Nuclear and chemical technologies, in particular, are felt as coming with unbearable great risks for human health, despite the experts' low estimates of risk regarding such technologies—hence, the great difficulties to find construction sites for industrial plants epitomized by the NIMBY (Not In My BackYard) attitude, even when dealing with relatively "green" technologies, e.g., renewable energy plants (van der Horst 2007).

New technologies, their function, and their associated risks are complex and obscure to most citizens, who perceive them as threatening and hard to manage. Given the low knowability of unfamiliar hazards, laypeople usually rely on intuition to make their risk judgments. This led to the creation of a specific interdisciplinary field of inquiry, *risk assessment*, devoted to obtain more reliable and comprehensive measures of risk factors and consequences. One of the aims was to reduce people's irrational fears by providing them with more accurate information on risk. When compared to "objective" risk assessments, citizens' risk perceptions turn out to be significantly different, even when they are presented with clear statistics (Yamagishi 1997). This marked inconsistency is a source of frequent conflict in crisis management: while experts tend to limit their analysis of risk to the amount of predictable negative outcomes (e.g., mortality rate, injuries, material damages), people's intuitive risk assessment seems to rely on a much broader array of aspects that characterize risk, such as uncertainty, dread, controllability, equitable distribution of risks and benefits, familiarity with the hazard, catastrophic potential, risk to future generations (Slovic 1987; Kasperson et al. 1988).

People's perception and acceptance of risk must not necessarily result from ignorance or irrationality; instead it can be the outcome of the—not always conscious—consideration of a series of attributes of the hazards that statistical analysis simply overlooks, along with personal psychological tendencies and cultural background. Failing to acknowledge the complex and socially determined nature of people's concept of risk may lead experts to unnecessarily polarize the debate on crisis management, with no added benefits. Indeed, careful estimation of mathematical probabilities should not exclude taking into account also a broader set of contextual factors, insofar as they have a demonstrable impact on subjective risk perception. Notable examples in the literature include gender (Steger and Witt 1989; Flynn et al. 1994) and worldview, especially in terms of fatalism, hierarchism, and individualism (Dake 1992; Jenkins-Smith 1993; Peters and Slovic 1996).

17.2.1 The Concept of Trust

Trust can promote compliance and cooperation, and is a fundamental construct for social interaction. Its relation with risk perception has been of increasing interest in the last years, as it is believed that enhancing citizens' trust in institutions will reduce their fears towards technological and natural risks (Kasperson et al. 1992).

Although there is still a lively debate among theorists on how to define trust, most approaches rely on some version of the conception proposed by Rousseau et al., for which trust is "a psychological state comprising the intention to accept vulnerability based upon positive expectations of the intentions or behavior of another" (1998, p. 395; see also Mayer et al. 1995).

There is still some confusion, though, on the breadth of the meaning of trust, as some authors refer to trust as limited to its causes or antecedents, or either to its outcomes or consequences. According to Castelfranchi and Falcone (2010), trust must be viewed as a layered notion: in its basic meaning of subjective trust, trust is "a belief, attitude, or expectation concerning the likelihood that the actions or outcomes of another individual, group or organization will be acceptable or will serve the actor's interests" (Sitkin and Roth 1993). Also labelled as general trust, trust in this sense corresponds to a trait variable, a general willingness of the trustor to rely on others.

The term trust can also be used to refer to the trust antecedents, that is, the factors that lead to trust. Among them, the most significant is trustworthiness, a property that is related to the trustee but that is also defined in relation with the context and the trustor's perception. The components of trustworthiness have been variously identified as ability, benevolence and integrity (Mayer et al. 1995); competence, objectivity, fairness, and consistency (Renn and Levine 1991); ability and intentions (Nooteboom 1996); responsibility and competence (Barber 1983). Yet nearly all these classifications can be traced back to the two dimensions of ability (competence) and willingness (commitment and care).

Finally, trust can be intended as the very act of relying upon somebody, that is the resulting manifestation of subjective trust, better identified with the label of "behavioral trust" (Castelfranchi and Falcone 2010). This aspect is especially crucial in crisis management, in particular in prevention and preparedness, since a key obstacle is often in translating good intentions into consequent deeds: thus trust in experts and institutions needs to be acted upon, rather than merely endorsed as an attitude.

17.2.2 The Role of Trust in Risk Perception

Whereas the existence of a strong relationship between trust and risk perception is widely accepted, full agreement on the nature of this relationship has not been reached yet. Trust and risk are related concepts by definition: many agree that risk, or the condition of having something at stake, is an essential element of trust (Deutsch 1958). As Mayer and colleagues noted, though, "it is unclear whether risk is an antecedent to trust, is trust, or is an outcome of trust" (1995, p. 711). Risk can be seen as an outcome of trust, corresponding to its behavioral manifestation, or to the act of becoming vulnerable.

In Das and Teng account (2004), trust and risk are mirror images of each other, as both hinge on uncertainty and probabilities: a perception of low trust will result in a perception of high risk, and vice versa. Citizens' risk acceptance seems to be highly dependent on the level of trust they have in risk managers. It has been proved that a key reason why the public often rejects experts' risk assessment is lack of trust (Nooteboom et al. 1997; Earle 2012). Lack of trust is also the reason why institutional risk communication often fails to achieve satisfactory results in allaying public's fears: as we shall discuss more thoroughly in Sect. 17.3, a distrusted source can hardly hope to persuade its audience, regardless of the quality of its message.

Indeed, the most influential view on the function of trust is that its purpose is to reduce complexity in the assessment and acceptance of risk in situations of low knowledge of the hazard. Laypeople usually have poor competence on the specifics of a crisis, so they resort to their level of trust in institutions to decide whether to accept or not the risks deriving from the hazards they have to face (Siegrist and Cvetkovich 2000; Visschers, and Siegrist 2008).

But how do people decide whether to trust risk management institutions? Many studies suggest that emotions play a fundamental role in that choice (Slovic 1993; Slovic et al. 2007).

Heuristics, or the strategy of resorting to emotions in order to manage uncertainty, have been typically analyzed in the field of reasoning and decision making (Tversky and Kahneman 1974; Damasio 1994), but have also been found to govern our processes of risk evaluation.

As far as common biases in risk assessment are concerned, Weyman and Kelly (1999) surveyed six different perspectives of research in risk perception: behavioral decision theory, the mental models approach, the value expectancy models, sociocultural theory, and safety culture. In their review on these approaches, they selected several recurring biases within the literature, which have specific impacts on how individuals perceive risk: perceived control, significant connections between psychological time and risk, effects of familiarity of the perceived risk, perceptions of vulnerability related to risks and hazards, biases such as the "framing effect," issues in the numerical representations of risk, the focus on "risky situation" vs. "risky individual" and issues in the perception of hazardous substances are all subject to fallacious instances of risk assessment.

With regard to trust judgments, we often rely on similarity and affect: that is, we tend to trust people that are similar to ourselves, who share our values, intentions, and group identity (Foddy et al. 2009)—so much so, that even mere physical resemblance has been found to affect trust attributions (DeBruine 2002). Moreover, closeness tends to produce positive emotions towards another agent: a study by Dunn and Schweitzer (2005) found an effect of irrelevant emotions on trust, such as anger or happiness. Finucane et al. (2000), tested people's judgment of risks and benefits under a time pressure condition designed to limit the use of analytic thought and enhance the reliance on affect. As expected, time pressure strengthened the inverse relationship (for the effects of time pressure on argumentation processes during a crisis, see also Sect. 17.4).

Risk managers should better acknowledge the fundamental, heuristic role that similarity and affect play in trust judgments and risk perception, especially in the context of crisis management. Heuristics and biases are often trotted out to explain why risks that are assessed by technical experts to be relatively minor, often elicit strong public worries, and to charge non-experts with irrationality. But people's personal perception of risk may not require any "adjustment." Actually, further research will be needed to deeply comprehend the mechanisms underlying it. A clear understanding of social trust and its determinants and its influence on risk perception could facilitate the difficult task of designing effective communication strategies, thus promoting cooperation between citizens and institutions.

17.2.3 The Risks of Risk Communication

Communication is a risky activity per se: it constantly involves the need to smooth the interactional management in order to reduce different types of risks for the interacting partners (e.g., risks of conflict, of losing face, or of refusal). When the communicated object is a threatening one (such as an environmental hazard or a public biological risk) and the illocutionary force is most likely aimed at addressing a point of both awareness and a reasonably harmless and self-protective behavior in certain target population, those responsible for this act must make sure that an even wider array of needs and dimensions are taken into account. Cognitive, affective, metapragmatic, sociocultural, and even stylistic dimensions are all relevant aspects of effective risk communication. As far as the cognitive dimensions are concerned, risk communication must especially gain insight into the qualitative and shared understandings of the lay public. It should also provide various levels and types of actions and messages that are rooted in—and which aim to shape adequately—the public's concerns, beliefs, attitudes, influences, and goals.

Although persuasive communication may rely on a remarkable research history within the fields of rhetoric, social psychology, and linguistics, risk communication constitutes a relatively new topic of investigation. Basic issues such as how it should be defined, how risk itself could be conceptualized, and the range of variables which should be taken into account in risk assessment are still a matter of discussion (Weyman and Kelly 1999). Especially because the frame of reference of the audience should appropriately be matched by the conveyed messages (Pidgeon 1992), precious insights both from cognitive psychology and from argumentation theory (Castelfranchi and Falcone 2010; Paglieri 2009, 2013a, b, 2015; Paglieri and Castelfranchi 2010; Hample et al. 2012) may specifically help bridging the gap between the representational level and the expressive one. The contribution of these two disciplines may also allow a more comprehensive theoretical framework for risk communication and a more effective practical one.

17.2.4 Possible Definitions, Loci of Interest, Objectives, and Models

Most generally, risk communication is intended as "an interactive process of exchange of information and opinion among individuals, groups, and institutions" which have a topical focus on a certain risk or hazard; it is a dynamic process that "involves multiple messages about the nature of risk and other messages, not strictly about risk, that express concerns, opinions, or reactions to risk messages or to legal and institutional arrangements for risk management" (Committee on Risk Perception and Communication, National Research Council 1989, p. 21; see also Covello et al. 2001; Infanti et al. 2013). Acknowledgments, considerations and emotions elicited with different levels of epistemic and affective commitment by that risk or hazard are not necessarily considered foregrounded features of risk communication. Secondary and tertiary effects triggered by the communication process are thought to be vital elements of the analysis of risk communication, as well as their psychological, social, and political repercussions (Kasperson et al. 1988). Although the effects that those messages have on psychological and social processes are not strictly included in the scope of analysis of risk communication (Renn and Levine 1991), they represent a precious corpus of information on the possible efficacy and felicity of the risk communication acts.

Covello et al. (1986) analyze risk communication as a deliberate exchange of information on health or environmental risks among interested parties concerning the degree, the significance and the policies aimed at managing or controlling those risks. Infanti and collaborators (2013), in line with Covello and Sandman (2001), distinguish risk communication from crisis communication, insofar as the latter does not necessarily involve the public in matters such as risk assessment, risk management, and decision making. Whereas the communication of a state of crisis is a reactive practice whose messages tend to focus on the spatiotemporal immediacy of a certain harmful event, and whose main aim is that to express or represent what is known and what is not known about it, risk communication is a planned practice. It emerges either long before the state of crisis occurs, or even if the latter represents a reasonable hypothesis in certain contexts. Risk communication is thus based on projections and calculations of the potential for future harm and it provides messages aimed at encouraging collaborative practices among all interested parties, including the public (Covello et al. 2001), instead of involving them in a reactive process. For these authors, while the source of a crisis communication in most cases simply reacts to possibly unexpected and contingent events, the source of risk communication deals with likely or expected hazards and posits itself as possibly directive and influential, regardless if these hazards actually happen. In so doing, the message source tends to downgrade the immediacy and possibly the intensity of the chosen communicative strategies, whereas the source of a crisis message tends to intensify them. However, this distinction is probably too rigid to do justice to the complex interplay between crisis management and risk communication:

a more nuanced approach would rather see these practices as continuous, with risk communication playing a key role in crisis prevention and preparedness (see Sect. 17.3).

The loci of interest involved in risk communication also need a shared identification. In a pure semiotic fashion, Covello and colleagues (1986) identified four components of risk communication: the message source, the message design, the delivery channel, and the target audience. In this model, risk communication is depicted as rather linear: a known, expert source delivers scientific information to specified recipients (usually, the lay public) through designated channels (see also Kasperson and Stallen 1991; Bell and Tobin 2007). Interested parties may include governments, agencies, corporations, industry groups and unions, scientists, and professional organizations as sources and public interest groups or individual citizens as target recipients (Covello et al. 1986), as well as various entities with a translatory role, such as insurance agents, planners or managers (Faulkner et al. 2011).

As far as the message production is concerned, risk communication may observe the following felicity conditions: the source believes that it is not obvious to the target audience that a certain hazard may occur, and that, if the hazard occurs, it will be detrimental to the target audience (preparatory precondition). The source genuinely believes that the hazard of reference is going to have harmful effects on the target audience (sincerity condition) and any communicative act counts as an attempt by the source to have the target audience recognize that a possible, future hazard will be truly detrimental (essential condition). The illocutionary force of risk messages may vary from directive acts (e.g., orders or instructions), to declarative acts, to representative acts (e.g., informative messages) or even to commissive acts (e.g., states of affairs that the government, industry, or organization promises to reach in order to protect the target audience or to gain trust from it). The illocutionary points may cover a great variety of purposes, instead.

Renn and Levine (1991) tried to summarize the main objectives and goals of risk communication in a list of functions, which contemplates:

- Risk understanding among target groups;
- Information disclosure about hazards to their potential victims;
- The legitimation of risk related decisions;
- The explanation and justification of risk management routines trust enhancing in the competence and fairness of the management process;
- Enhancing public protection through information about individual risk reduction measures;
- The encouragement of protective behavior or supportive actions toward the communicating source;
- The provision of guidelines for emergencies or behavioral advice during emergencies;
- The education of decision makers about public concerns and perceptions;
- Assistance in reconciling conflicts about risk-related controversies.

Many of these objectives rely on a certain amount of trust among the communicators to be effective. This gives further proof that trust and credibility constitute a fundamental feature of the risk communication process and that they should be co-constructed and maintained prior and during any act of risk management and announcement (see Sect. 17.3).

A widely accepted theory of risk communication is the one proposed by Covello and collaborators (2001). According to these authors, risk communication is based on four theoretical models: risk perception, mental noise, negative dominance, and trust determination. These models describe how risk information is processed, how risk perceptions are formed, and how risk decisions are made. Covello and colleagues (2001) identify at least 15 crucial factors: voluntariness, controllability, familiarity, equity, benefits, understanding, uncertainty, dread, trust in institutions, reversibility, personal stake, ethical nature, human origin vs. natural origin, victim identity, and catastrophic potential. These factors have a combined influence on how risks are perceived, influencing affective arousals on a variety of emotions (concern, worry, anger, anxiety, fear, hostility, outrage), which in turn impact significantly on attitudes and behavior, thus directly affecting risk communication outcomes. Specifically, an individual's perception of risk can be based on a combination of hazard and outrage. The latter especially takes on strong moral and emotional overtones, predisposing that individual to react emotionally, and thus significantly amplifying his or her levels of perceived risk.

To deal with these effects, Covello and collaborators suggest that risk communication should contemplate a series of risk perception related activities, especially aimed at gaining insight on how different stakeholders conceptualize and perceive risk. As we discuss in Sect. 17.3, the collective dialogue sustained by such activities is instrumental to achieve trust via rational argumentation, open-ended dialogue and attentive listening of the multiple public standpoints.

Since 2001, the four milestones of risk communication posited by Covello and collaborators have been corroborated by various findings, while at the same time new models and approaches have emphasized the importance of social and cultural factors to risk information processing and to the acceptance of risk messages. In particular, several recent frameworks conceptualize risk as a sociocultural process and highlight the importance of public trust in institutions and organizations in determining risk perception and preparedness (Lion 2001; Joffe 2003). Moreover, the perception of similar risks may vary or be assessed in significantly different manners across different groups or communities: as a case in point, consider the gender differences in risk perception mentioned before.

17.3 Argumentation in Crisis Prevention

In crisis prevention and preparedness, the focus of the relevant authorities is typically on persuading the public to change their mind: citizens are often assumed to be misinformed (their beliefs are mistaken) and/or misguided (their goals are wrong), and authorities see their mission as one of correcting their mistakes. As discussed in previous sections, this often generates poor results and a lot of paternalism in risk communication. Even more crucially, it leads to overlook another essential function of argumentation for crisis prevention: *building trust*.

Incidentally, it is worth stressing that a paternalistic attitude in risk communication is, quite frequently, epistemically justified: after all, "emergency managers (EMs) actually *are* often more knowledgeable about what publics need in emergency situations, and quicker to understand those needs, than the publics themselves. This is how they gained certification as emergency managers. Second, EMs are trained to deal with issues on probability (e.g., flood warnings, accidental chemical releases, pandemic flu risk) and to take precautions based on these probabilities" (Rowan et al. 2009, p. 181). Unfortunately, the reliable grounds upon which experts base their presumption of superior knowledge in matters of risk do not make paternalistic messages more effective at all. In other words, the fault of paternalistic risk communication is practical, not epistemic: it does not work as an effective communication strategy, regardless of whether experts do, in fact, "know better." Importantly, some of the reasons why people do not like to be simply told what to do can be ascribed to quirks of our psychology (e.g., my-side bias, or even plain pride); but other reasons for resisting one-sided risk communication may be perfectly sound, such as taking into account costs (e.g., what to do with one's kids when schools are closed due to a public alert) and dangers (e.g., the risk of looting in one's property during an evacuation) that experts either ignore or systematically downplay.

From an argumentative point of view, there are two distinct functions that risk communication must serve to be effective in crisis prevention and preparedness: persuading the public to accept a certain set of information and act accordingly, and building a climate of trust and mutual understanding among all relevant stakeholders. The problem is that there is a potential tension between these two functions of argumentation: they are not necessarily in conflict, but they can (and often are) contingently at odds. Not only because paternalistic and ineffective crisis communication can lead to lower trust in the sources of this communication, but also (and mostly) because trust in those sources is a *precondition* for the success of any communicative effort. The motto of risk communication should be: "trust first, content later." In a climate of trust, even a relatively poorly argued message is likely to be accepted and acted upon, precisely because of the credit the public is willing to give to its source. On the other hand, unfortunately, the opposite of trust, distrust,[1] is lethal to crisis communication, insofar as it poisons its wells. Once this happens, it

[1] Several attempts have been made to clarify the different meanings and interrelations between trust, distrust, and mistrust, both in cognitive psychology (Castelfranchi 2000), political studies (Lenard 2008), computer science (Marsh and Dibben 2005), and philosophical logic (Primiero and Kosolosky 2013). These taxonomies often do not align on the technical definitions of distrust and mistrust, although there is a tendency to consider the former stronger than the latter (a notable exception is Castelfranchi 2000, who claims the opposite): here we will adopt the convention of using distrust to indicate the active suspicion of the other's intention to cause harm (or at least

does not matter how effectively subsequent communication is designed: it will still be likely to fail, because the public is either no longer listening, or it has developed a bias against anything that (distrusted) source has to say.

As an illustration of the pernicious effects of distrust on crisis communication, consider the following example: the terroristic attack on the satirical journal Charlie Hedbo in Paris on January 7, 2015, where 12 people were killed, was partially captured on camera. In particular, a video showed the murder of the police agent Ahmed Merabet: a masked terrorist shot him in the head at close range with an AK-47 rifle, while the victim was lying wounded on the ground. In the days immediately after the attack, conspiracy theorists started suggesting that things were not as they seemed (of course), and the video of Merabet's assassination was paradoxically mentioned as evidence of the fact that "something was fishy." In particular, conspiracy theorists argued that, since in the video the head of the police agent was not blown to pieces by the bullet, then something strange must have happened, and perhaps the whole thing was staged or even fake. Another video soon became viral, "Chef Jeff's AK-47 Watermelon Fruit Salad," showing what happens when you shot a watermelon with an AK-47 from 10 ft away—not surprisingly, the watermelon disintegrates. Thus the "reasoning" of conspiracy theorists went as follows: since the head of the policeman did not explode as a watermelon, as it should have if he had been truly hit, then the policeman was never hit and the whole video must be a fake.

The whole idea is preposterous, from any point of view. In terms of the alleged evidence, the penetration force of an AK-47 bullet at point blank is such that it simply goes through, causing no explosion whatsoever in the head of the victim. More generally, it does not take a ballistic expert to realize that heads shot at close range explode only in a Tarantino's movie, not in real life—sadly, we have countless videos of terrorist executions attesting to this fact. As for the idea of considering a watermelon as a reliable substitute for a human head, it is not even worth criticizing. Not to mention the lack of plausibility of the alternative reconstruction proposed by conspiracy theorists: the dead body of Ahmed Merabet was seen by many people, and the unfortunate victim was publicly buried. Suggesting that he never died and is now living in secrecy is implausible to the point of idiocy. Finally, what should be the motive behind such a convoluted plot to fool the public? Who could possibly stand to gain by showing the world an Islamic terrorist killing a Muslim policeman?

The patent inconsistency of this conspiracy theory, combined with its widespread circulation at a time when the indignation for the attack was extremely high across the globe, raises an issue that is relevant to present purposes: what can possibly prompt more or less well-functioning human beings to put aside any decency towards the victims of a crime and endorse manifestly outlandish beliefs, based on a complete perversion of the rational standards of evidence ("they" are showing me X, therefore X must be false)? The answer is largely to be found in the fundamental

---

prevent good) to the interested party, as opposed to a merely cautious attitude towards someone or something that is not considered entirely trustworthy (mistrust, according to Lenard 2008).

distrust towards political institutions and media channels that fuel the vast majority of conspiracy theories. In the absence of a minimal level of trust towards political authorities and communication agencies, everything is fair game, and any evidence coming from a source of authority is automatically regarded with suspicion. The effects of such a climate of distrust are both evident and profound, especially during a time of crisis.

If endemic distrust is the nail in the coffin of risk communication, conversely taking into account the need for trust promotion as a precondition for effective crisis prevention and preparedness is the key to successful persuasive efforts. There is now ample of evidence of this fact in the literature, both concerning the value of public participation in risk assessment (Pelling 2007; Van Aalst et al. 2008) and risk communication (Kasperson 1986; Arvai 2003), and the effectiveness of messages that preemptively incorporate counterarguments into their own structure. The latter technique is known in communication studies as *two-sided persuasion* (Hovland et al. 1949; O'Keefe 1999, 2002; Eisend 2007), and it has long been known to produce much better effects than one-sided messages, even when the source and the aim of the messages are exactly the same. Moreover, these findings receive further corroboration from *inoculation theory* (McGuire 1961; Pfau 1992, 1997; Banas and Rains 2010), according to which repeated exposure to mildly persuasive attacks against one's views strengthen the ability to withstand subsequent, more serious criticism. By anticipating and neutralizing a variety of minor objections, a message thus comes to be perceived as better "immunized" against counterargument, and the public tends to be more ready to accept it. In argumentation theory, similar ideas are reflected in Ralph Johnson's idea of the *dialectical tier* (2000, 2003, 2007), which refers to the arguer's obligation of being ready to meet (at least some of) the potential objections raised by his/her argument. As noted by Rowan and collaborators (2009), the implications of this body of work on the argumentative mechanisms of persuasion for risk communication and crisis management are both obvious and far-reaching. While arguing to persuade and arguing to create a shared understanding are two very different activities, evidence suggest that, when it comes to communication between institutions and citizens, the latter is a *conditio sine qua non* of the former.

17.4 Argumentation in Crisis Intervention

An obvious and crucial constraint for argumentative processes at a time of crisis is the urgency to make a (good) decision: virtually all key choices in crisis intervention are time-sensitive, and typically delayed action is as bad as (or even worse than) any timely response, even a suboptimal one. In short, during a crisis time is a luxury decision-makers do not have, and this imposes a clear time pressure on their collective deliberation via argumentation. This is true at all levels where group decisions are required, from institutional authorities down to small groups of citizens trying to coordinate in the face of an ongoing emergency.

While there is a lack of studies on the effects of time pressure on argumentation, some useful implications can be extrapolated from the rich literature on group decision making under time constraints. As noted by Kelly and Lovig, "time pressure (. . .) seems to affect group performance by altering the type of information that is offered during group discussions" (2004, p. 187), consistently with the Attentional Focus Model of performance developed by Karau and Kelly (1992). In particular, past research shows that time pressure prompts group members to (1) attend a narrower set of cues, (2) be more focused towards task completion during group discussion, and (3) be less willing to tolerate dissenting voices (on the latter point, see Kruglanski and Webster 1991, Study 1). While the first effect is typically beneficial for performance, e.g., leading people to work at a faster rate (Kelly and McGrath 1985; Karau and Kelly 1992; Kelly and Karau 1999), the tendency towards "tunnel vision," groupthink (Turner and Pratkanis 1998) and heuristic reasoning (De Dreu 2003) characteristic of discussion under time pressure may be detrimental for performance, depending on a variety of factors. Clearly, the quality of the restricted set of cues on which the group focuses is crucial to determine the effects of time pressure on decision quality: Kelly and Karau (1999) demonstrated that a group engaged in an information sharing task will benefit from time pressure when the initial information distribution is optimal, whereas time pressure ruins performance under suboptimal initial conditions (e.g., when some key information is in possess of only a few group members, as it is often the case during a crisis). Also the type of task modulates the impact of time pressure: since discussion under time pressure tends to rely more on normative influence (e.g., "We should do X because it is the right thing to do according to a relevant norm") and less on information influence (e.g., "We should do X because it is the course of action best supported by available information"), this leads to improved performance on judgmental tasks, whereas time pressure becomes detrimental in intellectual tasks (Kelly et al. 1997). The take-home message of these studies is that, given the necessarily abbreviated and focused nature of group discussion under time pressure, it is crucial to enter deliberation well prepared—that is, with all the relevant information readily available and correctly distributed among group members.

Applied to argumentation in crisis intervention, this leads to identify two needs that have to be balanced against each other: minimizing the time devoted to discussion, due to the urgency implicit in a crisis, and making sure decisions are reached only after vetting all (or at least most) potential key objections. During a crisis, argumentation cannot be (and should not be) the open-ended exchange of reasons we discussed in relation with crisis prevention and preparedness. This is not to say that there is no room to argue during a crisis—on the contrary, the importance of reaching good group decisions makes intersubjective deliberation especially crucial, and argumentation is a necessary means to achieve it. But argumentation has to be efficiently streamlined, to be of use during a crisis: what we need are well-defined cross-examination pattern, aimed at rapidly and thoroughly checking the soundness of a proposed action plan.

Argumentation theory offers a promising tool in this respect: Walton's taxonomy of argumentation schemes, and the related notion of *critical questions* (Walton 1996;

Walton et al. 2008). Argumentation schemes are defined as "forms of argument (structures of inference) that represent structures of common types of arguments used in everyday discourse, as well as in special contexts like those of legal argumentation and scientific argumentation. They include the deductive and inductive forms of argument that we are already so familiar with in logic. However, they also represent forms of argument that are neither deductive nor inductive, but that fall into a third category, sometimes called defeasible, presumptive, or abductive. Such an argument may not be very strong by itself, but may be strong enough to provide evidence to warrant rational acceptance of its conclusion, given that its premises are acceptable" (Walton et al. 2008, p. 1). In order to establish whether the evidence mustered by a defeasible argumentation scheme is enough, each scheme is paired with a well-defined set of critical questions. As Walton and colleagues illustrate, "a set of critical questions attached to each scheme is the device for criticizing any argument fitting the structure of the scheme. The asking of a question, along with the response to it, implies a kind of dialogue structure in which two parties interact with each other. If an argument put forward by a proponent meets the requirements of a scheme, and the premises are acceptable to the respondent, then the respondent is obliged to accept the conclusion. But such an acceptance—or commitment, as it is often called—is provisional in the dialogue. If the respondent asks one of the critical questions matching the scheme and the proponent fails to offer an adequate answer, the argument defaults" (2008, p. 9).

While the notion of argumentation schemes was not originally designed to deal with crisis intervention, the stylized cross-examination implied by critical questions is precisely the kind of well-defined checklist we need for arguing effectively at a time of crisis. Instead of either bypassing completely group deliberation (which would be irresponsible and unethical) or getting entangled in prolonged debate (which would be impractical and ultimately disastrous), crisis management requires us to rapidly check whether current information sources are reliable and possible action plans are viable. Argumentation schemes and critical questions provide a valuable tool to accomplish this feat.

A recent taxonomy of schemes include no less than 60 entries (Walton et al. 2008): while not all of them are likely to be relevant for crisis intervention, some of them are clearly crucial. By way of example, here we will consider only two schemes: the argument from expert opinion, and the argument from consequences. The former is essential in a crisis to establish how much weight should be given to the opinion of experts, especially when dissenting views are voiced—as it often happens. The latter is crucial in determining what to do in dealing with a crisis, given the expected effects of various action plans.

Walton et al. (2008, p. 310) characterize arguments from expert opinion as follows (for further details, see Walton 1997, 2002):

Major Premise: Source E is an expert in domain S containing proposition A.
Minor Premise: E asserts that proposition A is true (false).
Conclusion: A is true (false).

Critical Questions

CQ1: *Expertise Question*: How credible is *E* as an expert source?
CQ2: *Field Question*: Is *E* an expert in the field that *A* is in?
CQ3: *Opinion Question*: What did *E* assert that implies *A*?
CQ4: *Trustworthiness Question*: Is *E* personally reliable as a source?
CQ5: *Consistency Question*: Is *A* consistent with what other experts assert?
CQ6: *Backup Evidence Question*: Is *E*'s assertion based on evidence?

Arguments from consequences, instead, are analyzed as follows (Walton et al. 2008, p. 332; see also Walton 1995, 2000):

Premise: If *A* is brought about, then good (bad) consequences will occur.
Conclusion: Therefore, *A* should (not) be brought about.

Critical Questions

CQ1: How strong is the likelihood that the cited consequences will (may, must) occur?
CQ2: What evidence supports the claim that the cited consequences will (may, must) occur, and is it sufficient to support the strength of the claim adequately?
CQ3: Are there other opposite consequences that should be taken into account?

The application of these schemes to the domain of crisis management is straightforward: with respect to reliance on expert opinion, for instance, all key aspects are brought into focus—the expertise of the source in question (CQ1), its relevance for the matter at hand (CQ2), the proper interpretation of the expert report (CQ3), his/her reliability in this specific instance, e.g., with respect to conflicts of interest (CQ4), the presence of conflicting views among experts (CQ5), and the soundness of the evidence upon which the expert opinion is based (CQ6). Similarly, the appeal to consequences is cross-examined in terms of their likelihood (CQ1), the evidence upon which that likelihood is based, in relation to the strength of the argument (CQ2), and the presence of other relevant effects that need to be considered (CQ3). The fact that all these questions are likely to sound as mere common sense is not surprising, since argumentation schemes are designed to capture familiar patterns of reasoning, nor does it detract from their usefulness, since argumentation during a crisis can greatly benefit from some ready-made common sense, especially when it comes to rapidly checking the reliability and viability of intervention options.

Articulating a comprehensive proposal for the application of argumentation schemes and critical questions to crisis management is beyond the aims of this chapter. Indeed, doing so would require a much closer look through the list of schemes, which contains many other instances that are bound to be relevant for arguing during a crisis: just to mention a few obvious candidates, argument from witness testimony, argument from values, practical reasoning, argument from fear appeal, argument from danger appeal, argument from need for help, argument from distress, argument from rules (for further details, see Walton et al. 2008, pp. 308–346). Hopefully, the brief remarks offered here have successfully established the potential value of using argumentation schemes as blueprints for concise critical deliberation for crisis intervention. This potential is further increased by the significant progresses made in encoding argumentation schemes in a variety of AI systems, such as Rationale™, Araucaria, Carneades, Compendium, and ArguMed, among others (for discussion on the use of argumentation schemes in computer science, see Walton et al. 2008, pp. 393–415).[2]

17.5 Conclusions

A quick survey of the literature on risk perception and risk communication (Sect. 17.2) helped us to identify two crucial roles that argumentation can play to support decision making at different stages of crisis management. In crisis prevention and preparedness, argumentation is needed to balance a participatory process of trust construction with the necessary (and typically legitimate) persuasive aims of authorities and experts: as discussed in Sect. 17.3, insights from persuasion studies (in particular, studies on two-sided persuasion and inoculation theory) can help structuring public dialogue in ways most likely to support positive results and best practices. In contrast, argumentation in crisis intervention is a time-sensitive process, and thus must be carefully designed in advance to yield maximum benefits with minimum delay: in Sect. 17.4, we outlined how argumentation schemes and critical questions can provide valuable guidance in this respect, offering a principled map of what are the critical "pressure points" of typical arguments used in crisis management, and practical suggestions on how to test them effectively.

Looking at the potential impact of argumentation in crisis prevention and preparedness, two aspects stand out:

- Participatory instruments (e.g., collective discussion and deliberation, focus groups, online forums) are quite effective, but they must be adopted only when the relevant authorities are *fully committed* to this form of crisis communication and management; otherwise they are bound to backfire, since there is nothing as damaging for public trust as the impression of being manipulated, ignored, and hastily dismissed by the authorities. To put it more bluntly: once citizens are given the chance of expressing their needs, beliefs, feelings, and suggestions regarding a potential crisis, then they must be given a proper degree of attention.
- Institutions need to design their communication strategies based on relevant results in *persuasion studies*: among other things, the style of the message, its

[2]The following are the URLs for each of these systems, in the order in which they are mentioned in the text: http://rationale.austhink.com/, http://araucaria.computing.dundee.ac.uk/doku.php, http://carneades.github.io/carneades/Carneades/, http://compendiuminstitute.net/about.htm, http://www.ai.rug.nl/~verheij/aaa/argumed3.htm

contents and its argumentative structure has to be attuned to the cognitive skills of the intended audience; potential objections to a proposed course of action must be anticipated and carefully assessed, especially when expecting opposition and resistance from the audience; the relevant goals that preventive measures are supposed to serve must be explicitly acknowledged and discussed.

As for the role of argumentation in crisis intervention, the following recommendations should be heeded:

- While principled abstract schemes of argument, like those proposed by Walton and collaborators (2008), can provide a valuable blueprint for time-effective cross-examination of crucial decisions, they should never become a straight-jacket. Argumentation schemes, by their very nature, are meant to capture broad generalities about recurring patterns of argument: this implies that the specifics of a concrete scenario may well elude them. Thus, crisis managers should supplement argumentation-based debugging with *domain-specific checklists*, in order to improve the rapid evaluation of key decisions at a time of crisis.
- Given the extreme time pressure characteristic of crisis intervention, most of the argumentative activity needed for it has to take place before the crisis even begins: this suggests the opportunity of a *preemptive approach* to argumentation in crisis intervention. In practical terms, this means that crisis managers should design in advance not only their decision procedures during a crisis, but also their corresponding argumentation protocols: when and how to argue, to face the various contingencies of a crisis. The use of argumentation schemes constitutes a first step in that direction, but other measures could and should be explored: assigning specific argumentative roles to members of a crisis unit (e.g., proponent, opponent, critic, advocate), setting an adequate standard of proof to establish when an opinion or course of action is warranted (how much evidence is required, and of what kind? what sort of intersubjective agreement? unanimity, absolute majority, relative majority, the single view of authoritative figures, or what else?), and devising protocols to handle the burden of proof (who has to prove what?).

Taken together, these considerations underscore the potential benefits that properly managed argumentative practices can yield for crisis management, at the interface between information fusion and decision making. Two caveats should temper our enthusiasm, though. First, as it is often the case in real life, harvesting such benefits is easier said than done: the kind of public argumentation required to engender trust and achieve persuasion in crisis prevention and preparedness is neither easy nor cheap, and massive efforts will be needed to effectively support it and avoid any "backfiring effect" (Cohen 2005; Paglieri 2009); similarly, the concrete application of argumentation schemes to define protocols for rapid deliberation in crisis intervention is still just a promising project, one that will require both theoretical and practical work to get off the ground. Second, precisely because the careful management of argumentative process can benefit much crisis management, a biased approach can wreak havoc on that very same process: we already evoked the poor track record of one-sided, narrow-minded, paternalistic forms of risk communication, as well as the inherent dangers in arguing too much, too minutely and too randomly at a time of crisis, without recognizing the need for quick and effective decisions. Thus, while the positive value of argumentation is contingent on sustained efforts to make it work, its ill-advised application remains a constant threat for effective crisis management, which is all the more reason to promote further research on this topic, if we want to steer clear of argumentative crisis and reap the benefits of critical arguments for rational deliberation.

Acknowledgments This research was financially supported by the PON research project *CLoud plAtform and smart underground imaging for natural Risk Assessment* (*CLARA*), funded by the Italian Ministry for Education, University and Research (MIUR). We are grateful to Rino Falcone, Cristiano Castelfranchi, and the members of the Trust: Theory & Technology group (T3) and the Goal-Oriented Agents Lab (GOAL) at the ISTC-CNR for insightful discussion on the topics of this chapter, and to Galina Rogova for providing useful feedback on a previous version of it. All remaining mistakes and omissions are our sole responsibility.

References

Arvai JL (2003) Using risk communication to disclose the outcome of a participatory decision-making process: effects on the perceived acceptability of risk-policy decisions. Risk Anal 23(2):281–289

Banas JA, Rains SA (2010) A meta-analysis of research on inoculation theory. Commun Monogr 77(3):281–311

Barber B (1983) The logic and limits of trust. Rutgers University Press, New Brunswick

Bell HM, Tobin GA (2007) Efficient and effective? The 100-year flood in the communication and perception of flood risk. Environ Hazards 7:302–311

Besnard P, Hunter A (2008) Elements of argumentation. MIT Press, Cambridge

Caminada M, Pigozzi G (2011) On judgment aggregation in abstract argumentation. Auton Agent Multi-Agent Syst 22(1):64–102

Castelfranchi C (2000) Affective appraisal versus cognitive evaluation in social emotions and interactions. In: Paiva A (ed) Affective interactions. Springer, Berlin, pp 76–106

Castelfranchi C, Falcone R (2010) Trust theory: a socio-cognitive and computational model. Wiley, Chichester

Cionea I, Hample D, Paglieri F (2011) A test of the argument engagement model in Romania. In: Zenker F (ed) Argumentation: cognition & community. Proceedings of OSSA 2011. CD-ROM. Windsor, OSSA, pp 1–15

Cohen D (2005) Arguments that backfire. In: Hitchcock D (ed) The uses of argument. Proceedings of OSSA 2005. OSSA, Hamilton, pp 58–65

Committee on Risk Perception and Communication, National Research Council (1989) Improving risk communication. National Academies Press, Washington

Covello VT, von Winterfeldt D, Slovic P (1986) Communicating risk information to the public. Risk Abstracts 3(4):1–14

Covello VT, Peters R, Wojtecki J, Hyde R (2001) Risk communication, the West Nile virus epidemic, and bioterrorism: responding to the communication challenges posed by the intentional or unintentional release of a pathogen in an urban setting. J Urban Health 78(2):382–391

Covello VT, Sandman PM (2001) Risk communication: evolution and revolution. In: Wolbarst A (ed) Solutions to an environment in peril. John Hopkins University Press, Baltimore, pp 164–178

Dake K (1992) Myths of nature: culture and the social construction of risk. J Soc Issues 48(4):21–37
Damasio AR (1994) Descartes' error: emotion, reason, and the human brain. Avon, New York
Das TK, Teng BS (2004) The risk-based view of trust: a conceptual framework. J Bus Psychol 19(1):85–116
De Dreu CKW (2003) Time pressure and closing of the mind in negotiation. Organ Behav Hum Decis Process 91(2):280–295
DeBruine LM (2002) Facial resemblance enhances trust. Proc R Soc Lond Ser B Biol Sci 269(1498):1307–1312
Deutsch M (1958) Trust and suspicion. J Confl Resolut 2:265–279
Dunn JR, Schweitzer ME (2005) Feeling and believing: the influence of emotion on trust. J Pers Soc Psychol 88(5):736
Dung PM (1995) On the acceptability of arguments and its fundamental role in nonmonotonic reasoning, logic programming and n-person games. Artificial Intelligence 7:321–357
Earle TC (2012) Trust in cooperative risk management: uncertainty and scepticism in the public mind. Routledge, London
Eisend M (2007) Understanding two-sided persuasion: an empirical assessment of theoretical approaches. Psychol Mark 24(7):615–640
Faulkner H, Mc Carthy S, Tunstall S (2011) Flood risk communication. In: Pender G, Faulkner H (eds) Flood risk science and management. Blackwell, Hoboken, pp 386–406
Finucane ML, Alhakami A, Slovic P, Johnson SM (2000) The affect heuristic in judgments of risks and benefits. J Behav Decis Mak 13(1):1–17
Flynn J, Slovic P, Mertz CK (1994) Gender, race, and perception of environmental health risks. Risk Anal 14(6):1101–1108
Foddy M, Platow MJ, Yamagishi T (2009) Group-based trust in strangers: the role of stereotypes and expectations. Psychol Sci 20(4):419–422
Gilbert M (1997) Coalescent argumentation. Lawrence Erlbaum Associates, Mahwah
Hample D (2005) Arguing: exchanging reasons face to face. Lawrence Erlbaum Associates, Mahwah
Hample D, Benoit P, Houston J, Purifoy G, Vanhyfte V, Wardwell C (1999) Naive theories of argument: avoiding interpersonal arguments or cutting them short. Argum Advocacy 35:130–139
Hample D, Dallinger J (1990) Arguers as editors. Argumentation 4:153–169
Hample D, Dallinger J (1992) The use of multiple goals in cognitive editing of arguments. Argum Advocacy 28:109–122
Hample D, Paglieri F, Na L (2012) The costs and benefits of arguing: predicting the decision whether to engage or not. In: van Eemeren F, Garssen B (eds) Topical themes in argumentation theory. Springer, Berlin, pp 307–322
Hample D, Werber B, Young D (2009) Framing and editing interpersonal arguments. Argumentation 23:21–37
Hovland C, Lumsdaine A, Sheffield F (1949) Experiments in mass communication. Princeton University Press, Princeton
Infanti J, Sixsmith J, Barry MM, Núñez-Córdoba J, Oroviogoicoechea-Ortega C, Guillén-Grima F (2013) A literature review on effective risk communication for the prevention and control of communicable diseases in Europe. ECDC, Stockholm
Jenkins-Smith HC (1993) Nuclear imagery and regional stigma: testing hypotheses of image acquisition and valuation regarding Nevada. University of New Mexico, Institute for Public Policy, Albuquerque, NM
Joffe H (2003) Risk: from perception to social representation. Br J Soc Psychol 42(1):55–73
Johnson RH (2000) Manifest rationality: a pragmatic theory of argument. Lawrence Erlbaum Associates, Mahwah
Johnson RH (2003) The dialectical tier revisited. In: van Eemeren FH, Blair JA, Willard CA, Snoeck Henkemans AF (eds) Anyone who has a view: theoretical contributions to the study of argumentation. Kluwer, Dordrecht, pp 41–53

Johnson RH (2007) Anticipating objections as a way of coping with dissensus. In: Hansen HV, Tindale CW, Blair JA, Johnson RH (eds) Dissensus and the search for common ground (CD-ROM). OSSA, Windsor, pp 1–16

Karau SJ, Kelly JR (1992) The effects of time scarcity and time abundance on group performance quality and interaction process. J Exp Soc Psychol 28:542–571

Kasperson RE (1986) Six propositions on public participation and their relevance for risk communication. Risk Anal 6(3):275–281

Kasperson RE, Golding D, Tuler S (1992) Social distrust as a factor in siting hazardous facilities and communicating risks. J Soc Issues 48(4):161–187

Kasperson RE, Renn O, Slovic P, Brown HS, Emel J, Goble R, Kasperson JX, Ratick S (1988) The social amplification of risk: a conceptual framework. Risk Anal 8(2):177–187

Kasperson RE, Stallen PM (1991) Risk communication: the evolution of attempts. In: Kasperson RE, Stallen PM (eds) Communicating risk to the public: international perspectives. Kluwer Academic Press, Dordrecht, pp 1–12

Kelly JR, Karau SJ (1999) Group decision making: the effects of initial preferences and time pressure. Pers Soc Psychol Bull 25:1342–1354

Kelly JR, McGrath JE (1985) Effects of time limits and task types on task performance and interaction of four-person groups. J Pers Soc Psychol 49:395–407

Kelly JR, Jackson JW, Hutson-Comeaux SL (1997) The effects of time pressure and task differences on influences modes and accuracy in decision-making groups. Pers Soc Psychol Bull 23:10–22

Kelly JR, Lovig TJ (2004) Time pressure and group performance: exploring underlying processes in the Attentional Focus Model. J Exp Soc Psychol 40(2):185–198

Kruglanski AW, Webster DM (1991) Group members' reactions to opinion deviates and conformists at varying degrees of proximity to decision deadline and of environmental noise. J Pers Soc Psychol 61:212–225

Lenard PT (2008) Trust your compatriots, but count your change: the roles of trust, mistrust and distrust in democracy. Polit Stud 56(2):312–332

Lion R (2001) Security or opportunity: the effects of individual and situational factors on risk information preference. Universal Maastricht, The Netherlands

Marsh S, Dibben MR (2005) Trust, untrust, distrust and mistrust—an exploration of the dark(er) side. In: Herrmann P, Issarny V, Shiu S (eds) Trust management. Springer, Berlin, pp 17–33

Mayer RC, Davis JH, Schoorman FD (1995) An integrative model of organizational trust. Acad Manage Rev 20(3):709–734

McGuire WJ (1961) The effectiveness of supportive and refutational defenses in immunizing defenses. Sociometry 24:184–197

Nooteboom B (1996) Trust, opportunism and governance: a process and control model. Organ Stud 17:985–1010

Nooteboom B, Berger H, Noorderhaven NG (1997) Effects of trust and governance on relational risk. Acad Manage J 40(2):308–338

O'Keefe DJ (1999) How to handle opposing arguments in persuasive messages: a meta-analytic review of the effects of one-sided and two-sided messages. In: Roloff ME (ed) Communication yearbook 22. Sage, Thousand Oaks, pp 209–249

O'Keefe DJ (2002) Persuasion: theory and research, 2nd edn. Sage, Thousand Oaks

Ouerdane W, Dimopoulos Y, Liapis K, Moraitis P (2011) Towards automating decision aiding through argumentation. J Multi-Criteria Decis Anal 18:289–309

Paglieri F (2007) No more charity, please! Enthymematic parsimony and the pitfall of benevolence. In: Hansen H, Tindale C, Johnson R, Blair J (eds) Dissensus and the search for common ground: Proceedings of OSSA 2007. CD-ROM. OSSA, Windsor, pp 1–26

Paglieri F (2009) Ruinous arguments: escalation of disagreement and the dangers of arguing. In: Hansen H, Tindale C, Johnson R, Blair J (eds) Argument cultures: Proceedings of OSSA 2009. CD-ROM. OSSA, Windsor, pp 1–15

Paglieri F (2013a) Choosing to argue: towards a theory of argumentative decisions. J Pragmat 59(B):153–163

Paglieri F (2013b) Argumentation, decision and rationality. In: Mohammed D, Lewinsky M (eds) Virtues of argumentation: Proceedings of OSSA 2013. CD-ROM. OSSA, Windsor, pp 1–12
Paglieri F (2015) Arguments, conflicts and decisions. In: Poggi I, D'Errico F, Vincze L, Vinciarelli A (eds) Conflict and negotiation: social research and machine intelligence. Springer, Berlin
Paglieri F, Castelfranchi C (2010) Why arguing? Towards a costs-benefits analysis of argumentation. Argum Comput 1(1):71–91
Paglieri F, Castelfranchi C, da Costa Pereira C, Falcone R, Tettamanzi A, Villata S (2014) Trusting the message and the messenger: feedback dynamics from information quality to source evaluation. Comput Math Organ Theory 20(2):176–194
Paglieri F, Woods J (2011a) Enthymematic parsimony. Synthese 178:461–501
Paglieri F, Woods J (2011b) Enthymemes: from reconstruction to understanding. Argumentation 25(2):127–139
Parsons S, Atkinson K, Li Z, McBurney P, Sklar E, Singh M, Haigh K, Levitt K, Rowe J (2014) Argument schemes for reasoning about trust. Argum Comput 5(2–3):160–190
Pelling M (2007) Learning from others: the scope and challenges for participatory disaster risk assessment. Disasters 31(4):373–385
Peters E, Slovic P (1996) The role of affect and worldviews as orienting dispositions in the perception and acceptance of nuclear Power. J Appl Soc Psychol 26(16):1427–1453
Pfau M (1992) The potential of inoculation in promoting resistance to the effectiveness of comparative advertising messages. Commun Q 40(1):26–44
Pfau M (1997) The inoculation model of resistance to influence. In: Boster FJ, Barnett G (eds) Progress in communication sciences, vol 13. Ablex, Norwood, pp 133–171
Pidgeon N (1992) The psychology of risk. In: Blockley DI (ed) Engineering safety. McGraw-Hill, Maidenhead, pp 167–185
Primiero G, Kosolosky L (2013) The semantics of untrustworthiness. Topoi. doi: 10.1007/s11245-013-9227-2
Rahwan I, Simari G (eds) (2009) Argumentation in artificial intelligence. Springer, Berlin
Reed C, Norman T (eds) (2004) Argumentation machines. Kluwer, Dordrecht
Renn O, Levine D (1991) Credibility and trust in risk communication. In: Kasperson RE, Stallen PM (eds) Communicating risk to the public: international perspectives. Kluwer Academic Press, Dordrecht, pp 175–217
Rousseau DM, Sitkin SB, Burt RS, Camerer C (1998) Not so different after all: a cross-discipline view of trust. Acad Manage Rev 23(3):393–404
Rowan KE, Botan CH, Kreps GL, Samoilenko S, Farnsworth K (2009) Risk communication education for local emergency managers: using the CAUSE model for research, education, and outreach. In: Heath RL, O'Hair HD (eds) Handbook of risk and crisis communication. Routledge, New York, pp 168–191
Siegrist M, Cvetkovich G (2000) Perception of hazards: the role of social trust and knowledge. Risk Anal 20(5):713–720
Sitkin SB, Roth NL (1993) Explaining the limited effectiveness of legalistic "remedies" for trust/distrust. Organ Sci 4(3):367–392
Slovic P (1987) Perception of risk. Science 236(4799):280–285
Slovic P (1993) Perceived risk, trust, and democracy. Risk Anal 13(6):675–682
Slovic P (1999) Trust, emotion, sex, politics, and science: surveying the risk assessment battlefield. Risk Anal 19(4):689–701
Slovic P, Finucane ML, Peters E, MacGregor DG (2007) The affect heuristic. Eur J Oper Res 177(3):1333–1352
Steger MAE, Witt SL (1989) Gender differences in environmental orientations: a comparison of publics and activists in Canada and the US. West Polit Q 42:627–649
Turner ME, Pratkanis AR (1998) Twenty-five years of groupthink theory and research: lessons from the evaluation of a theory. Organ Behav Hum Decis Process 73(2–3):105–115
Tversky A, Kahneman D (1974) Judgement under uncertainty: heuristics and biases. Science 185:1127–1131

Van Aalst MK, Cannon T, Burton I (2008) Community level adaptation to climate change: the potential role of participatory community risk assessment. Glob Environ Chang 18(1):165–179
Van der Horst D (2007) NIMBY or not? Exploring the relevance of location and the politics of voiced opinions in renewable energy siting controversies. Energy Policy 35:2705–2714
Van Eemeren F, Grootendorst R (2004) A systematic theory of argumentation: the pragma-dialectical approach. Cambridge University Press, Cambridge
Visschers VH, Siegrist M (2008) Exploring the triangular relationship between trust, affect, and risk perception: a review of the literature. Risk Manage 10(3):156–167
Walton D (1995) A pragmatic theory of fallacy. University of Alabama Press, Tuscaloosa
Walton D (1996) Argumentation schemes for presumptive reasoning. Lawrence Erlbaum Associates, Mahwah
Walton D (1997) Appeal to expert opinion. Pennsylvania State University Press, University Park
Walton D (1998) The new dialectic: conversational contexts of argument. University of Toronto Press, Toronto
Walton D (2000) Scare tactics. Kluwer, Dordrecht
Walton D (2002) Legal argumentation and evidence. Pennsylvania State University Press, University Park
Walton D, Krabbe E (1995) Commitment in dialogue: basic concepts of interpersonal reasoning. SUNY Press, Albany
Walton D, Reed C, Macagno F (2008) Argumentation schemes. Cambridge University Press, New York
Weyman AK, Kelly CJ (1999) Risk perception and risk communication: a review of literature. HSE Contract Research Report
Yamagishi K (1997) When 12.86 mortality is more dangerous than 24.14%: implications for risk communication. Appl Cogn Psychol 11:495–506

Chapter 18
Fusion Trust Service Assessment for Crisis Management Environments

Erik Blasch, Youakim Badr, Salim Hariri, and Youssif Al-Nashif

Abstract Future crisis management systems need resilient and trustworthy infrastructures to quickly develop reliable applications and processes, apply fusion techniques, and ensure end-to-end security, trust, and privacy. Due to the multiplicity and diversity of involved actors, volumes of data, and heterogeneity of shared information; crisis management systems tend to be highly vulnerable and subject to unforeseen incidents. As a result, the dependability of crisis management systems can be at risk. This chapter presents a cloud-based resilient and trustworthy infrastructure (known as rDaaS) to quickly develop secure crisis management systems. The rDaaS integrates the Dynamic Data-Driven Application Systems (DDDAS) paradigm into a service-oriented architecture over cloud technology and provides a set of resilient DDDAS-As-A Service (rDaaS) components to build secure and trusted adaptable crisis processes. One service presented includes the fusion of information from human observers and surveillance systems to assess the credibility (trust) of a crisis alert. The fusion trust service within the rDaaS also ensures resilience and security by obfuscating the execution environment and applying Behavior Software Encryption and Moving Technique Defense over the users, machines, and communication network. A simulation environment for a nuclear plant crisis management case study is illustrated to build resilient and trusted crisis response processes.

E. Blasch (✉)
Air Force Research Laboratory, Information Directorate, Rome, NY, USA
e-mail: erik.blasch1@us.af.mil

Y. Badr
CNRS, LIRIS, UMR5205, INSA-Lyon, Université de Lyon, Lyon, France
e-mail: youakim.badr@liris.cnrs.fr

S. Hariri
Center for Cloud and Autonomic Computing, University of Arizona, Tucson, AZ, USA
e-mail: hariri@ece.arizona.edu

Y. Al-Nashif
Old Dominion University, Norfolk, VA, USA
e-mail: yalnashi@odu.edu

G. Rogova, P. Scott (eds.), *Fusion Methodologies in Crisis Management*,
DOI 10.1007/978-3-319-22527-2_18

18.1 Introduction

Crisis management (CM) for critical infrastructures, natural disasters, or civil unrest caused by natural or malicious forces requires complex, dynamic, and heterogeneous responses. A comprehensive investigation of information fusion methods for crisis analysis was given by Scott et al. (Scott 2007) in 2007. Since that time, many issues have arisen in the information fusion community such as reasoning strategies, uncertainty analysis, and information fusion as a service.

Distributed CM systems should detect, recognize, and disseminate huge amounts of heterogeneous dynamic events that operate at different speeds and formats. Furthermore, the processing of crisis events and the development of real-time responses are challenges when the security and trust of every human and resource must be evaluated to maintain security, privacy, and trust among different stakeholders collaborating in the crisis management environment. For example, managing a nuclear crisis requires several actors to establish *response crisis processes* to coordinate their missions and securely exchange information based on their trusted levels. Multiplicity and diversity of involved actors, volume and heterogeneity of shared information, critical dependencies between actions, as well as the unpredictability of the situation evolution make the crisis environment complex and dynamic. Information fusion for crisis response includes traditional Bayesian or Dempster–Shafer methods (Florea and Bosse 2009) methods as well as coordination between situation awareness and decision making (Kankanamge 2010). In our analysis, we compare evidential reasoning methods with the proportional conflict redistribution (PCR) (Dezert and Smarandache 2009) rule for trust assessment in high-level information fusion decision making (Blasch et al. 2012a).

In this chapter, we present our approach to leverage information fusion methods, Dynamic Data-Driven Application Systems (DDDAS) techniques, a Service-Oriented Architecture (SOA) paradigm, and cloud services to develop a trusted crisis management environment that we refer to as resilient DDDAS-As-A-Service (rDaaS). Service-oriented computing aims at assembling software components to build distributed, agile applications and business processes that span organizational, communications, and computing platforms. Despite the agility that service-oriented architectures promise, many SOA-based processes are increasingly encountering difficult self-adaptation issues and fail to deal with continuous and unpredictable changes in dynamic environments. Changes may occur during runtime due to unavailable services or unreachable actors, resources degradation and security threats - just to mention a few. Changes may also evolve from unforeseen incidents and new decisions that, consequently, require rebuilding responses processes partially or fully. On the other hand, the DDDAS paradigm allows continuous monitoring, analysis, and adaptation of all software components involved in networked systems, including sensors, computational resources, and applications. In addition, cloud computing is particularly a promising technology that may support crisis management with computational and storage capabilities and quickly deploy crisis processes. Virtualization and elasticity of resources makes possible to deploy application servers and devices, and scale-up with large amounts of data collected from various sources. Consequently, the integration of information fusion, DDDAS, and SOA into one cloud environment, called rDaaS, leads to a novel architecture enabling the construction of adaptable SOA-based processes to deal with unpredictable changes in dynamic environments.

In this chapter, we present a DDDAS cloud-based resilient and trustworthy infrastructure to quickly develop crisis management applications and processes. Recent efforts for crisis response have focused on the need for rapid decision making support (Schneider et al. 2013) and services (Marhsall 2014) to support disasters analysis. Processes are built based on resilient SOAs driven by requirements and adaptable to internal and external changes. In rDaaS, the cloud environment tends to continuously design, develop, and execute crisis management response processes and services. To enable trusted collaboration among different actors involved in the crisis management environment, the rDaaS evaluates trust values of all actors and resources at runtime before enabling trustworthy collaborations based on different trust levels.

The remaining sections of the chapter are organized as follows. Section 18.2 describes related work and background research in information fusion, security, and resiliency. Section 18.3 describes the Cloud-based resilient and trustworthy Dynamic Data-Driven Application Systems for Crisis Environments (called r-DDDAS-as-a-Cloud Service (rDaaS)) to develop resilient crisis management applications. Section 18.4 presents the information fusion approach for trust assessment in the rDaas. Section 18.5 presents our experimental design and evaluation of the rDaaS approach for trust analysis. Section 18.6 presents a discussion and finally the chapter concludes in Sect. 18.7.

18.2 Related Work

Crisis management has received an abundant attention from a managerial perspective as they are often reduced to suitable strategies to prepare for handling the crisis, ensure a rapid and adequate response, and define clear roles and responsibilities. Critical infrastructure designs require cybersecurity (trust) which includes adaptability by continuously integrating design and runtime (design once–run forever) support. In this section, we review the cybersecurity detection techniques in general and then give a brief overview of resilient techniques. Details of connections between DDDAS and information fusion have shown consistency in similar developments (Blasch et al. 2014a).

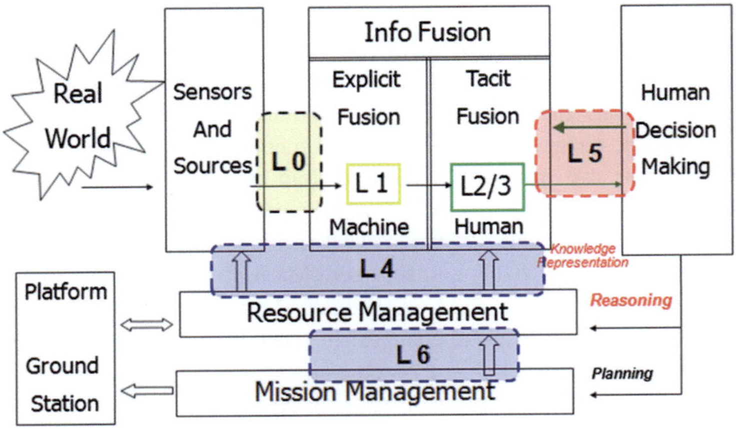

Fig. 18.1 Data Fusion Information Group (DFIG) model

18.2.1 High-Level Information Fusion

Building on the Boyd Observe, Orient, Decide, Act (OODA) model (Blasch et al. 2011a, b) and the Situation Awareness model (Endsley 1988); the Data Fusion Information Group model (Blasch and Plano 2005; Blasch 2015) was proposed, as shown in Fig. 18.1 along with the definitions.

In the DFIG model, the goal was to organize the information fusion and management functions. Management functions are divided into sensor control (L4), user selection (L5), and mission objectives (L6). Assessment (as traditional information fusion) includes object (L1), situation (L2), and impact (L3) analysis. The current definitions include:

Level 0—Data Assessment: estimation and prediction of signal/object observable states on the basis of pixel/signal level data association (e.g., information systems collection processing); [Examples: (Mendoza-Schrock et al. 2009)]

Level 1—Object Assessment: estimation and prediction of entity states on the basis of data association, continuous state estimation, and discrete state estimation (e.g., data processing); [Examples: (Yang et al. 2008; Liu et al. 2012)]

Level 2—Situation Assessment: estimation and prediction of relations among entities, to include force structure and force relations, communications, etc. (e.g., information processing); [Examples, (Blasch et al. 2012b; Abielmona et al. 2014)]

Level 3—Impact Assessment: estimation and prediction of effects on situations of planned or estimated actions by the participants; to include interactions between action plans of multiple players (e.g., assessing threat actions to planned actions and mission requirements, performance evaluation); [Examples: (Chen et al. 2007a; Blasch and Israel 2015)]

Level 4—Process Refinement (an element of Resource Management): adaptive data acquisition and processing to support sensing objectives (e.g., sensor management and information systems dissemination, command/control). [Examples: (Blasch et al. 2008; Yang et al. 2014)]

Level 5—User Refinement (an element of Knowledge Management): adaptive determination of who queries information and who has access to information (e.g., information operations) and adaptive data retrieved and displayed to support cognitive decision making and actions (e.g., human–computer interface). [Examples: (Blasch 2008; Blasch et al. 2009)]

Level 6—Mission Management (an element of Systems Management): adaptive determination of spatial–temporal control of assets (e.g., airspace operations) and route planning and goal determination to support team decision making and actions (e.g., theater operations) over social, economic, and political constraints. [Examples: (Blasch 2006; Chen et al. 2007b)]

A key element is Level 5 "user refinement" (Blasch 2003) which focuses on situational awareness as a user's *mental model*, to that of the fusion situation assessment display of a *geographical model* (Blasch et al. 2011c). The user has three elements of refinement (or analysis):

- *Physical*: Tasks including inputting a priori estimates, querying for results, and manipulating the data.
- *Cognitive*: Mental operations to infer meaning, utilize experiences, and modifying results based on context.
- *Information*: Dissemination and communication of data through reports and analysis.

As called out in issues and challenges (Blasch et al. 2012c), trends (Blasch et al. 2012d), and inquires (Ao et al. 2014): *high-level information fusion (HLIF)* requires evaluation of the techniques for user–machine coordination with low-level information fusion (LLIF) object assessment. An example is developing a semantic language (Wen 2011) so that a query (Aved and Blasch 2015) can be used to access situation assessment data. Currently, for information analysis, a semantic description requires a common ontology (Blasch et al. 2010a; Gruniger and Obrst 2014) for such things as the meaning of uncertainty (Costa et al. 2012).

18.2.2 Security Related Work

Intrusion detection can be broadly classified as signature-based and anomaly-based systems.

A *signature-based Intrusion Detection System* (IDS) uses pattern-matching algorithms to compare network traffic with an extensive library of attack signatures created by human experts (Pipkin 2000). A match indicates a probable attack. These techniques are extremely proficient in detecting known attacks because they can identify an attack as soon as it occurs. However, their foremost limitation is that it cannot detect new attacks. When a new attack is discovered, it takes time to develop signatures for the attack and deploy it into the existing IDSs. Some of the most commonly used signature-based intrusion detection techniques are introduced by SNORT (Roesch 1999), BRO (Paxson 1999), and others (Scarfone and Mell 2007). Anomaly-based detection techniques build a model of normal behavior and automatically classify statistically significant deviations from the normal profile as being abnormal (Al-Nashif et al. 2008; Ertöz et al. 2004; Denning 1987; Javitz and Valdes 1994). The advantage of this approach is that it is possible to detect unknown attacks. However, there is a potential for having a high rate of false positive alarms generated when the knowledge collected about normal behaviors is inaccurate. Examples of such techniques include Anomaly Behavior Analysis (ABA) methodology that has been successfully applied to a wider range of protocols (TCP/IP, DNS, WiFi, HTTP, etc.) (Alipour et al. DNS anomaly behavior analysis against cyber attacks; Alipour et al. IEEE 802.11 anomaly behavior analysis; Viswanathan et al. 2011), IDES (Lunt and Jagannathan 1988), NIDES (Anderson et al. 1995), EMERALD (Porras and Neumann 1997), and SPADE (Staniford et al. 2002).

An *anomaly-based IDS* builds a pattern based on a normalcy model and the compares incoming data to the trained model (Sequeira and Zaki 2002; Ye 2004; Yamanishi et al. 2000; Ye and Chen 2001; Eskin et al. 2002; Aggarwal and Yu 2001; Breunig et al. 2000; Knorr and Ng 1998; Ramaswamy et al. 2000; Sekar et al. 2002). The construction of the pattern, model, or database of normalcy can be coordinated with data mining techniques and machine learning (Shon and Moon 2007). For situations in which the actors are dynamic, game-theoretic approaches for cyber security over IDS incorporates learned behaviors for the normalcy of the situation (Chen et al. 2007c). The anomaly can be a threat to standard operations either measured or predicted (Blasch et al. 2006).

Cloud security suffers from a wide range of attacks such as those that target physical machines as well as the cloud virtualized environment (https://cloudsecurityalliance.org/research/secaas/). The dependency of cloud computing on the virtualized environment raises more security issues, like hypervisor exploitations (Schmidt et al. 2011; Goodin 2009). In addition, one of the main security issues in cloud computing is the insider attacks. With exchange of cloud data between different organizations, the risk of insider attacks increases.

Some previous works have presented classifications of Cloud Security (Subashini 2011; Bhadauria and Sanyal 2012; Modi et al. 2013). Cross-site scripting (Zeng 2013), Access control weaknesses, OS and SQL injection flaws, cross-site request forgery (Siddiqui and Verma 2011), cookie manipulation, hidden field manipulation, insecure storage, and insecure configuration are the threats to data stored in a SaaS cloud (Pék et al. 2013). Since, in the cloud model, the customers' data reside on the third-parties' data-centers, data security is a major concern for cloud consumers and providers and some researchers have addressed data security in their works (Abbasy and Shanmugam 2011; Feng et al. 2011; Kaufman 2009). Virtualization is one of the fundamental concepts of cloud computing. In a cloud system, multiple virtual guest machines share the same resources of a physical host machine. Recent advances have sought to combine cloud computing elasticity (Blasch et al. 2013a), DDDAS (Blasch et al. 2013b), and information fusion (Liu et al. 2014), for secure applications. There are some known security issues with common Virtual Machines (e.g., Xen, Microsoft HyperV) which can be exploited to threaten cloud services (Rosenblum and Garfinkel 2005).

18.2.3 Resilient Related Work

Moving Target Defense has been identified as a game changer approach to build self-defending systems (http://www.cyber.st.dhs.gov/docs/National_Cyber_Leap_Year_Summit_2009_Co-Chairs_Report.pdf). Some works presented a wide range of Moving target Defense (MTD) techniques to continuously change network configurations or parameters, firewall settings, operating systems, memory addresses, instruction sets, or application execution environments (Dunlop et al. 2011; Zhuang et al. 2012). For example, in Dunlop et al. (2011), the IP addresses are dynamically changed while maintaining existing connections. One can also randomize the configuration (Narain 2013) where the configuration variables of a system are randomized, while ensuring the availability of end to end services. In Kim (1998), the authors presented a survey of several software fault tolerance techniques. The fault tolerance techniques that are based on diversity include dual-node redundant operating stations with hardware or software result comparison, Recovery Block Station (Tyrrell 1996), Distributed Recovery Block with acceptance Test (Kim and Welch 1989), Voting Triple Modular Redundant Computing Stations (Toy 1987), N version programming (Avizienis 1985), and an Integrated Voting Algorithm (Latif-Shabgahi 2011). Also, in Evans et al. (2011), several diversity defense techniques in popular operating systems were discussed. These include Address Space Randomization (PaX 2000), Instruction Set Randomization (Barrantes et al. 2003), and Data Randomization (Cadar et al. 2008); where a comparison of these techniques is shown in Ge et al. (2014).

Some previous works have adopted diversity as a defense technique in a cloud environment. In Verissimo et al. 2012 the authors envision a Cloud of Clouds Architecture, which provides incrementally high levels of security and dependability to cloud infrastructures, in an open, modular, and versatile way. Their architecture employs diversity in deployment of cloud alternatives. However, they do not employ shuffling on these alternatives. In Vallee et al. (2008), a framework for Proactive Fault Tolerance is discussed that predict failures in nodes and migrate their processes away from the nodes that are about to fail. In Keromytis et al. (2012), the authors envision a cloud environment with continuous change in system configuration in order to create an unpredictable target for an adversary. In Luo and Wang 2010, the authors presented an intrusion tolerant cloud architecture that adopts the method of hybrid fault model, active and passive replicas, state update and transfer, proactive recovery and diversity. Future methods will use the MapReduce for cloud security for various crisis management applications such as object detection (Jeffrey and Sanjay 2008), movement (Cheng et al. 2014), and cybersecurity (Ge et al. 2015).

In our approach to implement rDaaS, we adopt diversity technique to the cloud application execution environment, redundancy in the resources is used to run the cloud services and randomly changing the versions, and resources are used to make it prohibitively expensive for attackers to figure out current execution environment, succeed in exploiting vulnerabilities, and launch attacks.

18.3 r-DDDAS-as-a-Cloud Service (rDaaS) Architecture

The r-DDDAS-as-a-Cloud Service (rDaaS) architecture aims at aligning cloud technology with the DDDAS paradigm. Figure 18.2 shows rDaaS main cloud services, which are used to develop resilient and trustworthy cloud-based event-driven crisis management systems. In this architecture, we leverage the DDDAS paradigm to combine the design stage with the runtime stage. The inseparability between design time and runtime makes it possible to deal with unpredictable events and enable prompt responses to confine and manage mitigation activities.

At the *design stage*, the resilient SOA Editor (RSE) helps decision makers to express their response plans in an abstract manner with desired metrics (Blasch et al. 2010b) and to develop effective and trusted and resilient responses even when they are experiencing cyber terrorisms or attacks. Likewise, the editor supports the

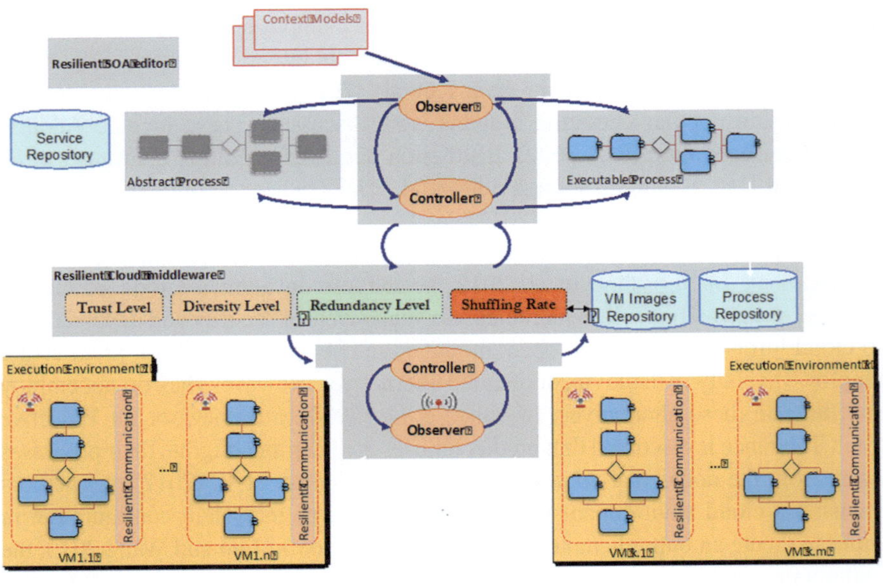

Fig. 18.2 rDaaS: Cloud-based resilient and trustworthy Dynamic Data-Driven Application Systems

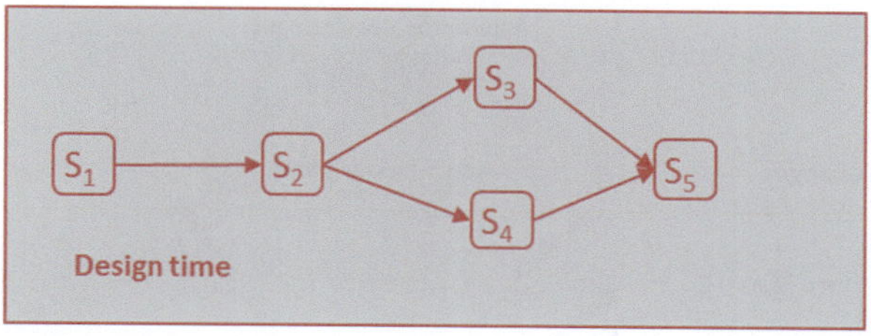

Fig. 18.3 Abstract response process

collection and management of information for decision makers (Blasch and Plano 2002). Information fusion techniques can be supported for DDDAS-based systems using cloud services (Liu et al. 2014) for unmanned aerial vehicle control (Peng et al. 2014). The RSE generates SOA-based executable processes from abstract response processes, taking into account global constraints expressed in contextual models (see Fig. 18.2).

At the *runtime stage*, the Crisis Runtime Manager (CRM) transforms executable response services into equivalent resilient and trustworthy services using the Resilient Cloud Middleware (RCM). In what follows, we discuss in further detail the RSE, the CRM, and the RCM functionalities.

18.3.1 Resilient SOA Response Service Design

The *resilient SOA Editor* (RSE) is a design assistant editor that assists decision makers to specify at the abstract level how the crisis response will be carried out by using several services provided by the SOA environment as shown in Fig. 18.3. Services, S_i $\{i = 1, \ldots, n)$ help crisis teams to communicate and interoperate with different types of internal and external resources. Examples include the integration of voice with imagery systems (Blasch 2014a; Blasch et al. 2014b).

In addition, the crisis coordinator can specify resilient requirements of services as well as the required trust level for all the actors involved in the execution of the crisis response services. The resilient requirement can be characterized by: (1) defining the required diversity level (how many different versions of an application and/or how many different platforms (e.g., operating system types) that are required to run the application); (2) defining the redundancy level (e.g., how many redundant physical machines are required); and (3) defining how often the execution environment needs to be changed (e.g., the number of application execution phases).

By using the resilient SOA Editor and the Resilient Cloud Middleware, we can continuously adopt the design of the crisis responses to match the security/trust policy, the context models, and any changes external or internal to the environment as shown in Fig. 18.4.

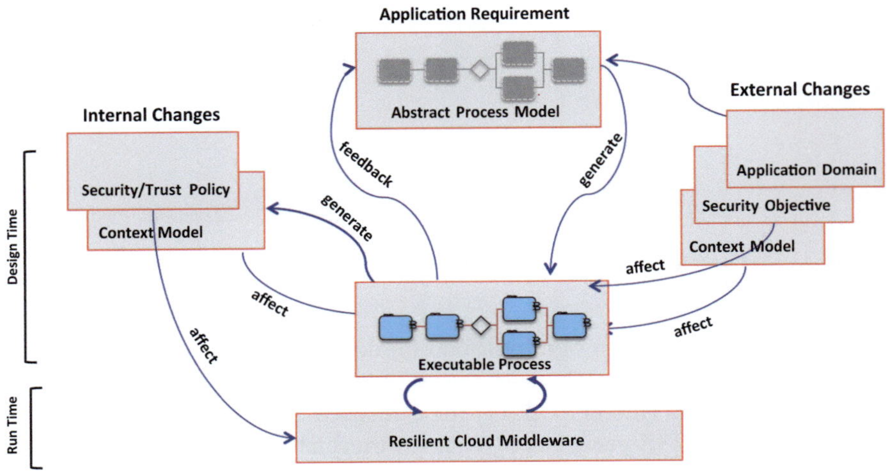

Fig. 18.4 Integrating design development stage of crisis response services with their runtime execution environment

The resilient SOA Editor's implementation is based on the *service farming-based composition* algorithm, which is developed by W. Li (Li et al. 2013). The service-farming algorithm generates optimal executable Web service-based processes from an abstract process by satisfying constraints on Web services (functional and nonfunctional properties) and models that describe internal and external changes (Fig. 18.4). Abstract processes are written with Business Process Model and Notation (BPMN) (Dijkman et al. 2008) to describe global functionalities independent from service implementations whereas models of external and internal changes are expressed with declarative rules that should be satisfy by the service farming algorithm and controlled by the Crisis Runtime Manager.

18.3.2 Crisis Runtime Manager

The Crisis Runtime Manager (CRM) provides control and management services to deploy and configure crisis response services and hardware resources that are required to achieve the response resilient requirements as specified by the Crisis Design Assistant. The CRM is implemented using our autonomic computing architecture shown in Fig. 18.5.

The CRM implementation is based on the autonomic computing environment (Autonomia) developed by the University of Arizona researchers (Tunc et al. 2014; Dsouza et al. 2013). The CRM functions will be implemented using two software modules: Observer and Controller modules. The Observer module monitors and analyzes the current state of the managed cloud resources or services. The Controller module is delegated to manage the cloud operations and enforce the

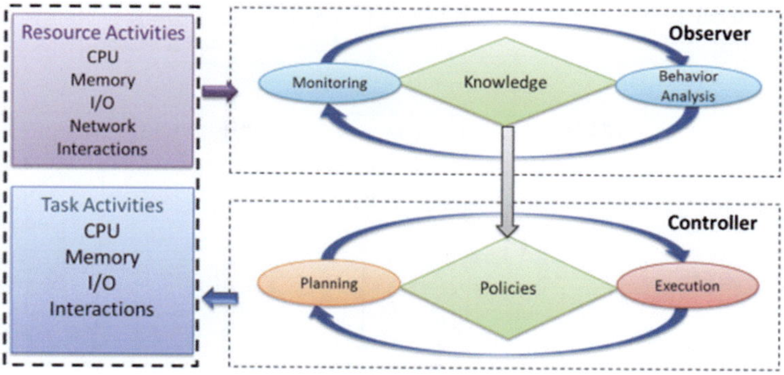

Fig. 18.5 CRM architecture

resilient operational policies. In fact, the Observer and Controller pair provides a unified management interface to support the CRM's self-management services by continuously monitoring and analyzing current cloud system conditions in order to dynamically select the appropriate plan to correct or remove anomalous conditions once they are detected and/or predicted. The Observer monitors and gathers data about the cloud resources and analyzes them to form the knowledge required by the Controller to enforce the ideal security/resilient management policies.

18.3.3 Resilient Cloud Middleware

The Resilient Cloud Middleware (RCM) provides tools and algorithms to implement desired resilient services as specified by the resilient SOA Editor. These capabilities are briefly described below:

- *Replication/Redundancy*: Redundancy is an approach that applies duplicated versions of the original system to tolerate hardware/software failures in one of the replicated components or systems. Redundancy can be implemented by using hardware, information, time or software redundant techniques.
- *Diversity and Automatic Checkpointing*: This capability enables us to generate multiple functionally equivalent, behaviorally different software versions (e.g., each software task can have multiple versions, where each version can be a different algorithm implemented in different programming language (e.g., C, Java, C^{++}) that can run on different computing systems.
- *Behavior Obfuscation* (*BO*): The BO method uses spatiotemporal behavior obfuscation to make active software components change their implementation versions and resources continuously and consequently evade attackers. The BO approach will significantly reduce the ability of an attacker to disrupt the normal operations of a cloud application. Also, it allows for adjusting the resilience level by dynamically increasing or decreasing the shuffling rate and scope of

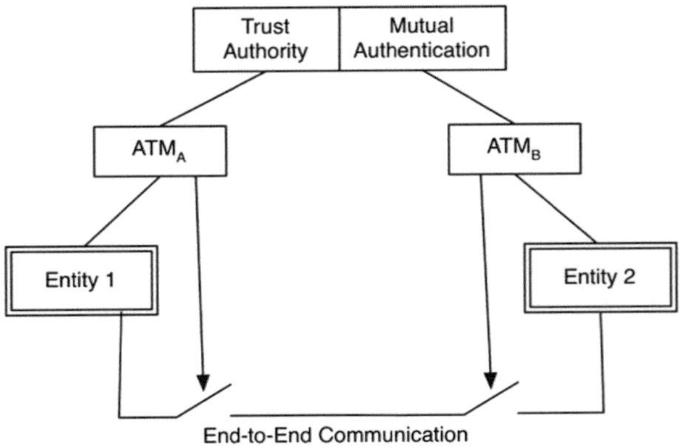

Fig. 18.6 Adaptive end-to-end trust

executing tasks' versions and their execution environments. A major advantage of this approach is that the dynamic change in the execution environment will hide the software flaws that would otherwise be exploited by a cyberattacker.

- *Trust Level*. The middleware provides the required algorithms to evaluate the trust of all resources and actors involved in the crisis management processes. When an entity communicates with another entity, the CRM obtains the trust level of the entity that needs to interact with from a Trust Authority (TA), see Fig. 18.6. If the trust level of the remote entity is below the minimum required trust level set in the policies, then the communication is dropped. By continuously checking with TA module, any interacting entities will not be able to communicate if they do not meet the end-to-end trust policies. Once the component trust level is verified, they can proceed and interact securely using the secure communications.

The resilient operation for any cloud service is achieved using RCM tools will enable us to use the Behavior Obfuscation (BO) algorithm that hides (analogous to data encryption) the execution environment by dynamically changing the number of versions used to run the service at each phase. The dynamic change in the service behavior makes it extremely difficult for an attacker to generate a profile with the possible flaws in the current implementation of the service. The decisions regarding when to shuffle the current variant, the shuffling frequency, and the variant selection for the next shuffle are guided by a continuous monitoring and analysis of current execution state of cloud applications and the desired resilience requirements. In addition to the shuffling of the execution of different application versions, we also apply hardware redundancy in order to tolerate the case when an insider discovers one of the physical machines used to run the application at one phase. The redundancy level determines how many of such scenarios can be tolerated. To speedup the process of selecting the appropriate resilient algorithms and execution environments, the CRM repository contains a set of BO algorithms and images of virtual machines that run in different operating systems (e.g., Windows, Linux) to implement supported crisis cloud services.

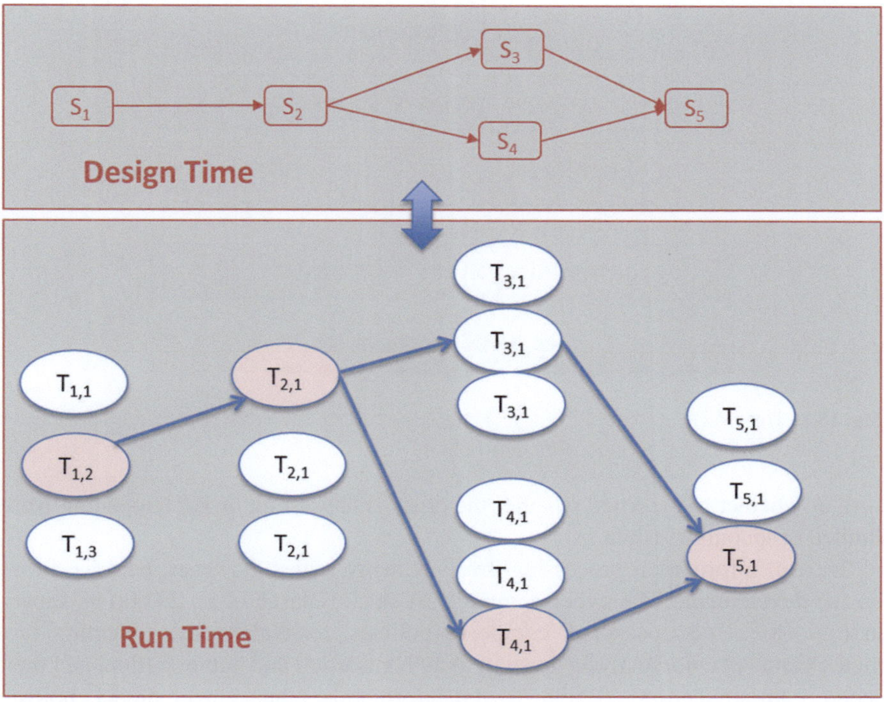

Fig. 18.7 Example implementation of five resilient crisis response services

As an illustrative example, let us assume that during the design stage, the decision maker develop a crisis response that consists of five services that were verified by the RSE module to be effective and meets the security and objective requirements. Once that is done, it will be passed to the CRM module that will use the RCM services and tools to build resilient execution of the selected services at runtime as shown in Fig. 18.7 that shows how the versions to be implement at each service at runtime.

18.4 Information Fusion for rDaaS Architecture

Inherent in the middleware is the ability to compute the trust level. To compute the trust level, we use evidential reasoning using the proportional conflict redistribution (PCR) rule. The PCR rules (e.g., PCR5 and PCR6) have been shown to reason faster than typical Bayesian or Dempster–Shafer methods with conflicting data. Given the uncertainty of crisis response, it is assumed that in the first phases of the response, that conflicting reports would be inherent as most human observers would not be

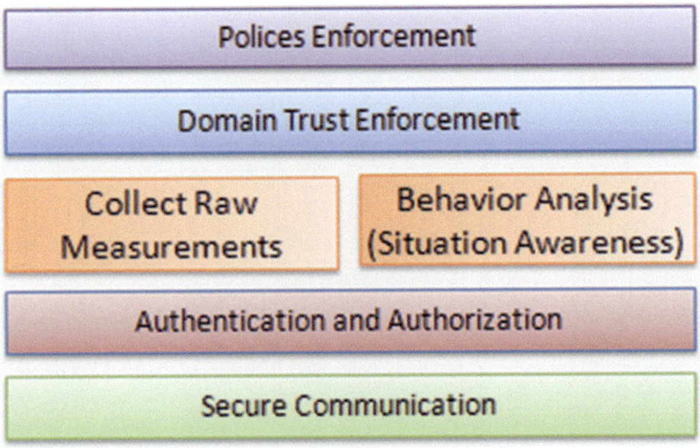

Fig. 18.8 Trust stack

sure of whether to ascertain whether the crisis is happening or not happening from limited or incomplete data.

Trust in information processing involves many issues; however, here we focus on the development of a cyber domain trust stack (Blasch et al. 2014a) as shown in Fig. 18.8. The trust stack composes policies, trust authority, collecting raw metrics and behavior analysis, leading to authentication and authorization, and then secure communications. Similar to the information management model, polices are important to determine whether data access is available. Likewise, sensor management gets access to raw metrics (Blasch et al. 2004) that need to be analyzed for situation awareness. The problem not being full addressed is the impeding results for secure communications. In what follows, we discuss the main functions to be provided by each layer in the trust stack shown in Fig. 18.8.

18.4.1 Behavior Analysis

Behavior analysis techniques apply statistical and data mining techniques to determine the current operating zone of the execution environment (situation awareness) and also project its behavior in the near future. The operating point (OP) of an environment can be defined as a point in an *n*-dimensional space with respect to well-defined attributes. An acceptable operating zone can be defined by combining the normal operating values for each attribute. At runtime, the operating point moves from one zone to another and that point might move to a zone where the environment does not meet its trust and security requirements. We use these movements in the OP to adjust the trust value of the current environment for the Domain Trust Authority. By continuously performing behavior analysis of the environment, we can then proactively predict and detect the anomalous behaviors from the reported crisis information. Furthermore, once it is determined that the environment's operating

Fig. 18.9 Trust metrics

point is moving outside the normal zone, it will adopt its trust value and then determine the appropriate proactive management techniques that can bring back the environment situation to a normal operating zone.

Figure 18.9 presents many trust metrics that need to be considered in a crisis response categorized into machine, user, application, connection, and security.

18.4.2 Domain Trust Authority

DTA evaluates the end-to-end trust over secure communications. It defines a tuple (machine, application, user, and data) to be an entity and all communications among entities has a certain context. Thus authentication is conducted per entity. Every entity has a trust level associated with it. In order to measure the trust, trust's metrics are introduced, and they take values between 0 and 1. Where 0 represents the distrust and 1 represent the blind or full trust. The trust measurements for all entities are stored in an entity call Trust Authority. The NIST standard SP 800-53 (NIST 2010) is used and it defines four levels of trust:

Level	Distrust	Low trust	Moderate	High trust
Trust value	0.00	0.33	0.66	1.00

Initially, a risk and impact analysis is performed to quantify the impact of each component on the overall operations of the network. Common Vulnerabilities and Exposures (CVE) and Common Vulnerability Scoring System (CVSS) are used to evaluate the initial impact for both software and the environment, and reputations of the users are used to assign their initial impacts. Based on the initial impact analysis, the initial trust values for each entity is determined. The risk and impact analysis performed is in consistence with the NIST "Recommended Security Controls for Federal Information Systems and Organizations" report. According to the NIST report, risk measures the extent to which entities are threatened by circumstances or events. The risk is a function of impact and its probability of occurrence. Risks arise from the loss of confidentiality, integrity, and/or availability of information and resources. Thus the initial trust T can be viewed as an inverse function of the risk R:

$$T = 1\,/R \tag{18.1}$$

Where the risk of an entity i is a function of the impact (imp):

$$\begin{aligned} R_i = {} & \mathrm{imp}_i\,(\text{confidentiality}) \bullet \Pr\ \mathrm{imp}_i\,(\text{confidentiality}) \\ & + \mathrm{imp}_i\,(\text{integrity}) \bullet \Pr\ \mathrm{imp}_i\,(\text{integrity}) \\ & + \mathrm{imp}_i\,(\text{availability}) \bullet \Pr\ \mathrm{imp}_i\,(\text{availability}) \end{aligned} \tag{18.2}$$

When a new entity is added, it has to register with the Mutual Authentication (MA) module and then its initial trust value can be quantified according to Eqs. (18.1) and (18.2).

When an entity communicates with another entity, an Autonomic Trust Management (ATM) agent obtains the trust level of the entity that needs to interact with from the Trust Authority (TA), see Fig. 18.6. If the trust level of the remote entity is below the minimum required trust level set in the policies, then the communication is dropped. By continuously checking with TA module, any interacting entities will not be able to communicate if they do not meet the end-to-end trust policies. Once the component trust level is verified, they can proceed and interact securely using the secure communications.

The trust value assigned to each component is not static and is updated continuously. The Trust Authority module is the one responsible for reevaluating the trust at runtime. As mentioned in the previous section, the trust is measured per entity and the trust levels are between 0 and 1.

$$T(E) \in [0,\ 1] \tag{18.3}$$

Each interaction between entities is governed by a context C. Thus, trust level for entities is computed per context:

$$T\,(E,\ C) \in [0,\ 1] \tag{18.4}$$

A Forgiveness Factor, F, is assigned to provide an adaptive mechanism for compromised entities to start gaining trust after all existing vulnerabilities have been fixed. Based on the impact of the entity on the overall operations, we can control the time it takes for that entity to recover its trust level. Monitoring, measuring, and quantifying trust metrics are required, and they are performed by the ATM. M_i will denote the collected trust metric, where i is the metric identifier. The function $m_i()$ is a quantifying function that returns a measurement between 0 and 1 for the metric M_i.

The overall trust for an entity is computed using two types of trust: (1) self-measured trust and (2) reputation-measured trust. The self-measured trust T_s is the trust that is evaluated based on the measurement performed by the ATM agent that manages the entity. While the reputation-measured trust, T_p is based on the trust metrics collected from peers based on a previous recent interaction with the entity for which the trust is being reevaluated. The T_s and T_p are given by following equations:

$$T_S(E, C) = T(\text{ATM}_E, C) \sum_{i=1}^{L} I_i(C) m_i(M_i)$$

$$T_P(E, C) = \frac{1}{K} \sum_{j=1}^{K} T\left(\text{ATM}_j, C\right) \sum_{i=1}^{L} I_i(C) m_i(M_i) \tag{18.5}$$

The values of the metric weight I_i for metric i is determined based on the feature selection technique, where:

$$\sum_{i=1}^{L} I_i(C) = 1 \tag{18.6}$$

Based on the context and the type of operations, the end-to-end trust is evaluated using three trust evaluation strategies: Optimistic, Pessimistic, and Average. The end-to-end trust for each strategy can be evaluated as follows:

Trust confidence	Trust evaluation strategy
Optimistic	$T(E, C) = \max\{T_S(E, C), T_P(E, C)\}$
Average	$T(E, C) = \text{ave}\{T_S(E, C), T_P(E, C)\}$
Pessimistic	$T(E, C) = \min\{T_S(E, C), T_P(E, C)\}$

Once $T(E,C)$ is computed, then it is mapped to the nearest of trust level: (High, Moderate, Low, and None).

The Trust Authority module continuously evaluates the trust for all components and their entities whenever new metrics are obtained from the ATM agents that require an update to entity trust evaluation above depending on the trust evaluation strategy. Various reasoning evaluation strategies exist, such as that of Bayesian, Evidential Reasoning, and Belief Functions (Blasch et al. 2013c) that can be used to evaluate trust.

In a DDDAS cyber environment, there are many levels of information fusion, but to build a trustworthy DDDAS environment, we need to check the trust of each level of information fusion. The Domain Trust Authority is the place to verify the trust of each entity passing information within the DDDAS environment. When the trust level drops below certain threshold; the incoming data can be dropped to enable secure communications. What follows are the DDDAS theory, simulations, measurements, and software analysis for Information fusion levels of cyber data, situation/behavior assessment, information management, and user refinement.

18.4.3 Information Fusion Evidential Reasoning

A fundamental technique for data fusion is Bayes Rule. Recently, Dezert (2012) has shown that Dempster's rule is consistent with probability calculus and Bayesian reasoning if and only if the prior $P(X)$ is uniform. However, when the $P(X)$ is not uniform, then Dempster's rule gives a different result. Yen (1986) developed methods to account for nonuniform priors. Assuming that we have multiple measurements $Z = \{Z_1, Z_2, \ldots, Z_N\}$ for cyber detection D being monitored, Bayesian and ER methods are developed next.

Assuming conditional independence, one has the Bayes method:

$$P\left(X \middle| Z_1 \cap Z_2\right) = \frac{P\left(X \middle| Z_1\right) P\left(X \middle| Z_2\right) / P(X)}{\sum_{i=1}^{N} P\left(X_i \middle| Z_1\right) P\left(X_i \middle| Z_2\right) / P\left(X_i\right)} \tag{18.7}$$

With no information from Z_1 or Z_2, then $P(X|Z_1, Z_2) = P(X)$. Without Z_2, then $P(X|Z_1, Z_2) = P(X|Z_1)$ and without Z_1, then $P(X|Z_1, Z_2) = P(X|Z_2)$. Using Dezert's formulation, then the denominator can be expressed as a normalization coefficient:

$$m_{12}(\varnothing) = 1 - \sum_{X_i;X_j \,|\, X_i \cap X_j} P\left(X_i \middle| Z_1\right) P\left(X_i \middle| Z_2\right) \tag{18.8}$$

Using this relation, then the total probability mass of the conflicting information is

$$P\left(X \middle| Z_1 \cap Z_2\right) = \frac{1}{1 - m_{12}(\varnothing)} \bullet P\left(X \middle| Z_1\right) P\left(X \middle| Z_2\right) \tag{18.9}$$

which corresponds to Dempster's rule of combination using Bayesian belief masses with uniform priors. When the prior's are not uniform, then Dempster's rule is not consistent with Bayes' Rule. For example, let $m_0(X) = P(X)$, $m_1(X) = P(X|Z_1)$, and

$m_2(X) = P(X|Z_2)$, then

$$m(X) = \frac{m_0(X)m_1(X)m_2(X)}{1 - m_{012}(\varnothing)} = \frac{P(X)P\left(X|Z_1\right)P\left(X|Z_2\right)}{\sum_{i=1}^{N} P(X_i)P\left(X_i|Z_1\right)P\left(X_i|Z_2\right)} \tag{18.10}$$

Thus, methods are needed to deal with nonuniform priors and appropriately redistribute the conflicting masses.

18.4.4 Proportional Conflict Redistribution

Recent advances in DS methods include *Dezert–Smarandache Theory* (DSmT). DSmT is an extension to the Dempster–Shafer method of evidential reasoning which has been detailed in numerous papers and texts: *Advances and applications of DSmT for information fusion (Collected works)*, Vols. 1–3 (Dezert and Smarandache 2009). Dezert et al. (Dezert 2002) introduced the methods for the reasoning and presented the hyper power-set notation for DSmT (Dezert and Smarandache 2003). Recent applications include the DSmT Proportional Conflict Redistribution rule 5 (PCR5) applied to target tracking.

The key contributions of DSmT are the redistributions of masses such that no refinement of the frame Θ is possible unless a series of constraints are known. For example, Shafer's model (Shafer 1976) is the most constrained DSm hybrid model in DSmT. Since Shafer's model, authors have continued to refine the method to more precisely address the combination of conflicting beliefs (Josang and Daniel 2006) and generalization of the combination rules (Smaradache and Dezert 2005; Daniel 2006). An adaptive combination rule (Florea et al. 2006) and rules for quantitative and qualitative combinations (Martin et al. 2008) have been proposed. Recent examples for sensor applications include electronic support measures (Djiknavorian et al. 2010), physiological monitoring sensors (Lee et al. 2010), and seismic-acoustic sensing (Blasch et al. 2011d).

Here we use the *Proportional Conflict Redistribution* rule no. 5 (PCR5). We replace Smets' rule (Smets 2005) by the more effective PCR5 to cyber detection probabilities. All details, justifications with examples on PCR*n* fusion rules and DSm transformations can be found in the DSmT compiled texts (Dezert and Smarandache 2009). A comparison of the methods is shown in Fig. 18.10.

In the DSmT framework, the PCR5 is used generally to combine the basic belief assignment (bba)'s. PCR5 transfers the conflicting mass only to the elements involved in the conflict and proportionally to their individual masses, so that the specificity of the information is entirely preserved in this fusion process. Let $m_1(.)$ and $m_2(.)$ be two independent bba's, then the PCR5 rule is defined as follows (see Dezert and Smarandache 2009, Vol. 2 for full justification and examples):

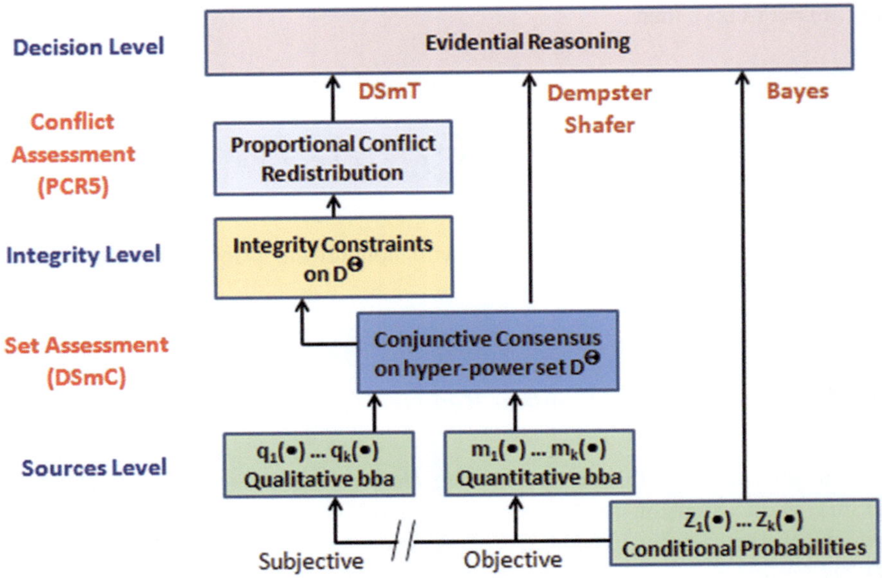

Fig. 18.10 Comparison of Bayesian, Dempster–Shafer, and PCR5 fusion theories

$m_{\mathrm{PCR5}}(\varnothing) = 0$ and $\forall X \in 2^{\Theta} \backslash \{\varnothing\}$, where $\varnothing$ is the null set and 2^{Θ} is the power set:

$$m_{\mathrm{PCR5}}(X) = \sum_{\substack{X_1; X_2 \in 2^{\Theta} \\ X_1 \cap X_2 = X}} [m_1(X_1) + m_2(X_2)] + \sum_{\substack{X_2 \in 2^{\Theta} \\ X_2 \cap X = \varnothing}} \left[\frac{m_1(X_1)^2 m_2(X_2)}{m_1(X_1) + m_2(X_2)} + \frac{m_1(X_1) m_2(X_2)^2}{m_1(X_1) + m_2(X_2)} \right] \tag{18.11}$$

where $\cap$ is the interesting and all denominators in the equation above are different from zero. If a denominator is zero, that fraction is discarded. Additional properties and extensions of PCR5 for combining qualitative bba's can be found in Dezert and Smarandache 2009 with examples and results. All propositions/sets are in a canonical form.

18.5 Case Study: Nuclear Plant Crisis Management

In order to experiment the Dynamic Data-Driven Application Systems (DDDAS) paradigm in managing crisis, we develop a crisis case study in which a large

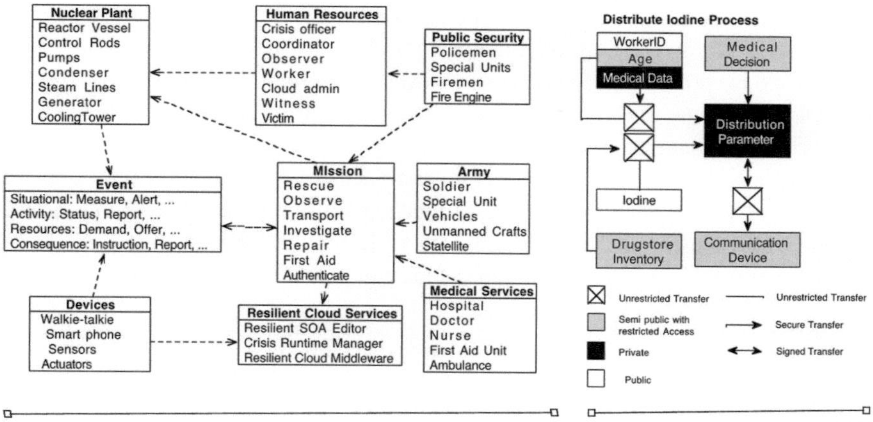

Fig. 18.11 Nuclear crisis feature diagrams and trust management

quantity of radioactive substance is accidentally evaded due to a critical incident in a nuclear plant. In response to this crisis, various parties at different levels and responsibilities have to collaborate together to reduce contamination risks. The volume and heterogeneity of exchanged information and critical dependencies between actions make crisis response processes vulnerable and subject to changes.

We build our crisis management system (CMS) based on the nuclear crisis management simulation in the SocEDA project in which 18 processes have been developed to validate a Cloud-based event-driven architectures (Barthe-Delanoë et al. 2013). As illustrated in Fig. 18.11 (left), the CMS involves identifying, assessing, and handling the crisis situation by orchestrating the communication between all actors involved in the crisis. It allocates and manages internal and external resources, and interoperates with different external services (e.g., military systems, police systems, government, medical services). It also provides access to authorized parties to exchange relevant information based on their trust levels.

18.5.1 Trust Management

In our scenario, a crisis is usually triggered by report from a witness and/or the surveillance system. A coordinator, who organizes all required resources and tasks, initiates the crisis response process. Observers with expertise in the nuclear field are assigned to the scene to observe the emergency situation. The tasks defined by the observer are crisis missions need to be processed to cope with the situation. Based on observer feedbacks from the scene, the coordinator is required to allocate suitable resources to each task. Human resources, for instance firemen, doctors, policemen, and technicians, act as first-aid workers, and resilient infrastructure, which may include resilient cloud resources and communication devices (e.g., PDAs or mobile phones). The workers are expected to perform their assigned tasks and report on the success or failure in carrying out the missions that allow the crisis to be concluded. Crisis response processes thus consists of tasks that should be executed in a specific orders to coordinate the communication between involved parties.

All crisis response processes should also keep track of trusted workers and enable rescue resources to provide or access location-sensitive information on the move. Figure 18.11 (right) shows a three-level mechanism (white, black and grey boxes) to indicate the trust level for accessing information with different trust levels (e.g., detailed maps, terrain data, and weather conditions) and routes leading to it.

18.5.2 Crisis Response Processes

As depicted in Fig. 18.12, the crisis response process involves three levels of tightly integrated functions: (1) *Strategic Level objectives* that will focus on protecting citizens, managing the situation, and recovering using the support of experts and scientific community; (2) *Operational Level tasks* that focuses on distributing, evacuating and confining activities and managing alerts and media; and (3) *Tactical Support Level situational assessments*, for managing resources and data (e.g., information fusion). These objectives, tasks, and assessments need to be performed as resilient and trusted cloud services such as Capture Incident, Assign Internal Resource, Assign External Resource, Execute Missions, Execute Observer Mission, Perform Rescue Missions, and Manage Adaptive end-to-end Trust. Details of tasks related to unforeseen situations (i.e., severe weather, risk of explosion, security attacks, etc.) that may affect the context in which the processes operates, and that require the adaptation to the environment are not included for space reasons. Keeping in mind that the rDaaS architecture deals with the environmental contextual changes and reacts in a certain way to ensure reliability and trustworthy.

The crisis management system contains server clusters and middleware that are hosted by the cloud and connected with different critical infrastructures, machines, and crisis team stakeholders. The crisis team members, such as coordinators and cloud administrators, use terminals or desktop machines to access the resilient Cloud services. External services and mobile workers with laptops are also connected to the cloud by means of Web services. The simulation of our CMS consists of:

1. The *simulated environment*, which comprises crisis response processes implemented as Web Services, each of which simulates the evolution of the crisis situation (e.g., the radiation propagation) or to receive feedbacks (events) from the crisis team (e.g., the evacuation status).
2. The *resilient SOA Response Service Development*, which enables the design of abstract processes and their implementation through distributed ESBs (Enterprise Service Bus). The ESB is used to simulate interactions between crisis actors by invoking Web service-based processes written in BPEL (Business Process Execution Language) and executed through a BPEL engine.

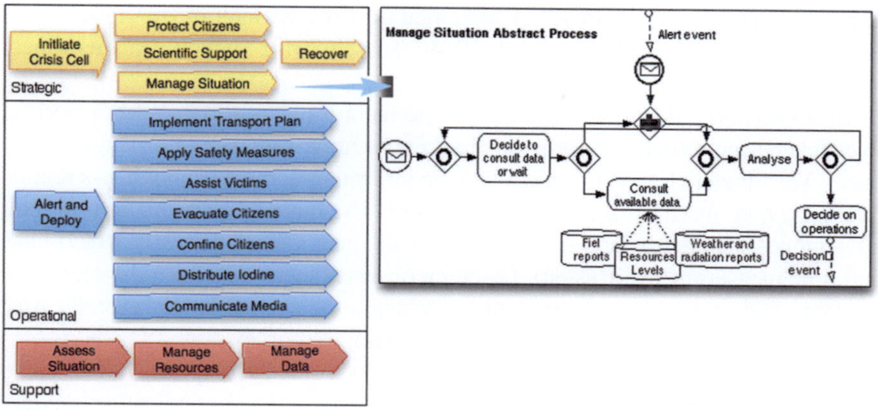

Fig. 18.12 Abstract crisis response processes (based on Tunc et al. 2014)

3. The *Cloud-based Resilient and Trustworthy infrastructure* provides services to access the resilient cloud middleware, ensure security, and establish end-to-end trust among various actors.

Since crisis management processes are event-driven, they particularly require an event-driven SOA (a.k.a SOA 2.0) that integrates complex event processing (CPE) and ESB. We use the Synapse/WSO2 ESB as a smart event broker that integrates events from Web services and redistributes them to end-points (e.g., actuators, terminals). The Sci-Flex plug-in makes possible to connect the Synapse with the Esper, a complex event processing (CEP). Esper rapidly processes large amount and high-frequency time-based event data. The Synapse/WSO2 ESB implements the WS-Eventing to provide simple publish-subscribe model in Web services and event subscription over SOAP. In our case-study, the ESB is considered as event consumer and Web Services are considered as event sources. An event is expressed in the WS-Eventing format and has an event type, which is part of an event topic.

18.5.3 Crisis Event Response Trust Simulation

In our case study, we focused on the development of the design environment in order to record and play streams of events and capture them by the event-driven SOA and processes. Interested readers can refer to our previous work on resilient cloud services using DDDAS and moving target defense. The metric-based evaluation regarding the simulation of our event-driven SOA in the resilient cloud supports trust in analysis. For crisis response, situation awareness is needed and more importantly trust in the data being circulated as to an emergency situation. Current challenges in managing the information is the presentation of the data to the user in normal and conflicting scenarios (Blasch et al. 2013d). We chose to use the proportional conflict redistribution (PCR) rule as compared to Bayesian and Dempster–Shafer rules. The uncertainty analysis using PCR has supported trust related metrics (for more details see Blasch et al. 2013e; Blasch et al. 2014c; Blasch 2014b) which supports timely mission response.

In our scenario, the crisis response process records the information from workers to enable rescue resources to provide or access location-sensitive information on the move. We simulate four sources of dynamic information, three from witness reports (workers) and one from a surveillance system. The human observers vary in their knowledge of a nuclear crisis, while the surveillance system typically would have real time measurements (trusted). Based on observer feedbacks from the scene, the coordinator is required to allocate suitable resources to each task. However, the trust in the report has to be determined before allocating resources.

The key issue is the witness reports from non nuclear experts, but close to the situation who utilize resilient cloud resources and communication devices (e.g., mobile phones). Thus, the trust of the observers (i.e., users), machines, and communication devices (with security) needs to be addressed before allocating resources. As a first step in the analysis, we are concerned with situations in which there is conflict in the reports in the first assessment of crisis. It all observers are reporting a similar situation of an impending crisis, information fusion should be obvious. However, that is not the case as the reporting is done from a normal state to a crisis state. Two cases are provided to demonstrate the use of the PCR rules and how the trust can be used for crisis assessment.

Figure 18.13 shows the case for high conflict (top) and low conflict (below) between crisis action teams for which the crisis of the disaster starts at $t = 50$ s.

High conflict between the crisis action teams in determining the extent of the crisis.

Low conflict between the crisis action teams in determining the extent of the crisis.

The difference between the scenarios which have high and low conflicts between observers has a big impact on the system trust in the crisis assessment. In Fig. 18.13 at the top, there is *high conflict* as different observers are unsure about whether a crisis is happening. The time to make a decision (say above 60 %) is not till (90 s − 40 s) = 50s after the crisis occurred for Bayesian and DS methods. PCR detects the conflict immediately but has a 5–10 s delay.

The trust metric is more revealing in the *low conflict* scenario. As with high conflict, the PCR evidential reasoning approach has a quicker reaction, while the Bayesian/DS have slightly longer delay times. The low conflict actually causes the Bayesian/DS processing steps to become complacent with large biases towards the status quo. The trust metric is higher for all methods in the low-conflict scenario versus the high-conflict scenario; however, the PCR method stays at a reasonable high trust value throughout the crisis analysis. Thus, variations in the scenario impact the trust metric results, but the higher trust is consistent with lower conflict. The higher trust has demonstrated a measure of resiliency in the data processing for emergency response situations which can be used in context modeling (Steinberg et al. 2014) for situation (e.g., crisis response) assessment.

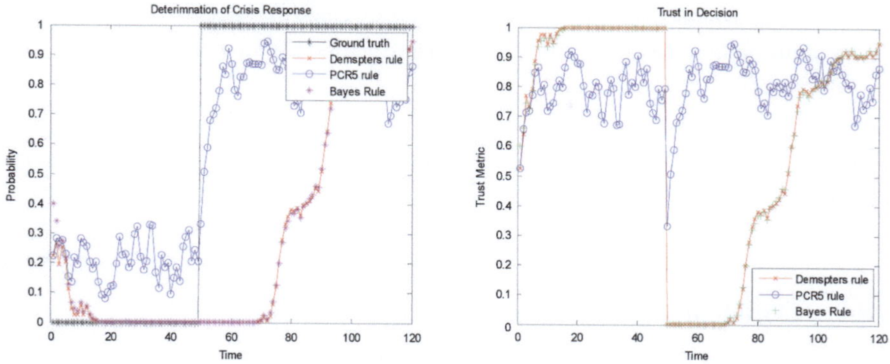

High conflict between the crisis action teams in determining the extent of the crisis.

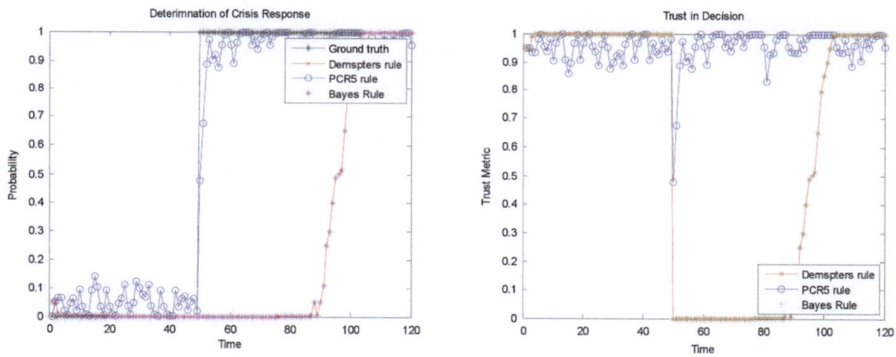

Low conflict between the crisis action teams in determining the extent of the crisis.

Fig. 18.13 High and low data conflict between crisis action teams for timely response

18.6 Discussion

There are many open questions being addressed by the simulation of which the infrastructure and information fusion techniques were described for the assessment of trust for an alert. The focus was on the real-time assessment of trust to determine the crisis alert as a first step to determine when a crisis was happening to then assign resources to alleviate the crisis. Three further issues of analysis include:

1. *Robustness Management*, to include the resilient cloud middleware testing when critical response nodes have a possible separation from the cloud services and the resulting change in the analysis based on the separation (see Fig. 18.2).
2. *Situation Management,* to include interaction among various (trusted) actors for further information to subsequently refine the situation (crisis) based on the content of the information from key actors (see Fig. 18.11).

3. 3-*Mission Management*, to include the web services that dictate whether or not the information gathered from the trusted actors aid in crisis reporting (see Fig. 18.12).

The combined elements of trust assessment of crisis and the resource allocation response would then support a complete response to mitigate the crisis through emergency response.

18.7 Conclusions and Future Work

Future trends for the integration of information fusion, DDDAS and SOA into one cloud environment include streaming big data, distributed information fusion, and detection of contextual changes in dynamic environments. In this chapter, we demonstrated the benefits of information fusion to enable resilience at the SOA-based process level for crisis response by assessing trust from distributed sources to enable end-to-end resilient services. Using the simulation framework for nuclear crisis management systems, we are conducting experiments and comparison with existing nuclear crisis management systems. We are also exploring rDAAS with a variety of multi-modal sensing and information fusion services to handle partial or missing contextual information while preserving SOA adaptability to changes and unforeseen situations. We are working on a simulation environment that record and play streams of events with temporal constraints that can extend to other dynamic situations.

Acknowledgements This work is partly supported by the Air Force Office of Scientific Research (AFOSR) Dynamic Data-Driven Application Systems (DDDAS), AFOSR FA95550-12-1-0241, National Science Foundation NSF IIP-0758579, NCS-0855087, and IIP-1127873, and Thomson Reuters in the framework of the Partner University Fund (PUF) project (PUF is a program of the French Embassy in the USA and the FACE Foundation and is supported by American donors and the French government).

References

16 September 2009. [Accessed January 2013]. http://www.cyber.st.dhs.gov/docs/National_Cyber_Leap_Year_Summit_2009_Co-Chairs_Report.pdf

Abbasy M, Shanmugam B (2011) Enabling data hiding for resource sharing in cloud computing environments based on DNA sequences. In IEEE World Congress

Abielmona R, Falcon R, Vachon P, Groza V (2014) Vessel tracking and anomaly detection using level 0/1 and high-level information fusion techniques. Workshop on soft computing applications

Aggarwal CC, Yu PS (2001) Outlier detection for high dimensional data. In SIGMOD Conference

Satam P, Alipour H, Al-Nashif Y, Hariri S (2015) DNS-IDS: Securing DNS in the Cloud Era, International Conference on Cloud and Autonomic Computing (ICCAC), pp 296–301

Alipour H, Al-Nashif YB, Satma P, Hariri S (2015) Wireless Anomaly Detection Based on IEEE 802.11 Behavior analysis. IEEE Transactions on Information Forensics and Security, 10(10), 2158–2170

Al-Nashif YB, Kumar A, Hariri S, Luo Y, Szidarovszky F, Qu G (2008) Multi-level intrusion detection system (ML-IDS). ICAC, pp 131–140

Anderson D, Lunt TF, Javitz H, Tamaru A, Valdes A (1995) Detecting unusual program behavior using the statistical component of the next-generation intrusion detection expert system nides. Technical Report SRI-CSL-95-06, Computer Science Laboratory, SRI International

Ao Z, Scholz J, Oxehham M (2014) A Scientific Inquiry fusion theory for high-level information fusion. International conference on information fusion

Aved AJ, Blasch E (2015) Multi-INT query language for DDDAS designs. Proc Comput Sci 51:2518–2532

Avizienis A (1985) The N-version approach to fault tolerant software. IEEE Trans Softw Eng SE-11(12):1491–1501

Barrantes E, Ackley D, Forrest S et al (2003) Intrusion detection: randomized instruction set emulation to disrupt binary code injection attacks. ACM conference on computer and communications security

Barthe-Delanoë A-M et al (2013) A platform for event-driven agility of processes: a delivery context use-case. Collaborative systems for reindustrialization. Springer, Berlin, pp 681–690

Bhadauria R, Sanyal S (2012) Survey on security issues in cloud computing and associated mitigation techniques. Int J Comput Appl 47(18):47–66

Blasch E (2003) Situation, impact, and user refinement. Proc SPIE 5096

Blasch E (2006) Level 5 (User Refinement) issues supporting Information Fusion Management. International conference on information fusion

Blasch E (2008) Chapter 19: Introduction to level 5 fusion: the role of the user. In: Liggins ME, Hall D, Llinas J (eds) Handbook of multisensor data fusion, 2nd edn. CRC Press, Boca Raton, FL

Blasch E (2014) Context aided sensor and human-based information fusion. IEEE National Aerospace and Electronics (NAECON)

Blasch E (2014) Trust metrics in information fusion. Proc SPIE 9091

Blasch E (2015) One decade of the Data Fusion Information Group (DFIG) model. Proc SPIE 9499

Blasch E, Israel S (2015) Situation/threat context assessment. International conference on information fusion

Blasch EP, Plano SB (2002) JDL Level 5 Fusion model 'user refinement' issues and applications in group tracking. Proc SPIE 4729

Blasch E, Plano S (2005) DFIG Level 5 (User Refinement) issues supporting Situational Assessment Reasoning. International conference on information fusion

Blasch E, Pribilski M, Daughtery B, Roscoe B, Gunsett J (2004) Fusion metrics for dynamic situation analysis. Proc SPIE 5429

Blasch E, Kadar I, Salerno J, Kokar MM, Das S, Powell GM, Corkill DD, Ruspini EH (2006) Issues and challenges in situation assessment (Level 2 Fusion). J Adv Inform Fusion 1(2):122–139

Blasch E, Kadar I, Hintz K, Biermann J, Chong C-Y, Salerno J, Das S (2008) Resource management coordination with level 2/3 fusion issues and challenges. IEEE Aerosp Electron Syst Mag 23(3):32–46

Blasch E, Valin P, Bosse E, Nilsson M, Van Laere J, Shahbazian E (2009) Implication of culture: user roles in information fusion for enhanced situational understanding. International conference on information fusion

Blasch E, Dorion É, Valin P, Bossé E, Roy J (2010) Ontology alignment in geographical hard-soft information fusion systems. International conference on information fusion

Blasch EP, Valin P, Bossé E (2010) Measures of effectiveness for high-level fusion. International conference on information fusion

Blasch E, Breton R, Valin P, Bosse E (2011) User information fusion decision making analysis with the C-OODA model. International conference on information fusion

Blasch E, Breton R, Valin P (2011) Using the C-OODA model for CIMIC analysis. IEEE National Aerospace and Electronics Conference

Blasch E, Deignan PB Jr, Dockstader SL, Pellechia M et al (2011) Contemporary concerns in geographical/geospatial information systems (GIS) processing. IEEE National Aerospace and Electronics Conference (NAECON)

Blasch E, Dezert J, Valin P (2011) DSMT Applied to Seismic and Acoustic Sensor Fusion. Proceedings of the IEEE National Aerospace Electronics Conference (NAECON)

Blasch EP, Bosse E, Lambert DA (2012a) High-level information fusion management and systems design. Artech House, Norwood, MA

Blasch E, Banas C et al (2012) Pattern Activity Clustering and Evaluation (PACE). Proc SPIE 8402

Blasch E, Lambert DA, Valin P, Kokar MM, Llinas J, Das S et al (2012c) High level information fusion (HLIF) survey of models, issues, and grand challenges. IEEE Aerosp Electron Syst Mag 27(9):4–20

Blasch E, Valin P, Jousselme A-L, Lambert DA, Bossé E (2012) Top ten trends in high-level information fusion. International conference on information fusion

Blasch E, Chen Y, Chen G, Shen D, Kohler R (2013a) Information fusion in a cloud-enabled environment. In: Han K, Choi B-Y, Song S (eds) High performance cloud auditing and applications. Springer, New York, pp 91–115

Blasch E, Seetharaman G, Reinhardt K (2013b) Dynamic data driven applications system concept for information fusion. Procedia Computer Science 18:1999–2007

Blasch E, Dezert J, Pannetier B (2013) Overview of Dempster-Shafer and belief function tracking methods. Proc SPIE 8745

Blasch E, Steinberg A, Das S, Llinas J, Chong C-Y, Kessler O, Waltz E, White F (2013) Revisiting the JDL model for information Exploitation. International conference on information fusion

Blasch E, Laskey KB, Joussselme A-L, Dragos V, Costa PCG, Dezert J (2013) URREF reliability versus credibility in information fusion (STANAG 2511). International conference on information fusion

Blasch E, Al-Nashif Y, Hariri S (2014) Static versus Dynamic Data Information Fusion analysis using DDDAS for Cyber Trust. International conference on computational science, Procedia Computer Science

Blasch E, Nagy J, Aved A, Pottenger WM, Schneider M, Hammoud R, Jones EK, Basharat A et al (2014) Context aided video-to-text information fusion. International conference on information fusion

Blasch E, Jøsang A, Dezert J, Costa PC, Laskey KB, Jousselme A-L (2014) URREF self-confidence in Information Fusion Trust. International conference on information fusion

Breunig MM, Kriegel H, Ng RT, Sander J (2000) LOF: identifying density-based local outliers. Proceedings of ACM SIGMOD International Conference on Management of Data, pp 93–104

Cadar C, Akritidis P, Costa, M, Martin J-P, Castro M, (2008) Data randomization. Technical Report MSR-TR-2008-120, Microsoft Research

Chen G, Shen D, Kwan C, Cruz JB et al (2007a) Game theoretic approach to threat prediction and situation awareness. J Adv Inform Fusion 2(1):1–14

Chen G, Tian Z, Shen D et al (2007) A novel framework for command and control of networked sensor systems. Proc SPIE 6578

Chen D, Chen G, Cruz J, Haynes L et al (2007) A Markov game theoretic data fusion approach for cyber situational awareness. Proc SPIE 6571

Cheng E, Ma L, Blaisse A et al (2014) Efficient feature extraction from wide area motion imagery by MapReduce in Hadoop. Proc SPIE 9089

Costa PC, Laskey KB et al (2012) Towards unbiased evaluation of uncertainty reasoning: the URREF ontology. International conference on information fusion

Daniel M (2006) Generalization of the classic combination rules to DSm hyper-power sets. Inform Security 20

Denning DE (1987) An intrusion-detection model. IEEE Trans Softw Eng 13(2):222–232

Dezert J (2002) Foundations for a new theory of plausible and paradoxical reasoning. Inform Security, An Int'l Journal, ed. by Prof. Tzv. Semerdjiev, vol 9

Dezert J (2012) Non-Bayesian reasoning for information fusion – a tribute to Lofti Zadeh. Submitted to J Adv Inform Fusion

Dezert J, Smarandache F (2003) On the generation of hyper-powersets for the DSmT. International conference on information fusion

Dezert J, Smarandache F (2009) Advances and applications of DSmT for information fusion (Collected works), vol 1–3. American Research Press, Rehoboth

Dijkman RM, Dumas M, Ouyang C (2008) Semantics and analysis of business process models in BPMN. Inf Softw Technol 50(12):1281–1294

Djiknavorian P, Grenier D, Valin P (2010) Approximation in DSm theory for fusing ESM reports. International workshop on belief functions

Dsouza G, Rodríguez G, Al-Nashif YB, Hariri S (2013) Building resilient cloud services using DDDAS and moving target defence. Int J Cloud Comput 2(2/3):171–190

Dunlop M, Groat S, Urbanski W, Marchany R, Tront J (2011) MT6D:a moving target IPv6 defense. In IEEE military communications conference (MILCOM)

Endsley MR (1988) Design and evaluation for situation awareness enhancement. Proc Hum Fact Ergon Soc Annu Meet 32:97–101

Ertöz L, Eilertson E, Lazarevic A, Tan P, Kumar V, Srivastava J, Dokas P (2004) Minds - Minnesota intrusion detection system. In: Data mining - next generation challenges and future directions. MIT Press, Cambridge

Eskin E, Arnold A, Prerau M, Portnoy L, Stolfo S (2002) A geometric framework for unsupervised anomaly detection: detecting intrusions in unlabeled data. In: Proceedings of applications of data mining in computer security. Kluwer Academics, Boston, pp 78–100

Evans D, Nguyen-Tuong A, Knight J (2011) Effectiveness of moving target defenses. In: Advances in information security. Springer, New York, pp 29–39

Feng J, Chen Y, Summerville D, Ku W, Su Z (2011) Enhancing cloud storage security against roll-back attacks with a new fair multi-party non-repudiation protocol. In Consumer communications and networking conference

Florea MC, Bosse E (2009) Crisis management using Dempster-Shafer theory: similarity measures to characterize source's reliability. RTO-MP-IST-086

Florea MC, Dezert J, Valin P, Smarandache F, Jousselme A-L (2006) Adaptive combination rule and proportional conflict redistribution rule for information fusion. COGIS '06 conference

Ge L, Yu W, Shen D, Chen G, Pham K et al (2014) Toward effectiveness and agility of network security situational awareness using moving target defense (MTD). Proc SPIE 9085

Ge L, Zhang H, Xu G, Yu W et al (2015) Towards MapReduce based machine learning techniques for processing massive network threat monitoring data. In: Networking for big data. CRC, Boca Raton

Goodin D (2009) Webhost Hack Wipes Out Data for 100,000 Sites. 8 June 2009 [Online]. http://www.theregister.co.uk/2009/06/08/webhost_attack/

Gruniger M, Obrst L (eds) (2014) Semantic Web and big data meets applied ontology. Ontology Summit

Javitz HS, Valdes A (1994) The nides statistical component: description and justification. Technical Report, SRI International Menlo Park, California

Jeffrey D, Sanjay G (2008) MapReduce: simplified data processing on large clusters. In Communications of the ACM

Josang A, Daniel M (2006) Strategies for combining conflict dogmatic beliefs. International conference on information fusion

Kankanamge RPR (2010) Information systems for supporting fire emergency. PhD Thesis, Loughborough University

Kaufman L (2009) Data security in the world of cloud computing. IEEE Security Privacy J 7(4):61–64

Keromytis A, Sethumadhavan GRS, Stolfo S, Junfeng Y, Benameur A, Dacier M, Elder M, Kienzle D, Stavrou A (2012) The MEERKATS Cloud Security Architecture. In 32nd international conference on distributed computing systems workshops

Kim K (1998) ROAFTS: a middleware architecture for real-time object-oriented adaptive fault tolerance support. In IEEE CS 1998 High-Assurance Systems Engineering (HASE) Symposium Washington, DC

Kim K, Welch H (1989) Distributed execution of recovery blocks: an approach for uniform treatment of hardware and software faults in real-time applications. IEEE Trans Comput 38(5):626–636

Knorr EM, Ng RT (1998) Algorithms for mining distance-based outliers in large datasets. In Proceedings of the 24th international conference on very large data bases, VLDB, pp 392–403, 24–27

Latif-Shabgahi S (2011) An integrated voting algorithm for fault tolerant systems. In Proceedings of the international conference on software and computer applications (IPCSIT), vol 9

Lee ZH, Choir JS, Elmasri R (2010) A static evidential network for context reasoning in home-based care. IEEE Trans Syst Man Cyber Part A Syst Hum 40(6):1232–1243

Li W, Badr Y, Biennier F (2013) Service farming: an ad-hoc and QoS-aware web service composition approach. Dans SAC 2013, Coimbra, Portugal, pp 750–756

Liu Z, Blasch E, Xue Z, Langaniere R, Wu W (2012) Objective assessment of multiresolution image fusion algorithms for context enhancement in night vision: a comparative survey. IEEE Trans Pattern Anal Mach Intell 34(1):94–109

Liu B, Blasch E, Chen Y, Aved AJ et al (2014) Information fusion in a cloud computing era: a systems-level perspective. IEEE Aerosp Electron Syst Mag 29(10):16–24

Lunt TF, Jagannathan R (1988) A prototype real-time intrusion-detection expert system. In Proceedings of the IEEE symposium on security and privacy, pp 18–21

Luo D, Wang J (2010) CC-VIT: virtualization intrusion tolerance based on cloud computing. In 2nd International conference on information engineering and computer science

Marhsall P (2014) Agencies push Geospatial data to the rank and file. GCN, Nov/Dec 2014

Martin A, Osswald C, Dezert J, Smarandache F (2008) General combination rules for qualitative and quantitative beliefs. J Adv Inform Fusion 3(2):67–89

Mendoza-Schrock O, Patrick JA et al (2009) Video image registration evaluation for a layered sensing environment. Proceedings of the IEEE National Aerospace Electronics Conference (NAECON)

Modi C, Patel D, Borisaniya B, Patel A, Rajarajan M (2013) A survey on security issues and solutions at different layers of Cloud computing. J Supercomput 63:561–592

Narain S (2013) Moving target defense with configuration space randomization. https://www.ncsi.com/nsatc11/presentations/thursday/emerging_technologies/narain.pdf. Accessed 30 Jan 2013

NIST (2010) Recommended security controls for federal information systems and organizations. NIST Special Publication 800-53, Revision 3

PaX Homepage 2000 [Online]. http://pax.grsecurity.net/. Accessed Oct 2012

Paxson V (1999) Bro: a system for detecting network intruders in real-time. Comput Networks (Amsterdam, Netherlands: 1999) 31(23–24):2435–2463

Pék G, Butty´an L, Bencsáth B (2013) A survey of security issues in hardware virtualization. ACM Computer Surv 45(3), Article 40 (July 2013), 34 pages

Peng L, Lipinski D, Mohseni K (2014) Dynamic Data Driven Application System for plume estimation using UAVs. J Intell Robot Syst 74:421–436

Pipkin DL (2000) Information security: protecting the global enterprise. Prentice-Hall, Inc., Upper Saddle River, NJ

Porras PA, Neumann PG (1997) Emerald: event monitoring enabling responses to anomalous live disturbances. In Proceedings of the National Information Systems Security Conference, pp. 353–365, 1997

Ramaswamy S, Rastogi R, Shim K (2000) Efficient algorithms for mining outliers from large datasets. Proceedings of the ACM SIDMOD international conference on management of data, pp 427–438

Roesch M (1999) Snort - lightweight intrusion detection for networks. 13th Systems Administration Conference - LISA

Rosenblum M, Garfinkel T (2005) When virtual is harder than real: security challenges in virtual machine based computing environments. In 10th conference on hot topics in operating systems, Berkeley

Scarfone K, Mell P (2007) Guide to Intrusion Detection and Prevention Systems (IDPS). Computer Security Resource Center (National Institute of Standards and Technology) (800–94). Retrieved 1 Jan 2010

Schmidt M, Baumgartner L, Graubner P, Bock D, Freisleben B (2011) Malware detection and Kernel Rootkit prevention in cloud computing environments. In 19th Euromicro international conference on parallel, distributed and network-based processing

Schneider J, Romanowski CJ, Stein K (2013) Decision making to support local emergency 960 preparation, response, and recovery, IEEE International Conference on Technologies for Homeland Security, pp 498–503

Scott PD (2007) Information fusion for natural and man-made disasters. AFOSR Grant F49620-01-1-0371, Jan 2007

Security as a Service. Cloud Security Alliance, [Online]. https://cloudsecurityalliance.org/research/secaas/. Accessed Jan 2013

Sekar R, Gupta A, Frullo J, Shanbhag T, Tiwari A, Yang H, Zhou S (2002) Specification-based anomaly detection: a new approach for detecting network intrusions. In CCS'02: Proceedings of the 9th ACM conference on computer and communications security, pp 265–274, New York, NY, USA

Sequeira K, Zaki M (2002) Admit: anomaly-based data mining for intrusions. Proceedings of ACM SIGKDD international conference on Knowledge discovery and data mining (KDD), pp 386–395

Shafer G (1976) A mathematical theory of evidence. Princeton University Press, Princeton, NJ

Shon T, Moon J (2007) A hybrid machine learning approach to network anomaly detection. Inform Sci 177(18):3799–3821

Siddiqui MS, Verma D (2011) Cross site request forgery: a common web application weakness. IEEE International Conference on Communication Software and Networks (ICCSN), pp 538–543

Smaradache F, Dezert J (2005) Information fusion based on new proportional conflict redistribution rules. International conference on information fusion

Smets P (2005) Analyzing the combination of conflicting belief functions. International conference on information fusion

Staniford S, Hoagland JA, McAlerney JM (2002) Practical automated detection of stealthy portscans. J Comput Security 10(1–2):105–136

Steinberg AN, Bowman CL, Haith G et al (2014) Adaptive context assessment and context management. International conference on information fusion

Subashini VS (2011) A survey on security issues in service delivery models of cloud computing. J Netw Comput Appl 34:1–11

Toy WN (1987) Fault-Tolerant Computing, *Advances in Computers,* Vol. 26, Academic Press, 201–279

Tunc C, Fargo F, Al-Nashif Y, Hariri S, Hughes J (2014) Autonomic Resilient Cloud Management (ARCM). ACM international conference on cloud and autonomic computing (CAC '14)

Tyrrell A (1996) Recovery blocks and algorithm based fault tolerance. In 22nd EUROMICRO conference

Vallee G, Engelmann C, Tikotekar A, Naughton T, Charoenpornwattana K, Leangsuksun C, Scott S (2008) A framework for proactive fault tolerance. International conference on availability, reliability and security

Verissimo P, Bessani A, Pasin M (2012) The TClouds architecture: open and resilient cloud-of-clouds computing. In IEEE/IFIP 42nd international conference on dependable systems and networks workshops

Viswanathan RP, Al-Nashif Y, Hariri S (2011) Application attack detection system (AADS): an anomaly based behavior analysis approach. ACS/IEEE international conference on computer systems and applications

Wen Y (2011) Heterogeneous sensor fusion in sensor networks: a language-theoretic approach. PhD Dissertation, Penn State University

Yamanishi K, Takeuchi J, Williams GJ, Milne P (2000) On-line unsupervised outlier detection using finite mixtures with discounting learning algorithms. In Knowledge discovery and data mining, pp 320–324

Yang C et al (2008) Fusion of tracks with road constraints. J Adv Inform Fusion 3(1):14–32

Yang C, Nguyen T et al (2014) Mobile positioning with fused mixed signals of opportunity. IEEE Aerosp Electron Syst Mag 29(4):34–46

Ye N (2004) A markov chain model of temporal behavior for anomaly detection. In Proceedings of IEEE 5th annual IEEE information assurance workshop

Ye N, Chen Q (2001) An anomaly detection technique based on a chi-square statistic for detecting intrusions into in-formation systems. Qual Reliab Eng Int 17:105–112

Yen J (1986) A reasoning model based on the extended Dempster Shafer theory. National conference on artificial intelligence

Zeng H (2013) Research on developing an attack and defense lab environment for cross site scripting education in higher vocational colleges. International Conference on Computational and Information Sciences

Zhuang R, Zhang S, DeLoach SA, Ou X, Singhal A (2012) Simulation-based approaches to studying effectiveness of moving-target network defense. In National symposium on moving target research

Part V
Decision Making

Chapter 19
Aggregation of Coherent Experts' Opinions: A Tractable Extreme-Outcomes Consistent Rule

Marcello Basili and Alain Chateauneuf

Abstract The paper defines a consensus distribution with respect to experts' opinions using a multiple quantile utility model. We show that the Steiner point (Schneider, Isr J Math 2:241–249, 1971) is the representative consensus probability. The new rule for aggregation of experts' opinions, which can be simply evaluated by the Shapley value, is prudential and coherent.

19.1 Introduction

Over the past few decades, there has been new interest in the aggregation of conflicting and not necessarily independent opinions of experts and scientists when the decision maker (DM) is faced with ambiguity. Examples of ambiguous events in which estimations are derived from empirical frequencies that induce not unique and not fully reliable assessments among experts are: the severity of global warming following a doubling of atmospheric CO_2 concentration with respect to preindustrial levels, the relationship among biodiversity loss, environmental service changes and transmission of re-emerging infectious diseases, the environmental safety and direct effects on human health of genetically modified organisms, and the morbidity and mortality of a pandemic flu in humans.

Because of difficulties and failure of the frequency theory due to incomplete, sparse, or unavailable data, the Bayesian theory emerged as the normative theory for solving opinion aggregation problems. In the Bayesian context, the concept of

M. Basili (✉)
DEPS and SEM, University of Siena, Piazza San Francesco 7, 53100 Siena, Italy
e-mail: marcello.basili@unisi.it

A. Chateauneuf
IPAG Business School and PSE-CES, University of Paris-I, Paris, France
e-mail: Alain.Chateauneuf@univ-paris1.fr

G. Rogova, P. Scott (eds.), *Fusion Methodologies in Crisis Management*,
DOI 10.1007/978-3-319-22527-2_19

opinion, which encompasses different notions, such as prior, vague prior, sequence of odds ratios, finitely additive measures finds a natural encode in subjective probability distribution.[1]

Nevertheless, in the Bayesian axiomatic approach to consensus distribution, there is no room for ambiguity. However, ambiguity attitude, which emerges when individuals face vague and incomplete statistical data, influences the perception of risky events and induces human beings to elicit probabilities and apply decision rules that violate the standard rationality paradigm. Recently, a stream of processes involving both modeling and behavioral aspects has been proposed to calibrate the aggregation-combination-composition of experts' opinions through the DM's ambiguity attitude. These methods differ from Bayesian pooling operators used to form a single consensus distribution and can be included in three main classes: (1) pooling methods based on Dempster's rule of combination or theory of evidence (Stephanou and Lu 1988; Bi et al. 2007; Ha-Duong 2008; Denoeux 2008); (2) combination rules based on possibility distributions and fuzzy measures (Sandri et al. 1995; Dubois and Prade 1994; Yu 1997), and (3) methods of aggregation based on multiple priors or capacities (Crès et al. 2011; Gajdos and Vergnaud 2013).[2]

We consider an aggregation scheme of opinions expressed through different probability distributions and a DM that adopts a multiple priors decision model. Following the standard literature, the set of probability distributions or lotteries of all experts can be considered a reflection of the DM's assessment of the reliability of available information about the underlying uncertainty, that is, her perception of ambiguity; the optimal aggregation rule incorporates the DM's attitude about scanty and vague information. Facing the set of all probability distributions attached by experts to possible events; the DM considers the mean value of their common probability set, indeed the mean value of their probability intersection. Such a mean value is the *Steiner point* of the convex capacity that emerges from the aggregation of experts' opinions. We introduce an original prudential rule to combine expert opinions: the DM combines the probability distribution summarizing the consensus experts' opinions with the most pessimistic and optimistic probability distributions on extreme events. Such a suggested procedure for eliciting expert opinions overcomes problem of calibration and informativeness among experts assessment.

[1]Whether opinions are expressed as probability measures, densities, mass functions, or odds, subjective probability distributions are used to form a consensus distribution through linear opinion pools (Stone 1961), linear opinion pools that only satisfy the marginalization property (DeGroot and Montera 1991), logarithmic opinion pools that satisfy external Bayesianity (Winkler 1968), or generalized logarithmic opinion pools (Genest et al. 1986) that use mathematical aggregation models.

[2]Crès et al. (2011) focus on the maxmin expected utility model so that "the decision maker's valuation of an act is the minimal weighted valuation, over all weight vectors of the experts' valuations" (a weighted pessimistic scenario). Gajdos and Vergnaud (2013) characterize preferences that exhibit independently aversion towards imprecision and conflict. They aggregate different evaluations through a multiple weights model where "a decision maker will satisfy the conflict aversion hypothesis whenever her degree of conflict aversion is higher than her degree of imprecision aversion."

Our new rule alludes to a simple measure of confidence of expert opinions based on agreement: the larger is the set of common distributions among all the panel (intersection of the all experts possible distributions) the greater the agreement is. If the common set is too narrow or empty, the DM could consider the panel not enough trustworthy to guide a choice. Our new approach is consistent and preserves stochastic dominance, then it can be used for sensitivity analysis that is strongly required.

This paper is organized as follows. Section 19.2 illustrates expert elicitation in complex issues. Section 19.3 introduces an aggregation rule based on a quantile utility model and a DM that is supposed to be pessimistic with respect to extreme negative outcomes (catastrophic losses), ambiguity neutral in an interval of more reliable outcomes (familiar results), and optimistic with respect to extreme positive outcomes (windfall gains). In Sect. 19.4, we translate in terms of attitude towards lower tails, upper tails, and intermediate quantiles, and define a less conservative criterion for eliciting a single consensus distribution. Moreover, we show that stochastic dominance holds. Concluding remarks follow.

19.2 Opinions Aggregation

Discussing the use of proper pooling methods to discover not only opinion of individuals who the DM regards as experts, but also to judge how well informed they are, Savage observes that "risks characterized by tiny probabilities may be difficult to have a reliable experts' assessment, that experts' opinions might be divergent, and, what is more relevant, you might discover which expert is optimistic or pessimistic in some respect and therefore temper his judgments. Should he suspect you of this, however, you and he may well be on the escalator to perdition" (Savage 1971). Combining experts opinions is a formal process for eliciting a common judgement in the form of subjective probability distribution, called consensus distribution, about the value of some decision-relevant quantities or event occurrences. Experts' judgments elicitation is a multidisciplinary process apt to fill data gaps or partial scientific knowledge. Mathematical and behavioral approaches for combining experts opinions have been proposed and used. Mathematical Bayesian aggregation models manage individual probability distributions to obtain a single combined probability distribution operating with different degrees of complexity: equal-weight, best-expert, copula, etc. (Genest and Remillard 2006; Cooke et al. 2007; Roman et al. 2008; Curtwright et al. 2008). Judgements in the form of subjective probability distributions can be obtained from subject-matter experts by interacting with them (sharing assessment) in elicitation protocols such as Delphi, Q-Methodology, Nominal Group Technique, Kaplans approach, etc. Even though elicitation protocols suffer from many problems such as polarization, strategic manipulation, overconfidence, self-censorship, pressure to conformity, and more extreme probability estimates in order to generate some kind of consensus distribution (Cooke 1991; Plous 1993), a number of algorithmic approaches based on Bayesian theory have been offered (Ayyub 2001; Hammitt and Zhang 2012).

If the DM does not have access to a reliability measure for each expert (e.g., likely loss or other measures of effectiveness), a measure of correlation between experts judgements (e.g., experts calibration and weighting) and faces ambiguous events, a situation may arise in which there is not enough information to form a unique reliable probability distribution. Therefore, the combination of experts opinions using a Bayesian statistical approach can lead to inconsistent and incoherent consensus distribution.

19.2.1 Expert Elicitation in Complex Issues

Expert Elicitation (EE) entails a process of formalizing and quantifying, generally in probabilistic terms, expert judgments about uncertain quantities and events that occur in real life. EE is defined in a decision problem with several alternatives and a panel of decision makers or experts that try to achieve a common solution. EE is currently used by US federal agencies, EU commission, the private sector, academia etc. (US Nuclear Regulatory Commission 1996; US Office of Management and Budget 2002; US Department of Transportation/Federal Railroad Administration 2003; US Environmental Protection Agency 2005, 2006, 2007, 2011; Guidance Note for Lead Authors 2010; Stocker et al. 2013; Kriegler et al. 2009; Zickfeld et al. 2010). Since 1980s the US Environmental Protection Agency's (EPA) uses EE to assess exposure-response relationships for lead/ozone, the Office of Air Quality, Planning and Standards (USOAQPS) uses EE to analyze uncertainty in the relationship between exposures to fine particles and the annual incidence of mortality and the Department of Energy (DOE) uses EE to evaluate nuclear waste and related issues. In 1990s the US Nuclear Regulatory Commission (USNRC) and the Commission of the European Communities (EU) conducted a seminal study on uncertainty analysis of accident consequences for nuclear power plants. A large part of the EU-USNRC joint report is devoted to analyze dependence modeling and dependence elicitation. The joint EU-USNRC report represents a benchmark in each of the submodeling areas; it involves 2036 elicitation variables, assessed by 69 experts spread over nine panels (15,422 individual expert-variable elicitations). The major conclusion pointed out is that there is no alternative but combine experts' judgements. In 2005, Guidelines for Carcinogen Risk Assessment revised and replaced the US EPA's Guidelines for Carcinogen Risk Assessment, published in 1986 and 1999. These guidelines emphasized the importance of weighing all of the evidence in reaching conclusions about the hazard, dose response, and exposure assessments of potential carcinogenic agents. The cancer guidelines were flexible enough to accommodate the use of EE to characterize cancer risks when the lack of knowledge about the lack of peer-reviewed methods and data leads to an ambiguous distribution of values for each model's parameter. In 2011 EPA published Expert Elicitation Task Force White Paper (US Environmental Protection Agency 2011) under the auspices of the former Science Policy Council (SPC) that analyzed EE and determined when and how expert judgments should be aggregated. In environmental studies about climate change, uncertainty is a crucial aspect and a particular cure is devoted to its consistent treatment and representation. Kriegler et al. (2009) studied major changes in the Atlantic meridional overturning circulation (AMOC), the Greenland ice sheet (GIS), the West Antarctic ice sheet (WAIS), the Amazon rainforest and the ElNino/Southern Oscillation (ENSO) by consulting a panel of 52 experts with a computer-based interactive questionnaire completed individually by participants. The intergovernmental panel on climate change (IPCC) released a Guidance Note for Lead Authors of Fifth Assessment Report on Consistent Treatment of Uncertainties. In such a note it is suggested that likelihood, that may be based on statistical or modeling analyses and elicitation of expert views, could be expressed through categories having fuzzy boundaries, i.e., "a statement that an outcome is *likely* means that the probability of this outcome can range from $\geq$66 % (fuzzy boundaries implied) to 100 % probability. This implies that all alternative outcomes are *unlikely* (0–33 % probability)." Zickfeld et al. (2010) elicited experts' probability, about different levels of radiative forcing could trigger some *basic* state change in the climate system, with linguistic interval fuzzy preference relations. Similarly, Stocker et al. (2013) pointed out that "the degree of certainty in key findings in this assessment is based on the author teams" evaluations of underlying scientific understanding and is expressed as a qualitative level of confidence (from very low to very high) and, when possible, probabilistically with a quantified likelihood (from exceptionally unlikely to virtually certain). Confidence in the validity of a finding is based on the type, amount, quality, and consistency of evidence (e.g., data, mechanistic understanding, theory, models, expert judgment) and the degree of agreement.

19.3 Consensus Distribution Through a Quantile-Function

In EE, possibility theory is considered as a flexible tool for representing incomplete, imprecise, or uncertain information. A possibility distribution can be viewed as a likelihood function in information fusion processes. Possibility theory can be considered as the membership function of the fuzzy set of possible value of an unknown variable. A possibility measure is a special case of capacity and it is well known that there exists at least one possibility measure that dominates any capacity. Then any capacity "can be viewed as a lower possibility measure and an upper necessity measure with respect to two distinct families of possibility distributions" Dubois (2011). It is well known that in cumulative prospect theory (CPT), the reference point originates a bipolar scale that is captured by bipolar capacities. Bipolar capacity and the proper Choquet integral (Grabish and Labreuche 2005) are natural generalization of capacities. We introduce an approach to form a consensus distribution that adopts a quantile-function (Basili and Chateauneuf 2011). Our approach combines a formalization of CPT under ambiguity, that relies on the idea that the DM has a set of outcomes called ordinary, because they are considered more reliable (*familiar* or closer to her life experiences), and two fat tails in which more ambiguous extreme (*unfamiliar*) events are included, with the fuzzy theory where expert opinions are represented by a set of probability distributions on possible events, none of which is considered fully reliable.

We consider m experts $j = 1, \ldots, m$ aiming at valuing the possible probability distributions P governing an uncertain situation $S = \{s_1, \ldots, s_i, \ldots s_n\}$, where one, and only one state, $s \in S$ will occur, but where it is assumed that there is a unique unknown probability distribution P_0 governing that situation. Formally, S is the finite set of states of the world and 2^S is the set of all subsets of S. Any given expert i, will be asked to give *lower* and *upper bounds* for the probability $p_0^i = P_0(\{i\})$. Therefore, the set of possible probabilities P_j considered by expert j will be $P_j = \left\{P = (p_1, \ldots, p_i, \ldots, p_n), a_i^j \leq p_i \leq b_i^j, i = 1, \ldots, n\right\}$.

We indeed assume $0 \leq a_i^j \leq b_i^j \leq 1$.

$$\text{It is straightforward that } P_j \neq \varnothing \text{ if and only if } \sum_i a_i^j \leq 1 \leq \sum_i b_i^j . \tag{19.1}$$

As proved in Chateauneuf and Cornet (2012) and de Campos et al. (1994), it follows that as soon as $P_j \neq \varnothing$, then P_j is the *core* $C(\upsilon_j)$ of a convex capacity υ_j, which can be defined easily as:

$$\forall A \in 2^S,\ \upsilon_j(A) = \text{Max}\left(\sum_{i \in A} a_i^j, 1 - \sum_{i \notin A} b_i^j\right).$$

Even if experts do not know P_0, because of their skill, they can conceive a set $\mathcal{P}_j$, such that $P_0 \in \mathcal{P}_j$. Therefore, one should expect that $\bigcap_j \mathcal{P}_j \neq \varnothing$.

The first test to validate the quality (competence and reliability) of the panel of experts should be to check that $\bigcap_j \mathcal{P}_j \neq \varnothing$.

From (19.1), it is immediate that $\bigcap_j \mathcal{P}_j \neq \varnothing$ is equivalent to $\sum_i a_i \leq 1 \leq \sum_i b_i$, where: $a_i = \underset{j}{\text{Max}} a_i^j$, $b_i = \underset{j}{\min} b_i^j$, and indeed $a_i \leq b_i\ \forall i$.

Once these conditions were checked, or if $\bigcap_j \mathcal{P}_j = \varnothing$, the experts were asked to revise their opinions by enlarging their considered initial $\mathcal{P}_j$, in order to satisfy the consistency requirement $\bigcap_j \mathcal{P}_j \neq \varnothing$. One could summarize the *consensus opinion* $P = \bigcap_j \mathcal{P}_j$ through a convex capacity[3] υ with the known formula

[3]A capacity υ is convex if $\upsilon(A \cup B) + \upsilon(A \cap B) \geq \upsilon(A) + \upsilon(B), \forall A, B \in 2^S$.

$$\upsilon(A) = \text{Max}\left(\sum_{i\in A} a_i, 1 - \sum_{i\notin A} b_i\right).$$

In other words, this convex capacity υ will now be considered the aggregation of the multiple prior opinions.

19.3.1 Multiple Quantile Utility Model

We now suggest the use of the multiple quantile utility model (as considered in Basili and Chateauneuf 2011) with respect to the previous convex capacity. In this way, we show that the Steiner point $\Pi^{\vartheta} \in C(\upsilon)$ can be considered the representative probability of the consensus experts' opinions.

As a matter of fact, the Steiner point[4] is defined as the *center of* $C(\upsilon)$ and a meaningful probability summarizing the consensus experts' opinions. Moreover, for convex capacities, the Steiner point is nothing more than the famous *Shapley value*, the computation of which is quite simple.[5]

Let us recall (i.e., Owen 1968) that Shapley's Π^{Sh} (i.e., $\Pi^{\text{Sh}} = \Pi^{\vartheta}$) is defined by:

$$\forall i \in [1,n] \;\; \Pi_i^{\vartheta} = \sum_{\{i\}\subset A\subset S} \frac{(|A|-1)!(n-|A|)!}{n!}\left[\upsilon(A) - \upsilon(A\backslash\{i\})\right].$$

Example 1. Computation of the Shapley value for the following "probability-interval" capacity υ

$$S = s_1 \; s_2 \; s_3$$
$$b_i = \tfrac{6}{12} \; \tfrac{5}{12} \; \tfrac{7}{12}$$
$$a_i = \tfrac{2}{12} \; \tfrac{3}{12} \; \tfrac{4}{12}$$

Therefore υ is given by

A	$\{s_1\}$	$\{s_2\}$	$\{s_3\}$	$\{s_1, s_2\}$	$\{s_1 s_3\}$	$\{s_2 s_3\}$	S
$\upsilon(A)$	$\frac{2}{12}$	$\frac{3}{12}$	$\frac{4}{12}$	$\frac{5}{12}$	$\frac{7}{12}$	$\frac{7}{12}$	$\frac{12}{12}$

and

[4]The Steiner point of a polytope, also called Steiner curvature centroid, is the weighted average of its vertices, in which the weight for each vertex is proportional to its outer angle. Details are in Schneider (1971, 1993).

[5]The Shapley value is an operator that assigns an expected marginal contribution to each player in a coalitional game with respect to a uniform distribution over the set of all permutations on the finite set of players. Details are in Gajdos et al. (2008).

$$\Pi_1^{\vartheta} = \frac{1}{6}\left\{2.\frac{2}{12} + \frac{1.(5-3+7-4)}{12} + \frac{2.(12-7)}{12}\right\} = \frac{19}{72}$$
$$\Pi_2^{\vartheta} = \frac{1}{6}\left\{2.\frac{3}{12} + \frac{1.(5-2+7-3)}{12} + \frac{2.(12-5)}{12}\right\} = \frac{27}{72}$$
$$\Pi_3^{\vartheta} = \frac{1}{6}\left\{2.\frac{4}{12} + \frac{1.(7-2+7-4)}{12} + \frac{2.(12-7)}{12}\right\} = \frac{26}{72}$$

hence, $\Pi_i^{\vartheta} = (\frac{19}{72}, \frac{27}{72}, \frac{26}{72})$.

19.4 A Coherent Prudential Aggregation Rule for Experts Opinions

According to the assumption, the DM's ambiguity attitude is modeled through a convex capacity υ. For an act $X : S \to \mathbb{R}$ and $(\alpha, \beta) \in [0, 1]^2, \alpha \leq \beta$, such that $[\alpha, \beta]$ determines the interval of cumulative probability between which outcomes can be considered ordinary, the DM values outcomes in $[\alpha, \beta]$ in an ambiguity neutral way by Π^{ϑ}. With respect to ambiguity attitude on extreme outcomes, we propose a prudential aggregation rule: namely, one that assumes pessimism on the lower tail $[0, \alpha]$ and optimism on the upper tail $[\beta, 1]$.

For any act, the common pseudo-inverse F_X^{-1}, indeed the quantile-function, is defined.[6]

As a result, the DM chooses $\alpha, \beta \in [0, 1]$, where $\alpha \leq \beta$, and computes the value of $X \in \mathbb{R}^S$ through $I(X) = I_1(X) + I_2(X) + I_3(X)$, where, defined $\overline{\upsilon}$ the conjugate of υ[7]: $I_1(X) = \int_0^{\alpha} F_X^{\upsilon^{-1}}(p)dp$; $I_2(X) = \int_{\alpha}^{\beta} F_X^{\Pi^{\vartheta^{-1}}}(p)dp$, and $I_3(X) = \int_{\beta}^{1} F_X^{\overline{\upsilon}^{-1}}(p)dp$.

Once the probability distribution $\Pi^{\mathrm{Sh}} = \Pi^{\vartheta} \in \mathrm{core}(\upsilon)$ has been selected, it is possible to define the DM pessimism with respect to outcomes in the lower tail, and optimism with respect to outcomes in the upper tail. In fact, pessimism and optimism are defined with respect to the probability distribution that expresses ambiguity neutrality, i.e., Π^{ϑ}.

Definition 2. The DM is pessimistic with respect to the lower tail if she overestimates losses and underestimates gains in this tail with respect to the probability $\Pi^{\vartheta} \in C(\upsilon)$, i.e., if $I_1(X) \leq \int_0^{\alpha} F_X^{\Pi^{\vartheta^{-1}}}(p)dp$.

Definition 3. The DM is optimistic with respect to the upper tail if she underestimates losses and overestimates gains in this tail with respect to the probability $\Pi^{\vartheta} \in C(\upsilon)$, i.e., if $I_3(X_i) \geq \int_{\beta}^{1} F_X^{\Pi^{\vartheta^{-1}}}(p)dp$.

Proposition 1. *Under the prudential rule, the DM is pessimistic with respect to the lower tail and optimistic with respect to the upper tail.*[8]

[6]Details are in Basili and Chateauneuf (2011).

[7]The dual or conjugate capacity $\overline{\upsilon}$ is defined by $\overline{\upsilon}(A) = 1 - \upsilon(A^C)\ \forall A \in 2^S$.

[8]Proposition 1 is a direct consequence of Proposition 3 in Basili and Chateauneuf (2011).

Interestingly, this rule could be particularly useful for describing potential global temperature increases and climate sensitivity after recent evidence showed that the 5-year mean global temperature has been flat for a decade. In case this rule appears prudential or cautious since it mitigates the overestimated high temperatures and aggravates the underestimated low temperatures (Stocker et al. 2013; Kriegler et al. 2009; Zickfeld et al. 2010).

Stochastic dominance provides a powerful method for act analysis since it does not require assumptions concerning the shape of the probability function or utility function and utilizes every point in the set of probability distributions.

Furthermore, it is known that under risk, one of the rare rules upon which decision theorists agree, is the respect of first order stochastic dominance (Hadar and Russell 1969; Levi and Wiener 1998). Therefore, we introduce the following definition.

Definition 4. A rule $I : \mathbb{R}^S \to \mathbb{R}$ is said to be coherent with the probabilistic information ϑ if given $X, Y \in \mathcal{F}$, X. First Order Stochastically Dominates Y for any probability $P \in C(\upsilon)$, i.e., if $P(X \geq t) \geq P(Y \geq t) \quad \forall t \in \mathbb{R}$ and $\forall P \in C(\upsilon)$ implies $I(X) \geq I(Y)$.

Proposition 2. *The prudential rule is coherent.*

Proof. Let $X, Y \in \mathcal{F}$ such that X first order stochastically dominates Y for any probability $P \in C(\upsilon)$. Let us first show that $I_1(X) \geq I_1(Y)$. Recall that for $p \in [0, 1]$, $F_X^{\bar{\vartheta}^{-1}}(p) = \inf\left\{t \in \mathbb{R}, 1 - \vartheta(X > t) \geq p\right\}$ and $F_X^{\bar{\vartheta}}(t) = \bar{\vartheta}(X \leq t) = 1 - \vartheta(X > t)$ $\forall t \in \mathbb{R}$. Take $\underline{P} \in C(\upsilon)$ then $\underline{P}(X > t) \geq \underline{P}(Y > t)$ so $\min_{P \in C(\upsilon)} P(X > t) \geq \min_{P \in C(\upsilon)} P(Y > t)$. Since ϑ is convex, hence exact, this implies $\vartheta(X > t) \geq \vartheta(Y > t)$, hence $F_X^{\bar{\vartheta}}(t) \leq F_Y^{\bar{\vartheta}}(t)$, and therefore, $F_X^{\bar{\vartheta}^{-1}}(p) \geq F_Y^{\bar{\vartheta}^{-1}}(p)$, $\forall p \in [0, \alpha]$, which gives $I_1(X) \geq I_1(Y)$.

Similar proofs apply for I_2 and I_3, which completes the proof. ∎

19.5 Concluding Remarks

We developed a new approach to form a consensus distribution by considering the composite inverse cumulative function. We provided evidence that a rational DM would aggregate experts' probability distributions in a functional that combines her different attitude with respect to likely and extreme events. The new functional allows the representation of the DM ambiguity attitude about experts competence regarding the uncertain events being examined. The functional form overcomes misevaluation induced by cognitive insensitivity to small probability outcomes. Finally, our prudential rule preserves stochastic dominance.

References

Ayyub B (2001) A practical guide on conducting expert-opinion elicitation of probabilities and consequences for corps facilities. IWR report 01-R-01

Basili M, Chateauneuf A (2011) Extreme events and entropy: a multiple quantile utility model. Int J Approx Reason 52:1095–1102

Bi Y, Bell D, Wang H, Guo G, Guan J (2007) Combining multiple classifier using Dempster's rule for text categorization. Appl Artif Intell 21:211–239

Chateauneuf A, Cornet B (2012) Computing the cost spread of a financial structure with bid/ask. Working paper, Universite Paris 1 and University of Kansas

Cooke R (1991) Experts in uncertainty: opinion and subjective probability in science. Oxford University Press, Oxford

Cooke R, Wilson A, Toumisto J, Morales O, Tanio M, Evans J (2007) A probabilistic characterization of the relationship between fine particulate matter and mortality: elicitation of European experts. Environ Sci Toxicol 41:6598–6605

Crès H, Gilboa I, Vieille N (2011) Aggregation of multiple prior opinions. J Econ Theory 146:2563–2582

Curtwright A, Morgan M, Keith D (2008) Expert assessment of future photovoltaic technology. Environ Sci Technol 42:9031–9038

de Campos L, Huete J, Moral S (1994) Probability intervals: a tool for uncertain reasoning. Int J Uncertain Fuzziness Knowl Based Syst 2:167–196

DeGroot M, Montera J (1991) Optimal linear opinion pools. Manag Sci 37:546–558

Denoeux T (2008) Conjunctive and disjunctive combination of belief functions induced by non distinct bodies of evidence. Artif Intell 172:234–264

Dubois D (2011) Fuzzy measures on finite scales as families of possibility measures. EUSFLAT-LFA

Dubois D, Prade H (1994) Possibility theory and data fusion in poorly informed environments. Control Eng Pract 2:812–823

Gajdos T, Vergnaud J (2013) Decisions with conflicting and imprecise information. Soc Choice Welf 41:427–452

Gajdos T, Hayashi T, Tallon JM, Vergnaud JC (2008) Attitude toward imprecise information. J Econ Theory 140:27–65

Genest C, Remillard B (2006) Comments on T. Mikosh's paper "Copulas: tales and facts". Extremes 9:27–36

Genest C, McConway K, Schervish M (1986) Characterization of externally Bayesian pooling operators. Ann Stat 14:487–501

Grabish M, Labreuche C (2005) Bi-capacities II: the Choquet integral. Fuzzy Set Syst 151: 237–259

Guidance Note for Lead Authors (2010) IPCC 5th assessment report on consistent treatment of uncertainties. Intergovernmental panel on climate change (IPCC), IPCC cross-working group meeting on consistent treatment of uncertainties, Jasper Ridge, CA. https://www.ipcc.ch/pdf/supporting-material/uncertainty-guidance-note.pdf

Hadar J, Russell W (1969) Rules for ordering uncertain prospects. Am Econ Rev 59:25–34

Ha-Duong M (2008) Hierarchical fusion of expert opinion in the transferable belief model, application to climate sensitivity. Int J Approx Reason 49:555–574

Hammitt J, Zhang Y (2012) Combining experts' judgments: comparison of algorithmic methods using synthetic data. TSE working paper, pp 12–293

Stocker F, Qin D, Plattner G, Tignor M, Allen S, Boschung J, Nauels A, Xia Y, Bex V, Midgley PM (2013) Climate change 2013: The physical science basis. Contribution of working group I to the fifth assessment report of the intergovernmental panel on climate change. Cambridge University Press, Cambridge/New York, NY

Kriegler E, Hall J, Held H, Dawson R, Schellnhuber H (2009) Imprecise probability assessment of tipping points in the climate system. Proc Natl Acad Sci USA 106:5041–5046

Levi H, Wiener Z (1998) Stochastic dominance and prospect dominance with subjective weighting functions. J Risk Uncertain 16:147–163
Owen G (1968) Game theory. Saunders Company, Philadelphia
Plous S (1993) The psychology of judgment and decision making. McGraw-Hill, New York
Roman H, Walker K, Walsh T, Conner L, Richmond H, Hubbell B (2008) Expert judgment assessment of the mortality impact of changes in ambient fine particulate matter in the US. Environ Sci Technol 42:2268–2274
Sandri S, Dubois D, Kalfsbeek H (1995) Corrections to elicitation, assessment, and pooling of expert judgments using possibility theory. IEEE Trans Fuzzy Syst 3. doi:10.1109/TFUZZ.1995.4819
Savage L (1971) Elicitation of personal probabilities and expectations. J Am Stat Assoc 66:783–801
Schneider R (1971) On Steiner points of convex bodies. Isr J Math 2:241–249
Schneider R (1993) Convex bodies: the Brunn-Minkowski theory. Cambridge University Press, Cambridge
Stephanou H, Lu S (1988) Measuring consensus effectiveness by a generalized entropy criterion. IEEE Trans Pattern Anal Mach Intell 10:544–554
Stone M (1961) The opinion pool. Ann Math Stat 32:1339–1342
US Department of Transportation/Federal Railroad Administration (2003) Chapter 2: approach to estimation of human reliability in train control system studies. In: Human reliability analysis in support of risk assessment for positive train control. DOT/FRA/ORD- 03/15. Office of Research and Development, Washington, DC
US Environmental Protection Agency (2005) Guidelines for carcinogen risk assessment. EPA/630/P-03-001B. Risk Assessment Forum, Office of Research and Development, Washington, DC
US Environmental Protection Agency (2006) Regulatory impact analysis of the final PM national ambient air quality standards, Washington, DC
US Environmental Protection Agency (2007) Regulatory impact analysis of the proposed revisions to the national ambient air quality standards (NAAQS) for ground-level ozone
US Environmental Protection Agency (2011) Expert elicitation task force white paper, EPA-452/R-07-008, Washington, DC
US Office of Management and Budget (USOMB) (2002) Guidelines for ensuring and maximizing the quality, objectivity, utility, and integrity of information disseminated by federal agencies. Fed Regist 67(36):8452–8460
US Nuclear Regulatory Commission (USNRC) (1996) Branch technical position on the use of expert elicitation in the high-level radioactive waste program. NUREG-1563. Division of Waste Management, Office of Nuclear Material Safety and Standards, Washington, DC
Winkler R (1968) The consensus of subjective probability distributions. Manag Sci 15:61–75
Yu D (1997) Expert opinion elicitation process using a fuzzy probability. J Korean Nucl Soc 29: 25–34
Zickfeld K, Morganb M, Frame D, Keithd D (2010) Expert judgments about transient climate response to alternative future trajectories of radiative forcing. Proc Natl Acad Sci USA 107:12451–12456

Chapter 20
Decision Making Under Ignorance

Phan H. Giang

Abstract Ignorance is an extreme form of uncertainty. In most narrow and technical sense, it means inability to assign a meaningful probability to the phenomena of interest. In more general sense, the state of ignorance is the result of the absence of knowledge about structural factors that influence the issues, the lack of reliable information or inability to completely determine the space of alternatives and consequences. We argue that the practice of casually papering over the ignorance with subjective judgments and analytic assumptions can have serious consequences. This chapter provides a structured survey (and necessarily selective) of significant ideas and proposals for decision making under ignorance, from the ground breaking work by Hurwicz and Arrow to the latest result of τ-anchor utility theory. A careful analysis and isolation of ignorance in the system of knowledge about a subject or a problem is of particular importance in the context of risk assessment and risk management.

20.1 Introduction

In the article *Rumsfeld's Knowns and Unknowns: The Intellectual History of a Quip* appeared in *The Atlantic* magazine on March 2014, journalist David Graham (2014) tells a fascinating story about the memorable quip that Donald Rumsfeld, who served as Secretary of Defense in George W. Bush's administration, is remembered for:

> As we know, there are known knowns; there are things we know we know. We also know there are known unknowns; that is to say we know there are some things we do not know. But there are also unknown unknowns - the ones we don't know we don't know.

The statement was a reply to a question raised by reporter Jim Miklaszewski of *NBC News* in an official news briefing by US Department of Defense in February 2002,

P.H. Giang (✉)
George Mason University, MS 1J3, 4400 University Dr., Fairfax, VA 22030, USA
e-mail: pgiang@gmu.edu

G. Rogova, P. Scott (eds.), *Fusion Methodologies in Crisis Management*,
DOI 10.1007/978-3-319-22527-2_20

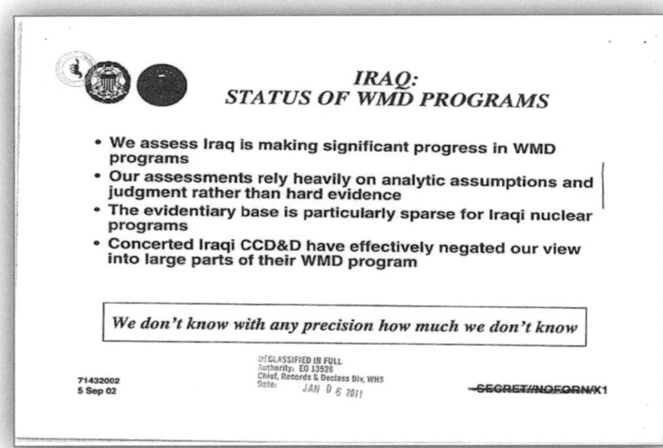

Fig. 20.1 Rumsfeld's memo to Gen. Myers

demanding the evidence for the claims that Iraq's dictator Saddam Hussein had weapons of mass destruction (WMD) and was willing to transfer them to terrorist networks.[1]

It was a difficult question for Rumsfeld to answer and, to give him credit, he was smart and in this case at least, truthful. We now know his true state of knowledge from a recently declassified secret memo he sent to the Chairman of the Joint Chiefs of Staff, Gen. Richard Myers, dated September 5, 2002.[2] In that memo, Rumsfeld included a slide (Fig. 20.1) from the top secret presentation prepared on his specific request by the Chief of Military Intelligence. The presentation summarized the information that US Government knew about the status of WMD programs [of Iraq]. The salient points of the slide were that the assessments about significant progress of Iraq's WMD program relied mostly on *analytic assumptions* and *judgments* (emphasis added) not the hard evidence; the evidentiary base for the assessments was sparse and finally, the concerted efforts of Iraq's government to hide their intention effectively blocked US intelligence view into WMD program. The slide concluded "We don't know with any precision how much we don't know." The state of knowledge that Mr. Rumsfeld described in such a convoluted way, fits the technical definition of ignorance.

We all know how the war that was sold to the public mostly on supposed WMD dangers, ended. Admittedly, it was quite possible that if the WMD were not an issue, something else could be used to sell the war. Everyone can draw a lesson from this

[1]For the context of the quip, the entire transcript of the news briefing can be found here: http://www.defense.gov/transcripts/transcript.aspx?transcriptid=2636.

[2]http://library.rumsfeld.com/doclib/20110202_20020909toMyersreWMD.pdf.

tragic history. One lesson that we can draw from this story is that substituting so called analytic and judgmental assumptions for the hard evidence may have adverse, potentially disastrous, consequences.

Most applications in science, engineering, medicine, business, and military express uncertainty in the language of the probability theory. Probability estimation is based on data and/or human opinions. Accurate estimation of probability requires a well-defined state space and availability of reliable and relevant data in sufficient quantity. Reliability and relevancy are essential for data to be useful. Subjective or personal probability is estimated on the basis observed or hypothetical behavior of individuals. An essential condition for estimation of subjective probability is that the behaviors must be internally consistent. Without sufficient data or behavioral observations, probability can be arrived by invoking the philosophical principle of insufficient reason (or its close cousin the principle of maximum entropy) that assigns equal probability to alternative states. An important but casually dismissed implication of using probability is that an individual commits herself to very strong epistemic assumption according to which anything that can be learned about a phenomenon is known to the individual. In that state of knowledge, the outcomes are random events which, from the individual's standpoint, are quantitatively exchangeable with the outcomes of coin tossing, roulette spinning, or radioactive decay.

However, in many practical situations, the real rationale behind the adoption of probability is convenience rather than justification. Almost a century ago, Knight (1921) and Keynes (1921) came to the conclusion that probability theory is not a universally appropriate language of uncertainty. In particular, the principle of insufficient reason is problematic because the same real situation can be modeled in different ways resulting in different sets of alternatives. The rationality of imposing the internal consistency, the critical condition for the estimation of subjective probability, is also debatable. Gilboa, Postlewaite, and Schmeidler argue that rationality requires a compromise between internal coherence and justification, the same way that the compromises are forced on moral dilemmas (Gilboa et al. 2009). When the requirements for internal coherence and justification contradict each other, they wrote "it is more rational to admit that one does not have sufficient information to generate a prior than to pretend that one does." More mundane challenges in probability estimation that everyone can bear witness to are routine unavailability of the data necessary for estimation, the cost of collecting data is too expensive for one's budget, the relevant data are scarce or conflicting, the data that you can trust are irrelevant and finally in competitive games, the information about the opposing party can be intentionally misleading.

The opposite epistemic state of knowing everything is the state of ignorance. This is an extreme form of uncertainty when an individual has no reliable information about the phenomenon of interest and therefore is not able to produce, in any meaningful way, a probability distribution.

In the context of risk assessment and risk management (Aven and Steen 2010), the term ignorance is used to describe the situations characterized by the lack of knowledge about the subject, a poor basis for probability assignments (data are scarce, unreliable, conflicting, etc.) and inability of the decision makers to fully determine all possible consequences. A more structural view held by the health and safety executive (HSE) in the UK (Health and Safety Executive 2001) highlights the key difference between ordinary uncertainty and ignorance is the knowledge about the factors that influence the issues. Uncertainty refers to a state of knowledge in which influencing factors are known, but the likelihood of consequences or effects cannot be precisely described. Ignorance, on the other hand, refers to the lack of knowledge about the factors influencing an issue. Another condition often associated with the term ignorance is the sample space ignorance (SSI) (Pushkarskaya et al. 2010) where the decision maker has difficulty in determining the set of possible alternative states. Daily situations that can be identified with ignorance occur without much attention perhaps because the stakes involved are quite low. For example, you are deciding whether or not to buy warranty for a piece of electronics made by a new to the market manufacturer but the information necessary to set up an optimization problem such as the rate of failure and the cost of fixing it is not known. Practically, in this situation you have to make decision under ignorance (Hogarth and Kunreuther 1995).

It is necessary to emphasize that, strictly speaking, the situations of pure or complete ignorance are rare. The rationale for studying decision making under ignorance is not based on the argument that pure or complete ignorance occurs frequently in practice. The reason that makes study of decision making under ignorance necessary is that an adequate description of uncertainty for most practical problems almost always includes some form of the singular state of knowledge. If anyone still dismisses the relevance of ignorance in the age of Big data, Rumsfeld's episode should be enough to refute that argument. Instead of making the ignorance disappear, modeling efforts should focus on separating the parts of the problem for which reliable data are available from those where reliable evidence is absent.

This chapter is structured as follows. Section 20.2 offers a review of literature on decision under ignorance. In Sect. 20.3, a new utility theory under ignorance is developed by imposing a condition for the certainty equivalent operator on Hurwicz–Arrow's decision theory under ignorance. Finally, Sect. 20.4 contains the discussion and examples.

20.2 A Brief Review of Decision Under Ignorance

Before going into technical presentation, we list the basic set of notations used in this paper. Our framework includes *variables* which are denoted by upper case letters I, X, Y, and so on. A variable (X) has a *domain* (Ω_X). A *decision* or *act* involves one or more variables. It is a mapping from the state space formed by the domains of its variables to the set of prizes. The set of prizes denoted by $\mathcal{O}$. Acts are denoted by lower case letters such as d, f, g. The decision maker's behavior is described by preference relation $\succeq$ on acts. In this chapter $\succeq$ is assumed to be a weak order with some exceptions which will be made explicitly.

With the development of formal decision theory under risk in the 1940s spurred by pioneering work of von Neumann and Morgenstern (1953), economists like Shackle (1949), Hurwicz and Arrow (1977) started pondering the question how an individual makes decision if she cannot associate any probability distribution to consequences of an act.

20.2.1 Hurwicz–Arrow's Theory of Choice Under Ignorance

A short paper (Arrow and Hurwicz 1977), reprinted in 1977, outlines the theory of decision under ignorance developed by Hurwicz and Arrow in early 1950s. The result was first obtained by Hurwicz and then improved by Arrow. The axiomatic approach they employed laid the foundation for later studies of decision making under ignorance.

We review the basic results of HA theory in our setting. Consider a collection of variables $\{I_1, I_2, \ldots\}$ whose domains are sets Ω_{I_i} . A *decision* or *act* defined on variable I_i is a mapping $f : \Omega_{I_i} \rightarrow \mathcal{O}$ where $\mathcal{O}$ is the (normalized) *prize space*. The domain of f and is denoted by $\Omega(f)$. A *decision problem* is a non-empty set of decisions that have the same domain. Denote the set of acts defined on variable I by $\mathcal{D}_I$ and the set of all acts by $\mathcal{D}$. Further we make technical assumptions: Ω_{I_i} are finite subsets of the set of natural numbers $\mathbb{N}$ and $\mathcal{O}$ is the (normalized) *prize space* which is assumed to be the real unit interval $\mathbb{R}_{0,1}$.

In the HA paper, an optimal operator $\hat{.}$ that maps each decision problem A to a subset of optimal acts $\hat{A}$. In this review, the optimal operator construct is replaced by a preference relation $\succeq$ on the set of acts $\mathcal{D}$. We assume that $\succeq$ is a weak order, i.e., it is reflexive, transitive, and complete. Formally, for any $f, g, h \in \mathcal{D}$ $f \succeq f$, if $f \succeq g$ and $g \succeq h$ then $f \succeq h$; either $f \succeq g$ or $g \succeq f$. From preference relation $\succeq$, a strict preference ($\succ$) and indifference ($\sim$) are defined as follows. $f \succ g$ means $f \succeq g$ and $g \not\succeq f$. $f \sim g$ means $f \succeq g$ and $g \succeq f$. For act f and prize $x \in \mathcal{O}$ if $f \sim x$ then x is called *certainty equivalent* of f.

The correspondence between HA operator $\hat{.}$ and $\succeq$ is as follows: For a decision problem $A, f \in \hat{A} \Leftrightarrow \forall g \in A, f \succeq g$. From the definition it follows that if $f, g \in \hat{A}$ then $f \sim g$.

The lasting impact of HA theory is due to the axioms that elegantly capture the essence of the notion of ignorance. Formally, HA theory presupposes that $\succeq$ must satisfy four axioms, originally named as the properties A–D as follows. Property A requires that $\succeq$ is a weak order. Property D or the weak dominance property, stipulates that if f_1 and f_2 are acts on the same domain $\Omega(f)$, and $\forall w \in \Omega(f), f_1(w) \geq f_2(w)$ then $f_1 \succeq f_2$. Properties A and D are standard axioms of preference which are not specific for preference under ignorance but hold for preference under risk and uncertainty. Properties B and C on the other hand are specific requirements that make sense only in the case of ignorance.

Property B is called the invariance under relabeling. Formally, if acts f_1, f_2 are isomorphic in the sense that there is an one-to-one mapping $h : \Omega(f_1) \to \Omega(f_2)$ such that $\forall s \in \Omega(f_1), f_1(s) = f_2(h(s))$ then f_1 and f_2 are indifferent $f_1 \sim f_2$.

The acts f_1, f_2 in (B) have domains of the same cardinality. An act can be viewed as a vector of its prizes. Thanks to the one-to-one mapping h, vector f_2 can be seen as obtained from vector f_1 by a permutation. Property B requires that the (preferential) valuation of a vector does not depend on the positions of its values. The states in the domain of Ω_{f_i} are treated symmetrically. No state is considered more likely than another. If the uncertainty is described a probability function only the uniform distribution among all distributions satisfies this symmetry property.

Property C, invariance under deletion of duplicate states, is a requirement unique for the state of ignorance. Suppose f_1, f_2 are two decisions, f_2 is said to be *derived from* f_1 by deleting duplicate states if (1) $\Omega(f_2) \subset \Omega(f_1)$ f_1 and f_2 are coincide on $\Omega(f_2)$ and (2) for each $w \in \Omega(f_1) - \Omega(f_2)$, there exists $w' \in \Omega(f_2)$ such that $f_1(w) = f_1(w')$. Property C requires that if f_2 is *derived from* f_1 then f_1 and f_2 are indifferent. Viewed as vectors of prizes, this property requires that if vector f_1 has two equal components $f_1(w_i) = f_1(w_j)$ then one of the component can be deleted. The newly obtained vector f_2 is indifferent to the original vector f_1. The term "duplicate states" refers to the states that f-map to the same prize.

This property addresses one of the epistemic challenges that create the state of ignorance, namely, the inability to determine and justify the complete set of alternative states [for example, the SSI (Pushkarskaya et al. 2010)]. Let us consider a simple example.

Example 1. Suppose that you are standing in front of two identical urns that contain 100 balls each. You are told that the first urn may include balls of two different sizes (small and big) but no information about their proportions. You need to decide how much to pay for a bet that pays \$0 if a ball drawn from the urn is small and \$1 if the ball is big. In the second urn, you have the same information about the sizes and proportions as in the first urn. But on top of it, you are told that small balls have only one color—white, while big balls can be painted in one of three colors: red, green, or blue. Again, you do not know the proportions of balls by colors. You have to decide how much you are willing to pay for a bet that pays \$0 if the ball is small-white, and \$1 if the ball is big-red or big-blue or big-green.

Let us formally describe the urns. For the first one, the domain of the variable "size" is $\{\text{small}, \text{big}\}$ and the bet is $f_1(\text{small}) = 0$ and $f_1(\text{big}) = 1$. For the second one, the state space has elements $\{\text{small.white, big.red, big.green, big.blue}\}$ and the bet is $f_2(\text{small.white}) = 0, f_2(\text{big.red}) = 1, f_2(\text{big.green}) = 1, f_2(\text{big.blue}) = 1$. Property C requires that bets f_1 and f_2 are indifferent, that is, you would pay the equal prices for f_1 and f_2. It is quite easy to see the reasonableness of the indifference in this case because the color is clearly irrelevant to bet f_2.

To make color variable relevant, let us modify f_2 into bet f_2' as follows $f_2'(\text{small.white}) = 0$, $f_2'(\text{big.red}) = 0$, $f_2'(\text{big.green}) = 1$, $f_2'(\text{big.blue}) = 1$. Property C still requires that $f_1 \sim f_2'$. Thus, under property C, the price you pay for a bet depends on the *set of distinct prizes*, not the vector of prizes where a prize can appear multiple times. The rationale of this axiom is that if it is difficult for an individual to determine and justify the space of alternative states then the evaluation should not depend on the notational form tied to a particular state space. If you are ignorant about the proportions of balls by sizes, colors then it does not matter if you model the act using only size variable or color variable or both. One can say that the domain of f_2 is obtained from the domain of f_1 by splitting states or vice versa domain of f_1 is obtained from merging states in the domain of f_2 (big into big.red, big.blue, big.green). ■

Property C is unique for ignorance. No probability distribution has this property because the uniformity of a distribution is destroyed when states are splitted/merged. Thus, properties B and C formulated by Hurwicz and Arrow capture the essence of ignorance. Much later in the context of statistical inference, Walley (1996) proposes two criteria that a ignorance belief must satisfy. The *embedding* principle requires that plausibility of an event A should not depend on the sample space in which A is embedded. The *symmetry* principle says that all elements in the sample space should be assigned the same plausibility. These two principles are reincarnations of properties B and C in HA theory. The main result of HA theory is a theorem that characterizes the preference relation that satisfies properties A through D.

Theorem 1 (Hurwicz–Arrow). *The necessary and sufficient condition for preference relation $\succeq$ on the set of acts $\mathcal{D}$ satisfies properties A through D is that there exists a weak ordering $\geq^2$ on the space of ordered pairs of real numbers $\mathcal{Z}^2 = \{\langle a, b\rangle \mid 0 \leq a \leq b \leq 1\}$ that satisfies the following properties: (1) if $a \geq a'$ and $b \geq b'$ then $\langle a, b\rangle \geq^2 \langle a', b'\rangle$; (2) for actsf, g of the same domain Ω_I,*

$$f \succeq g \text{ if } \left\langle \min_w f(w), \max_w f(w) \right\rangle \geq^2 \left\langle \min_w g(w), \max_w g(w) \right\rangle. \tag{20.1}$$

Let us call the preference relation that satisfies the conditions of HA theorem *HA preference relation*. Few implications can be drawn from HA theorem. First, the preferential comparison between two acts reduces to comparing its extreme values. Intermediate prize values do not matter. Under ignorance the state space is not important, therefore, acts can be identified with their prizes. The order of prizes is not important either. The set of finite non-empty *bags*[3] of $\mathcal{O}$ is denoted by $\mathcal{F}(\mathcal{O})$. Thus, two symbols denote the same thing $\mathcal{D} \equiv \mathcal{F}(\mathcal{O})$. From now on, we have a flexibility to denote, at our convenience, an act under ignorance either by a vector or a set of its prizes. Finally, HA theorem does not fully describe the preference relation $\succeq$. It does not specify the preference between two acts when the comparisons of their minimal elements and maximal elements point to opposite directions.

Facing the same informational ignorance and the same set of prizes, the prices that different individuals would be willing to pay for the bet are different. Intuitively, a pessimistic person would pays more attention to the negative side, i.e., the worst

[3] A bag is different from a set in the sense that a bag can contain the same item multiple times while all the elements in a set are distinct.

prize in the set while an optimistic person would look more into the positive side, i.e., the best prize in the set. Hurwicz proposed to quantify that attitude by a parameter α to control the weights attached to the worst and best prizes in a set. According to Hurwicz's α-rule, for acts $f_1, f_2 \in \mathcal{F}(\mathcal{O})$

$$f_1 \succeq_\alpha f_2 \text{ iff } \alpha \min(f_1) + (1-\alpha)\max(f_1) \geq \alpha \min(f_2) + (1-\alpha)\max(f_2). \quad (20.2)$$

That is, comparing two acts, an individual must calculate the combinations of the worst and the best prizes with α coefficient for each act and then compares the calculated values. For example, acts that have the same combined value are indifferent. In particular, the value $\alpha \min(f) + (1-\alpha)\max(f)$ is the certainty equivalent of act f.

Hurwicz's rule or criterion belongs to the family of preferences sanctioned by HA theorem. Moreover, it is intuitively appealing because α parameter allows a smooth variation of the degree of pessimism in different individuals. An extremely pessimistic individual who focuses exclusively on the worst possibility has $\alpha = 1$ while on the opposite end, an extremely optimistic individual who focuses on the best prizes has $\alpha = 0$. Despite its popularity and appeal, Hurwicz's rule suffers from a drawback—it is sequentially inconsistent. The following example clarifies the problem.

Example 2. Consider a urn of 100 balls. A ball has two characteristics: size (small/large) and color (black/white). The composition of the balls is not known (on both size and color). A bet is offered whose rewards depend on both size and color as follows. $f(\text{small.black}) = 1, f(\text{small.white}) = 0.4, f(\text{large.black}) = 0.7,$ and $f(\text{large.white}) = 0$. Figure 20.2 presents two ways of viewing the bet.

Suppose an individual has $\alpha = 0.6$ (slightly pessimistic). First, she can reason as follows (Fig. 20.2a). Suppose the size of the ball is small, she has a bet on color only. The set of rewards in this case is $\{1, 0.4\}$. Using Hurwicz's rule the set is

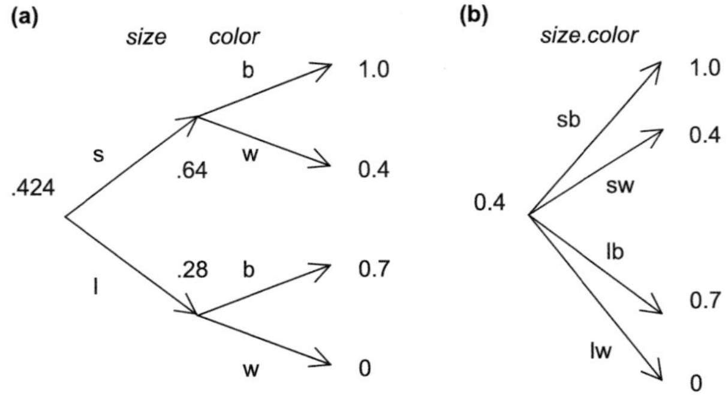

Fig. 20.2 Hurwics's rule: sequential inconsistency illustration: (**a**) "size" is followed by "color" and (**b**) "size" and "color" are merged

indifferent to $0.64 = 0.4 * 0.6 + 0.4 * 1$. With similar reasoning, if the ball turns out to be large ball then she gets a set of prizes $\{0, 0.7\}$ which is indifferent to 0.28. Under the ignorance about the proportions of small and large balls, the set of rewards $\{0.28, 0.64\}$ is indifferent to 0.424.

The individual wants to verify the value she calculated using a different line of reasoning. Because she has no information about both size and color, she lumps two variables into one called "size.color". Under this view, the set of rewards is $\{1, 0.7, 0.4, 0\}$ (Fig. 20.2b). Using $\alpha = 0.6$ she finds the set of rewards is indifferent to 0.4.

The difference is puzzling because no new information is added in case (b). The difference between (a) and (b) is due to purely subjective view of the individual. ∎

The notion of sequential consistency is closely related to an important identity in probability theory—the law of iterated expectation. Basically, if X and Y are random variables then

$$\mathbb{E}[X] = \mathbb{E}[\mathbb{E}[X|Y]]. \tag{20.3}$$

The unconditional expectation of X equals the expectation of the conditional expectation of X given Y. In particular, this law allows a divide-and-conquer strategy to compute $\mathbb{E}[X]$. One can divide the state space of X into several subclasses and create a variable Y to be the class ID, i.e., $\mathbb{E}[X|Y = j]$ is the mean of class j. Compute the mean for each class and then compute the mean of those conditional expectations. From a conceptual point of view, the law is assuring because no matter how the state space of X is divided the final value of expectation remains the same.

As the example shows, Hurwicz's criterion does not have this property. That is, the value of the certainty equivalent of a set of prizes calculated by Hurwicz's rule (denoted by $\mathcal{CE}_\alpha(A)$) depends on the way the set is partitioned.

Formally, suppose A is a set of prizes and $\{B_i|1 \leq i \leq m\}$ is a partitions of A, i.e., $A = \cup_{i=1}^m B_i$ and $B_i \cap B_{i'} = \emptyset$ for $i \neq i'$. Equation (20.4) below is referred to as the *law of iterated certainty equivalence*. The law says that the certainty equivalent of a set of prizes is equal to the certainty equivalent of a set of certainty equivalents calculated for each subset of the partition. In other words, the certainty equivalent is invariant of the partition of the set.

$$\mathcal{CE}(A) = \mathcal{CE}(\{\mathcal{CE}(B_i)|1 \leq i \leq m\}). \tag{20.4}$$

In general, Hurwicz's rule does not satisfy that law. But there are two exceptions: for $\alpha = 0$ and $\alpha = 1$. The fact that (20.4) holds if $\mathcal{CE}_0(A) = \max(A)$ and $\mathcal{CE}_1(A) = \min(A)$ is not difficult to prove. Later we will show a sufficient statement that (20.4) holds only if $\alpha = 0$ or $\alpha = 1$.

20.2.2 Decision Under Ignorance: Alternatives to HA Theory

One of complaints about HA theory is its reliance only on two extremal values and ignore the intermediate values in the set even the result is derived from reasonable axioms. There are many attempts to relax the HA axioms to account for non-extremal values. We review representative works and ideas in the following.

Cohen and Jaffray (CJ) (1980) described a system of axioms for rational behavior under complete ignorance. CJ theory assumes a state space Ω and acts are mappings from the state space to the set of prizes. The basic object in CJ theory is the strict preference relation P, i.e., fPg is the notation for "f is strictly preferred to g." P is assumed to be asymmetric and transitive. From relation P two relations R and I are defined. fRg means not fPg (it is not the case that f is strictly preferred to g). fIg means not $(fPg \vee gPf)$. Intuitively one can view R as (non-strict) preference and I as indifference but the difference between R and $\succeq$ in HA theory is that R is not transitive and neither is I even as P is. Giving up the transitivity requirement for non-strict preference, CJ were able to add a weak dominance axiom. A relation D is defined between two acts f, g as follows:

$$fDg \Leftrightarrow \forall w \in \Omega, f(w) \geq g(w) \quad \text{and} \quad \exists w_0 \in \Omega, f(w_0) > g(w_0). \tag{20.5}$$

That is for all states f is as good or better than g and there is at least in one state f is strictly better than g. The weak dominance axiom stipulates that weak dominance implies strict preference, i.e., $fDg \Rightarrow fPg$. This axiom implies that states are non-null. In addition to that, CJ also introduced an axiom "increase" (Axiom 6) which is less intuitive. They define a class of "rational decision criteria" consisting of those that satisfy their system of axioms. The central result is that CJ rational decision criteria in a "first-order approximation, depend on the sole comparison between the extremal values of acts, the taking into account of weak dominance which is required of criteria, or of other interactions between acts bringing in events, which remains a possibility, can only have a second-order influence on choices" (Cohen and Jaffray 1980). An example of CJ rational criterion is

$$fPg \Leftrightarrow \begin{cases} (m_f + M_f > m_g + M_g) \text{ or} \\ (m_f + M_f = m_g + M_g, M_f - m_f < M_g - m_g) \text{ or} \\ (m_f + M_f = m_g + M_g, M_f - m_f = M_g - m_g \\ \min_{\Omega'} f > \min_{\Omega'} g) \end{cases}, \tag{20.6}$$

where $m_f = \min_\Omega f$, $M_f = \max_\Omega f$, and $\Omega' = \{w \in \Omega | f(w) \neq g(w)\}$. That is a type of lexicographic criterion. The first condition used to compare two acts is the sum of their min and max elements. If that condition does not resolve in a strict preference, the second condition used is the difference between the max and min elements. If the second condition does not resolve the comparison then the minimal elements among the states where f and g are different are used.

In Congar and Maniquet (2010), Congar and Maniquet (CM) investigated axiomatic system for a rational decision under ignorance which is defined as "no available information regarding plausible probability distributions over the possible outcomes." An act is a vector of outcomes which are von Neumann–Morgenstern utilities. CM's assumption that the outcome is vNM utility makes it clear that the ignorant variable precedes the risk variable in their model.

CM theory includes five axioms: quasi-transitivity (transitivity required only for the strict preference relation), Savage's independence, Duplication (split/merge of states with the same outcomes do not change preference), Strong dominance (dominance holds for all permutations of outcomes), and Scale invariance (linear transformation of the utilities does not affect the preference). The quasi-transitivity property is inspired by Cohen–Jaffray work reviewed earlier. Only three decision criteria satisfy these requirements. The first criterion, the *protective* criterion, ignoring the common outcomes of acts u and v, compares the minimal elements among remaining outcomes. This criterion reflects extreme pessimism. The second decision criterion, *hazardous* criterion, is the dual version of protective criterion. Instead of comparing the minimal elements of u and v excluding the common part, it compares the maximal elements. Finally, the third criterion, *neutral* criterion, is the conjunction of both protective and hazardous criteria. A prominent property is that all three criteria compare acts by restricting the attention to the states in which the outcomes are different. This feature, due to Savage's axiom, is also present in the example of CJ rational behavior as in Eq. (20.6). The Savage's independence axiom differentiates CM theory from Hurwicz–Arrow's decision criterion (Theorem 1). Consider acts u and v which have the same minimal and maximal outcomes m, M but are different on other outcomes. Hurwicz–Arrow's decision criterion would make u and v indifferent but u and v may not be indifferent according to the protective or the hazardous or the neutral criteria. The scale invariance axiom requiring that adding a constant to outcomes or multiplying the outcomes with a constant do not change the preference between two acts. Among five axioms, the rationale for this axiom is far from convincing. An often voiced critique for decision criteria including the protective, hazardous or neutral or the original HA criterion is that they do not permit individualization of attitude toward uncertainty. CM argue that three attitudes toward uncertainty, namely pessimism, optimism, and neutral, are implemented by the protective, hazardous, and neutral decision criteria. However, if individual A and individual B who are both pessimistic (optimistic) and have the same vNM utility function and then they have identical preference under ignorance. It is impossible to express the idea that while both are pessimistic (optimistic), one is less so than the other. The adoption of Savage's independence axiom for ignorance is also problematic. Perhaps, the key distinction between decision under risk and under uncertainty has to do with Savage's independence (sure-thing principle). A cornerstone in the theory of subjective probability, the axiom has been conclusively shown in many studies beginning with Ellsberg's ground breaking work (Ellsberg 1961), to be violated in decision under uncertainty. Because ignorance is the extreme form of uncertainty it would require truly compelling arguments, which are not there, to convince the validity of the independence axiom in the case of ignorance.

A proposal to the problem of decision under ignorance by Gravel et al. (2012) is based on the principle of insufficient reason and expected utility theory. They define a "completely uncertain" decision as the finite set of its consequences and an "ambiguous decision" as a finite set of possible probability distributions over a finite set of consequences. They would then apply the principle of insufficient reason to assign to every consequence (probability distribution) an equal probability and as comparing decisions on the basis of the expected utility of their consequences (probability distributions) for some utility function. This proposal avoids dealing with difficulties caused by ignorance altogether. A similar proposal was found in the works by Smets (2005) in the context of decision making with Dempster–Shafer belief functions. This family of proposals does not satisfy a basic property of ignorance in Hurwicz–Arrow sense namely the invariance under splitting/merging states. For example, adding a small random noise to the outcomes of an act would have a dramatic effect on its utility because it changes the set of different outcomes and hence the probability distribution derived from the principle of insufficient reason.

Viewing an act under ignorance as a set of prizes naturally suggests using a familiar descriptive statistic as the certainty equivalent of the set. Among three basic statistics, mean, mode, and median, the calculations of mean and mode rely on the frequency information which is not known in the state of ignorance and therefore excluded. Nitzan and Pattanaik (1984) describes an system that characterizes the median criterion according to which the certainty equivalent of an act under ignorance is equal to the median of its set of prizes. The most important property of the median statistic is its insensitivity to the outliers, i.e., the extreme (max and min) values of a set can change dramatically without affecting its median value. In this sense, the median criterion is radically different from the prescription of HA theory. The most obvious problem with the median criterion is its violation of HA property C. The certainty equivalent of an act is dependent on the way state space is modeled or merging/splitting of states.

The axiomatic approach to decision under ignorance has been discussed, among others, by Maskin (1979), Nehring and Puppe (1996), and more recently Puppe and Schlag (2009). A literature survey on the topic by Barbera, Bossert, and Pattanaik is given in Barbera et al. (2004).

20.3 A Utility Theory Under Ignorance

We have seen in the previous section the approaches to decision under ignorance that reject some of the assumptions in HA theory. In this section we presents a theory that accepts all the HA assumptions and further imposes the law of iterated certainty equivalence that Hurwicz's α-rule, the most famous criterion in the family of criteria sanctioned by HA theory, violates. The result was first reported in Giang (2011) where the proofs for the propositions in this section are found.

We assume a preference relation $\succeq$ which is a weak order and satisfies HA axioms A–D (Sect. 20.2.1). We make two technical assumptions which are not part of HA theory. First, when the prize set of an act is a singleton (mapping all states into a single prize) we have a constant act. In such a case the uncertainty (ignorance included) about the states does not matter. We assume that the preference among constant acts is exactly the arithmetic order.

Assumption 1 (Constant Acts—CA). *For $x, y \in \mathcal{O}$, $x \succeq y$ iff $x \geq y$.*

For $z \in \mathcal{F}(\mathcal{O})$ define $z^{\uparrow} = \{x \in \mathcal{O} \,|x \succeq z\}$ and $z^{\downarrow} = \{x \in \mathcal{O} \,|z \succeq x\}$. Clearly, both $z^{\downarrow}$ and $z^{\uparrow}$ are non-empty ($0 \in z^{\downarrow}$ and $1 \in z^{\uparrow}$) and because of completeness of $\succeq$, $z^{\downarrow} \cup z^{\uparrow} = \mathcal{O}$. By (CA) property, there is a unique $x \in \mathcal{O}$ such that $x \sim z$. Thus, for each act $A \in \mathcal{F}(\mathcal{O})$ there is a unique prize $c \in \mathcal{O}$ which is the *certainty equivalent* of A. It follows from HA Theorem 1 that $\min(A) \leq c \leq \max(A)$. To see that $\min(A) \leq \mathcal{CE}(A) \leq \max(A)$, suppose the contrary that either $\mathcal{CE}(A) < \min(A)$ or $\mathcal{CE}(A) > \max(A)$. Choose a value $z \in \mathcal{O}$ such that $\mathcal{CE}(A) < z < \min(A)$. On the one hand, $\{\min(A), \max(A)\} \succeq z$ because the min and max of the constant act z are less than those of act A. On the other hand, $z \succ \mathcal{CE}(A)$ by CA. That contradicts the fact that $A \sim \mathcal{CE}(A)$. $\mathcal{CE}$ will be referred to as *certainty equivalent operator*. Because a HA preference relation $\succeq$ completely determines (and is completely determined by) its certainty equivalent operator, we can interchangeably discuss the properties of $\succeq$ and $\mathcal{CE}$.

The second technical assumption is about continuity of preference relation (equivalently of ce operator). Viewing acts as vectors of prizes naturally leads to the concept of convergence of a sequence of acts. Suppose $(f_i)_{i=1}^{\infty}$ is a sequence of acts, in the vector form $f_i = \{x_{i1}, x_{i2}, \ldots x_{in}\}$. The sequence $(f_i)_{i=1}^{\infty}$ is said to converge to act $f = \{x_1, x_2, \ldots x_n\}$ (notation $\lim_{i\to\infty} f_i = f$) if $\lim_{i\to\infty} x_{ij} = x_j$ for $1 \leq j \leq n$.

Assumption 2 (Continuity of Certainty Equivalence Operator—C). *If* $\lim_{i\to\infty} f_i = f$ *then* $\lim_{i\to\infty} \mathcal{CE}(f_i) = \mathcal{CE}(f)$.

This assumption simply says that a small change in the prizes of an act under ignorance would not lead to a jump in the certainty equivalent. If $(f_i)_{i=1}^{\infty}$ converges to f and every member of that series is preferred to g then f is preferred to g. Because $f_i \succeq g$, $\mathcal{CE}(f_i) \geq \mathcal{CE}(g)$. So $\lim_{i\to\infty} \mathcal{CE}(f_i) \geq \mathcal{CE}(g)$. By (C), $\lim_{i\to\infty} \mathcal{CE}(f_i) = \mathcal{CE}(f)$. Thus, $\mathcal{CE}(f) \geq \mathcal{CE}(g)$ or equivalently $f \succeq g$.

The following lemma summarizes the properties of $\mathcal{CE}$ operator.

Lemma 1. *Suppose $\succeq$ is a HA preference relation on $\mathcal{F}(\mathcal{O})$ that satisfies properties (CA) and (C) then operator $\mathcal{CE}$ is well defined and satisfies:*

1. *Unanimity. For $x \in \mathcal{O}$, $\mathcal{CE}(x) = x$.*
2. *Range. $\forall A \in \mathcal{F}(\mathcal{O}), \mathcal{CE}(A) = \mathcal{CE}(\{\min(A), \max(A)\})$.*
3. *Monotonicity. If $a \geq a'$ and $b \geq b'$ then $\mathcal{CE}(\{a, b\}) \geq \mathcal{CE}(\{a', b'\})$.*
4. *Continuity. $\lim_{x\to a} \mathcal{CE}(\{x, b\}) = \mathcal{CE}(\{a, b\})$; $\lim_{x\to b} \mathcal{CE}(\{a, x\}) = \mathcal{CE}(\{a, b\})$.*

The ce operator $\mathcal{CE}$ is a function that maps from $\mathcal{F}(\mathcal{O})$, the set of finite subsets of reals in the unit interval, to the unit interval $\mathcal{O}$. Because $\mathcal{CE}$ is a certainty equivalent operator for HA preference, only the maximal and minimal elements of its argument matter. So we can define a two-placed function $\gamma : \mathcal{Z}^2 \to \mathbb{R}^1_0$ where $\mathcal{Z}^2 = \{\langle a, b\rangle \,|0 \le a \le b \le 1\}$ that carries all information of $\mathcal{CE}$.

$$\forall A \in \mathcal{F}(\mathcal{O}),\ \mathcal{CE}(A) = x \ \Leftrightarrow\ \gamma(\min(A), \max(A)) = x. \tag{20.7}$$

The following lemma details the properties of γ.

Lemma 2. *Suppose $\mathcal{CE} : \mathcal{F}(\mathcal{O}) \to \mathcal{O}$ satisfies Unanimity, Range, Monotonicity, Continuity and Iterated certainty equivalence and γ is defined via (20.7) then*

(1) *If for some $x \in \mathbb{R}^1_0$, $\gamma(x, 1) = a > x$ then $\forall y \in [x, a]$, $\gamma(y, 1) = a$.*
(2) *If for some $z \in \mathbb{R}^1_0$, $\gamma(0, z) = b < z$ then $\forall y \in [b, z]$, $\gamma(0, y) = b$.*

It is easy to verify that γ is continuous in each argument and satisfies: (i) For $0 \le x \le 1$, $\gamma(x, x) = x$; (ii) if $x \ge x', y \ge y'$ then $\gamma(x, y) \ge \gamma(x', y')$; and (iii) For $0 \le x \le y \le 1$, $\gamma(x, y) = \gamma(\gamma(x, x), \gamma(x, y)) = \gamma(\gamma(x, y), \gamma(y, y))$. By properties (i) and (iii), we have $a = \gamma(x, 1) = \gamma(\gamma(x, x), \gamma(x, 1)) = \gamma(x, a)$ and $a = \gamma(x, 1) = \gamma(\gamma(x, 1), \gamma(1, 1)) = \gamma(a, 1)$. It follows from property (*ii*) that for any y in the interval $[x, a]$, $a = \gamma(x, a) \le \gamma(y, 1) \le \gamma(a, 1) = a$. Symmetrically, (2) can be proved.

It turns out that γ function corresponding to a HA preference relation that satisfies the law of iterated certainty equivalence must have a special functional form.

Lemma 3. *Suppose $\gamma : \mathcal{Z}^2 \to \mathbb{R}^1_0$ two following statements are equivalent*

(1) γ satisfies (i) $\gamma(x, x) = x$ for $0 \le x \le 1$; (ii) $\gamma(x, y) \ge \gamma(x', y')$ if $x \ge x', y \ge y'$; (iii) $\gamma(x, y) = \gamma(\gamma(x, x), \gamma(x, y)) = \gamma(\gamma(x, y), \gamma(y, y))$ for $0 \le x \le y \le 1$.
(2) There exists a value $\tau \in [0, 1]$ such that

$$\gamma(x, y) = \begin{cases} y \text{ if } y \le \tau \\ \tau \text{ if } x \le \tau \le y \\ x \text{ if } x \ge \tau \end{cases}. \tag{20.8}$$

Combining Lemmas 2 and 3 leads to a representation theorem.

Theorem 2. *A HA preference relation $\succeq$ on $\mathcal{F}(\mathcal{F}(\mathcal{O}))$ that satisfies (C), (CA) and the law of iterated certainty equivalence iff there exists a value $\tau \in \mathbb{R}^1_0$ such that the certainty equivalent operator $\mathcal{CE}$ of $\succeq$ has the form: for $A_i \in \mathcal{F}(\mathcal{O})$, $1 \le i \le m$*

$$\mathcal{CE}(A_i) = \begin{cases} \max(A_i) & \text{if } \max(A_i) \le \tau \\ \tau & \text{if } \min(A_i) \le \tau \le \max(A_i) \\ \min(A_i) & \text{if } \min(A_i) \ge \tau \end{cases}, \tag{20.9}$$

$$\mathcal{CE}(\cup_{i=1}^m A_i) = \mathcal{CE}(\{\mathcal{CE}(A_i) | 1 \le i \le m\}). \tag{20.10}$$

While Eq. (20.10) is the law of iterated certainty equivalence, Eq. (20.9) describes the special form that ce operator must have. The central role is the value τ. The behavior of $\mathcal{CE}$ operator has three cases depending on the position of τ relative to the set (actually its minimal and maximal values). If the entire set of prizes lies above τ then $\mathcal{CE}$ behaves like min() function. If the entire set of prizes lies below τ then $\mathcal{CE}$ behaves like max() function. If τ is in between the minimal and maximal elements of the set then the entire set of prizes is indifferent to τ. For that reason, τ in (20.9) is called the *characteristic value* of a preference relation. Since the behavior of an individual decision maker under ignorance is described by a preference relation, τ can also be viewed as the characteristic value of the individual.

Another interesting property of Eq. (20.9) is that the ce of a set of prizes is the value between the min and max elements of the set that minimizes the distance to the characteristic value. For this reason, the function in (20.9) will be called the *τ-anchor* utility function. Imagine an ideal rubber cord, one end of it is fixed to the anchor τ. The other end of the cord is fixed to a movable point which is allowed to move within interval $[m_A, M_A]$ where $m_A = \min(A)$ and $M_A = \max(A)$. The ce of a set is the point where equilibrium is attained.

To get an interpretation for the characteristic value one can use the equality $\tau = \mathcal{CE}(\{0, 1\})$. τ is the value that the decision maker would give in exchange for the entire set of prizes [0, 1]. This is a situation of total ignorance. Not only the decision maker is ignorant about the likelihood of variable realization but also the consequences. No prize in the set of possible prizes is excluded.

We can see how the value of τ can be used to classify individual's qualitative attitude toward ignorance. As ignorance is an extreme form of uncertainty one can reasonably argue that an uncertainty averse (seeking) individual must demonstrate the ignorance averse (seeking) attitude. Suppose a "probabilistically sophisticated" individual has a vNM utility function under risk $u(x)$. Facing the prize set $\{0, 1\}$ with no probability distribution, she would adopt the uniform distribution. In effect, she converts the act under ignorance to a fair coin lottery $(H{:}0, T{:}1)$ that brings reward of \$0 if the coin lands *H*ead and \$1 if the coin lands *T*ail. The ce of this lottery is $c_u^{un} = u^{-1}(0.5 * u(0) + 0.5 * u(1))$. The probabilistically sophisticated person is by definition uncertainty neutral (Epstein 1999). A uncertainty averse (seeking) attitude implies that the certainty equivalent under ignorance is less (more) than the certainty equivalent under the uniform probability. This leads to the following classification of attitude toward ignorance. For an individual with utility function under risk u, her attitude toward ignorance is averse (seeking) if $\tau \leq (\geq)\ c_u^{un}$. For example, if an individual is risk neutral with vNM utility function $u(x) = x$, then $c_u^{un} = 0.5$. $\tau = 0.4 < 0.5$ is classified as ignorance averse. For another individual with vNM utility function $u(x) = x^{0.5}$ (risk averse), hence, $c_u^{un} = 0.25$. $\tau = 0.4 > 0.25$ is classified as ignorance seeking.

For two individuals with the same utility under risk, the comparison of their characteristic values tells their relative attitude toward ignorance, i.e., $\tau_1 < \tau_2$ means that individual number 1 is more averse toward ignorance than the individual number 2.

Finally, we note that min and max decision criteria are special cases of (20.9) with $\tau = 0$ and $\tau = 1$, respectively. On the other hand, (20.9) excludes Hurwicz's α-criterion as well as the median rule. There is no nontrivial $0 < \alpha < 1$ that makes Hurwicz's α-criterion equivalent to (20.9).

There is experimental evidence to support the model of decision making under ignorance based on to the τ-*anchor* utility. In one of few works dedicated specifically to find evidence about human decision making under ignorance, Hogarth and Kunreuther (1995) designed series of experiments on human subjects to understand the difference between decision making under ignorance and that when probabilities and consequences are known. They found that under ignorance, a significant number of subjects would switch to "meta-strategy" and use some higher-order rules to solve the choice problem. An example of such high order rules is "buy warranty to for the peace of mind" or "I would regret not buying the warranty should a breakdown occur." Hogarth and Kunreuther observed that "An important feature of a meta-strategy is that, although the use of the strategy is triggered by the stimulus encountered, it is not responsive to detailed features of the stimulus." This observation fits well with τ-anchor utility model. In our model, τ represents the "peace of mind" level of an individual. Although the certainty equivalence of a decision under τ-anchor utility depends on its prizes, it is quite stable when the prizes vary.

In Fig. 20.3 is the plot of the τ-utility function with $\tau = 0.4$. The coordinates on *XY* surface are the extremal values of sets of prizes. The ce of the sets are read on *Z* axis. For example points $(0.2, 0.5)$ or $(0.5, 0.2)$ represent sets of prizes whose minimal element is 0.2 and the maximal element is 0.5. The ce in these cases are 0.4 (τ). The origami-like surface is continuous but not differentiable at the folding lines.

In literature, a result that closest to (20.9) was established by Nehring and Puppe (1996). They derived the functional form from two axioms: a continuity axiom (C) and a strong independence axiom (SI) and other technical assumptions. Strong independence condition requires that adding a new prize that was not included in either of two sets of prizes would not change the preference between the sets.

$$\forall A, B \in \mathcal{F}(\mathcal{O}), x \in \mathcal{O}, x \notin A, x \notin B, \text{ if } A \succeq_{NP} B \Rightarrow A \cup x \succeq_{NP} B \cup x. \quad (20.11)$$

SI condition is often contrasted with the condition named Independence (I) which requires that preference between two sets made by adding two new prizes to a set is the same as the preference between added prizes. $A \in \mathcal{F}(\mathcal{O})$, $x, y \in \mathcal{O}$ and $x, y \notin A$, if $x \geq y$ then $A \cup \{x\} \succeq A \cup \{y\}$. (SI) is stronger than (I). For example both Hurwicz's α-criterion with $\alpha \neq 0, 1$ and the median rule satisfy (I) but not (SI). Theorem 2, on the other hand, is based on imposing the law of iterated certainty equivalence on HA preference. Taken together, Nehring–Puppe's result and Theorem 2 provide independent supports for the conceptual foundation of τ-anchor utility function.

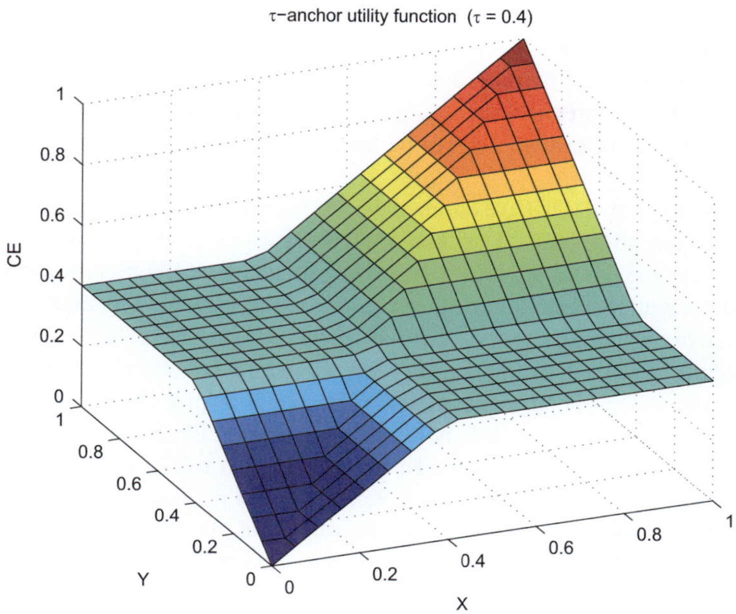

Fig. 20.3 τ-Anchor utility function

20.4 Related Literature and Discussion

On this chapter the topic of decision in the condition of ignorance has been examined mostly from a mathematical and economic point of view. Understanding of the topic would not be possible without contributions from researchers in behavioral and neurobiological fields. Hogarth and Kunreuther (1995) argued that many practical situations where individuals have to make decision lack the basic features of a gamble, namely, the outcomes and probabilities. For example, an individual decides whether to buy warranty for an electronic device without knowing probability of its breakdown and the repair cost. They designed a series of experiments on human subjects to understand the difference between behavior under probability and that when no probability or cost are known. They found clear evidence for difference of behaviors when the subjects have and do not have probability (ignorance). They found that under ignorance the subjects use two types of strategies to arrive at a choice. One of those is using a "principle that resolved the choice conflict and was insensitive to the particular features of different options." The authors expressed some degree of surprise at the finding and wrote "It is perhaps ironic that, under ignorance, when people should probably think harder when making decisions, they do not. In fact, they may be swayed by the availability of simple arguments that serve to resolve the conflicts of choice." We hold a different opinion. We think that the observed behavior is perfectly rational. Under ignorance, there is no reliable information as input for one to think harder. Thinking harder in this case often means filling the void left by lack of hard evidence with personal analytic or judgmental assumptions which may be false. The casual practice of acting upon those assumptions as if they were facts is misleading and sometime dangerous. As it was argued in the previous section, the τ-anchor utility fits the description of this type of strategy. For example, τ can be interpreted as "peace of mind" level for a subject.

Pushkarskaya et al. (2010) examined the neurological evidence of decision under ignorance using fMRI scan. They consider two types of missing information (MI): ambiguity (vague probabilities) and SSI. They found that different types of MI activate distinct neural substrates in the brain. A popular view held by neuroscientists, the reductive viewpoint, suggests that individuals reduce SSI (extreme uncertainty) to ambiguity and then to risk by forming subjective beliefs, about the partition of a sample space and then about the corresponding probability. However, the prediction of the reductive view holds only for ambiguity averse individuals and not for ambiguity-tolerant individuals. They concluded with a key suggestion that theories of decision making under uncertainty should include individual tolerance for missing information. The characteristic value that plays a central role of in the utility theory in Sect. 20.3 provides an answer to this challenge.

Finally, we illustrate with an example based on Ellsberg's paradox (Ellsberg 1961) on how ignorance and probability can be used together.

Example 3 (Ellsberg's Urn). A urn of 90 balls of three colors: red ($\boldsymbol{r}$), white ($\boldsymbol{w}$), and yellow ($\boldsymbol{y}$). The proportion of red is $\frac{1}{3}$. The proportions of white and yellow are unknown. $\bar{\boldsymbol{r}}$ or $\sim \boldsymbol{r}$ denotes not red, i.e., white or yellow. The same way $\bar{\boldsymbol{w}}, \bar{\boldsymbol{y}}$ are defined. A ball is drawn from the urn. A bet on proposition α where α can be $\boldsymbol{r}, \boldsymbol{w}, \boldsymbol{y}$ or their negations, pays \$1 if α holds and nothing otherwise.

In our framework, the given information about the urn is modeled by two variables X, Y. The domain of X has two propositions $\{\boldsymbol{r}, \bar{\boldsymbol{r}}\}$. The conditional domain of Y given $X = \boldsymbol{r}$ has only one state denoted by $\top$ (on this domain ignorance and certainty are the same). Y conditional on $X = \bar{\boldsymbol{r}}$ is an ignorant variable of two states $\{\boldsymbol{w}, \boldsymbol{y}\}$ (Fig. 20.4).

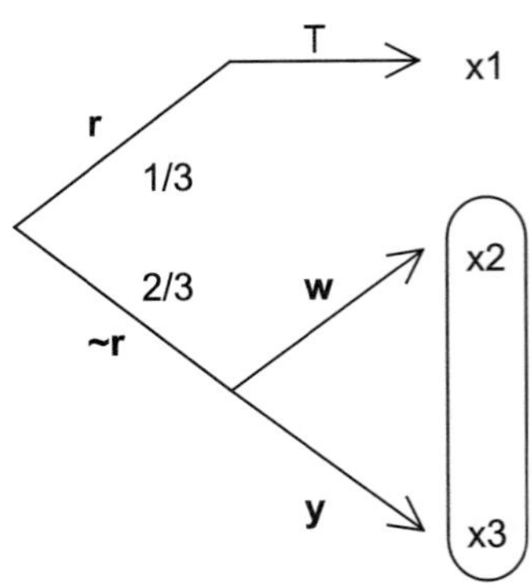

Fig. 20.4 Ellberg's urn

Let assume that an individual has a risk averse (concave) utility function $u(x) = \sqrt{x}$ and characteristic value under ignorance $\tau = 0.20$. Note that under the uniform distribution and with utility function $u(x) = \sqrt{x}$, the lottery $(0.5{:}0, 0.5{:}1)$ has the certainty equivalent of 0.25. So for a uncertainty averse individual, her characteristic value under ignorance must be $\tau < 0.25$. For a bet on red, the prizes are $x_1 = 1, x_2 = x_3 = 0$. If the drawn ball is red, the prize is 1. If the color is not red then the prize is 0. The expected utility is $1/3 * \sqrt{1} + 2/3 * \sqrt{0} = 1/3$. The certainty equivalent is 0.1111.

The bet on $\boldsymbol{w}$ has the following prizes $x_1 = x_3 = 0$ and $x_2 = 1$. The prize if the ball is red is 0. If the ball is not red, the individual has set of prizes $\{0, 1\}$ under ignorance. The ce of that act is the characteristic value $\tau = 0.20$. The expected utility is 0.2981 and the ce of entire act is 0.0889.

The bet on $(\boldsymbol{w} \vee \boldsymbol{y})$ has the following prizes $x_1 = 0, x_2 = x_3 = 1$. The expected utility is 0.6667 and the ce is 0.4444.

The bet on $(\boldsymbol{r} \vee \boldsymbol{y})$ has the following prizes $x_1 = 1, x_3 = 1$, and $x_2 = 0$. If the ball is not red the individual has set of prizes $\{0, 1\}$ under ignorance. The ce of this ignorant act is 0.20. The expected utility is 0.6315 and the ce of entire bet is 0.3988. Thus, $\boldsymbol{r} \succ \boldsymbol{w}$ and $(\boldsymbol{w} \vee \boldsymbol{y}) \succ (\boldsymbol{r} \vee \boldsymbol{y})$. ∎

20.5 Conclusion

Our goal in this chapter is to convince readers that the state of ignorance does occur in the real world, almost always within a bigger problem. A careful analysis and isolation of ignorance in the system of knowledge about a subject or a problem is of particular importance in the context of risk assessment and risk management. We argue that the practice of casually papering over the ignorance with subjective judgments and analytic assumptions can have serious consequences. We provide a structured survey (and necessarily selective) of significant ideas and proposals for decision making under ignorance, from the ground breaking work by Hurwicz and Arrow to the latest result of τ-anchor utility theory.

References

Arrow KJ, Hurwicz L (1977) An optimality criterion for decision making under ignorance. In: Arrow KJ, Hurwicz L (eds) Studies in resource allocation processes. Cambridge University Press, Cambridge

Aven T, Steen R (2010) The concept of ignorance in a risk assessment and risk management context. Reliab Eng Syst Saf 95:1117–1122

Barbera S, Bossert W, Pattanaik PK (2004) Ranking sets of objects. In: Barbera S, Hammond P, Seidl C (eds) Handbook of utility theory: extensions, vol 2. Springer, Berlin, pp 893–978

Cohen M, Jaffray J-Y (1980) Rational behavior under complete ignorance. Econometrica 48(5):1281–1299

Congar R, Maniquet F (2010) A trichotomy of attitudes for decision-making under complete ignorance. Math Soc Sci 59:15–25
Ellsberg D (1961) Risk, ambiguity and the Savage's axioms. Q J Econ 75(4):643–669
Epstein LG (1999) A definition of uncertainty aversion. Rev Econ Stud 66:579–608
Giang PH (2011) Dynamic consistency and decision making under vacuous belief. In: Cozman FG, Pfeffer A (eds) Uncertainty in artificial intelligence: proceedings of the 27nd conference (UAI–2011). The Association for Uncertainty in Artificial Intelligence Press, pp 230–237
Gilboa I, Postlewaite A, Schmeidler D (2009) Is it always rational to satisfy Savage's axioms? Econ Philos 25:285–296
Graham DA (2014) Rumsfeld's knowns and unknowns: the intellectual history of a quip. Atl Mag. http://www.theatlantic.com/politics/archive/2014/03/rumsfelds-knowns-and-unknowns-the-intellectual-history-of-a-quip/359719/
Gravel N, Marchant T, Sen A (2012) Uniform expected utility criteria for decision making under ignorance or objective ambiguity. J Math Psychol 56:297–315
Health and Safety Executive (UK) (2001) Reducing risks, protecting people: HSE's decision-making process. HSE Books, London
Hogarth RM, Kunreuther H (1995) Decision making under ignorance: arguing with yourself. J Risk Uncertain 10:15–36
Keynes JM (1921) A treatise on probability. Macmillan, London
Knight FH (1921) Risk, uncertainty and profit. Houghton Mifflin, Boston
Maskin E (1979) Decision-making under ignorance with implications for social choice. Theor Decis 11:319–337
Nehring K, Puppe C (1996) Continuous extensions of an order on a set to the power set. J Econ Theory 68:456–479
Nitzan S, Pattanaik PK (1984) Median-based extensions of an ordering over a set to the power set: an axiomatic characterization. J Econ Theory 34:252–261
Puppe C, Schlag KH (2009) Choice under complete uncertainty when outcome spaces are state dependent. Theor Decis 66:1–16
Pushkarskaya H, Liu X, Smithson M, Joseph JE (2010) Beyond risk and ambiguity: deciding under ignorance. Cogn Affect Behav Neurosci 10(3):382–391
Shackle GLS (1949) Expectations in economics. Cambridge University Press, Cambridge
Smets P (2005) Decision making in the TBM: the necessity of the pignistic transformation. Int J Approx Reason 38(2):133–147
von Neumann J, Morgenstern O (1953) Theory of games and economic behavior, 3rd edn. Princeton University Press, Princeton, NJ
Walley P (1996) Inferences from multinomial data: learning about a bag of marbles. J R Stat Soc Ser B 58(1):3–57

Chapter 21
Modeling Extreme Events Using Heavy-Tailed Distributions

Mathukumalli Vidyasagar

Abstract Typically, in constructing a model for a random variable, one utilizes available samples to construct an empirical distribution function, which can then be used to estimate the probability that the random variable would exceed a pre-specified threshold. However, in modeling extreme events, the threshold is often in excess of the largest sampled value observed thus far. In such cases, the use of empirical distributions would lead to the absurd conclusion that the random variable would never exceed the threshold. Therefore it becomes imperative to fit the observed samples with some appropriate distribution. For reasons explained in the paper, it is desirable to use the so-called stable distributions to fit the set of samples. In most cases, stable distributions are heavy-tailed, in that they do not have finite variance (and may not even have finite mean). However, they often do a very good job of fitting the data. This is illustrated in this paper via examples from various application areas such as finance and weather.

21.1 Problem Formulation

Suppose X is a real-valued random variable, and as is customary, define

$$\Phi_X(u) = \Pr\{X \leq u\}, \bar{\Phi}_X(u) = \Pr\{X > u\} = 1 - \bar{\Phi}(u)$$

to be the cumulative distribution function (cdf) and complementary cumulative distribution function (ccdf) of X, respectively. An "extreme" event can be defined as one for which $\bar{\Phi}_X(u)$ is very small. Estimating the probabilities of extreme events can arise in one of two ways. First, a threshold θ is specified, and one is asked to estimate $\bar{\Phi}_X(\theta)$, that is, the likelihood that the random variable X exceeds the threshold θ. The second is that a level parameter α is specified, and one is asked to determine the $1-\alpha$ "value at risk" (VaR), which is the smallest real number u such

M. Vidyasagar (✉)
The University of Texas at Dallas, 800 W. Campbell Road, Richardson, TX 75080, USA

Indian Institute of Technology Hyderabad, Kandi, Medak 502285, TS, India
e-mail: M.Vidyasagar@utdallas.edu

G. Rogova, P. Scott (eds.), *Fusion Methodologies in Crisis Management*,
DOI 10.1007/978-3-319-22527-2_21

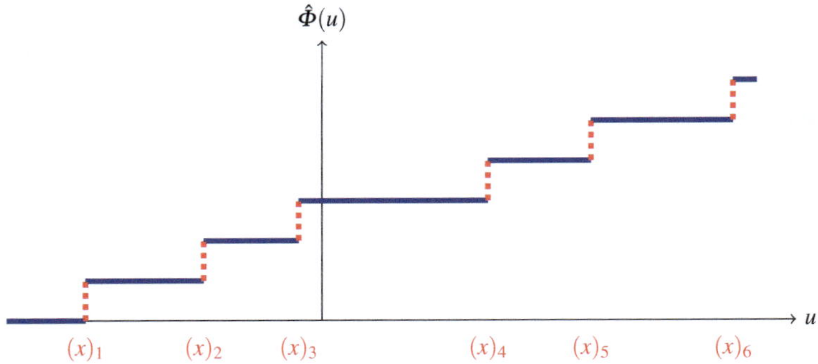

Fig. 21.1 Empirical cumulative distribution (staircase) function

that $\bar{\Phi}_X(u) \leq \alpha$. Note that the VaR is associated with $1 - \alpha$ and not α. Thus one speaks of a 99 % VaR when $\alpha = 0.01$, and not a 1 % VaR.

In practice, the true cdf $\Phi_X(\cdot)$ is not known. Instead, one has access to independent samples $x_1, \ldots, x_m$ of the random variable X. One can construct an "empirical" cdf $\hat{\Phi}(\cdot)$ as follows:

$$\hat{\Phi}_X(u) = \frac{1}{m} \sum_{i=1}^{m} I_{\{x_i \leq u\}},$$

where I denotes the indicator function. Thus $\Phi_X(u)$ is just the fraction of samples that are less than or equal to u. If the samples are sorted and denoted by (x), so that $(x)_1$ denotes the smallest sample and $(x)_m$ denotes the largest sample, then the empirical distribution is a "staircase" function as shown in Fig. 21.1.

Well-known results such as the Glivenko–Cantelli lemma assure us that the empirical distribution $\hat{\Phi}(\cdot)$ converges uniformly in probability[1] to the true cdf $\Phi(\cdot)$ as the number of samples $m \to \infty$. Thus one might be tempted to use the empirical distribution as a proxy for the true cdf in order to estimate the complementary cdf $\Pr\{X > u\}$.

Estimating the probability of extreme events can be challenging when one of two situations arises: (1) the threshold θ is larger than the largest sample $(x)_m$, and (2) the level parameter α is less than $1/m$, where m is the number of samples. In each case, using $\hat{\Phi}$ in the place of Φ leads to absurd results. If the threshold θ is larger than the largest sample recorded, then clearly $\hat{\Phi}(\theta) = 1$, so that the empirically estimated ccdf evaluated at θ equals zero. This would lead to the conclusion that the probability of X exceeding the largest sample $(x)_m$ is precisely zero. Similarly, if a level parameter α is specified, the VaR at level $1 - \alpha$ is the $(1 - \alpha)$-th percentile value of X. But if $\alpha < 1/m$, then once again the VaR at the level $1 - \alpha$ would equal the largest sample $(x)_m$. In essence therefore, the challenge in estimating the

[1] In fact, the convergence is almost sure.

probability of extreme events arises when the threshold is larger than the largest sample witnessed thus far, and/or the level parameter α is smaller than the reciprocal of the number of samples. Both situations can arise in practice, and are illustrated via the two case studies in the paper.

Case Study 1: Rainfall in Houston Data from the NOAA for the city of Houston contains the rainfalls (in tenths of millimeters) for 2039 days, out of which it did not rain at all for 1469 days, leaving 570 days with nonzero (at least a tenth of a millimeter) rainfall. Suppose it is desired to estimate the threshold such that the probability of exceeding that threshold is 0.001, given that it rains. In this case there are only $m = 570$ samples, whereas as $\alpha = 0.001 < 1/m$. Therefore the empirical distribution cannot be used to predict this value.

Case Study 2: Dow-Jones Industrial Average Data is available for the daily closing price of the Dow-Jones industrial average (DJIA) for several years. In the present example, the data is from September 2007 through December 2012. It is desired to estimate the 99 % VaR of the DJIA for the next day, using this data. In this instance, the number of samples is rather large. However, it would not be prudent to use historical data stretching over many years to compute the distribution function of the DJIA. It would be preferable to use only the most recent data, say from the preceding 30 trading days, to fit a distribution function. In this case it not the shortage of data that causes the difficulty; rather, it is the fact that much of the data is not relevant to the problem at hand.

21.2 Fitting Gaussian Distributions to Data

If one wishes to estimate the probabilities of extreme events, one has to construct a distribution function to the observations or measured samples. The fitted distribution function cannot simply be the empirical distribution function shown in Fig. 21.1, because that leads to absurd estimates for the probabilities of extreme events, as pointed out above. The success of estimation methods depends on how well one can fit the observations. In turn this depends on the families of distributions used to fit the observations.

By far the most commonly used distribution is the Gaussian distribution, which is characterized by just two parameters, namely the mean and the standard deviation (or variance, which is the square of the standard deviation). The Gaussian density function with mean μ and standard deviation σ is defined by

$$\phi_G(x) = \frac{1}{\sqrt{2\pi}\sigma} \exp(-(x-\mu)^2/2\sigma^2).$$

A special case of the Gaussian random variable is the normal random variable, which has zero mean and standard deviation of one. The normal density function is defined by

$$\phi_N(x) = \frac{1}{\sqrt{2\pi}} \exp(-x^2/2).$$

Given a set of samples $x_1, \ldots, x_m$, the maximum likelihood Gaussian estimator is obtained by choosing μ and σ to be the sample mean and sample standard deviation, respectively; thus

$$\hat{\mu} = \frac{1}{m}\sum_{i=1}^{m} x_i, \hat{\sigma} = \left[\frac{1}{m}\sum_{i=1}^{m}(x_i - \hat{\mu})^2\right]^{1/2}. \tag{21.1}$$

21.3 Central Limit Theorem

One of the main justifications for using the Gaussian to fit observations is the central limit theorem. Since this theorem is "central" to the contents of the present paper, it is described in detail. Suppose X is a random variable with cdf Φ,[2] and let $X_1, \ldots, X_m$ be independent copies of X. Define the m-fold sum and the m-fold average of these random variables in the natural manner, namely

$$S_m = \sum_{i=1}^{m} X_i, \; A_m = \frac{S_m}{m} = \frac{1}{m}\sum_{i=1}^{m} X_i. \tag{21.2}$$

At this stage, one can investigate the behavior of these random variables as m becomes large. Suppose X has a finite expected value $E(X)$. Then, under mild conditions, it can be established that the average A_m converges in probability to $E(X)$. In other words, for each fixed $\epsilon > 0$,

$$q(m, \epsilon) = \Pr\{|A_m - E(X)| > \epsilon\} \to 0 \text{as} m \to \infty.$$

This is usually referred to as "the law of large numbers." However, this property gives only a very crude picture of the limit behavior. One can ask whether there exist sequences of real numbers $\{a_m\}, \{b_m\}$ such that the sequence of "centered and scaled" random variables

$$Y_m = \frac{S_m - b_m}{a_m} \tag{21.3}$$

converges to a limiting random variable. Now the central limit theorem states that if the random variable X has finite standard deviation σ, then we can choose

$$a_m = \frac{\sigma}{\sqrt{m}}, b_m = E(X),$$

[2]The subscript X is dropped in the interests of clarity.

and the resulting sequence $\{Y_m\}$ converges in the distributional sense to a normal random variable; that is, as $m \to \infty$, the cdf of Y converges uniformly to that of a normal random variable.

21.4 Stable Distributions

At this point, one can ask: What are *all possible* limit distributions of random variables Y_m that can be achieved by choosing the constants a_m and b_m appropriately? If X has a finite standard deviation (which in turn implies that it has a finite mean), then it turns out that the only possible limits are Gaussian random variables, as asserted by the central limit theorem. However, if one were to remove the restriction that X has finite variance, then a broader class of limits results, known as stable distributions. This section is devoted to a discussion of this very useful class of distributions. An excellent reference for this material is Breiman (1992).

A cdf Φ of a random variable X is said to be a **stable distribution** if the following is true: If Y, Z are independent random variables, each of which has the cdf Φ, and a, b are real numbers, then there exist real numbers c, d such that $aY + bZ$ and $cX + d$ have the same probability distribution (i.e., they are "distributionally equivalent"). Φ is **strictly stable** if $d = 0$, and p-**strictly stable** if in addition $c = (|a|^p + |b|^p)^{1/p}$. Observe that a Gaussian distribution is two-strictly stable.

It can be shown that all possible limit distributions of sequences $\{Y_m\}$ defined in (21.3) must be stable distributions. If Φ_S is a stable distribution, and Φ_X is the cdf of a random variable X such that the cdf of the centered and scaled random variable Y_m defined in (21.3) converges to Φ_S as $m \to \infty$, then the cdf Φ_X is said to be **in the domain of attraction** of the stable distribution Φ_S. The central limit theorem assures us that every cdf corresponding to a random variable with finite standard deviation is in the domain of attraction of a Gaussian distribution. Interestingly, the domain of attraction of a Gaussian distribution also contains some distributions without a finite standard deviation.

The characteristic functions of all stable distributions can be characterized by just four parameters, referred to here as $\alpha, \beta, \gamma, \delta$, as follows: If $\alpha = 2$, then the characteristic function is a Gaussian (which in turn implies that the cdf is also a Gaussian). If $\alpha < 2$, then

$$\psi_X(u) = \exp\left(\mathbf{i}\delta u - \gamma^\alpha |u|^\alpha \left[1 - \mathbf{i}\beta \tan\left(\frac{\pi\alpha}{2}\right) \frac{|u|}{u}\right]\right), \text{ if } \alpha \neq 1,$$

$$\psi_X(u) = \exp\left(\mathbf{i}\delta u - \gamma u \left[1 + \mathbf{i}\beta \frac{2}{\pi} \frac{|u|}{u} \log|u|\right]\right) \text{ if } \alpha = 1.$$

In the above, $\mathbf{i}$ denotes $\sqrt{-1}$.

The most interesting and important among these parameters is α, which is referred to as the "exponent." Actually, there are *two* exponents, one for large positive values and another for large negative values. This is made precise next. Stable distributions are everywhere differentiable. Suppose $\Phi(\cdot)$ is a non-Gaussian stable distribution function, and let $\phi(\cdot)$ denote its density. Then there exist constants M_1, M_2, not both zero, and constants α_1, α_2 with $\alpha_i \in (0, 2)$ if $M_i \neq 0$, such that

$$\phi(x) \sim L_1(x)x^{-\alpha_1} \text{ as } x \to -\infty \text{ if } M_1 \neq 0,$$

$$\phi(x) \sim L_2(x)x^{-\alpha_2} \text{ as } x \to \infty \text{ if } M_2 \neq 0,$$

where L_1, L_2 are "slowly varying functions," i.e.,

$$\lim_{x\to\infty} \frac{L_i(tx)}{L_i(x)} = 1, \ \forall t > 1.$$

Moreover, the exponent α equals $\min\{\alpha_1, \alpha_2\}$. Therefore, as x approaches $\pm\infty$, the non-Gaussian stable density functions exhibit "scale-free tail behavior." Further, the smaller the exponents α_1 or α_2, the more slowly the corresponding tail decays.

Unfortunately, closed-form formulas for the actual cdf, which is the inverse Fourier transform of the characteristic function, are available in only a few simple cases. The number $\alpha \in (0, 2]$ is called the "exponent" of the stable distribution. The case $\alpha = 2$ corresponds to the Gaussian distribution, and the Gaussian is the only stable distribution with finite standard deviation. All other stable distributions, corresponding to $\alpha < 2$, are "heavy-tailed" random variables with infinite standard deviation. If $1 < \alpha$, the random variable X has finite mean, whereas if $0 < \alpha \leq 1$, then X does not even have a finite expected value.

Besides the central limit theorem, one of the appeals of the Gaussian distribution is that there are explicit formulas for the maximum likelihood estimates, as shown in (21.1). There are now systematic procedures available for computing the maximum likelihood estimates of $\alpha, \beta, \gamma, \delta$ in order to fit a stable distribution to a given set of data. However, these are difficult to describe within the space available.

21.5 Gaussian vs. Stable Fits

Because the Gaussian is a special case of a stable distribution, it is obvious that a stable distribution would *always* achieve at least as good a fit to observations as does a Gaussian. However, it is surprising to note that the fit is substantially better when stable distributions are used. Figure 21.2 shows the outcomes of fitting a Gaussian and a stable distribution to the annualized daily returns of the DJIA over the period from September 2007 through December 2012. The stable fit is virtually indistinguishable from the actual empirical distribution function, whereas the Gaussian fit is rather poor. The optimal parameter values for the stable

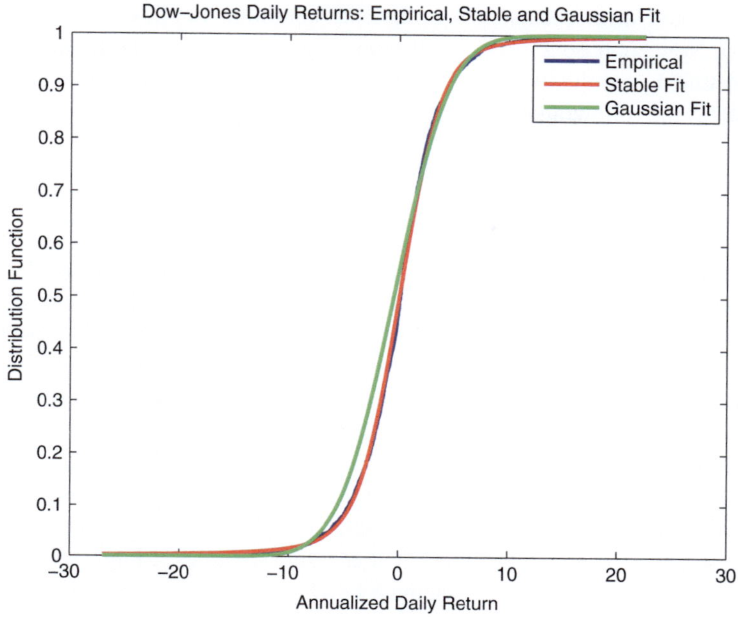

Fig. 21.2 Dow-Jones industrial average: stable and Gaussian fits

distribution are

$$\alpha = 1.6819,\ \beta = -0.0651,\ \gamma = 2.2345,\ \delta = -0.0150.$$

It is interesting to note that the exponent parameter α is considerably smaller than the Gaussian exponent of 2. This suggests that the annualized daily returns of the DJIA are anything but Gaussian.

21.6 Extremal Behavior of Sums of Independent Random Variables

In this section, we discuss one of the more dramatic differences between stable distributions with $\alpha < 2$ and the Gaussian distribution (which can be thought of as a stable distribution with $\alpha = 2$).

Suppose Φ_X is within the domain of attraction of a stable distribution with exponent $\alpha < 2$. Note that Φ_X itself need not be a stable distribution. Find constants t_m such that $m\bar{\Phi}_X(t_m) \to 0$ as $m \to \infty$. For instance, we can choose any sequence $\{\epsilon_m\} \downarrow 0$, and define

$$t_m = \bar{\Phi}_X^{-1}(\epsilon_m/m)$$

if the indicated inverse exists. Note that the condition implies that $t_m \to \infty$ as $m \to \infty$. Now suppose $X_1, \ldots, X_m$ are independent copies of the random variable X, and define the cumulative sum S_m as in (21.2). Then a result due to Nagaev (1969, 1979) states the following:

$$\lim_{m \to \infty} \frac{\Pr\{\max\{X_1, \ldots, X_m\} > t_m\}}{\Pr\{S_m > t_m\}} = 1. \tag{21.4}$$

What this equation means is this. The denominator is the probability that the sum of all m random variables exceeds a specified threshold t_m, while the numerator is the probability that *just one* among the random variables exceeds t_m. The result states that, as the number of samples grows, the likelihood of the sum of all m random variables assuming a really large value is the same as the likelihood that *just one* random variable assumes a really large value.

What makes this result interesting is the fact that, if Φ_X is within the domain of attraction of a Gaussian, for example, if Φ_X corresponds to the standard deviation being finite, then the indicated limit equals zero. In this case, as the number of samples grows, it is far more likely that the sum assumes a large value because many of the m random variables assume relatively large values, and not because just one random variable assumes a really large value.

The result can be stated in another way. Suppose Φ_X is within the domain of attraction of a stable distribution with exponent $\alpha \leq 2$. Then the limit indicated in (21.4) equals zero if $\alpha = 2$, and equals one if $\alpha < 2$. These are the only possible values for the limit. Moreover, it does not matter how close α is to 2—the limit value in (21.4) is still going to be one. Thus there is a sharp discontinuity in the extremal behavior of sums of random variables, depending on whether the exponent of the associated limit stable distribution exactly equals two, or is less than two.

21.7 Case Studies

21.7.1 Houston Rainfall

The rainfall data for Houston consists of 2309 records of daily rainfall measured in tenths of millimeters. In 1439 of those days, there was no measurable precipitation, while some precipitation was recorded only on 570 days. The maximum recorded rainfall was 2002 tenths of millimeters, or roughly eight inches.

Figure 21.3 shows the empirical ccdf of the rainfall *assuming that there is some rain*. In other words, this empirical ccdf is based on only the 570 records corresponding to days on which there was measurable precipitation. The horizontal axis is plotted on a logarithmic scale. It can be seen from this figure that, for intermediate values of α, say $0.1 \leq \alpha \leq 0.9$, the ccdf can be approximated by a linear function on a log scale; that is,

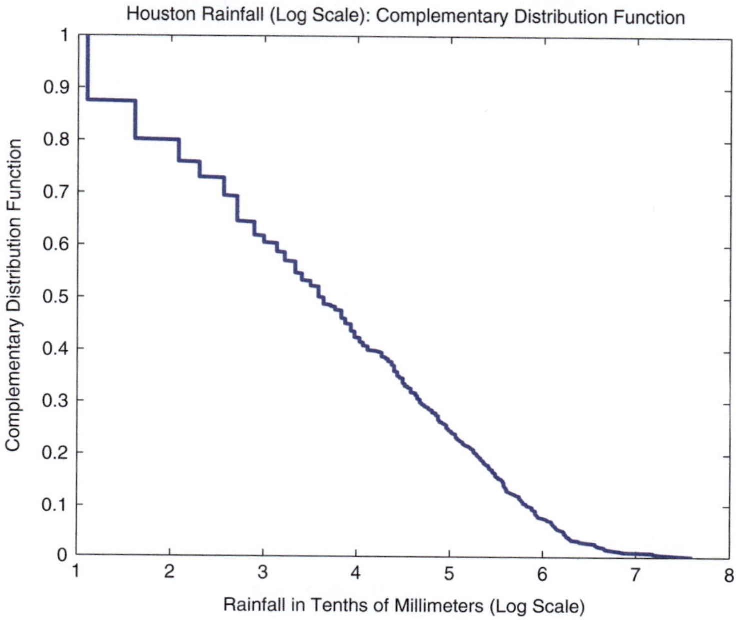

Fig. 21.3 Houston rainfall on a log scale

$$\bar{\Phi}(\alpha) \sim b_1 - b_2 \log \alpha.$$

However, for values of α smaller than 0.1, the empirical ccdf is no longer linear. Instead, if the empirical ccdf is plotted on a log–log scale, then it can be seen from Fig. 21.4 that

$$\log(\bar{\Phi}(\alpha)) \approx c_1 - c_2 \log \alpha,$$

where $c_1 = 11.4241, c_2 = 2.3376$. This can be written as

$$\bar{\Phi}(\alpha) \approx \exp(c_1)\alpha^{-c_2}.$$

Therefore, for extreme values of rainfall (or equivalently, for extremely unlikely scenarios), the rainfall likelihoods follow a power law. Because $c_2 > 2$, the corresponding random variable has a finite standard deviation, and is thus not heavy-tailed. Using the above approximation to compute $\bar{\Phi}^{-1}(0.001)$ leads to an estimated VaR of 2545.5 tenths of millimeters, which is higher than the highest rainfall ever recorded in Houston during the period under study.

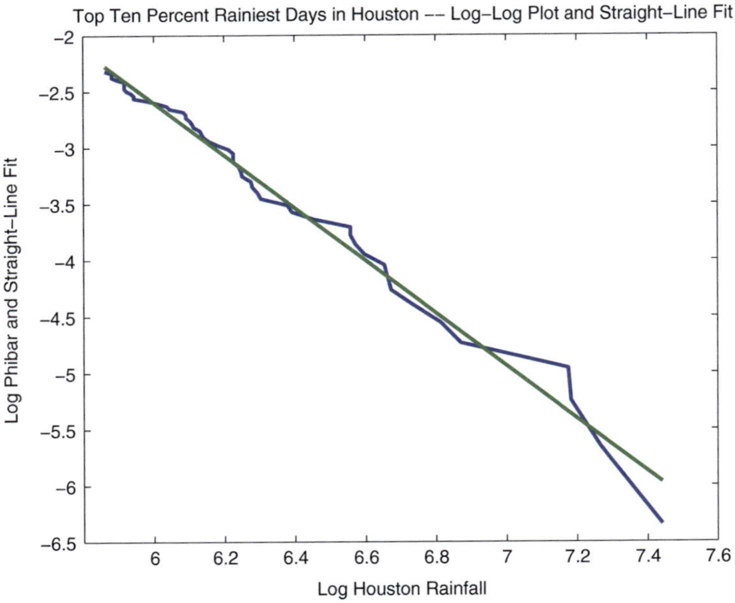

Fig. 21.4 Houston rainfall on a log–log scale and a linear fit

21.7.2 Dow-Jones Industrial Average

The objective of this case study is to determine the 99 % VaR of the DJIA, based on the annualized daily returns of the last 30 trading days in 2012. A stable distribution fit to these 30 days' returns yields the following parameters:

$$\alpha = 1.3897,\ \beta = 0.2680,\ \gamma = 1.3478,\ \delta = 0.3548.$$

Thus the stable fit to the last 30 days' daily returns (on an annualized basis) is even more heavy-tailed than the stable fit of the daily returns of five and a half years. This is because the former has $\alpha = 1.3897$ whereas the latter has $\alpha = 1.6819$. Figure 21.5 shows the empirical distribution of the daily return (on an annualized basis), together with the stable cdf corresponding to the above parameter vector. The VaR at the 99 % level can be computed by inverting the stable cdf at 0.99. This yields an annualized return of 10.6815, or 0.0293 on a daily basis. The interpretation is that, based on the stable fit to the data, one would predict that the next day's DJIA would not be more than 1.0293 times the last day's closing average, with a probability of 99 %. It is interesting to note that the largest upward movement within the 30-day period is 4.7301 on an annualized basis, or a daily return of 0.0130. Thus the VaR predicted using the stable fit is more than twice the largest upward movement observed. In contrast, if one were to use the empirical distribution blindly, the predicted 99 % VaR would be the largest observed upward movement.

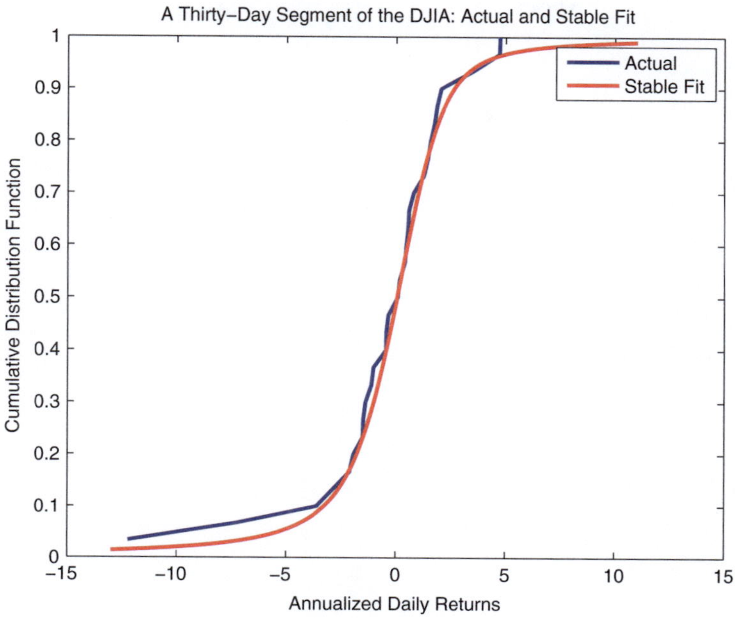

Fig. 21.5 Dow-Jones industrial average: stable and Gaussian fits

List of Symbols

Φ_X: Cumulative distribution function (cdf) of a random variable X.
$\bar{\Phi}_X$: Complementary cumulative distribution function (ccdf) of a random variable X.

References

Breiman L (1992) Probability. SIAM, Philadelphia, PA

Nagaev AV (1969) Integral limit theorems taking large deviations into account when Cramér's condition does not hold. Theory Probab Appl 14(1):51–67

Nagaev SN (1979) Large deviations of sums of independent random variables. Ann Probab 7(5):745–789

Part VI
Case Studies

Chapter 22
A General Framework for Using Social and Traditional Media During Natural Disasters: QuOIMA and the Central European Floods of 2013

Gerhard Backfried, Christian Schmidt, Dorothea Aniola, Christian Meurers, Klaus Mak, Johannes Göllner, Andreas Peer, Gerald Quirchmayr, Gerald Czech, and Markus Glanzer

Abstract Traditional media have a long history in covering natural disasters and crises. In many instances, these media remain major providers of information about an event. In recent years, however, information about natural disasters has increasingly been disseminated on a significant scale via Social Media platforms. These media provide new, additional and complementary angles on events and, combined with traditional media, produce a more complete spectrum of coverage. We present an approach, combining information from across the different kinds of media—traditional as well as social—and also across multiple languages, providing opportunities for first responders and decision makers to gain improved situational awareness and allowing for improved disaster relief, support, mitigation and resilience measures. The approach is put into context by relating it to a long-term strategic model including horizon-scanning and risk-management activities and a 5-phase disaster model forming the basis for information gathering and dissemination activities. To illustrate the research efforts the QuOIMA (Quelloffene

---

G. Backfried (✉) • C. Schmidt • D. Aniola
SAIL LABS Technology, Vienna, Austria
e-mail: Gerhard.Backfried@sail-labs.com; Christian.Schmidt@sail-labs.com; Dorothea.Aniola@sail-labs.com

C. Meurers • K. Mak • J. Göllner • A. Peer
National Defence Academy, Ministry of Defence and Sports, Vienna, Austria
e-mail: Christian.Meurers@bmlvs.gv.at; klaus.mak@gmx.at; johannes.goellner@meinesteuerberatung.at; peer.a@gmx.at

G. Quirchmayr
Multimedia Information Systems, Faculty of Computer Science, University of Vienna, Vienna, Austria
e-mail: Gerald.Quirchmayr@univie.ac.at

G. Czech • M. Glanzer
Austrian Red Cross, Vienna, Austria
e-mail: Gerald.Czech@roteskreuz.at; Markus.Glanzer@roteskreuz.at

G. Rogova, P. Scott (eds.), *Fusion Methodologies in Crisis Management*,
DOI 10.1007/978-3-319-22527-2_22

Integrierte Multimedia Analyse) project, based on the pillars of cross-media, multimedia, and multilingual processing and representing major aspects of the general framework is presented. QuOIMA focuses on the information gathering aspects from the point of view of a first responder and crisis manager or -communicator rather than the management of active (outgoing) communication. Initial findings on data collected during the 2013 Central European floods are reported and discussed.

22.1 Towards Decision Support Systems for Crisis Management

Crisis and Disaster Management (CDM) typically takes place in a highly dynamic, complex and elusive environment. In this context it is often difficult for decision makers to make sense of all assets, circumstances and information for their decision making- and communication processes. Decision Support Systems meet that challenge and integrate the dynamics of systems as well as sophisticated risk management and modeling techniques to enhance the decision making process itself and especially to improve the decisions based on unsecured information and information which has not been validated yet. For an organization it is fundamental to understand that this approach not only concerns operational tasks but has to be implemented on a strategic level as well. This covers not only decision support for crisis management but also a comprehensive long-term planning procedure integrating all relevant approaches covering horizon scanning, scenario-development, and risk management techniques with their underlying specific models, technologies, tools, and processes. The overall approach is outlined in the following model (the so-called *Z-model* (Klerx et al. 2014)) presenting a generic and comprehensive approach for strategic long-term horizon scanning and risk management (Fig. 22.1).

This model provides the basis for an enhanced and sophisticated research and development initiative to implement a comprehensive and specific planning framework for strategic risk, crisis, and disaster management. The first set of *Horizon Scanning* processes deals with the scope of big data analysis and the development of new technologies and systems to gather information from a variety of sources like social media, traditional media etc. and subsequent analyses to identify topics, relations, trends, and sentiments. Results are integrated into a scenario development process and a first risk evaluation of identified assets. The challenge of this research is not only to develop new and more efficient approaches but also to automate the processes as far as possible in a robust manner. The second and final set of *Risk Management* processes improves situational awareness and supports a common understanding of the CDM-situation before risk analysis is conducted. The results are used to compute ratings and set up portfolios to be able to develop or extend knowledge models in the subsequent step, which are then applied to enhance decision processes.

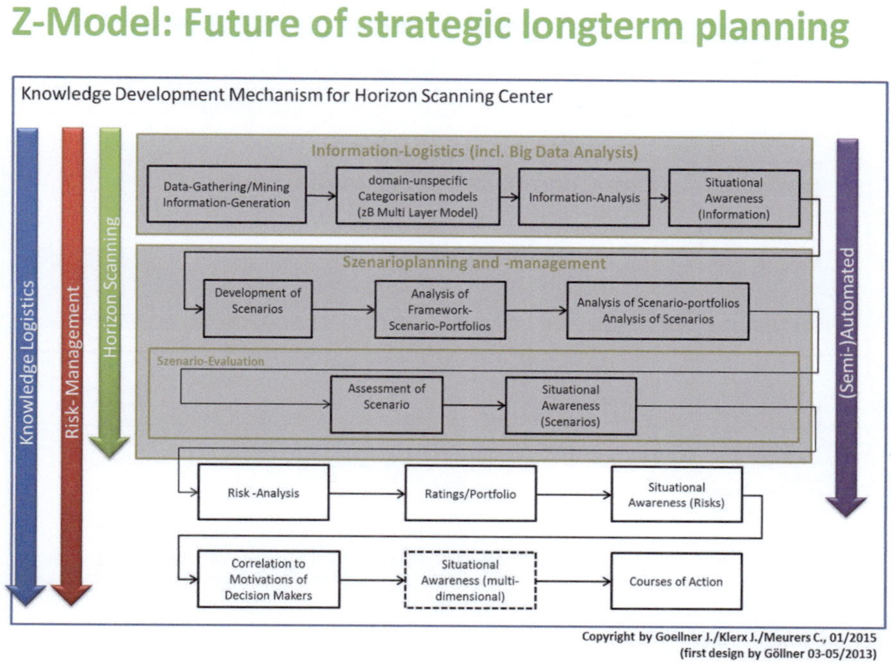

Fig. 22.1 Z-Model (Klerx et al. 2014)

Crisis management on an operational level covers three phases:

- Pre-execution phase: Initial signals of a natural disaster are detected and monitored. First responders explore on-site circumstances and build up operational staffs. The affected population is informed. First recommendations and actions are implemented.
- Execution phases: Integrated execution of all first responders' activities with a focus on the actual intervention for protection of the at-risk population and mitigation of the disaster event.
- Follow-up phase: Recovery and reconstruction, immediate help measures, handover of first responders to other organizations like local authorities, first analyses of the experiences of the intervention take place.

Crisis Management on a strategic level above all requires a transfer of the operational crisis model to a strategic point-of-view.

The common 3-phase-model can be expanded by two additional phases—a *prevention and initialization phase* and a *postvention phase*. These additional phases allow a feedback-cycle of planning, control and development activities integrating the results, experiences, and lessons learned from the *postvention* phase. (Backfried et al. 2013a, b) The resulting 5-phase-model is shown in Fig. 22.2 below.

A major benefit of this model is that it allows defining the five different stages of disasters and consequently identifying the point at which certain emergency

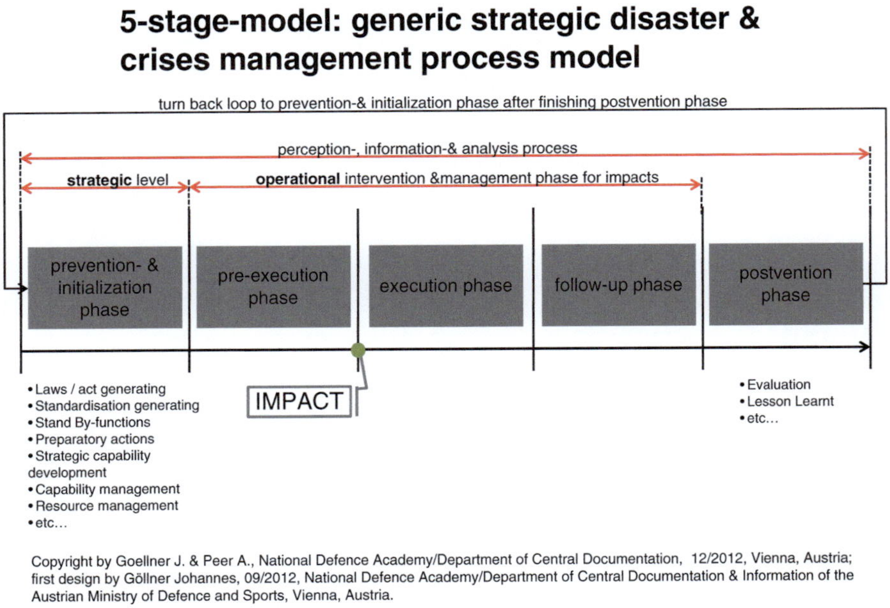

Fig. 22.2 5-phase-model for Crisis Management (Backfried et al. 2013a, b)

legislation, general standardization, and regulations can be applied. Likewise, different strategies for data-collection, analysis, visualization as well as different patterns of cooperation and information sharing might be associated with the individual phases. The 5-phase-model was developed as a generic model for the development or adaptation of innovative architectural systems consisting of Management Information Systems (MIS), Executive Information Systems (EIS) and Decision Support Systems (DSS) for the execution of operational, tactical, and strategic integrated management tasks and constitutes a comprehensive reunification and networked consideration of various models and processes of different provenances (Backfried et al. 2013a, b)

For information and data analysis purposes, it is necessary to specify the requirements of first responders for each of the above phases and to categorize the gathered information. A categorization system like the so-called *Multi-Layer Multiple Vector Model* (Langer et al. 2014; Göllner et al. 2014) developed at the Austrian Ministry of Defence (MoD) can be applied to classify information as well as an event itself. This model allows to document relevant content concerning events and incidents for further analysis as well as to differentiate events referring to factor such as:

- Organization
- Causers/initiators
- Type of the event
- Timely framework
- Dimension/influence
- Abstractional level

Every incident or event can be categorized, documented and be related to actors and knowledge role models. It allows recognizing correlations and dependencies in order to provide information to analyze or to identify additional information demands in the decision making processes.

22.2 Analysis Support in the QuOIMA Project

Against the background of the above theoretical models, the QuOIMA project (Backfried et al. 2012) was launched in November 2012 to create a practical framework and apply it against real-world phenomena and actual use-cases in the area of natural disasters. As it is a nationally funded Austrian security research project,[1] the emphasis lies on incidents involving Austria directly or indirectly. Related to the Z-model, it covers the initial phases of the *Horizon Scanning* activities. Related to the 5-phase disaster model, the *execution phase* following the impact forms the core of project activities within QuOIMA. The system itself is disaster-phase agnostic—i.e., all functionality is required to be available across all disasters phases. Only open (freely available) sources are considered. Validation and interpretation must be carried out by experts and in relation to the decision making process.

Traditional media such as TV, radio, and web-sources have long been employed to provide information about crises and disasters. Even today, depending on the infrastructure and social factors, these media may still be an important and highly additive source of information. In certain regions of the world, they even remain the primary source of information for large parts of the population. However, under certain circumstances, e.g., cross-border or international settings of crises, official and traditional channels may not be available to an organization or only be able—or willing—to provide partial information. Social media allow professionals as well as affected individuals to provide and retrieve further insights, different angles, and unfiltered (and to some extent unbiased) information on events covered by traditional media and are able to add valuable dimensions like detailed micro-perspectives and (near) real-time images of dynamically changing situations to an already multilingual, multi-source, multimedia environment.

Their value in the scope of disaster and crisis communication is widely accepted and has led to an increasing number of systems and approaches integrating their strengths (Holmes 2011; Nilsson et al. 2012; American Red Cross 2010; Lindsay 2011; DHS 2013). Especially multimedia data in the form of images and video is becoming more common-place with the ubiquity of portable devices. Individuals carrying such devices will often be on-site, delivering visual content and meta-data associated with short comments in real-time as an event unfolds. In many cases, Social Media might be ahead of traditional media in reporting about a certain

[1] Research Program KIRAS: http://www.kiras.at/

incident. Based on these observations, QuOIMA aims at fusing the diverse, mixed, and complementary open sources of information—in particular Social Media and Traditional Media—and at allowing decision makers to quickly make sense of a large, multilingual, and inhomogeneous body of data. In particular, the approach is based on the following assumptions:

- Both, Traditional and Social Media play fundamental and complementary roles and will continue to do so in the foreseeable future
- Their respective strengths can be increased by combination of sources and channels. Links detected between different media provide insights not available by approaches based on single media only
- Multimedia processing capabilities are required to ingest the complete spectrum of sources covering an event such as robust and extensible image-, text-, and audio-processing technologies
- Crises and disasters often take place in multinational, cross-border, multilingual, and multicultural settings. As a consequence, media in different languages and dialects need to be collected and processed (Fig. 22.3).

Users of the QuOIMA platform are enabled to access data from different modalities typically just minutes after they have been published and perform various types of visualization and analysis activities by connecting and inter-linking information from different sources and media. The early detection of structures and patterns of communication are expected to allow responders to react earlier, swifter and more adequately in situations of crisis or disaster. QuOIMA focuses on the harvesting and collection aspects of the communication process from the point of view of a first responder and crisis manager or -communicator. Emphasis is placed on the gathering and analysis of (incoming) information, rather than the

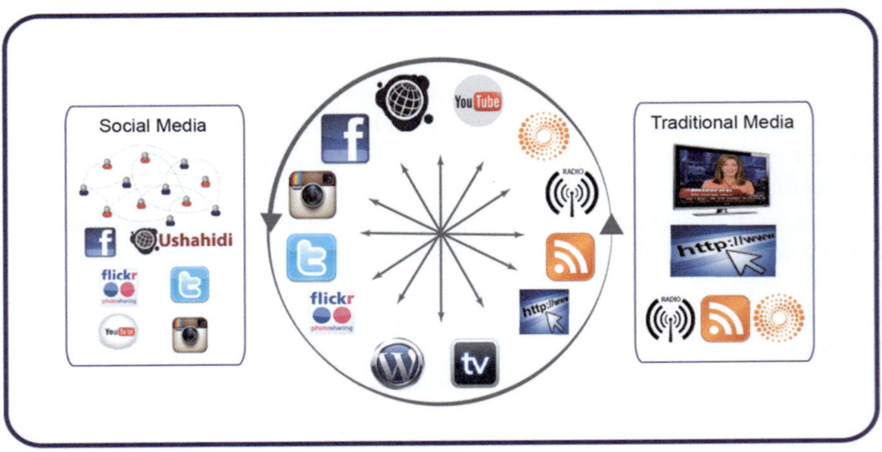

Fig. 22.3 Schematic overview of types of cross-media links occurring between and within traditional and Social Media sources (taken from Backfried et al. 2013a, b)

management of active (outgoing) communication.[2] Cross-media communication and -patterns occurring at different stages of a disaster form one core area of interest within the project. Textual processing, audio-processing, and visual-processing are combined; an underlying multilingual ontology is employed to allow cross-lingual clustering and processing. In particular, the connections between the terminology and vocabulary used at various stages of a disaster (as outlined by the 5-phase disaster model) in combination with cross-media analysis and the identification of communication patterns and their relation to stages of a disaster are investigated. The integral inclusion of Social Media and their correlation with Traditional Media also presents opportunities for crowd-sourcing and crowd-tasking activities.

The overall architecture is based on the Sail Labs Media Mining System[3] designed for real-time processing of open-sources. All individual components, such as the automatic speech recognition engine, are available for a number of languages and built in a way to make them reusable and extensible in a straightforward manner regardless of the actual incident.

In collaboration with a first responder, the Central European Floods of 2013 were selected as the use-case of QuOIMA. Further data-collection is ongoing and will continue throughout the project duration (November 2013), allowing for continuous evaluation of methods and results.

22.3 Central European Floods

The rainfalls in Central Europe of May and June of 2013 caused some of the worst flooding recorded in the affected regions in the last 500 years. Some locations received more than 400 l of rain per square meter in the period of only a few days. This caused severe flooding along the Danube and Inn rivers in Southern Germany and Austria (and subsequently down-river in the Slovak Republic, Hungary and the Balkans) as well as along the Elbe and Saale rivers in Eastern Germany, the Czech Republic, and Poland. This natural disaster is also estimated to have been one of the most expensive catastrophes of 2013 with an overall loss exceeding 12 billion Euros (CEDIM 2013) and causing casualties in several of the affected countries.

Figure 22.4 below shows the areas most affected by the flood.

For Austria, the most extensive flooding was observed along the Inn and Danube rivers in the period from the end of May to the end of June 2013. For the QuOIMA project, the period starting on May 20th—the time when the severe floods started receiving increasing attention by the media—to June 23rd—the time when the peak of the flood had moved downstream from Austria—was chosen as the use-case for

[2]This aspect is certainly of equal importance and will be added to the framework in a follow-on project

[3]www.sail-labs.com

Fig. 22.4 Map of 2013 European floods, Alexrk2 (licensed under CC BY-SA 3.0)

data collection and information analysis activities. Data was collected before, during and following this period in a number of sources, languages and in different types of media, focusing on events associated with the flood.

22.3.1 Data Collection and Corpus Creation

To allow the analysis of cross-media communication phenomena, a parallel corpus spanning various types of media, sources and languages all pertaining to a single event is required. To this end, a corpus covering the floods on various types of media has been created. Data-sources and amounts collected are listed in Table 22.1.

Traditional sources (TV, web, press agencies) were selected according to geographical and content-related coverage. TV sources include public and private stations from Germany and Austria (e.g., ZDF, ORF, NTV) as well as TV-stations broadcasting on a global scale (e.g., CNN, BBC). Web-sources include international, national (Germany and Austria) as well as local newspapers and

Table 22.1 Collected data sources

Medium	Amount	Comment
TV	218 h	13 TV programs, 9 TV stations, 4 countries, German and English
Web	3500 pages	102 German and English sources
Tweets	470 k tweets	German, English, Dutch
Facebook	9800 posts/comments	16 public pages, German and English posts and comments from more than 1000 users
Press Agencies	750 press releases	German

magazines. Facebook data includes posts and comments from public-pages of organizations and individuals involved in disaster-relief. Twitter data was generated in an incremental approach by starting from generic disaster-related terminology and finally producing an active set of actual hash-tags and user-accounts which were used with Twitter's streaming-API to collect tweets. All data collected were processed (transcribed, cleaned, tokenized, . . .) annotated and enriched with meta-data for subsequent analysis purposes.

The resulting corpus spans the period from May 20th 2013 to June 23rd 2013 including the period of the execution phase in Austria. The data collected covers many affected areas; however, due to the project's focus, the following analyses have been carried out with a focus on Austria. Further data-collection activities will continue until the end of the project, allowing reevaluation and verification of findings made during analysis of the first use-case. Further models to be able to cope with languages spoken in countries along the Danube will be added to the QuOIMA system in 2014.

22.4 Observations/Insights

22.4.1 Use of Social Media in Times of Crisis and Disasters in Austria

The topic of Social Media in times of crisis and disasters in the context of Austria has not been researched extensively. Episodic evidence exists, that Social Media—in particular messaging- and image- and video-sharing services like Twitter or YouTube or social networks like Facebook—do get employed in the process of disaster communication by the population as well as by first responders. The *Twitter population* of Austria is currently estimated to be approximately 117,000 persons/accounts—the overall population of the country being approximately 8.5 million. This results in less than 1.5 % of individuals present on Twitter.

In comparison, more than 38 % of all Austrians are estimated to be on Facebook.[4] However, these numbers do not reflect the presence of first responders and organizations as well as of foreigners living in Austria. In spite of the relatively small number of users, analysis of the data collected during the 2013 floods indicates that Twitter is indeed actively used by both, the population as well as organizations. Furthermore, Facebook (we only investigated public pages) was also used actively during the flooding incident. A substantial number of links to further Social Media sites (such as Instagram or Flickr) provides further evidence of the existing (and possibly increasing) importance of Social Media for the disaster context in Austria. It can thus be assumed that these media indeed play a role for crisis- and disaster-communication in the Austrian context.

22.4.2 Coverage Patterns (Timing Behavior)

Over the observed period from May 20th to June 23rd 2013, all data sources, traditional as well as social, generally follow the same trend. Activity is low and rises slowly towards the final days of May. Within the first days of June, however, the picture is quite different with a surge of activity across all sources. TV coverage displays a first peak on May 31st during the ramp-up leading to the height of the flood. As the situation was gradually getting worse with forecasts indicating extreme conditions, these forecasts received substantial attention across all TV stations monitored. Social Media coverage yields most documents relating to the event on June 3rd and 4th (following the impact of flooding by a couple of days) the activity on Twitter leading Facebook posts and comments by 1 day. Web-sources peak 1 day later on June 5th.

All sources display multiple activity peaks corresponding to the relative peaks of the Danube- and Elbe-related floods (whose peak occurred a few days later than the Danube-related one) within a few days only. TV-coverage quickly drops after the initial peaks (potentially due to other, international events replacing the flood-topic swiftly) whereas social media and web-sources continue with high coverage of the events. Figure 22.5 below displays the activity over all observed media by plotting the percentage of documents relative to the total volume within a medium, over the observed period.

22.4.3 Cross-Media Communication

Various forms of cross-media communication may be expected during a major disaster-event: TV programs may list hash-tags and accounts to be used for

[4]Social Media Radar: http://www.socialmediaradar.at/

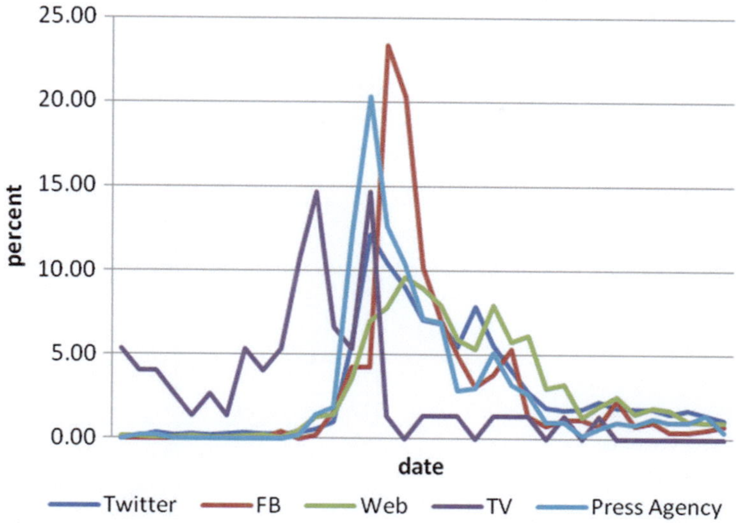

Fig. 22.5 Percentage of documents per medium compared to total volume over observed period of time

communication on Social Media. Posts on Social Media platforms may link to newspaper feeds or image-sharing platforms, newspapers may link to video-sharing sites etc. These interconnections form just a few examples of the many possible combinations and links between and within media types and sources. Not all connections may occur with equal intensity during all phases of a disaster—e.g., links to photo-sharing sites might occur more frequently during the initial phases whereas links to websites might prevail during later stage. All of the above links provide ample opportunities to extend the data collection mechanism—by adding the appropriate sources and by downloading and processing the actual document—as well as for analysis of actors, their messages and agendas. Furthermore, feedback on relief measures and the perception of first responders can readily be made visible through this data and supporting or contradicting information be highlighted.

Table 22.2 below provides an overview of the most frequently linked-to sites from tweets referring to the flood. A large proportion of links emanating from tweets points to other social networking sites. Links to Facebook feature among the most frequent ones. Links to public pages—as private ones are not considered within our project—can be followed and the target pages added to the data-collection component of QuOIMA. Video- and image-sharing websites such as YouTube, Instagram, Twitpic, and Flickr likewise feature prominently. A portion of the targeted videos as well as images can be downloaded and processed using audio- or visual-processing components such as OCR, scene detection or the detection of near duplicates. Visual objects receiving most links (and/or rated on their respective platforms according to meta-information available via APIs) can be identified and presented to analysts. Using this approach, the most-shared image on a popular

Table 22.2 Frequently linked-to Websites from tweets

Domain	Count	Social	News	Weather	Org
www.facebook.com	10,030	x			
www.seismoblog.de	5703			(x)	
Youtube	4946	x			
instagram.com	3524	x			
www.spiegel.de	3187		x		
www.welt.de	2552		x		
www.focus.de	2263		x		
www.mz-web.de	2170		x		
www.tagesschau.de	1778		x		
www.n24.de	1717		x		
www.n-tv.de	1642		x		
www.wetterblogger.de	1598			x	
www.sueddeutsche.de	1476		x		
www.br.de	1458		x		
twitter.com	1420	x			
www.mdr.de	1331		x		
twitpic.com	1152	x			
stuke2.piratenpartei-bayern.de	1031				x
fluddhilfe.de	1022			x	
www.epochtimes.de	1009		x		
www.wetter.com	931			x	
www.dnn-online.de	890		x		
www.faz.net	848		x		
www.morgenpost.de	813		x		
www.zeit.de	812		x		
www.ndr.de	781		x		
www.deutscher-warndienst.de	748			x	x
www.bild.de	741				

Facebook page could be identified as having been taken during a flood several years ago (also along the Danube, so location and flood condition might be genuine, however, the time is not). A further group of link-destinations are news-outlets associated with newspapers, magazines, and TV-stations which can, in turn be added to the list of sites providing information about the event. Finally, weather stations posting river-levels and specific weather information as well as political entities such as parties and local governments are linked to. Conspicuously missing from the table are first responders, such as the Austrian Red Cross[5] or the "Technisches Hilfswerk"

[5]www.roteskreuz.at

(the Federal Agency for Technical Relief, THW) in Germany.[6] This could be an indicator that the presence of such organizations on Twitter is still weak or that information posted is not referenced and passed-on frequently. Both cases could prompt a more active stance of such organizations on Social Media platforms.

22.4.4 Linking Behavior Over Time

Among other factors, the linking behavior of Social Media posts might vary depending on the phases of a disaster and the amount of news-coverage an incident receives. The ubiquity of smart-phones and hand-held devices allows individuals witnessing events to post images in real-time (sometimes even to the point where people have to be reminded to seek shelter before tweeting). These could subsequently be passed-on and linked-to by the recipients or the general public. Later phases might be characterized by news agencies or organizations (such as first responders) providing in-depth and broader coverage which might then be disseminated along the same channels. Figure 22.6 below shows a graph of links to Social Media sites, news-outlets and sites for donations over the core-period of the flooding in Austria. The graph depicts the percentage of links to the particular kind of media as compared to the overall volume per day of links to the same kind of medium.

No clearly discernible pattern of difference in the linking behavior can be observed over time. That is links to Social Media as well as to news outlets follow a similar trend with neither one seemingly being preferred at any stage of the observed period. Links to sites dealing with donations, however, do display this time-related behavior with links increasing dramatically in the days following the highest flood levels (peaking around June 12th, several days into the incident). This observation indicates that information gathering cannot assume any staged

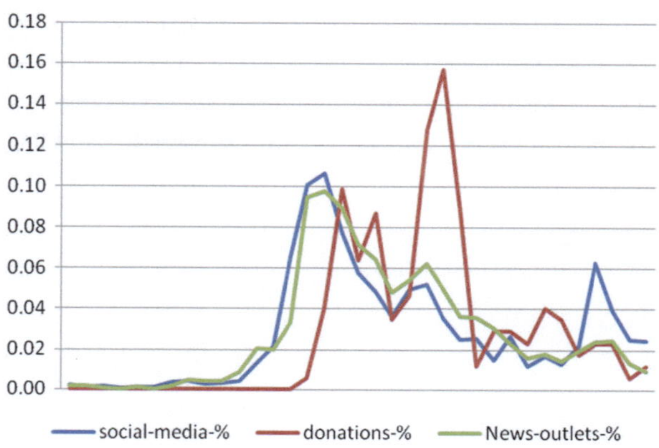

Fig. 22.6 Social Media Sites in the core-period of floods in Austria

[6]www.thw.de

time-distribution over the different types of media but rather should address all types of media—traditional and social—simultaneously and in parallel, i.e., in a phase-agnostic manner

22.4.5 Sentiment Analysis and Bots on Twitter

As part of the evaluation, we also apply sentiment analysis to the different types of media in order to detect common and diverging tendencies, regularities and exceptions (Shalunts et al. 2014).

Applying sentiment analysis to tweets over the period shows predominantly negative sentiment, which—given the nature of the incident—is not surprising. However, specific users yielded a pattern different to the general trend. Several of these accounts turned out to be *bots*—automatically operated accounts on Twitter. Tweets generated by bots do not necessarily have identical content or structure, but may vary considerably over time. In order to detect such *bots*, a bot-detection mechanism much as the one described in Chu et al. (2012) was implemented, taking into account results from sentiment-analysis. Using a conservative parameter-setup for bot-detection shows that approximately 1 % of all accounts involved in tweeting about the natural disaster can be assumed to be bots. However, the resulting amount of tweets corresponds to roughly 22 % of all tweets collected. Including these tweets in analysis and statistics might thus lead to biasing of results. Being able to detect bots and their tweets thus forms a crucial task for data collection. Table 22.3 shows the proportions of users and tweets (for tweets in German pertaining to the flood).

22.4.6 Communication on Facebook

Only so-called public pages on Facebook are considered for data-collection and input within QuOIMA. All posts and comments made on these pages form the documents ingested and analyzed. Accounts include several NGOs from Germany (THW), Austria (Team Österreich[7]), Switzerland (Red Cross[8]) as well

Table 22.3 Proportion of tweets and users for *humans* and *bots*

	Total tweets	Percent	Total users	Percent
Humans	182, 423	78	22, 574	99
Bots	51, 043	22	215	1
Combined	233, 466		22, 789	

[7] www.teamoesterreich.at

[8] www.redcross.ch

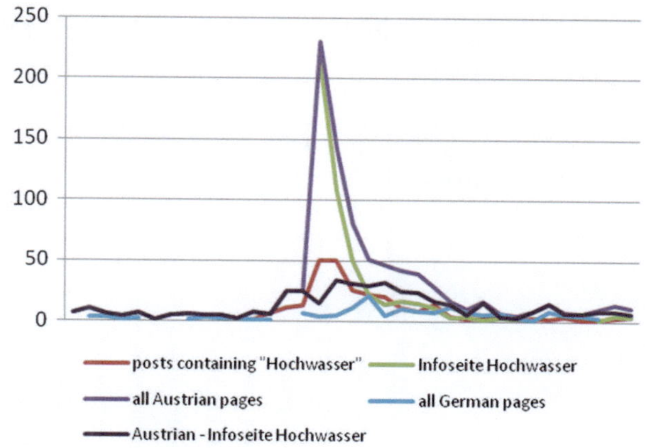

Fig. 22.7 Number of posts and comments on public Facebook pager over time

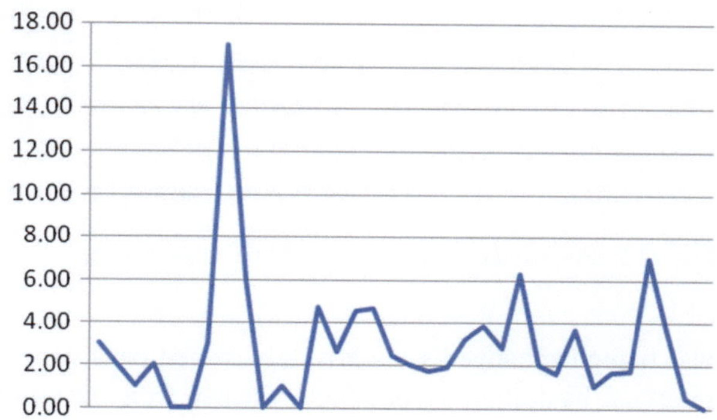

Fig. 22.8 Ratio of comments over posts over time

as international organizations such as the Global Disaster Alert and Coordination System (GDACS[9]) or the Disasters Emergency Committee (DEC[10]) and one particular account created by an individual shortly after the flood hit Austria (Fig. 22.7).

Viewed over time, posting and commenting-volume largely coincide. However, as Fig. 22.8 shows, certain days (and posts) incurred more active commenting (as well as *linking* and *sharing*) making these posts particularly interesting to first responders.

Figure 22.9 shows that approximately 57 % (5583 out of a total of 9801 posts and comments) of all recorded activity relates to a particular account created by an individual shortly after the flood hit Austria (within only the Austrian pages more than 66 %).

Furthermore, the account managed to absorb this share of communication within only a few days after its creation. Requests made by the Austrian Red Cross did not result in cooperation, which yielded a situation where a large share of information

---

[9]www.gdacs.org

[10]www.dec.org.uk

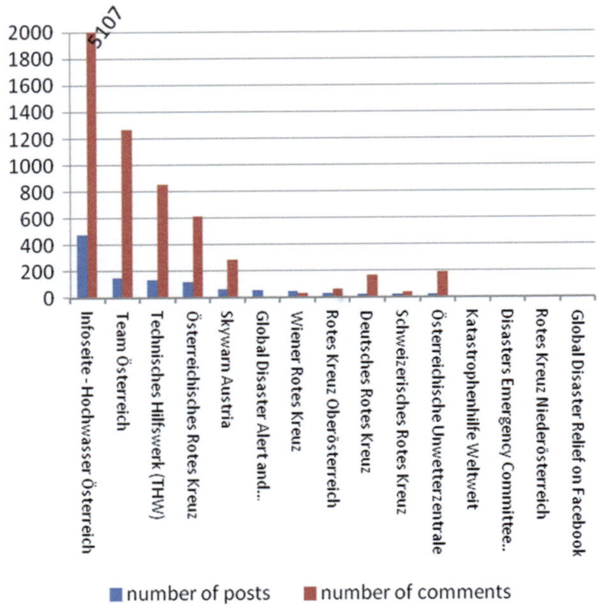

Fig. 22.9 Posts and comments over public Facebook pages observed

was being exchanged via a privately owned page without the possibility to directly intervene by authorities or NGOs. Clearly, from the point of view of a first responder this situation is not optimal, especially with rumors and incorrect information also occasionally being spread through the page. In addition, the owner of a public page may not be obvious and lead to false assumptions of people commenting on posts. This environment clearly demonstrates the need for first responders to monitor media across a wide spectrum including public Facebook pages.

22.5 Expected Positive Impacts for Decision Makers

First responders have a distinct need for a large amount of information and a wide spread of communication channels especially in the domain of crisis and disaster management. Disaster events like flooding scenarios are always a race against time and require therefore a sophisticated and specific management of resources and personnel. Financial, human, relational and structural capital have to be employed and protected effectively and efficiently for all first responders involved in disaster management, be they civil or military.

An analysis of communication content and patterns might lead not only to improved targeted information for decision makers but also to the identification of structures indicating a specific situation, such as a certain phase already having started.

From a practical point of view, it is a strategic advantage for decision makers sending in first responders to know how a situation is developing, especially how people affected by a disaster act and react. Being able to identify certain forms of stress, disappointment, chaos and maybe even a developing tendency to resort to violence against first responders are essential information for team configuration and mission preparation and execution. Furthermore first responders have the possibility to assess concerned persons' needs, alert citizens and other organization of changing conditions, new threats, and the development of the crisis event itself. The result is an improved and more targeted communication with the affected population and a more efficient and effective operational development (Backfried et al. 2013a, b).

Basically the combination of social and traditional media has two immediate effects for first responders.

- Gathering of information emerging in nearly real-time from traditional and social sources allows to improve their situational awareness and arrive at a more complete, varied, and richer picture, enhancing the basis for decision making processes.
- Disseminating of quality-assured, trustworthy information via social and traditional media allows to communicate in a more precise manner with the affected population to coordinate evacuation and disaster relief activities and to guide affected people through the crisis in general.

The QuOIMA System establishes a common information space for all phases of the disaster, where decision makers can retrieve phase-relevant pre-analyzed and quality assured information not only to integrate this information in their processes and management activities but to implement a shared awareness between all organizations (and the particular representatives in the affected areas) involved in a relief effort.

22.6 Conclusions

Following the above mentioned Z-model, the actual project is one more step on the way towards a more comprehensive and generic strategic long-term planning based on risk management and horizon scanning systems.

The analysis of and work with Social and Traditional Media have led to the question, how information and communication in crisis and disaster management will look like in the future. The research brought up not only the benefits but also the problems of such a system following a comprehensive and generic approach. Whereas these insights concern mainly the information gathering aspects, outbound communication requires similar solutions and methods to be investigated in follow-on projects.

Such a comprehensive system must be able to provide all types of information in every phase of a crisis or disaster. Disasters like flooding events are not only critical in terms of time but often cover a broad geographic area crossing national, international and cultural borders. Considering that the peak of the use-case flooding, e.g., moves from East to West, the first responders in the affected areas are in different stages or phases within the event development. This fact demands different information for different people who are following different operational processes. Additionally first responders have not only to coordinate and synchronize activities beyond their operational processes, limitations, and structures but often to deal with language differences as well.

The QuOIMA System meets these challenges. We described the relevance and importance of the usage of Social Media in crisis and disaster management and implemented a multilingual, real-time information and communication system, which provides information fast, phase-agnostic, relevant, and quality-assured to improve not only decision making processes but communication and disaster relief measurements as well.

Acknowledgements The research work leading to these results is funded by the Austrian Federal Ministry for Transportation, Innovation and Technology[11] (BMVIT) within the KIRAS[12] program.

References

American Red Cross (2010) The case for integrating crisis response with social media. White Paper

Backfried G, Göllner J, Quirchmayr G, Rainer K, Kienast G, Thallinger G, Schmidt C, Pfeiffer M, Meurers C, Peer A (2013) Cross-media analysis for communication during natural disasters, IAIT 2013. International conference on advances in information technology, Bangkok, Thailand, Dec 2013

Backfried G, Schmidt C, Quirchmayr G, Rainer K, Kienast G, Thallinger G, Göllner J, Peer A (2013) Integration of media sources for situation analysis in the different stages of disaster management, EISIC 2013, European Intelligence and Security Informatics Conference, Uppsala, Sweden, Aug 2013

Backfried G, Schmidt C, Pfeiffer M, Quirchmayr G, Glanzer M, Rainer K (2012) Open source intelligence in disaster management, EISIC 2012, European Intelligence and Security Informatics Conference, Odense, Denmark, Aug 2012

CEDIM, Juni-Hochwasser 2013 in Mitteleuropa - Fokus Deutschland, http://www.cedim.de/download/FDA_Juni_Hochwasser_Bericht2.pdf on 2014/06/23

Chu Z, Gianvecchio S, Wang H, Jajoda S (2012) Detecting automation of twitter accounts: are you a human, Bot or Cyborg? IEEE Transactions on Dependable and Secure Computing 9(6):811–824

DHS Report (2013) Lessons learned: social media and hurricane sandy. https://www.llis.dhs.gov/sites/default/files/Lessons%20Learned%20-%20Social%20Media%20and%20Hurricane%20Sandy.pdf, 06/25/2014

Göllner J, Meurers C, Peer A, Langer L, Kammerstetter M (2014) Bedeutung des Risikomanagements für die Sicherheit von Smart Grids. Proceedings of 13 Symposium Energieinnovation, Graz

[11] http://www.bmvit.gv.at/

[12] http://kiras.at/

Holmes W (2011) Crisis communication and social media: advantages, disadvantages and best practices. Univ. of Tennessee, CCISymposium, 2011, Knowville, USA

Klerx J, Göllner J, Mak K (2014) Horizon scanning for emerging risks in supply chains. Presentation 24.04.2014, EMCSR 2014, conference European meetings on cybernetics and systems research, Vienna

Langer L, Göllner J, Meurers C, Peer A, Kammerstetter M, Bleier T (2014) Importance of risk management for the security of smart grids. EMCSR 2014, conference European meetings on cybernetics and systems research, Vienna, 2014

Lindsay B (2011) Social media and disasters: current uses, future options, and policy considerations. Congressional Research Service

Nilsson J et al (2012) Making use of new media for Pan-European crisis communication. ISCRAM 2012, Vancouver, Canada

Shalunts G, Backfried G, Prinz K (2014) Sentiment analysis of German social media data for natural disasters. 11th International conference on information systems for crisis response and management, ISCRAM 2014, University Park, PA, USA

Chapter 23
Coordination of Decision-Making in Crisis Management

John Dowell

Abstract Catastrophic civil events give rise to large and complex response operations involving many agencies and individuals. Coordination of this response operation has been a long standing problem that cannot be solved by simply creating better procedures; on the contrary, it is an emergent and transmuting phenomenon that arises in the interactions between multiple agents as they confront high risks, short time-scales and poor data. This chapter examines coordination by considering crisis management in terms of distributed multi-agent decision-making. Coordination is then identified separately with the planning, actions, communications and knowledge of this multi-agent system. This framework is used to examine decision-making coordination at the scene of a major railway accident at Clapham UK in 1988. Decisions affected by the particular difficulty of ensuring the electrical isolation of the accident scene are a focus for the case study. Factors influencing the quality of coordination are assayed from the case study and this analysis informs our understanding of crisis management preparedness and training.

23.1 Introduction: The Issue of Coordinating the Managed Response to a Crisis

Catastrophic accidents, natural disasters and equivalent kinds of disruptive event are as unpredictable as they may be inevitable. Nevertheless, when civil crises occur, an effective response is expected and this turns on the coordination achieved in managing the response (Quarantelli 1988; Auf der Heide 1989).

The difficulties of managing a crisis response operation increase significantly with the involvement of greater numbers of responders and with multiple agencies. As well as the emergency services, major emergency operations can involve local authorities, specialist expert groups and commercial organisations. Consequently, there will be complex inter-dependencies between the activities of different agents; for example, ambulance crews arriving at the scene of a catastrophe will expect

J. Dowell (✉)
University College London, London, UK
e-mail: j.dowell@cs.ucl.ac.uk

G. Rogova, P. Scott (eds.), *Fusion Methodologies in Crisis Management*,
DOI 10.1007/978-3-319-22527-2_23

directions from the police who themselves will need to be informed by the fire service about the location of people needing treatment and about how to access the site. This is the problem of coordination and is both widely recognised and longstanding. For example, it was a significant factor in the response to the Harrow rail accident in 1952, the Kings Cross Station fire in 1988 and the aftermath of Hurricane Katrina (Fennell 1988; Townsend 2006), with one public enquiry report recording that "*the court was left with the impression that there had been a breakdown of communications at command level between the emergency services. Each diligently pursued its own duty but there was a lack of liaison between them*" (Fennell 1988). The persistence of the problem of coordination is an indication of its emergent and transmuting nature.

Recommendations for coordinating crisis management have been developed and are reflected in contingency plans and training, a representative example being the plans published by the UK government (CCS 2013). This guidance acknowledges that inter-agency coordination arises from the essential need for different agencies to work together, rather than working under instruction from one or other. It recommends that coordination at the operational level "on the ground" should be facilitated by a lead responder agency, typically the Police. An ad hoc multi-agency coordination group should also be convened to determine coordination requirements and solutions at a tactical level. When there is a distinct scene for the incident, this tactical group will meet face-to-face at the local and temporary incident command room. The UK guidance acknowledges the essential non-monotonic nature of crisis management and advocates a principle of subsidiarity wherein "*decisions should be taken at the lowest appropriate level with coordination at the highest necessary level*". In sum, these recommendations for coordination are focused on establishing organisational structures capable of recognising, negotiating and effecting coordination, rather than on recommending definite processes or mechanisms by which coordination would be achieved. Formulating procedures for coordination requires a better understanding of the substantive nature of coordination.

This chapter examines coordination by considering crisis management in terms of distributed multi-agent decision-making. This supports an analysis of decision-making in the response operation at the Clapham railway accident. Factors determining the quality of coordination of that decision-making are assayed from the case study.

23.2 Crisis Management as Multi-agent Decision-Making

The operation mounted in response to a civil crisis will progress through phases of containment, rescue and restoration (Partridge et al. 2012). Whilst the priorities of the response operation change through these different phases, the decisions made are all fundamentally concerned with allocating resources (personnel, supplies, equipment and facilities) to meet demands (threat neutralisation; movement of victims to safety; medical care; protection of evidence, etc.). Usually the demand for resources exceeds and may overwhelm their availability; this is what marks out a crisis from a routine emergency. The decisions made in managing a crisis are therefore fundamentally about resource allocation (Fiedrich et al. 2000; Hick et al. 2012) and are particularly characterised by risks, uncertainties and limited time.

The risks are that losses may be un-controlled, that lives may not be saved, and that suffering, damage and disruption may continue; there are risks too for the responders themselves and for preserving their own assets.

Decisions will be made under considerable uncertainty about the scale and nature of the demands and about constraints on the availability of the resources. For example, it is vital that the first responders correctly assess the situation and determine whether to report it as a major incident, so triggering a scaled crisis response operation. In many incidents the scale of the situation will be immediately apparent, but not so in all and a delay in initiating the correct scale of response will significantly affect outcomes.

Decisions will be made under strong time constraints because a crisis possesses its own tempo as the situation evolves and as the effect of the operation itself is felt (Samurçay and Rogalski 1993). The opportunities for deploying resources effectively are time limited, for example, most lives that can be saved will be saved in the so-called 'golden hour' following a catastrophic event.

Crisis management is also decision-making that is distributed over multiple agents (Mendonca et al. 2001; Van Veelen et al. 2006; Fiedrich and Burghardt 2007). Decisions about the allocation of resources to demands in managing the response to a crisis are shared by multiple agents within and across agencies. Typically, knowledge about those demands and resources will be distributed amongst these multiple agents. In spite of the complexity created by this shared decision-making and distributed knowledge, the coordination of crisis management can be associated with the separate factors of planning, action, communication, and knowledge

23.2.1 Coordinating Multi-agent Planning

Each agency participating in an emergency operation will have contingency plans drawn up in readiness for the different kinds of catastrophe that may occur. Those plans will be known to each agent and will be learnt and refined through training exercises. Yet because catastrophes are fundamentally unpredictable, contingency plans must be partial, high level, and conditional; accordingly, for the occasion of each catastrophe, contingency plans will be interpreted by each agency and each agent, regardless of level of authority. If this plan interpretation is conducted implicitly and informally, coordination will deteriorate. In particular, the recognised partitioning between the priorities of different agencies may become dislocated, leading to priorities that are duplicated, in conflict, or which neglect important concerns. Because of limited time and uncertainty, planning will need to be opportunistic but this too will likely confound coordination of decision-making.

Coordination requires that contingency plans are explicitly interpreted and that conflicts can be recognised and resolved. It requires that the priorities of all agents, whilst necessarily varying, should be coherently aligned. It also requires opportunistic planning to be made explicit.

23.2.2 Coordinating Multi-agent Activities

Since plans cannot be reduced to simple procedures (Quarantelli 1988), the actions of different agents during the crisis response are determined by mutual constraints established between those agents and not by aggregation of individual contingency plans. In this way, individuals create behavioural dependencies between each other. Action becomes reactive to the effects of other agents' actions and the operation emerges in the interaction between actions rather than simply implementing a master plan (Mendonca et al. 2001).

Coordination relies on the recognition and negotiation of dependencies between activities and it requires that the effects of the actions of each agent are visible to those who will need to act on them.

23.2.3 Coordinating Communications

Information disseminates progressively but often only slowly and indirectly during crisis response operations (Hale et al. 2005; Reddy et al. 2009). Information channels will often be limiting, making communications asynchronous and reducing feedback (Kapucu 2008).

Coordination requires authoritative, validated and accurate information to be disseminated rapidly. It relies on the timely passing on of information directly to those who may need it. In the absence of feedback, coordination requires an understanding of how others will respond to messages and actions.

23.2.4 Coordinating Role and Task Knowledge

If coordination requires each agent to know how others will respond to their messages and actions, then different agents must have knowledge of each other's tasks. There must be at least a partial redundancy in knowledge and skills (Emery and Trist 1973). This redundancy of task knowledge is a prerequisite for modifying the real divisions of labour, whether the modification is achieved through violation or negotiation (Hutchins 1991).

Coordination requires agencies and their agents to have redundant knowledge and skills. It requires differences in the division of labour to be negotiated. The cognitive horizon that delimits the uniqueness and redundancy of an agent's knowledge is critical to recognising which agents need extra support: giving extra support where it is needed is a form of coordination and means that cognitive horizons must be sufficiently broad.

23.3 Making Decisions About Track Isolation at the Clapham Railway Accident

Faulty wiring in the railway signalling system at Clapham in December 1988 caused an automatic signal to start showing green instead of red. A commuter train travelling into rush-hour London then collided with a stationary commuter train waiting ahead in a deep, curved cutting. The first train then veered to its right, striking a third, oncoming train. As a result, 36 people died, and some 500 were injured, including seriously injured and trapped people; in all some 1500 passengers were involved in the accident. It was the worst accident on British railways in the last half century and would significantly change how the system is operated thereafter (Vaughan 2000).

A detailed account of the emergency relief operation is contained in the report of the public inquiry into the disaster (Britain and Hidden 1989) and finds that the operation was highly effective. This was achieved in spite of significant difficulties confronted, some of which arose from the structures and processes of that operation itself, including the coordination of the different agencies involved. By examining just one of the difficulties facing the rescuers—the electrical isolation of the track—the nature of coordination of multi agent decision-making in crisis management is revealed.

All railway lines through Clapham have three rails, the third rail providing the 750 V traction current for electric trains. It was a foremost priority of the emergency operation to remove the current from the third rail and ensure the electrical isolation of the accident. Electrically isolating the emergency would prevent other trains becoming involved in the accident. At 8:10 am, the time of the accident, trains were running with 2 min headway between them and—as will be apparent, the risks to other trains were considerable. Electrical isolation would also remove the risk of electrocution to passengers escaping from the wreckage and to rescuers coming to their aid. So how was the isolation of the accident achieved?

Immediately following the collision, Inspector Forster of the British Transport Police who had been travelling on the second train and was uninjured in the accident, climbed onto the track. Using a trackside phone, Inspector Forster told Signalman Cotter at the Clapham signal box: "*You have a major train crash: summon the emergency services*"; the signalman immediately set to red all the manual signals in his area of responsibility. Inspector Forster then ran to the rear of the wrecked trains, shouting to the passengers to remain on-board. Once behind the crash, he placed a short-circuiting bar on the rails behind the wrecked trains; this bar shorted out the traction current. Because there was no electrical arcing when he attached the bar, Inspector Forster concluded that the current was already off, most likely shorted out by the collision itself.

The traction current was actually controlled from the Earlsfield sub-station some miles away from the accident scene. Working at the sub-station, a Mr Reeves had noticed the indicators on his "diagram" start to move at 8:10 am. The alarm bell had then rung and flashing lights appeared for all the Up and Down main lines at Clapham. At the same time, the sub-station current had tripped at the National Grid supply, from which Mr Reeves concluded there had been a severe derailment at the "Up Cross-over". Mr Reeves first action was to restore the supply to allow trains to continue to run and to be signalled safely. When that was done, he proceeded to create neutral electrical sections around Clapham. Given the apparent severity of the situation at the Up crossover, Mr Reeves isolated the Up and Down tracks of the local as well as the shorted out main lines. This was because isolating all four lines would ensure the greatest protection to whatever may have occurred. Mr Reeves did not attempt, as the rules instruct, to at intervals reopen the circuit breakers to the shorted lines.

Mr Reeves completed this work by 8:12 am; his next task was to use the internal telephone to inform Signalman Cotter at the Clapham signal box of what had been done. Mr Reeves found the line to the signal box constantly engaged, and so tried the traffic controllers at Waterloo station: they were also engaged. It was some 10 min before Mr Reeves could inform Signalman Cotter at the signal box, and then only by informing the Waterloo controllers.

Yet during this time, a train was heading towards Clapham on the Up main line. Driver Pike in the cab of this fourth train sensed the disconnection of traction power. Because the faulty automatic signal was still set to "all clear", Driver Pike assumed there was an electrical fault on his locomotive, and so decided to coast as far as possible on to Clapham; without radio or telephone, he was oblivious to the accident ahead. Rounding the bend on the approach to Clapham and at about 60 mph, Driver Pike was able to see the wreckage of the accident and applied the emergency brakes, bringing his train to rest with 60 yards to spare.

Together with the police patrol, two fire tenders arrived at the scene at 8:17 am, Station Officer Mills (fire brigade) taking charge of the emergency. SO Mills immediately summoned eight more fire tenders, eight ambulances and a surgical unit; when he had made a more thorough assessment, SO Mills declared a "Major Incident", bringing into action a massive rescue operation. SO Mills then directed the rescuers as they arrived, before being replaced by more senior officers.

Amongst those arriving at the scene was the Area Manager from Clapham Junction, having been hastily briefed by Signalman Cotter. The rule book states that on arrival, the Area Manager must "*check that the traction current has been switched off and that Electrical Control is aware.*" However, the Area Manager had not been told, or could not recall being told, if the current was removed. Rather than dealing with this issue as a priority, he instead became involved in organising a diesel train to evacuate passengers; evidently traumatised by the scene he was witnessing he was shortly relieved by other arriving railway management.

So in deciding whether to enter the crash site and make the critical assessment of the scale of the disaster and response needed, SO Mills was confronted with the considerable uncertainty about whether the track was still live. He had not spoken with Inspector Forster, the Area Manager could not tell him, and he had no communication with either Signalman Cotter or Mr Reeves. SO Mills became convinced—correctly, but for reasons which are now unknown that the traction current was removed. He therefore ordered that walking passengers be evacuated immediately and allowed the medical teams to enter the wreckage. Very severe injuries were found requiring immediate life-saving procedures, to the extent of setting up blood transfusions for some trapped passengers even before they could be freed from the wreckage. SO Mills ignored the procedure of posting lookouts on the track to warn other trains, presumably because he believed the crash site was now isolated and safe from other traffic. At 8:32 am a message was sent to British Rail at Waterloo requesting confirmation that the current was off and that confirmation was returned at 8:35 am.

23.4 Coordination of Decision-Making at the Clapham Accident

This single aspect of the crisis response operation—the removal of the traction current, is richly woven with the raw phenomena of decision-making coordination. By viewing the operation in terms of distributed multi-agent decision-making, the coordination of decisions relating to the track isolation can be recognised. Claims about factors affecting the quality of coordination are also exemplified through the Clapham case.

Even when individual responses are well trained, ineffectiveness can result if the responses are not coordinated. Inspector Forster, for instance, knew to tell passengers to stay on the train and knew to place the shorting bars on the track. He also could tell that the current was already shorted out. Mr Reeves knew to isolate all the lines around the Up crossover; he even informed Signalman Cotter and the Waterloo controllers that the lines were isolated. SO Mills knew that the lines must be isolated before ordering the evacuation, yet he was not informed—when he most needed to know, that the traction current was indeed off.

Contingency plans are necessarily partial and conditional; because they must be interpreted by each agent for each crisis, coordination problems may occur. Mr Reeves, for instance, interpreted his contingency plan for track isolation as requiring the isolation of the local lines as well as the Up and Down main lines. As a consequence, it is possible that the Area Manager did not know, until 8:35 am, whether the local lines were isolated, and so whether he should procure only a diesel train for evacuating passengers. *Contingency plans may even be ignored altogether, with both beneficial and detrimental consequences for coordination.* Mr Reeves, for instance, ignored the contingency plan requiring him to attempt, at intervals, to restore the current. His action allowed rescuers at the accident scene to continue their work in safety. By contrast, the failure of the Area Manager to obtain track isolation made fraught the decision by SO Mills to proceed with the evacuation.

Individuals and agencies may have different priorities, and without consultation important concerns may be ignored: coordination requires individuals and agencies to have a common structure of priorities. SO Mills, for instance, concentrated attention on evacuating passengers and attending to the injured. These are proper priorities of the fire brigade, but nevertheless, as overall manager of the crisis, SO Mills should also have posted lookouts to remove the risk to other trains; containment of the crisis has an equal priority with the rescue of victims.

The sequencing of different agents' actions during the emergency operation follows from the mutual constraints between agents; it does not come from individual contingency plans. In this way, individuals create behavioural dependencies between each other: coordination relies on the recognition and negotiation of those dependencies. Mr Reeves, for example, established a mutual constraint with Signalman Cotter on which would depend the sequence in which the traction current was restored to different track sections away from the scene of the accident. Mr Reeves would restore the current to the sections at the request of Signalman Cotter—who would redirect the halted train traffic through other parts of the Clapham network. Mr Reeves would maintain a level of isolation in negotiation with Signalman Cotter. The crowded telephone line did not enable this dependency.

The actions of one agent are reactive to the effects of other agents' actions. In this way, the response operation emerges in the interaction between decisions about actions, it does not follow some master plan. Coordination requires that the effects of actions can be recognised by those who will need to act on them. Driver Pike, for instance, sensed the removal of the traction current and, since the signals were still showing all clear, incorrectly inferred a fault with his locomotive. The effects of Mr Reeve's actions were not transparent to Driver Pike, even though they were immediately felt. *Again, feedback may be poor or absent: coordination relies on knowing how others will respond to your messages or actions.* For instance, in the absence of feedback, Signalman Cotter would have inferred that all trains had been halted. If Signalman Cotter had known how Driver Pike would react to the disconnection of traction current, he might have urged the posting of a lookout on the line and the early placing of warning detonators.

The opportunity for some actions will be short, limiting the time for communications and decisions: coordination depends on the right decisions and communications at the right time. Inspector Forster, for instance, had to inform the signal box of the accident as soon as possible, since Signalman Cotter could best prevent another train from colliding with the wreckage, but only if informed sufficiently quickly. Inspector Forster phoned Signalman Cotter before installing the shorting bar.

Authoritative, validated and accurate information disseminates progressively but often only slowly. Coordination requires the rapid dissemination of authoritative, validated and accurate information. Mr Reeves, for instance, was only "informed" of the accident 10 min after it occurred; meanwhile, he had acted on the basis of assumed and ambiguous information (the mimic board indicators moving, and so on). As regards the authority of information, Inspector Forster, told Signalman Cotter there had been a "major train crash" which brought the first emergency service units to the scene. When SO Mills made his assessment, he declared a "Major Incident", a precise term reserved for the most severe of accidents and understood by all agencies to demand the fullest response. Only an officer of SO Mill's rank could use the term; his message carried the greatest authority.

Communications are asynchronous: coordination depends on "passing on" information to those who may need it. The Area Manager, for instance, failed to recognise that he should be informed by Signalman Cotter about the state of the track isolation before leaving for the accident scene because on arrival at the scene he would need to brief SO Mills.

The nominal and real divisions of labour may vary, either through violation or negotiation. Coordination requires differences in the division of labour to be negotiated. The Area Manager, for instance, departed from his nominal role by not obtaining electrical isolation and by allowing SO Mills to take on this task. Mr Reeves also violated the nominal division of labour, but with positive effects, by deciding to isolate the Clapham south track areas without receiving instruction from the signal box. *Further, if the division of labour is to be negotiated, then there must be at least a partial redundancy in knowledge and skills: coordination requires a redundancy of knowledge and skills between agents and agencies.* This redundancy might only be inferred in the actions of the agents at the Clapham accident, for instance, in SO Mills requesting the track isolation, in Inspector Forster attaching the shorting bar.

The cognitive horizon which delimits the uniqueness and redundancy of an agent's knowledge is critical to recognising which agents need extra support: if coordination requires that extra support be given where needed, then cognitive horizons must be sufficiently broad. Mr Reeves, for instance, recognised that if a major derailment had occurred, whoever was working in the Clapham Signal Box would be needing all the help they could get to stop the other traffic. Isolating the area without waiting for a discussion was the greatest help Mr Reeves could provide: his cognitive horizon was sufficiently broad to allow him to recognise the help Signalman Cotter would be needing.

23.5 Developing Coordination of Crisis Management

When viewed as distributed multi-agent decision-making, crisis management has systemic coordination characteristics that can be separately associated with the coordination of planning, of action, of communications, and of knowledge. Each of these factors gives rise to primary coordination requirements:

- Coordination requires that contingency plans are explicitly interpreted and that conflicts can be recognised and resolved. It requires that the priorities of all agents, whilst necessarily varying, should be coherently aligned. It also requires opportunistic planning to be made explicit.
- Coordination relies on the recognition and negotiation of dependencies between activities and it requires that the effects of the actions of each agent are visible to those who will need to act on them.
- Coordination requires authoritative, validated and accurate information to be disseminated rapidly. It relies on the timely passing on of information directly

to those who may need it. In the absence of feedback, coordination requires an understanding of how others will respond to messages and actions.

- Coordination requires agencies and their agents to have redundant knowledge and skills. It requires differences in the division of labour to be negotiated. The cognitive horizon that delimits the uniqueness and redundancy of an agent's knowledge is critical to recognising which agents need extra support: giving extra support where it is needed is a form of coordination and means that cognitive horizons must be sufficiently broad.

These coordination requirements were evident in the aftermath of the Clapham accident. The potential consequences of poor coordination included another train colliding with the wreckage, the possible electrocution of rescuers and delays in evacuating passengers and attending to the injured. Those consequences were avoided and in fact the public inquiry recorded that the emergency services had worked together in an "exemplary manner". Examining the particular question of the track isolation has exposed the processes within the operation and the elements of coordination associated with them. They reveal an organisation convened and fashioned to the particular exigencies of the crisis where the dependencies between decisions made within and across agencies is only partly managed by reference to plans and procedures and is very significantly managed by mutual knowledge and shared inference. Even when individual responses are well trained, ineffectiveness can result if agencies and their agents are not well coordinated.

Joint-planning between agencies was recommended by the Clapham inquiry (Britain and Hidden 1989), to be reinforced by training including multi-agency table-top simulation exercises (Dowell 1995). Such exercises are now an established, routine part of crisis preparation. This analysis of the Clapham accident identifies features of coordination that need to be incorporated into those exercises and which should be the focus for the development of processes, structures and technologies to better support coordination.

References

Auf der Heide E (1989) Disaster response: principles of preparation and coordination. CV Mosby Company, St. Louis

Britain G, Hidden A (1989) Investigation into the Clapham Junction railway accident. HMSO, London

Civil Contingencies Secretariat (CCS), 2013, Emergency Response and Recovery. Cabinet Office: London. https://www.gov.uk/government/publications/emergency-preparedness

Dowell J (1995) Coordination in emergency operations and the tabletop training exercise. Trav Hum 58(1):85–102

Emery FE, Trist EL (1973) Towards a social ecology: contextual appreciation of the future in the present. Plenum Press, London

Fennell D (1988) Investigation into the King's Cross underground fire. HMSO, London

Fiedrich F, Burghardt P (2007) Agent-based systems for disaster management. Commun ACM 50(3):41–42

Fiedrich F et al (2000) Optimized resource allocation for emergency response after earthquake disasters. Saf Sci 35(1):41–57
Hale JE et al (2005) Crisis response communication challenges building theory from qualitative data. J Bus Commun 42(2):112–134
Hick JL et al (2012) Allocating scarce resources in disasters: emergency department principles. Ann Emerg Med 59(3):177–187
Hutchins E (1991) The social-organization of distributed cognition. In: Perspectives on socially shared cognition. American Psychological Association, Washington, DC, pp 283–307
Kapucu N (2008) Collaborative emergency management: better community organising, better public preparedness and response. Disasters 32(2):239–262
Mendonca D et al (2001) Decision support for improvisation during emergency response operations. Int J Emerg Manag 1(1):30–38
Partridge RA et al (2012) Oxford American handbook of disaster medicine. Oxford University Press, Oxford
Quarantelli EL (1988) Disaster crisis management: a summary of research findings. J Manag Stud 25(4):373–385
Reddy MC et al (2009) Challenges to effective crisis management: using information and communication technologies to coordinate emergency medical services and emergency department teams. Int J Med Inform 78(4):259–269
Samurçay R, Rogalski J (1993) Cooperative work and decision making in emergency management. Trav Hum 56(1):24
Townsend FF (2006) The federal response to Hurricane Katrina: lessons learned. The White House, Washington, DC
Van Veelen B et al (2006) Effective and efficient coordination strategies for agile crisis response organizations. Proceedings of ISCRAM 2006
Vaughan A (2000) Tracks to Disaster. Ian Allan

Chapter 24
HAZMAT Tracking: Compatibility Organizational Theory Case Study

Nicolas A. Valcik

Abstract Universities, colleges and research centers have always been incubation centers for ideas, intellectual property and research on new technologies in the United States. Prior to Patriot Act, the only mandates universities, colleges and research centers faced were with radioactive material (e.g. U-235) and environmental protection guidelines (e.g. waste). After the passage of the Bioterrorism Preparedness 256 and Response Act of 2002 and the Homeland Security Chemical Facility Anti-Terrorism Standards passed in 2007, higher education institutions as well as research centers now are having to control for biological agents and chemical elements that are now mandated to be reported to the federal government under specific criteria. That being stated, organizational decision-making in universities, colleges and research centers have organizational culture to take into account when new procedures and policies are to be implemented. The case study seen in this chapter analyzes the decision-making process in context of the organizational culture when the new mandates were passed by the federal government for homeland security and disaster response purposes.

24.1 Introduction

The purpose of this research is to explore a research organization (which will remain anonymous) and how it inventoried its hazardous materials (HAZMAT) before and after the year 2005. While the effort to track HAZMAT should ideally be tightly coupled with the federal mandates, this case study illustrates how the ideal operational practices are at times not tightly coupled with the operational aspects of the organization. This research will demonstrate how technological solutions are often decoupled from organizational culture and behavior. The author gathered data from interviews with staff members, archival documentation, unobtrusive observation and participant observation. In this manner the author was able to triangulate the data in a qualitative study which can be utilized for critical assessment in the

N.A. Valcik, Ph.D. (✉)
Director of Institutional Research at West Virginia University, Public Administration, College of Arts and Sciences West Virginia University, Morgantown, WV, USA
e-mail: Nicholas.Valcik@mail.wvu.edu

G. Rogova, P. Scott (eds.), *Fusion Methodologies in Crisis Management*,
DOI 10.1007/978-3-319-22527-2_24

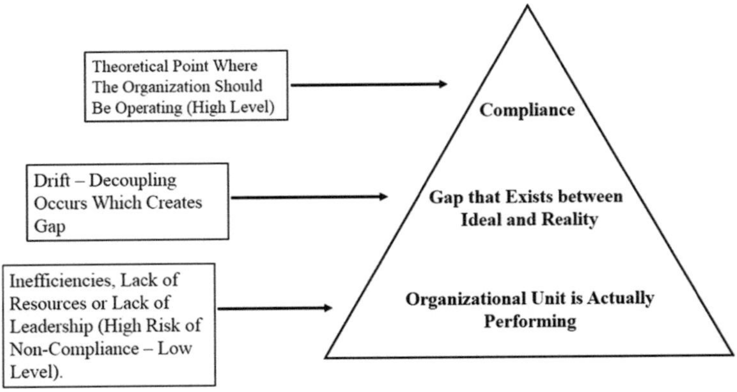

Fig. 24.1 Organizational drift—reality versus ideal

case study. The case study organization will be known as Research Institution, for the purposes of this paper and the identities of participants in the case study are identified only by letters to represent individuals. This case study will focus on the gap between what should be practiced (theory) and what actually occurred within a research organization which resulted in organizational drift (Meyer and Brian 1977) (Fig. 24.1).

24.2 Existing Management Information Systems (MIS) Organizational Theories

There are several organizational theories which address MIS in relation to the productivity or use by a particular organization (i.e., Policy Studies, Stages of Growth Theories and Political Process Models). There are very few MIS organizational theories however which take a "layered" approach to a set of organizational theories used to explain how new technology does or, more importantly, does not ultimately impact an organization's output.

Policy studies focusing on technology implementation and information resources management never appear to address the issues of the political aspects or existing personnel's agendas regarding implementation of newer technology (Caudle 1987; Morell 1989; Vasu et al. 1998). MIS economic theories are primarily focused on financial profit models (Janson et al. 2013). In this context, information technology systems/software vendors are expected to attempt to maximize their profit when selling to public or nonprofit organizations in accordance to economic theories.

MIS public goods theories focus on free markets driving information systems usage and adoptions (Boettke 2010). Neither of these theories consider the recipient institution's organizational culture, or for that matter, the organizational structure of public organizations whose primary objective is other than attempting to maximize profit for personal gain. These theories also do not address the issue of public organizations having an "ethical duty" to be as efficient as possible in serving the public interest and to keep costs down since public funds are used to finance the organization.

Stage of Growth Theories make an assumption that organizations change based on an evolving technology basis (Nolan 1979; King and Kraemer 1984). While the theory has plausible implications for MIS in public organizations, it does not take into account internal resistance to moving away from older technology. Operational personnel (or senior administrators for that matter) in decision-making positions have the ability to forestall implementation of new technology, sometimes indefinitely. These individuals may potentially be either actively resisting change or quietly sabotaging efforts progressing to newer technology by simply not using the technology. This resistance to change may actually cause certain improvement stages to be omitted or deferred as administrators or employees that resist a certain stage over a period of years retire, transfer, are terminated, or otherwise quit the organization.

Political process models (i.e., Socio-technical Models, Planning Process Models, and Political Process models), uniformly make an assumption that conflict with interested stakeholders is eventually be resolved (Keen 1981). The resolution of the conflict will then allow for progression of technology within the organization (Keen 1981). These theories also, however, do not address the issues of intransigent resistance to new technology by stakeholders within the organization that has no resolution. These theories similarly assume that an organization operates in an integrated fashion as opposed to departments internal to an organization that are in "silos" each with its own internal political power base.

If an organization is operating in a dated business model with silos for each department, then implementation of technology which relies on an integrated organizational business model necessarily will be incompatible with the existing organizational culture. An administrator or researcher could certainly use a Grounded Theory approach to assess the success or failure of such an implementation (Glaser and Strauss 1967). Such investigative techniques will allow an administrator to work backwards to identify specific factors that led to a particular IT solution implementation outcome.

Another omission from traditional MIS theories is that they do not substantially consider the situation of weak central leadership in regard to information technology issues as related to the entire organization. This incompatibility can potentially result in a failure to adopt and implement new technology in the absence of strong central leadership in the organization that exhibits such a siloed organizational structure.

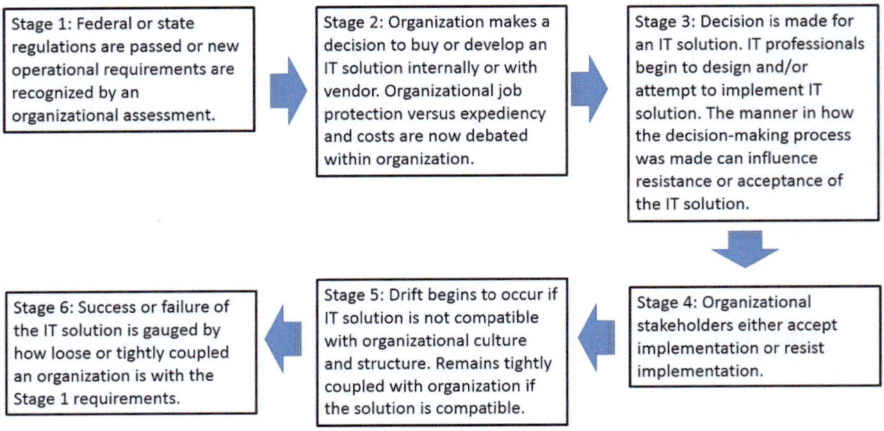

Fig. 24.2 Stage of organizational IT interactions

24.3 Proposed MIS Organizational Theory

This research focuses specifically on HAZMAT tracking, yet it could just as easily be focused on billing, accounting, personnel/HR records or generic inventory control systems. Organizational culture has been unintegrated with MIS organizational theories, which is a shortcoming of organizational theories since the 1960s. Allison and Zelikow state:

> Organizational culture emerges to shape the behavior of individuals within the organization in ways that conform with informal as well as formal norms. The result becomes a distinctive entity with its own identity and momentum (Allison and Zelikow 1999, p. 145).

Ignoring or omitting organizational culture as a component of an MIS theory causes a large portion of how an organization interacts with new technology to be unexplored and unexplained within an organizational context. Keen implicitly criticized the generally accepted relative inconsequence of organizational culture on MIS when he stated:

> The traditional view of MIS as a staff function ignores the pluralism of organizational decision-making and the link between information and power. (Keen 1981, p. 25)

The identity and momentum an organization exhibits behaviorally and culturally may indeed be just as powerful a force for implementing as for impeding a new form of technology. The IT solution process passes through multiple stages in a public or nonprofit organization (see Fig. 24.2).

The present case study in this research reveals common aspects which potentially exist in any public sector organization in relation to information technology or a technological business process. This research describes and utilizes a new organizational theory, Compatible Organizational Theory, to frame the interaction

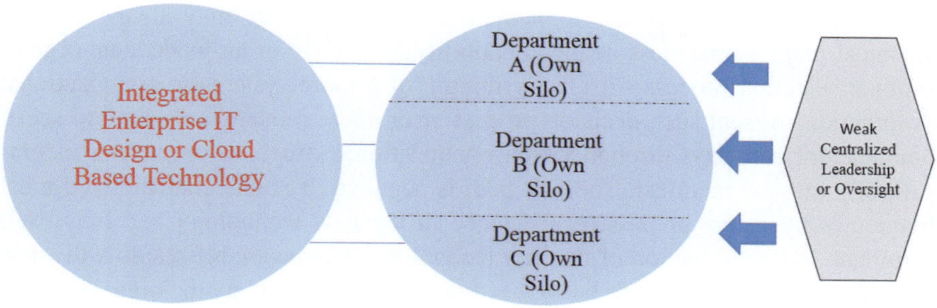

Fig. 24.3 Ideal IT solution versus compromised organizational structure

or possible absence of interaction, between an organization's structure, culture and technology implantation when resistance to technology change occurs.

Current IT trends are leading toward enterprise information systems designed as an integrated approach (i.e., PeopleSoft) or server-based Cloud technology. Neither approach is necessarily always compatible with how an organization is structured internally or with an organization's culture (see Fig. 24.3).

These internal conflicts become more evident upon further evaluation of how an organization implements (or fails to implement) newer technological solutions. Generally MIS organizational theories make the assumption that technology will systematically evolve over time and become integrated within public organizations but this perspective infrequently yields an accurate or complete depiction of the process or the impact (or lack thereof) of technological change on public sector and not-for-profit organizations.

Organizations able to adapt and to implement new technologies successfully exhibit an organizational culture and structure which both values and seeks out integrated technological solutions. Organizations struggling in the attempt to implement newer types of integrated systems and experiencing varying levels of difficulty are likely being hindered by an incompatible organizational culture and/or organizational structure that is fundamentally incompatible with the technological change being attempted. Any organization pursuing an integrated technology system must first uncritically accept the ultimate value of such a system to the organization and then be willing and able to adapt to such technology from both the culture and structural perspective. Institutional inertia contributory intangibles such as obdurate mid-level individuals' perceived reduction or loss of power/control and fear of obsolesce of current knowledge, skills, and abilities are primary drivers of internal resistance to technology change, and ultimately resistance to change in general. Strong centralized leadership, in combination with compatible mission and organizational objectives, is a requisite for successful implementation of new technologies. Burke summarized it thus:

> There can be leaders anywhere in an organization. But if the organizational change is large scale and transformational in nature, requiring significant change in mission, strategy, and culture, then leadership must come from the top of the organization, from executives, particularly the chief executive. (Burke 2002, p. 16)

The failure to successfully implement a new technology may always include technical inappropriateness or incompatibility, indicative of an inadequate organizational selection process—itself a symptom of a more systematic organizational dysfunction. Absent such decision process error, there remains only limited set of fundamental pathways through which organizational structural or cultural incompatibility may be manifest. The first path is negative. It is one characterized most frequently by active or passive resistance to the new technology and may even culminate in total rejection of the new technology. The invariable result is the new technology ultimately fails to be successfully operationalized by the organization.

The second path is one where leadership instills new organizational cultural precepts and structures by appropriately restructuring the organization up to and including significant personnel reassignments. The total cost of implementation is increased beyond the direct cost of the technology however the probability of successful operationalization is significantly enhanced. The potential for individuals to place spurious impediments to implementation is reduced as the responsibility for successful implementation is devolved to operational personnel whose newly defined positions are dependent on successful implementation.

The third and most costly path is extra organizational where a new, parallel public sector or nonprofit organization is created and delegated compliance and operational efficiency responsibilities for the primary agency requiring the use of the new technologies. Operational control of the new technology is therefore external to the primary organization which resisted the change in technology while executive control may remain unchanged.

Relevant to the issues constituting the focus of the current research, it will become clear that there are more than simply Homeland Security concerns related to HAZMAT inventory control and management. Theft of valuable materials, accidental exposure to HAZMAT, natural or man-made disasters, and regulatory compliance mandates all provide institutional imperatives for research organizations to compile and maintain accurate HAZMAT inventories using state of the art technologies enhancing both accuracy and scope.

24.4 Historical Context of HAZMAT Regulation

The virtually simultaneous, coordinated attacks on the Pentagon and the World Trade Center towers on September 11, 2001 initiated a large-scale government restructuring of most federal law enforcement, domestic security and emergency management organizations in the USA. Gone forever was the politically popular and publicly comfortable perception that a large scale foreign terrorist attack could never occur inside the United States. Laws were passed and many executive organizations were either merged from several small agencies (i.e., Homeland Security on November 25, 2002) or were given new powers to contend with potential threats of terrorist activities. It should be realized however these types of threats had always been a possibility even before 9/11, i.e., the incident at Haymarket Square in

1886 (Gross 2001), the first World Trade Center bombing in 1993 (FBI 2008), and the Oklahoma City bombing in 1995 (Michel and Herbeck 2001).

Terrorist willingness to use unconventional weapons against civilian targets as a means to gain an advantage over an opponent through either physical destruction or psychological warfare is evident throughout history. Biological warfare was used by the Mongols in the 1340s when attacking the city of Caffa by catapulting cadavers infected with the plague into the city with the intent to infect the residents (Wheelis 2002). The Imperial German Army during World War I used anthrax on the enemy's food supply and all belligerents used chemical weapons such as chlorine, phosgene, mustard gas, and other chemical agents in an attempt to gain a military advantage (Frischknecht 2003). The Imperial Japanese Army's Unit 731, operating from 1932 and throughout World War II, developed and used chemical agents and biotoxins as weapons against civilians in China and had plans to use them against the USA during the war (Frischknecht 2003). The Soviet Union and the USA both experimented with the use of biotoxins and chemical warfare during the "Cold War" (Frischknecht 2003). Iraq used chemical weapons during its protracted war with Iran (1980–1988) and on their own population of Kurdish citizens (1988) who were perceived to pose a potential threat to Saddam Hussain's regime (BBC 2013).

Iraqi insurgents in 2007 began using trucks loaded with chlorine gas (Kratovac 2007). These attacks represent an example of an unconventional weapon used indiscriminately against a large number of civilian non-combatants which in one attack killed 150 villagers (Kratovac 2007). Chlorine was used because it was readily available to the insurgents and it had the potential for devastating effects, both physically and psychologically (Kratovac 2007). Similarly, Sarin gas has been used on civilians in the Syrian civil war and has resulted in hundreds or thousands of people killed (Lederer 2013).

Unconventional warfare brings with it a higher probability that unconventional weapons will be used; ever increasing the need for legitimate users to keep accurate inventories of biological toxins, chemicals, radiological materials, and radioactive waste, all of which have the possibility to be weaponized.

Past attacks have proven that biological and chemical weapons use can be carried out even with inexpensive, low technology delivery systems. Two examples of these unconventional approaches by criminals were the biological attack carried out by the Rajneeshee Cult in 1984, in which salmonella was introduced into restaurant salad bars to poison potential voters (Ayers 2006) and the Tokyo Subway attacks in 1995 where the Aum Shinrikyo cult released Sarin gas (Bellamy 2008). These two terrorist organizations successfully attacked large civilian populations by primarily employing their own members as the delivery mechanism. The American anthrax attacks of 2001 employed the US Postal Service as the delivery mechanism for a biotoxin which not only affected the intended victims but also paralyzed the country's mail system for a number of days (Riamondo 2005).

Terrorist individuals, groups or organizations intent on using biotoxins and/or chemical agents in attacks need to acquire the agents, typically by theft or purchase through illegal channels. Therefore, organizations with legitimate uses for such materials must maintain very tight inventory control systems to prevent the misdirection of materials which may be weaponized from falling into the wrong hands. An accurate inventory system is the very first control if materials are to be secured and accounted for accurately.

Threats from terrorists are not the only reason to adhere to government mandates as other incidents have occurred inadvertently at public research organizations.

Researchers at Texas A&M University were exposed to select agents in two separate incidents in 2006. The institution failed to report the incidents as required by law to the Center for Disease Control (CDC), which resulted in significant fines being assessed against the university (Schnirring 2008).

Texas Tech University in 2010 experienced an accidental explosion in a chemistry laboratory which injured a graduate student (United States Chemical Safety Board 2011).

Ten people were injured at Villanova University in 2013 when they were exposed to chemicals and had to be medically treated (Moton 2013). The extensive level of research that occurs daily in such fields such as biology, chemistry, engineering, geology, and physics at public research organizations creates virtually innumerable situations, absent stringent controls, where potential HAZMAT might be accessed, misused, or misappropriated for criminal or terrorist intent. This is coupled with the simple increased probability of accident occurring at such organizations due to forces of nature (i.e., tornado, hurricane, or flood) or caused by a man-made factor (i.e., fire).

Radiological material has largely been controlled and inventories have been closely monitored since the Manhattan Project (1942–1946) was subsumed under the US Atomic Energy Commission on January 1, 1947 which has been succeeded by the Nuclear Regulatory Commission (National Regulatory Commission 2013). Biotoxins for weapons use has largely existed in the USA without extensive federal oversight or inventory control until recently. Polluting, noxious and/or hazardous chemicals must be reported for disposal purposes to the Environmental Protection Agency (EPA) (Environmental Protection Agency 2013) and inventoried if the chemical is categorized as a controlled substance by the Drug Enforcement Agency (DEA) (Drug Enforcement Agency 2013).

Two federal legislative initiatives passed after the Patriot Act in 2001, specifically to address deficiencies the inventory management of biotoxins and chemicals that previously did not fall under either EPA or DEA. The Bioterrorism Preparedness and Response Act of 2002 requires all entities that use or store any of 63 listed biotoxins designated as "Select Agents" to report their location and quantities to the federal government (US Congress 2002). The Homeland Security Chemical Facility Anti-Terrorism Standards passed in 2007 focuses on entities that use or store chemicals to report their inventories to the Federal government (Department of Homeland Security 2007). These two federal mandates link the Biosafety Microbiological and Biomedical Laboratories Manual (BMBL): 5th Edition (Center for Disease Control and Prevention 2009) and the requirement for an inventory control methodology to any organization that possesses Select Agents or a certain amount of chemicals (threshold limits for each type of chemical). These two statutes now provide federal agencies with a new mechanism to control and audit organizations that possess biotoxins or chemicals for safety, security, and regulation purposes.

24.5 Case Study: Organizational Culture and Structure of "Research Institution"

The Research Institution which provides the basis for this case study is typical of most research universities. It is a public university with a significant educational mission and maintains an open campus environment housed in approximately 50 buildings. Research Institution is ranked by the National Science Foundation among the top institutions in terms of federal research expenditures and performs unique, ground-breaking research in science and engineering. The organization employs over 2000 individuals, approximately 500 of whom are tenured or tenure-track faculty. The institution is located within a large metropolitan urban area. There are many very similar such organizations throughout the USA and the rest of the world.

Research Institution differs from other organizations in a few key areas. It is a comparatively young organization with a research capability that grew considerably over a short period of time. This growth in research activity occurred with sporadic executive direction and support and primarily ad hoc administrative planning and oversight to ensure the contemporaneous establishment of effective inventory management of HAZMAT. Research Institution still faces challenges which have not been resolved as of this writing.

Research Institution was founded as a high-tech research organization. The business model reflected a "siloed" management model featuring a central administration with little authority delegated among the various operating subunits. Throughout the history of the institution, the information technology department had not changed in structure and had developed an insular organizational culture mentality. Consequently, there was insignificant motivation within the IT department to upgrade its central administrative systems throughout the decades of the 1990s and 2000s. Research Institution's information technology department was heavily invested in personnel that relied upon (and perhaps even relished) mainframe technology, specifically an IBM 3390 that used DB2 and COBOL, for all of the university's critical data systems. When Y2K occurred, Research Institution purchased a new mainframe and continued to code in COBOL instead of taking the opportunity to upgrade to a new platform and exploring web-based technologies for its enterprise administrative systems (Fig. 24.4).

This was the critical point in time where technology and organizational culture intersected at Research Institution. Personnel outside the information technology department who suggested new techniques or new technologies were received with hostility, challenged or discredited, and ultimately ignored. Becker addressed the generality of this situation by writing:

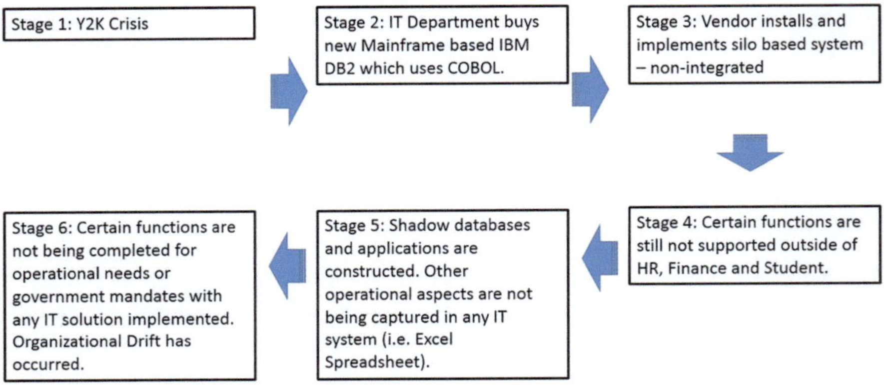

Fig. 24.4 Y2K IT solution

> Social rules define situations and the kinds of behavior appropriate to them, specifying some actions as "right" and forbidding others as "wrong". When a rule is enforced, the person who is supposed to have broken it may be seen as a special kind of person, one who cannot be trusted to live by the rules agreed upon by the group. He is regarded as an outsider. (Becker 1973, p. 1).

The information technology department was essentially allowed to define the institution's organizational culture inasmuch as it established the norms pertaining to which data processing systems and technologies were acceptable and which were not acceptable. The successful effort to retain an aging computer system that the IT department understood well, in time caused the IT department to decouple its mission from the needs of Research Institution. Personnel within the institution who possessed the means, capability, and authority to utilize contemporary computer systems and programming languages felt it necessary to do so autonomously in order to accomplish their work, thus circumventing Research Institution's IT department. Meanwhile, the IT department's information security group concerned itself with "discovering" these "unauthorized parallel" systems and worked almost exclusively toward shutting them down, claiming such systems posed an imminent security threat to the institution's administrative processes systems.

This decoupling of the objectives of the IT department from the much broader needs of the organization created a situation that eventually led Research Institution to be noncompliant with federal, state and local regulations regarding effective tracking of HAZMAT. Furthermore, this created a research environment in which misuse or mishandling of HAZMAT carried with it an ever-increasing potential for destruction of property and possibly personal injury—or worse—to staff, faculty, students, or the larger community. The standardized information systems supported and required by the IT department were technologically incapable of effectively tracking and reporting HAZMAT and were unable to provide adequate data reports to satisfy mandatory reporting for the organization's facilities. A completely new data processing approach was required.

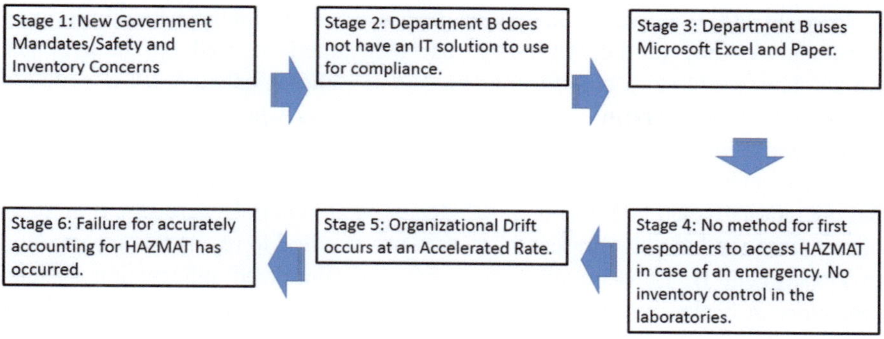

Fig. 24.5 HAZMAT tracking before Project Alpha

In 2001, executive management of Research Institution instructed "Department A" to develop an enterprise software application, designated as Project Alpha, which would become the data repository for facility information and would utilize Geospatial Information Systems (GIS) to generate accurate floor plans. Project Alpha was designed to produce accurate state and federal reports crucial to enabling the organization to successfully lobby for new resources for expansion, specifically financing for the construction of new facilities. Project Alpha's initial rollout caused ripples through Research Institution's information technology department. Project Alpha was technologically more advanced, less expensive, and orders of magnitude more capable than the dated and expensive systems the IT staff had advocated and required to be used. Project Alpha also used technology that the IT staff were not trained to use and did not understand.

24.5.1 *Lack of Hazardous Materials Inventory*

Research Institution, for much of its history, had been a comparatively small facility with few buildings. Research laboratories were housed adjacent to an administrative departments or teaching classrooms within the same building. Often laboratories were constructed in a building that did not contain specific systems often required for research activities, such as a proper ventilation system or secure HAZMAT storage apparatus. Often researchers and students would transport HAZMAT through open corridors, over carpeted areas, and in close proximity to staff and other students who did not wear protective gear.

Research Institution grew and the leadership's philosophy toward facility use changed. Increasingly, the benefits of designating new, purpose-built buildings solely for research activities were realized. Research Institution began a program to remove all research activities from areas occupied by non-research students and staff to specially constructed facilities that could be secured from open access foot traffic.

Prior to 2003, Research Institution had no real inventory beyond shipping documents of any of the HAZMAT that was stored or used on the grounds. Researchers could store biotoxins, chemicals or radiological samples in ordinary household refrigerators in common areas or hallways where any passerby could have stolen an item or been accidentally exposed. "Department B" (for the purposes of this paper) was charged with the responsibility to maintain a HAZMAT inventory but operationally had little or no knowledge of what materials were stored in many places since the individual researchers were not required to report this information to Department B. Consequently, there was no centralized information system, repository or business process to keep track of such items unless the federal DEA or NRC specifically required Research Institution to do so. Even when Research Institution tracked specific HAZMAT in order to remain compliant with federal regulations, the data were kept on a Microsoft Excel spreadsheets located on a staff member's personal computer, virtually inaccessible other Research Institution staff and wholly inaccessible to first responders in the event of an emergency (Fig. 24.5).

In one particular emergency instance, the local fire department answered a call in one of Research Institution's laboratories. The first responders arrived but refused to enter the building because no one was able to inform them of precisely what hazardous materials and the quantities were contained in the facility. A subsequent incident occurred when Research Institution's Police Chief off-handedly volunteered the suggestion to first responders conducting a routine facilities inspection to not enter particular areas in the event of a fire because there was no HAZMAT inventory for the location. This almost casual lack of an appropriate HAZMAT inventory created a potentially very dangerous environment for employees, students, first responders, and the community at large.

Agency theory states that people in an organization have and promote their own agenda and within Research Institution this pattern was observed (Downs 1967). Department B was more concerned about managing EPA-regulated waste than compiling and keeping current an inventory of HAZMAT. Researchers were more concerned with conducting their work than complying with intrusive and time-consuming edicts from the central administration. This situation led to slow compliance with the new federal requirements. Additionally, attempting to comply with the new requirements revealed that Research Institution was incapable of doing so because Department B was significantly understaffed and lacked the necessary data management systems and tools to do the work. Simply, Research Institution, at that time, had no effective, centralized computer system that could track and maintain an inventory of hazardous materials.

Furthermore, it was discovered that if a department ordered a research chemical, for instance, the shipment often bypassed the central receiving department and was delivered directly to the ordering department's laboratory or offices. Circumventing central receiving meant the chemical could arrive and be used without any institutional record of its existence, save for the accounts payable invoice, outside of that particular laboratory unless an inspection was carried out to find these wayward chemicals. Lack of a mandatory centralized receiving process created a gap in

Research Institution's business processes that could obviate an accurate inventory system even if such a system existed.

Department B was formally tasked with creating a HAZMAT inventory but was allocated no additional resources nor had the departmental leadership sufficient experience to conduct such a program. However, Department A came to recognize that Project Alpha could be readily modified to actively track all HAZMAT via a barcode system that used GPS locator tags or RFID tags on each container and then map the locations in real time to the floor plans using GIS. Department A submitted its proposal to Department B's director but the director was unwilling to commit personnel toward implementing such a tracking system with central receiving. Development of the conceptual HAZMAT tracking capability of Project Alpha was stalled at that point.

24.5.2 *Change in Focus and Change in Personnel*

An investigation performed in 2002, resulted in the demotion of the director of Department B and a decision to hire a replacement whose primary mission was to bring Research Institution into compliance with federal HAZMAT mandates. In 2003, the new director was hired. Unfortunately, the new director inherited Department B's personnel who collectively lacked the motivation, nor saw the need, to create, perform, and maintain an accurate HAZMAT inventory list. Department B's staff appeared to lack the knowledge, skills and abilities to create or maintain anything other than Microsoft Excel worksheets. Consequently, Department B's new director contacted Department A and requested that Project Alpha be modified to include HAZMAT tracking as soon as practical so that Department B personnel could begin to record HAZMAT characteristics such as ownership, type of chemical or toxin, quantities, and so on for each container.

HAZMAT containers were to initially receive a barcode upon delivery to the central receiving department and be logged into the central receiving information system. The central receiving system would then electronically send a file to Project Alpha which would lock all assets to the floor plans that were in GIS format. Items that did not come through central receiving were located and affixed with a barcode which allowed the items' characteristics to be added into Project Alpha. At this point the HAZMAT materials were being actively tracked in Project Alpha, and Research Institution could provide an accurate report for compliance purposes (Fig. 24.6).

Subsequently, Department A and the director of Department B jointly earned a national award for innovation in 2006.

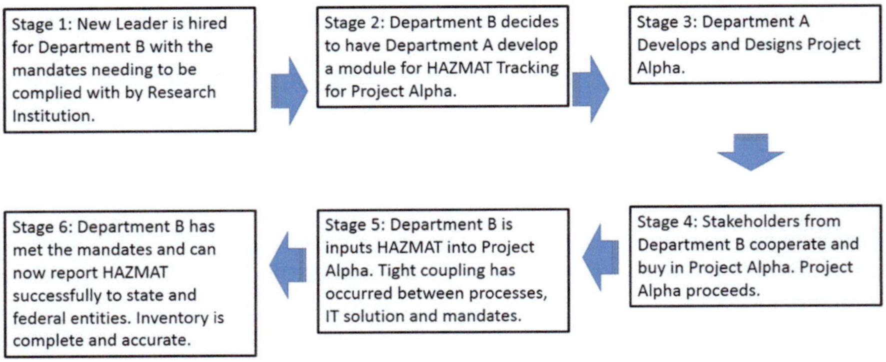

Fig. 24.6 HAZMAT tracking implementation of Project Alpha

Another Department B Leadership Change and New Obstacles

In 2007, the Department B Director left Research Institution and a new director was hired. Initially, the new Department B Director expressed interest in continuing the use of Project Alpha for tracking HAZMAT through central receiving to its destination and seemed to appreciate the reporting capability Project Alpha provided. However, the new director's staff again expressed resentment and disdain at the additional work generated by active HAZMAT tracking. In time, the central receiving information system ceased transmitting files to Project Alpha. Department B's Director seemed to lose interest in the effort as resistance from staff persisted. One of new director's subordinates contacted a colleague in the information security department and expressed "concerns" that Project Alpha was an insecure "parallel" system. Information security contacted Department A and immediately conducted a series of security audits on Project Alpha. Upon conclusion of the audits, it was determined that Project Alpha was indeed a secure system; however, the momentum for maintaining the HAZMAT tracking system was lost. The inventory effort stalled and eventually fell into nonuse (Fig. 24.7).

Research Institution continues to use and expand Project Alpha successfully in many other areas of the organization. Department B returned to their old methodology of using Microsoft Excel spreadsheets since this method requires a minimum amount of work, and which suffices, per regulatory definition, for compliance purposes. Nonetheless, for functional reasons, this method does not accurately record or reflect the chemical, biological, and radiological HAZMAT inventory at Research Institution at any given time—which was and is the intent of the various HAZMAT inventory mandates. The institution's current central executive administration perceives no palpable risks inherent with this "spreadsheet system" in maintaining compliance with federal mandates or the institution's ability to inform first responders of the locations and quantities of hazardous materials.

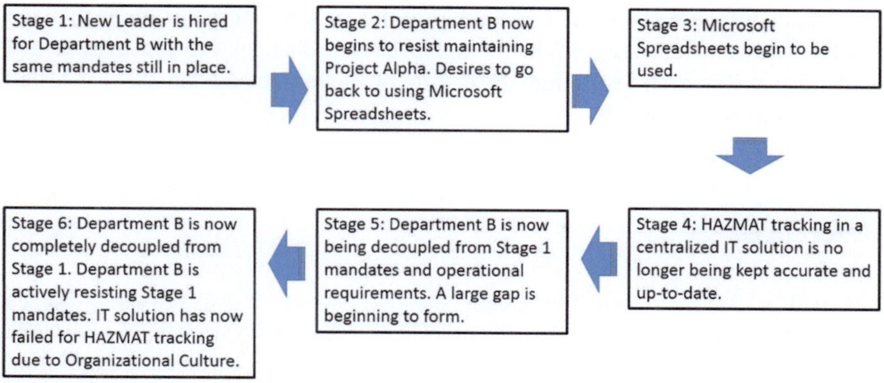

Fig. 24.7 HAZMAT tracking after new leadership in Department B

24.6 Conclusions

This case study is an example of leadership that failed at the upper and middle management levels. There appeared to be a significant disconnect between upper-level administrators and mid-level managers where the upper-level administrators have abdicated their leadership responsibilities. There was no evidence of any significant executive administrative pressure applied to ensure compliance with federal mandates regarding HAZMAT tracking, even after the successful development, implementation, and operationalization of a system that could make such tracking attainable and highly effective. This clearly indicates that the absence of HAZMAT tracking at Research Institution is a direct consequence of organizational and cultural issues and not a technological issue. Fluctuations in mid-level leadership led to changes in departmental priorities which, in turn, led eventually to a return to old, familiar methods for tracking hazardous materials.

The series of events related in this case study set the stage for two possible events to occur. The first possibility is that a crisis or emergency will occur, which may force a leadership change and with it a renewed interest in technology applications that can successfully promote safety compliance. The second possibility is that, through retirement and natural attrition, new leadership will assume control at in Research Institution's executive administration which will then mandate best technologies and procedures that will bring the Research Institution into compliance. Either way, these possibilities suggest that the tools for compliance are available but the current organizational culture is disconnected from implementation of appropriate and effective technological solutions. Simply, there has not been an emergency of sufficient proportions to command an effective response from Research Institution's management.

Finally, this case study also raises the question that if such a disconnect between technology applications and organizational culture exists regarding something as seemingly self-evident as the need to track hazardous materials at a public research institution, are there other functional areas in this or similar organizations that also exhibit similar disconnection? Until organizational changes are made by a new leadership at Research Institution, the culture of stagnation and "business as usual" will continue to exist.

References

BBC (2013) Saddam's Iraq key events: Iran-Iraq War 1980–1988 chemical warfare. BBC News. http://news.bbc.co.uk/2/shared/spl/hi/middle_east/02/iraq_events/html/chemical_warfare.stm. Retrieved 11 Nov 2013

Becker HS (1973) The outsiders: studies in the sociology of deviance. The Free Press of Glencoe, New York, NY. ISBN 0-684-83635-1

Boettke P (2010) The two social philosophies of Ostroms' institutionalism. George Mason University. Policy Stud J 39(1), MPRA Paper No. 33425, Posted September 15, 2011. http://mpra.ub.uni-muenchen.de/33425/. Retrieved 25 Nov 2013

Burke WW (2002) Organizational change: theory and practice. Sage Publications Ltd., London. ISBN 0-7619-1482-X

Caudle SL (1987) High tech to better effect. Bureaucrat. Spring 16(1):47–52. http://www.thepublicmanager.org/docs_articles/archive/Vol16,1987/Vol16,Issue01/V16N1_HightTech_Caudle.pdf. Retrieved 25 Nov 2013

Center for Disease Control and Prevention (2009) Biosafety Microbiological and Biomedical Laboratories Manual (BMBL), 5th edn., http://www.cdc.gov/biosafety/publications/bmbl5/BMBL.pdf. Retrieved 7 Jan 2014

Department of Homeland Security (2007) Department of Homeland Security Appropriations Act 2007: Section 550, DHS-2006-0073, RIN 1601-AA41, 6 CFR Part 27 – Chemical Facility Anti Terrorism Standards. http://www.dhs.gov/xlibrary/assets/IP_ChemicalFacilitySecurity.pdf. Retrieved 16 Jan 2009

Downs A (1967) Inside bureaucracy. Library of Congress Catalog Card No. 67-18259. Little Brown Publishing Company, Boston, MA

Drug Enforcement Agency (2013) History of DEA. U.S. Drug Enforcement Agency. http://www.justice.gov/dea/about/history.shtml. Retrieved 8 Nov 2013

Environmental Protection Agency (2013) Our history. Environmental Protection Agency. http://www2.epa.gov/aboutepa/epa-history. Retrieved 8 Nov 2013

Frischknecht F (2003) The history of biological warfare. EMBO Rep 4(Suppl 1). United States National Library of Medicine, National Institutes of Health, June 2003. http://www.ncbi.nlm.nih.gov/pmc/articles/PMC1326439/. Retrieved 8 Nov 2013

Glaser BG, Strauss AL (1967) The discovery of grounded theory: strategies for qualitative research. Aldine Publishing Company, New York, NY. ISBN 0-202-30260-1

Janson M, Cecez-Kecmanovic D, Ann B (2013) Information systems development: a study in communicative action. Center for International Studies, University of Missouri-St. Louis, Missouri. http://www.umsl.edu/~jansonma/myarticles/MariusDubravka.pdf. Retrieved 25 Nov 2013

Keen PGW (1981) Information systems and organizational change. In R Kling (Editor) Social impacts of computing. Communications of the ACM 24(1), January 1981. http://itu.dk/~rasesp/davesGuf/p24-keen.pdf. Retrieved 26 Nov 2013

King JL, Kraemer KL (1984) Evolution and organizational information systems: an assessment of Nolan's single stage model. Commun ACM 27(5):466–475

Kratovac K (2007) Iraq chlorine attacks raise new concerns. USA Today. February 22, 2007. http://usatoday30.usatoday.com/news/world/iraq/2007-02-22-chlorine-iraq_x.htm. Retrieved 8 Nov 2013

Lederer EM (2013) UN: 'Convincing evidence' of Syria chemical attack. Associated Press, September 16, 2013. http://news.yahoo.com/un-convincing-evidence-syria-chemical-attack-131653691.html. Retrieved 8 Nov 2013

Meyer JW, Brian R (1977) Institutionalized organizations: formal structure as myth and ceremony. Am J Sociol 83(2):340–363

Morell JA (1989) The impact of information technology on research: an agenda for social science investigation. Chapter 14. In Nagel and Garson (eds)

Moton K (2013) 10 injured in hazmat incident at Villanova University. ABC Action News. http://abclocal.go.com/wpvi/story?id=8990169. Retrieved 8 Nov 2013

Nolan RL (1979) Managingthecrisesindataprocessing. Harvard Bus Rev. 57(2) (March–April 1979)

Nuclear Regulatory Commission (2013) Our history. U.S. Nuclear Regulatory Commission. http://www.nrc.gov/about-nrc/history.html. Retrieved 8 Nov 2013

Schnirring L (2008) Texas A&M fined $1 million for lab safety. February 21, 2008. http://www.cidrap.umn.edu/news-perspective/2008/02/texas-am-fined-1-million-lab-safety-lapses. Retrieved 8 Nov 2013

United States Chemical Safety Board (2011) CSB Releases Investigation into 2010 Texas Tech Laboratory Accident; Case Study Identifies Systemic Deficiencies in University Safety Management Practices. October 19, 2011. http://www.csb.gov/csb-releases-investigation-into-2010-texas-tech-laboratory-accident-case-study-identifies-systemic-deficiencies-in-university-safety-management-practices/. Retrieved 8 Nov 2013

United States Congress (2002) Homeland Security Act 2002. H.R. Bill 5005, 23 Jan 2002

Vasu ML, Stewart DW, Garson DG (1998) Organizational behavior and public management, 3rd edn. Taylor and Francis, New York. ISBN 0-8247-0135-6

Wheelis M (2002) Biological Warfare at the 1346 Siege of Caffa. Centers for Disease Control and Prevention. Emerg Infect Dis 8(9), September 2002. Historical review. http://wwwnc.cdc.gov/eid/article/8/9/01-0536_article.htm. Retrieved 8 Nov 2013

Chapter 25
Decision Support for Wide Area Disasters

Alexander Smirnov, Tatiana Levashova, Nikolay Shilov and Alexey Kashevnik

Abstract Information integration processes utilized in a context-aware decision support system for emergency response are considered. The system supports decision making by providing fused outputs of different sources. The chapter demonstrates advantages of ontology-based context to integrate information and to generate useful decisions. A case study concerning a fire response scenario illustrates the system operation. This study focuses on planning fire response actions and evacuation of people in danger using the ride-sharing technology.

25.1 Introduction

The purpose of information fusion is to facilitate situation awareness and improve decision making (see, e.g., Appriou et al. 2001; Blasch et al. 2012; Steinberg and Bowman 2013; Kennewell and Ba-Ngu 2013). The main result of information fusion is a new meaningful matrix of information that is beneficial to decision making in less uncertainty, more precision, and/or more comprehensibility than the contributing parts (Dasarathy 2001; Haghighat et al. 2011; Waltz and Llinas 1990). High-level information fusion process turns information into actionable knowledge by capturing complex relations, reasoning over past and future events and discerning the usefulness of results to meet system-level goals (e.g., Masse et al. 2008; Holsapple and Whinston 1986; Scherl and Ulery 2004).

Intelligent context management can improve the quality of context-based information fusion results and ultimately quality of decision making since the quality of decision making depends upon the quality of information at hand. In this chapter we

A. Smirnov (✉)
St. Petersburg Institute for Informatics and Automation of the Russian Academy of Sciences, 39, 14th line, St. Petersburg 199178, Russia

ITMO University, 49, Kronverkskiy av., St. Petersburg 197101, Russia
e-mail: smir@iias.spb.su

T. Levashova • N. Shilov • A. Kashevnik
St. Petersburg Institute for Informatics and Automation of the Russian Academy of Sciences, 39, 14th line, St. Petersburg 199178, Russia
e-mail: tatiana.levashova@iias.spb.su; nick@iias.spb.su; alexey@iias.spb.su

G. Rogova, P. Scott (eds.), *Fusion Methodologies in Crisis Management*,
DOI 10.1007/978-3-319-22527-2_25

present an intelligent decision support system, which integrates dynamic data and information coming from different sources. This system has to be context-aware to dynamically adapt to the current situation, and to reduce information overload.

"Context is any information that can be used to characterize the situation of an entity. An entity is a person, place, or object that is considered relevant to the interaction between a user and an application, including the user and applications themselves" (Dey 2001, p. 5). Context utilization enables reduction of the volume of available information and knowledge to the information and knowledge relevant to, or useful in the current situation, that is to reduce information overload. Here we define information as data organized to produce meaning (Landry et al. 1970), and knowledge as information that can be used to make decisions (Zins 2007). Ontology building is a well recognized method of context formalization for higher level fusion (see, e.g., Baumgartner et al. 2010; Garcia et al. 2014; Gomez-Romero et al. 2011; Kokar et al. 2009). Ontology-based context representation is able to identify information needed to describe the current situation, make this information sharable and interpretable by the environmental sources and decision support systems, and support ontological reasoning over the integrated information.

This chapter demonstrates advantages of applying context for integration of information and enabling dynamic decision support as applied to the problem of emergency response.

The rest of the chapter is structured as follows. Section 25.2 gives an overview of the context-aware decision support system (CADSS) for emergency response and discusses the place of information integration in it. Utilization of this system for fire response actions planning is described in Sect. 25.3. Conclusions summarize the main results of the presented research in Sect. 25.4.

25.2 Context-Aware Decision Support System for Emergency Management CADSS

The CADSS for emergency management is intended to support decisions on planning emergency response actions[1]. The system scenario follows two main phases: preliminary and execution (Fig. 25.1). These phases comprise several stages; at some of them information integration processes occur.

25.2.1 Preliminary Phase

At the preliminary phase, an application ontology describing knowledge of the emergency management domain is built. This ontology represents non-instantiated

[1]This subsection presents a brief description of the conceptual framework of the CADSS. The detail description can be found in (Smirnov et al. 2005; Smirnov et al. 2015)

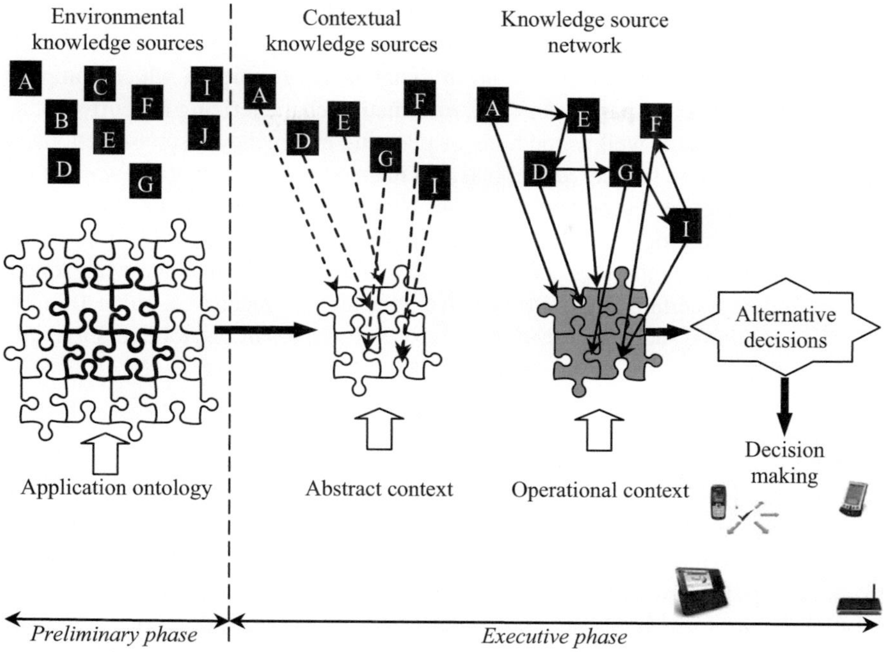

Fig. 25.1 Context-aware decision support

knowledge of two types: domain and problem-solving. The domain part of the application ontology contains various types of emergency events and knowledge that can be used to describe situations caused by such events. The problem-solving part represents the knowledge that may be required to solve the problem of planning emergency response actions. The application ontology is the result of the integration of multiple pieces of different ontologies provided by various ontology libraries and is semiautomatically created by domain experts. It is the result of combining domain and problem-solving knowledge.

The application ontology is formalized through constraints, which has proven successful for planning problems. This problem specified by constraints is solved as a constraint satisfaction problem (Tsang 1995). For the ontology representation, the formalism of object-oriented constraint networks (Smirnov et al. 2003) is used. This formalism supports an object-oriented knowledge specification. According to the formalism, the application ontology is described through sets of classes, class attributes, attribute domains, and constraints. The ontology is specified as a (non-instantiated) object-oriented constraint network.

In the ontology, the domain and problem-solving knowledge are related as follows. Any problem is represented by a hierarchy of methods that can be used to solve this problem. The methods are considered classes. Input and output arguments of the methods are represented as attributes. Functional constraints relate the input and output method arguments to class attributes contained in the representation of the domain knowledge. These constraints indicate what attribute values serve as values of the method input arguments or what attributes' values are assigned values received in a result of the methods' execution.

The CADSS is intended to be used in dynamic environments, which comprise multiple environmental parameters, i.e., information characterizing the current state of the environment as well as methods of planning emergency response actions. It is assumed that these sources are represented in an ontology-compatible way. At the preliminary stage, correspondence relations between the environmental sources and the application ontology are searched. An ontology matching model (Smirnov et al. 2010) is used for seeking the correspondences. The relations indicate what source can instantiate a certain class attribute. An attribute can receive its value from an information source such as a sensor or database or as a result of problem-solving.

25.2.2 Execution Phase

At the execution phase, a set of alternative decisions (emergency response plans) is generated and offered to the decision maker to provide him/her with an opportunity to choose a plan. Also at this phase emergency responders included in the plan are asked to confirm and implement it, and the contexts, in which decisions are made, are archived. It is in this phase that the context-aware information integration takes place.

In the framework being discussed, context is used to represent knowledge about an emergency situation (the settings in which decisions occur and the problems requiring solutions). A two-level context, *abstract context* and *operational context* are used to represent the situation (Fig. 25.1).

Abstract Context Construction

The application ontology serves as the basis for construction of abstract context. This context uses the application ontology to capture the knowledge pieces relevant to the emergency situation under consideration and significantly reduce the amount of knowledge represented in the ontology. The abstract context is a non-instantiated object-oriented constraint network similar to the application ontology. Both components (domain and problem-solving knowledge) comprising the application ontology are presented in the abstract context. From the information integration perspective the abstract context is a new knowledge source created through integration of multiple pieces of the application ontology.

The ontology inference supports the procedure of the integration. This inference is based on the relationships of the formalism of the object-oriented constraint network. Consequently, new ontology items (constraints, classes, attributes) may appear in the abstract context. This is the process of refinement of the existing situation representation as the result of information integration.

Building Operational Context

The operational context is an instantiation of the abstract context with the real-time information coming from available sources. A subset of all available environmental sources able to provide values for construction of instances of the classes represented in the abstract context or solve the problems is utilized for the abstract context instantiation. All these sources are organized in a source network, in which nodes are sources providing data values and/or solving the problems; network arcs signify the order of the node execution.

Initially, the operational context is a copy of the abstract context where variables of the constraint network representing this context are empty or take default values. The instantiation is carried out by using the constraint satisfaction method. When a variable in the object-oriented constraint network is satisfied, it is mapped on the corresponding attribute value in the operational context. The operational context reflects any changes in information coming from the source network. In the CADSS the operational context is a map, on which instances (emergency responders, buildings, etc.) are presented to help the users (decision makers) better understand it.

The operational context is the result of intelligent integration of heterogeneous data and information from contextual sources. This context is a new knowledge source created from the environmental information and represents a foundation for problem-solving and decision making.

Problem-Solving

The source network solves the planning problem embedded in the operational context as a constraint satisfaction problem. The solution is a set of alternative emergency response plans with each corresponding to a decision that can be made in the current situation. An emergency response plan is a set of emergency responders with required supporting services, schedules for the responders' activities, and transportation routes for the mobile responders. The process of problem-solving is a process of integration of information from various sources.

Decision Making and Decision Implementation

A decision maker is presented with a set of alternatives ordered by some predefined efficiency criteria (e.g., minimal time of the response actions or minimal cost of these actions, minimal time of transportation of the injured people to hospitals, etc.). He/she can choose criteria for alternatives ordering and select one plan from the set. Evidently, time is the most important criteria in many emergencies. Nevertheless, the CADSS provides multiple criteria options, which can be considered more important than response time in certain situations. For example, sometimes insignificant increase of response time can cause significant cost saving, which is important if the budget of a response provider is too small. In this case, the decision maker can compare plans produced for different criteria (for instance, optimization by time and optimization by costs), assess what the time loss for optimization by costs, and make the appropriate decision. One more example concerns several emergencies occurring at the same time at the same controlled territory. The decision maker can choose which response actions have the highest priority and chose the optimization by time for them.

The decision is delivered to the emergency responders included in the plan, i.e., to the actors responsible for the plan implementation. They have to approve the decision provided by the system by confirming their readiness to participate in the response actions (decision implementation). The emergency responders are enabled to participate in the approval procedure by using any Internet-accessible devices. The approval of the decision directly by the actors allows to avoid hierarchical decision making, which is time-consuming and therefore is not good for emergencies.

If some of the emergency responders are not able to participate due, for example, unexpected changes in the situation occurred between the time when decision is made and the emergency responders receive the plan then, in some cases, the plan can be adjusted. The plan adjustment implies a redistribution of the emergency functions corresponding the selected by decision maker functions between the other plan participants.

In the CADSS, emergency responders are represented by their profiles. If an emergency responder agrees to participate in some additional activity his/her profile is extended with a new capability.

Archival Knowledge Management

Decisions, the abstract context, and the operational context along with the source network are saved in a context archive. The operational context and the source network are saved in their states at the instant of the alternatives' generation. The archived components are the objects of archival knowledge management.

The archival knowledge management pursues several goals (e.g., revealing user preferences, grouping users with similar interests, decision mining). With reference to information integration, the main intention of the archival knowledge management is inference of new information based on the accumulated knowledge. For instance, new relations between the knowledge represented in the operational contexts can be discovered based on a comparative analysis of these contexts accumulated in the context archive. Finding the same instance in different operational contexts may lead to revealing new relations for this instance.

Archival knowledge management enables the user to reveal new relations between originally unrelated knowledge. This outcome allows refining the existing representations.

Abstract Context Reuse

The abstract contexts are reusable components of the CADSS. Reuse of an abstract context in settings when the available sources are not intended to solve the problems specified in this context is the reason to search for a new configuration of the source network so that this new configuration would be appropriate to solve the specified problems. Search for a new configuration implies search for sources that can support problem-solving by using the methods specified in the context as well as search for alternative methods. A basic condition for finding alternatives is an availability of sources that provide methods that can be used to support analysis of the specified problems.

If alternative methods have been found, they get specified in the abstract context. That is, the abstract context gets extended with the new knowledge about the conceptual scheme. The source network is reconfigured accordingly. The new configuration gains new capacities.

25.2.3 Information Integration in the CADSS for Emergency Management

Figure 25.2 shows the stages of information integration in the CADSS scenario. Some explanations to clarify the Figure are presented below. The ontology-based representation of the emergency situation comprises non-instantiated representation of the emergency management domain knowledge and abstract specification of the problem of emergency response planning. The real-time representation of the emergency situation is a dynamic view of this situation. The input shown for the block “archival knowledge management” is limited to the input relevant to the present discussion.

At the stage of application ontology building, domain and the problem-solving knowledge are integrated into one ontology. The combined ontology presents a new knowledge source representing knowledge about the application domain.

At the stages of abstract context construction, abstract context reuse, and archival knowledge management, the information integration processes bring refined knowledge about the problem at hand and about the application domain.

The stages of abstract context reuse and decision implementation enable an object (actor or source network) to gain new capabilities/competencies. They may appear as the result of knowledge reuse in new scenarios. It is, for instance, the case when an existing source network is reconfigured so that it achieves a new configuration with new capabilities or competencies, or when an actor takes a new practice.

Additionally, new settings in which the abstract context is reused may offer a new problem-solving method. Here, successful utilization of an existing method or

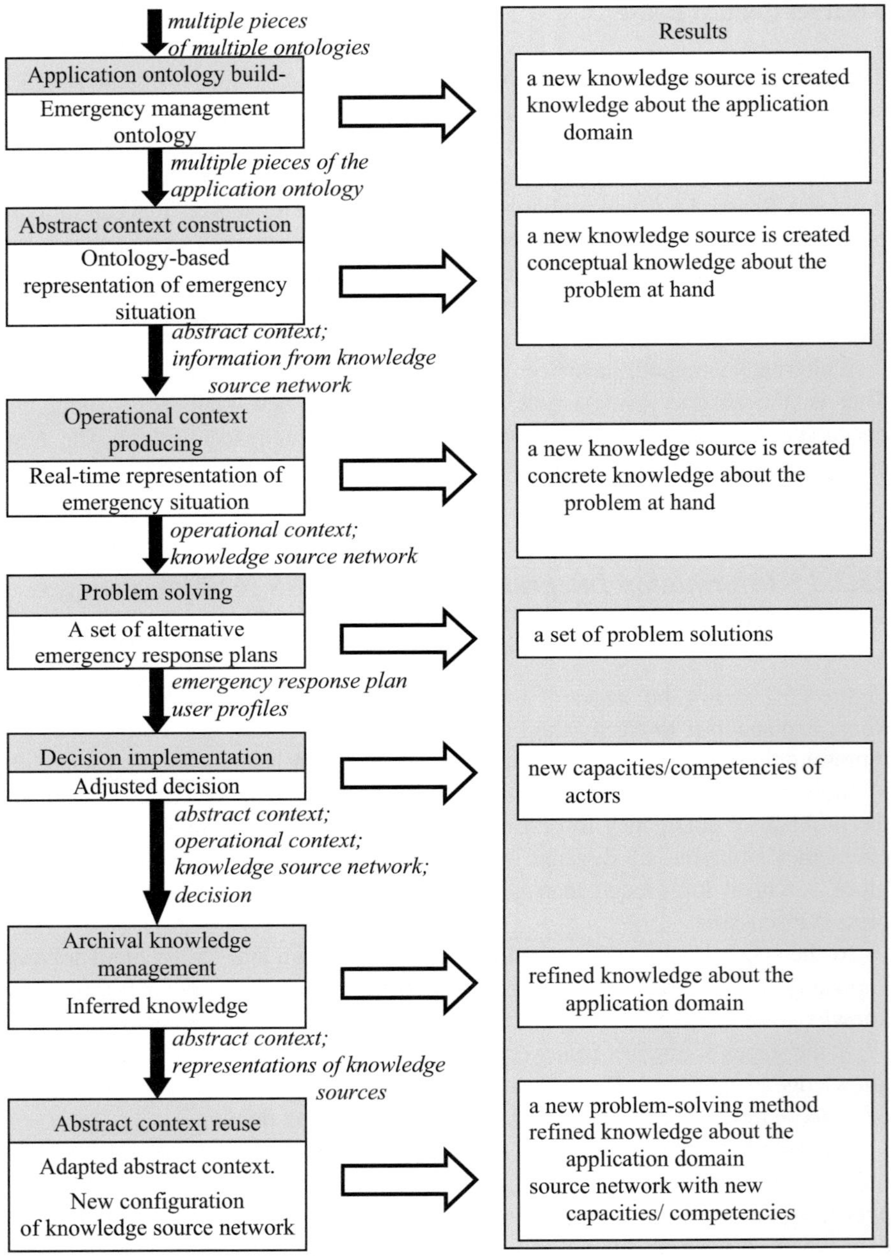

Fig. 25.2 Information integration in CADSS

a sequence of existing methods to solve a problem different from the problem these methods usually solve reviles a new problem method.

The operational context is the result of integration of heterogeneous information. It represents specific dynamic knowledge about the current situation and the problem at hand (planning emergency response actions).

At the stage of problem-solving the problem of planning emergency response actions specified in the operational context is treated as a constraint satisfaction problem. The result of problem-solving is a set of emergency response plans. Problem-solving integrates heterogeneous information to produce this set.

The following Section demonstrates information integration in the CADSS scenario on planning fire response actions.

25.3 Fire Response

Emergencies bring about highly dynamic distributed environments characterized by a vast amount of information coming from heterogeneous sources. Efficient decision support in such environments is heavily depends on the system ability to integrate information coming from these sources. This Section focuses on decision support for fire response as one of the possible emergency response scenarios.

The purpose of the CADSS in a fire situation is to produce a fire response plan for emergency responders, i.e., a plan for first response actions and an evacuation plan for potential victims to support decision making for the parties involved in the plans.

The fire event was simulated using an internal platform that supports a GIS-based simulation. The platform is able to generate random locations of emergency responders and their failures, random route availabilities, and random flows of cars. It allows ones to input information on types of emergency events, number of victims, etc.

In the simulated scenario it is assumed that a fire event has happened in Helsinki, Finland in a city's building; injured people have to be transported to hospitals; 26 persons need to be evacuated from the scene of fire. In the territory adjacent to the fire place there are available seven fire brigades, eight emergency teams, five hospitals having free capacities for four, four, two, three, and three beds available; six fire trucks and one fire helicopter assigned to the fire brigades, seven ambulances, and one rescue helicopter assigned to the emergency teams; one fire brigade is calculated to be required to extinguish the fire. The plan for actions designed for the emergency teams supposes that one vehicle can house one injured person.

The application ontology of the emergency management domain contains seven taxonomy levels, more than 600 classes, 100 class attributes, and 120 constraints of different types

Figure 25.3 presents abstract context created for the fire situation. This context specifies among other requirements that in the fire situation services provided by car drivers, hospitals, emergency teams and fire brigades are required. Car drivers,

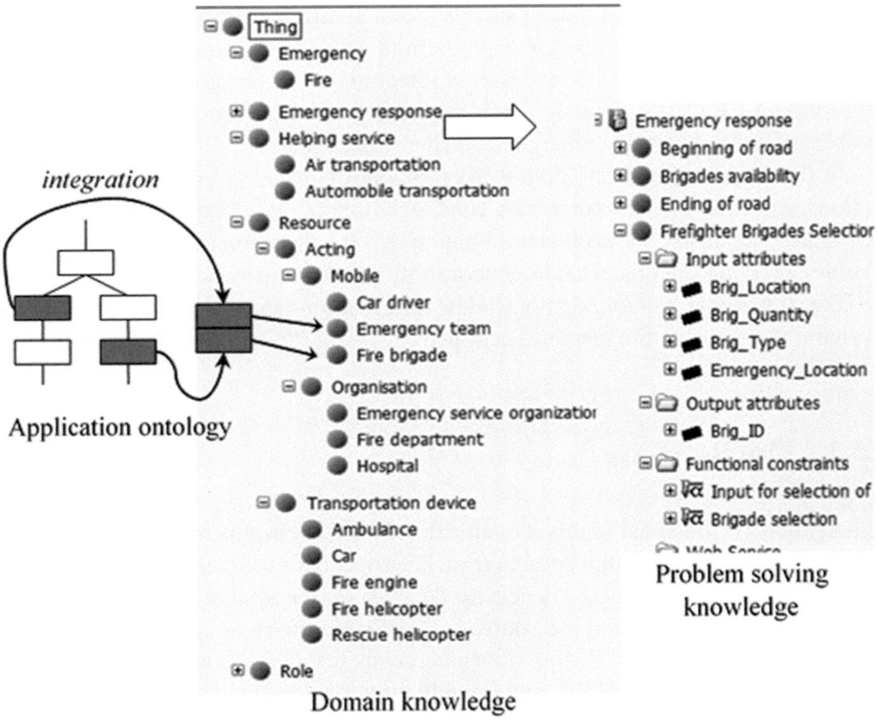

Fig. 25.3 Abstract context (a fragment) 120 constraints of different types

emergency teams, and fire brigades are mobile acting resources; they can use cars, ambulances, fire engines, and special-purpose helicopters for transportation.

In Fig. 25.3, the problem-solving knowledge specified in the abstract context is represented by "Emergency response" class. This class specifies the following problems:

- Selection of feasible hospitals, emergency teams, fire brigades, and car drivers;
- Determination of feasible transportation routes for ambulances and fire engines depending on the transportation network and traffic situation;
- Calculation of the shortest routes for transportation of the emergency teams by ambulances, fire brigades by fire engines, and evacuees by cars;
- Finding a set of feasible response plans for emergency teams, fire brigades, and hospitals;
- Finding a set of feasible ridesharing routes.

In the course of the abstract context construction a new relationship between the knowledge unspecified in the application ontology is inferred (Fig. 25.4). Namely, the application ontology specifies that a value for the variables representing the current location of a transportation device serves as an input variable of the routing method (1). The class "mobile" representing a mobile resource and the class

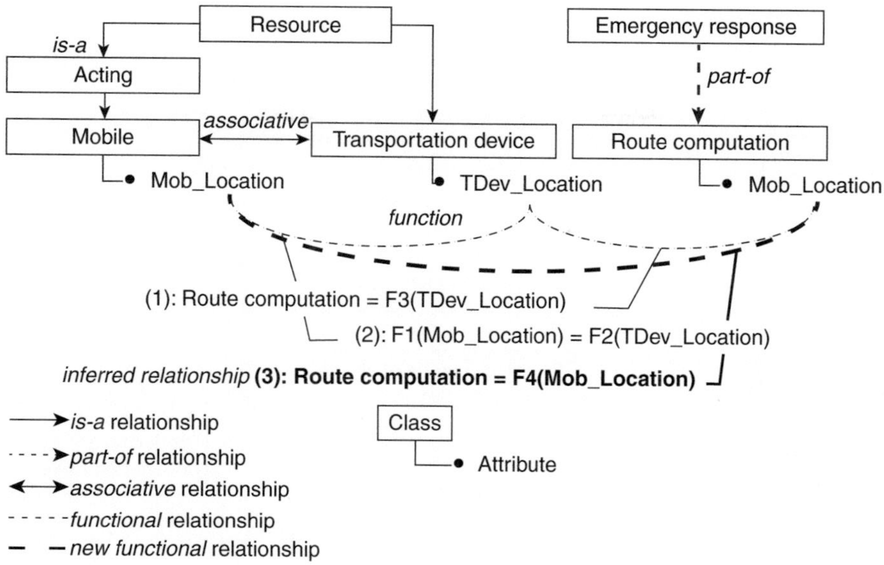

Fig. 25.4 Inferred relationship

"transportation device" are linked by a functional relationship (2) stating that the location of a mobile resource is the same as the location of the transportation device this resource goes by. In the abstract context a new functional relationship (3) is inferred. This relationship means that a value of the attribute representing the current location of a mobile resource serves as an input argument of the routing method. In other words, values for the both attributes representing the current location of a transportation device or the current location of a mobile resource can be used as one of the input arguments by the routing method.

Figure 25.5 presents a source network organized to produce operational context and solve the problems specified in the abstract context. In the Figure, arrows indicate the execution sequence of methods provided by the sources constituting the network.

Figure 25.6 presents the operational context based on the abstract context (Fig. 25.3) at some time instant. The operational context representation reflects any changes in information coming from the source network (e.g., the decision maker observes that mobile resources are changing their locations). In Fig. 25.6 the bold dot indicates the fire location.

For the situation presented by the operational context, the set of feasible plans for first response actions is generated for the criteria of minimal time and cost of transportation of all the victims to the hospitals, and minimal number of mobile emergency responders involved in the response actions. The set of plans comprises four plans. The efficient plan (Fig. 25.7) is selected based on the key indicator of minimal time of victim transportations. In the figure, the dotted lines show the routes proposed for the transportations of the emergency teams and fire brigades.

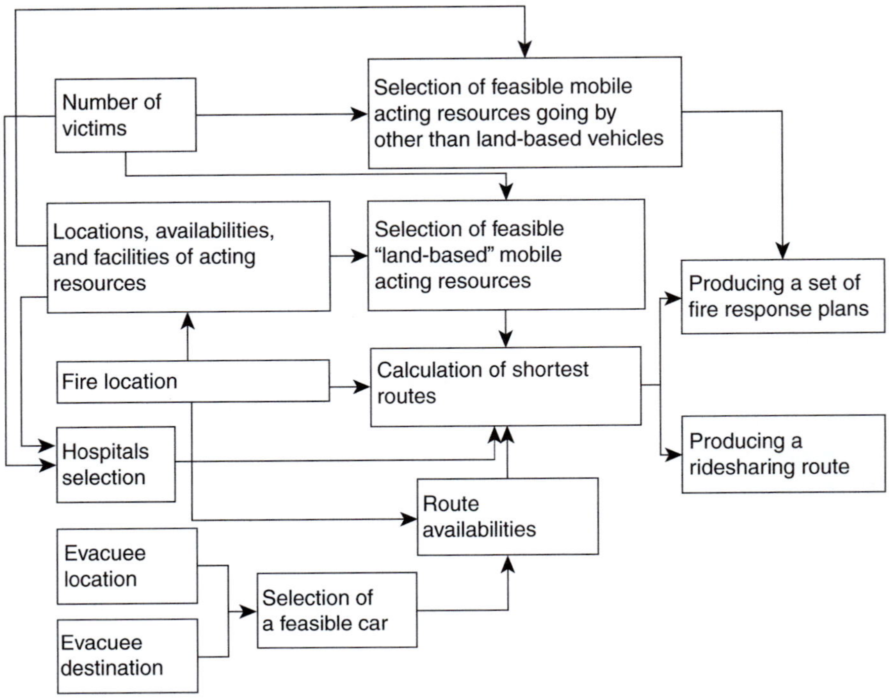

Fig. 25.5 Source network

Fig. 25.6 Operational context

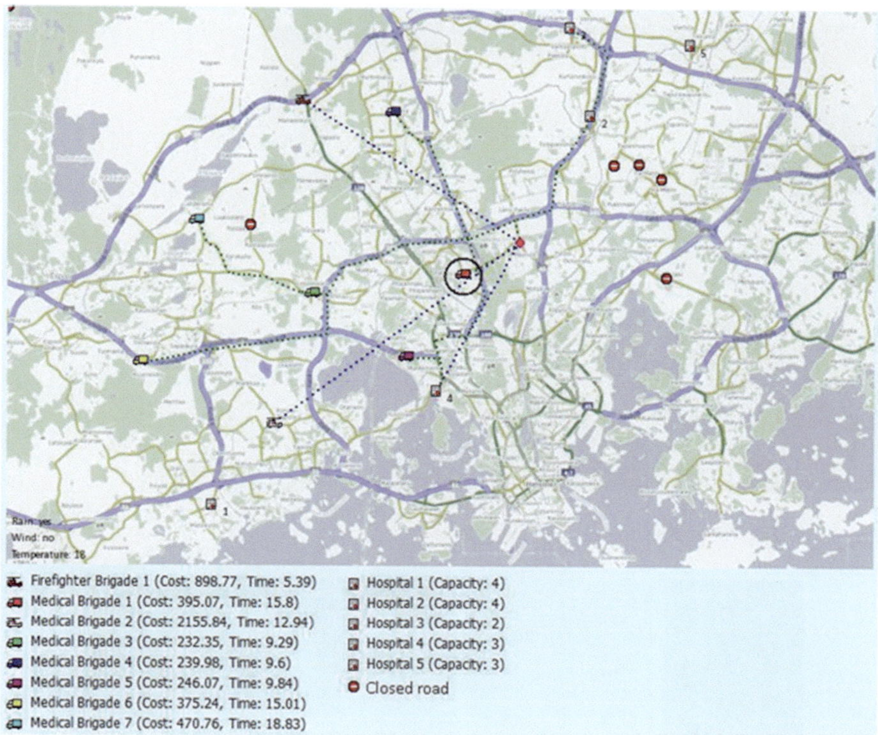

Fig. 25.7 Plan for first response actions

One ambulance (encircled in the figure) and the rescue helicopter go from the fire location to hospitals twice. The estimated time of the operation of transportations of all the victims to hospitals is 1 h 25 min.

The decision maker decides on the presented plan. This decision is delivered to the emergency responders for their approval. Figure 25.8 shows a part of the plan displayed on the Tablet PC belonging to the leader of an emergency team going by ambulance. In the considered scenario, all the responders of the team agree to participate in it. Therefore, no plan adjustment is needed.

Simultaneously with planning first response actions, evacuation activities are planned. Features provided by the ridesharing technology are used to organize the evacuation of people who might be in danger when the fire spreads.

Persons who need to be evacuated invoke a ridesharing service responsible for the evacuation. Clients of this service are supposed to be installed on the Internet-accessible devices of car drivers and other people involved in the fire situation. The persons enter the locations they would like to be driven to. The ridesharing service determines the persons' locations and searches for cars going to or by the same or close destinations that the person would like to be. It searches the cars among the vehicles passing the persons' locations. This service reads information about the destinations that the car drivers are going to from the navigators that the drivers

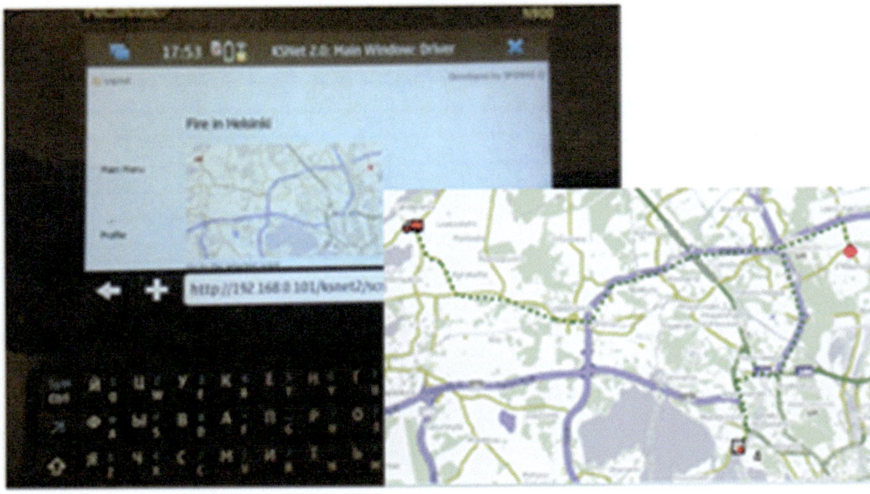

Fig. 25.8 Plan for actions for an emergency team

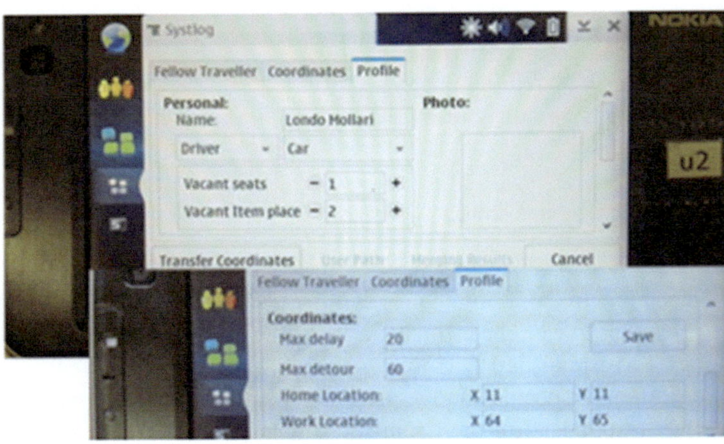

Fig. 25.9 Driver's profile

use or from the drivers' profiles. The profiles store periodic routes of the drivers and cars' properties as the number of passenger seats, the availabilities of baby car seats, etc. (Fig. 25.9).

Based on the information about the locations and destinations of the person and the found cars, a set of feasible routes for the person transportations is generated. An efficient route is determined based on the criterion of minimum transportation time.

The ridesharing service sends specific signals to the drivers included in the ridesharing routes and displays on the drivers' devices the routes each driver is selected for. The points where the driver is expected to pick up the passenger(s) is indicated in the routes. The ways the passengers have to walk to these points are routed for them as well. Besides the routes, the passengers are informed of the model, color, and license plate number of the car intended for their transportation. The persons that cannot be evacuated by passing cars are informed that they can be evacuated by taxi. If they agree, the ridesharing service makes orders for taxi.

Evacuation plan includes agreement between the driver and the evacuee to go according to the scheduled ridesharing route. If there is no agreement between a driver and an evacuee, another car for evacuation of this passenger is sought for. At that, the confirmed routes are not revised.

In the considered scenario, the results of simulations of evacuation by ridesharing are as follows: of 26 persons to be evacuated from the scene of fire 22 have been driven directly to the destinations by 16 cars whereas for 4 persons no cars have been found. Examples of ways routed for a driver and a passenger are given in Figs. 25.10 and 25.11. The encircled car in the Figures shows the location where the driver is offered to pick up the passenger.

The plans for first response actions and the evacuation plan are the results of problem-solving, during which information from heterogeneous sources is fused.

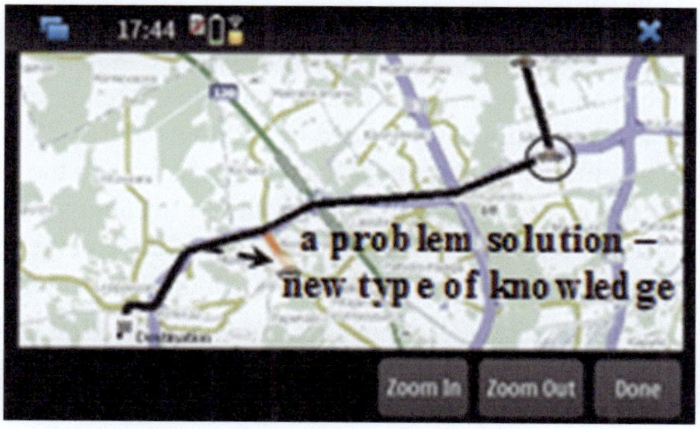

Fig. 25.10 Ridesharing route: driver's view

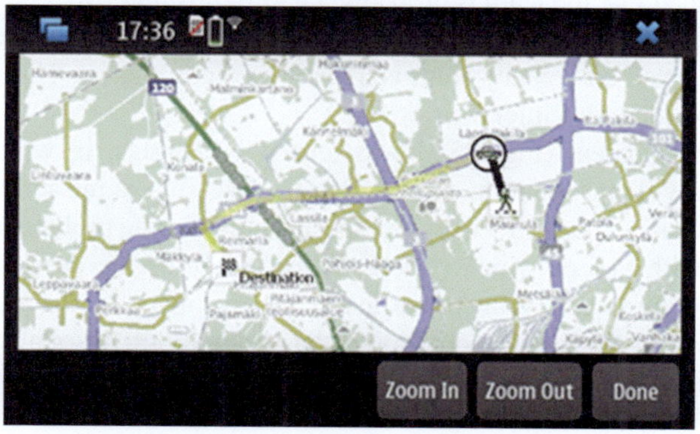

Fig. 25.11 Ridesharing route: passenger's view

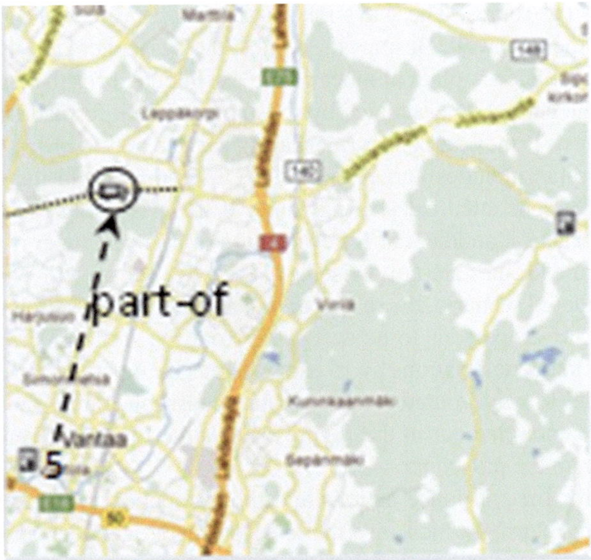

Fig. 25.12 History for emergency team

For instance, information about the availabilities of emergency responders, their locations, and road availabilities is provided by different sources. This information is fused to select an emergency responder that can participate in the joint actions.

The results of the analysis of operational contexts accumulated in the archive enable to produce additional knowledge concerning an emergency team. The emergency team encircled in Fig. 25.7 participated in different emergency response actions. Some operational contexts, in which this team appeared and then participated in corresponding actions do not represent any instances of the class *Emergency response organization* specified in the abstract context. This suggests that the emergency team is a part of one of the hospitals represented in the operational contexts together with this team. Based on the operational context (Fig. 25.12) it can be concluded that the team is probably a part of hospital five represented in this context since it is the only hospital represented. *Part-of* relation between the hospital 5 and the encircled emergency team is a new revealed relation. This relation is an outcome of the inductive inference.

The abstract context (Fig. 25.3) was reused in settings, in which the source intended to provide information about hospitals' locations is missing. Figure 25.13 illustrates this case. The abstract context specifies the routing problem as a hierarchy of methods, with one of which ("GetLocation") returns the current locations of objects in the format of point coordinates on the map. In the example under consideration it is required to determine the locations of hospitals. The method "GetLocation" uses data from sensors.

The set of environmental information sources does not include sensors dealing with static objects like hospitals but include other sources. One of them (*A*)

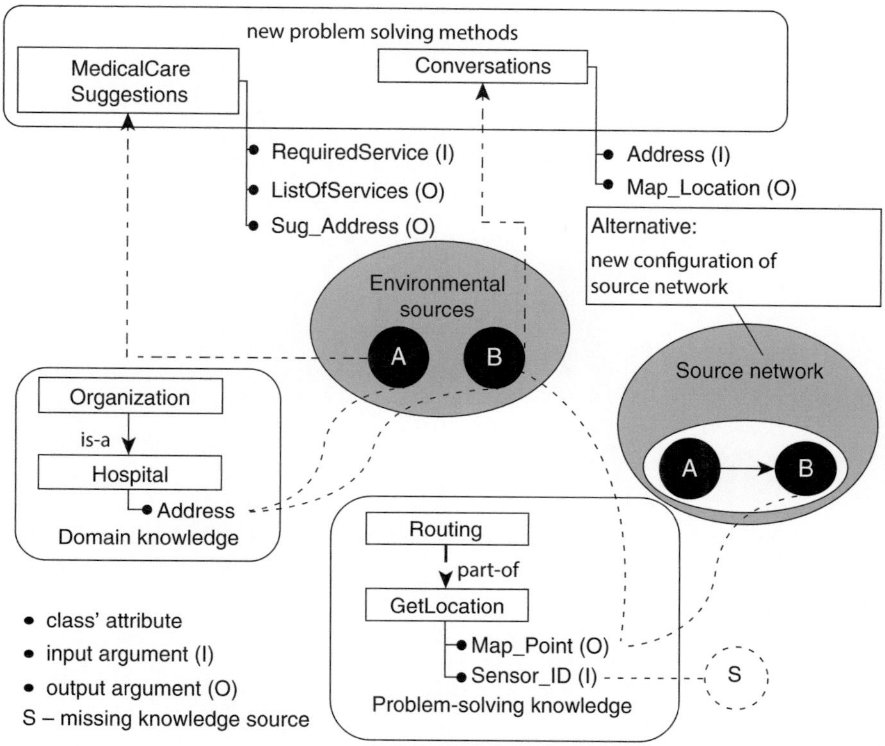

Fig. 25.13 Discovery of alternative problem-solving methods

implements the "MedicalCareSuggestions" method for making recommendations what medical care organizations can be used to access some specific medical service. This source contains a database with information about the hospitals. The source *A* returns the hospitals' addresses in an address format. The other source (*B*) implements the "Conversions" method that converts the address formats into the format of coordinates. The successive execution of the methods "MedicalCareSuggestions" and "Conversions" is an alternative way to calculate the hospital locations represented by their coordinates.

In the abstract context, the methods "MedicalCareSuggestions" and "Conversions" are not specified as an alternative to the method "GetLocation." A set of constraints have to be introduced to get this alternative explicitly specified. This introduction leads to the extension of the abstract context by new knowledge representation items.

The Smart-M3 platform (Honkola et al. 2010) was used in the simulation of the fire response scenario. Tablet PC Nokia N810 (Maemo4 OS), smart phone N900 (Maemo5 OS), and different mobile phones served as the user devices. Personal PCs based on Pentium IV processors and running under Ubuntu 10.04 and Windows XP were used for hosting other services.

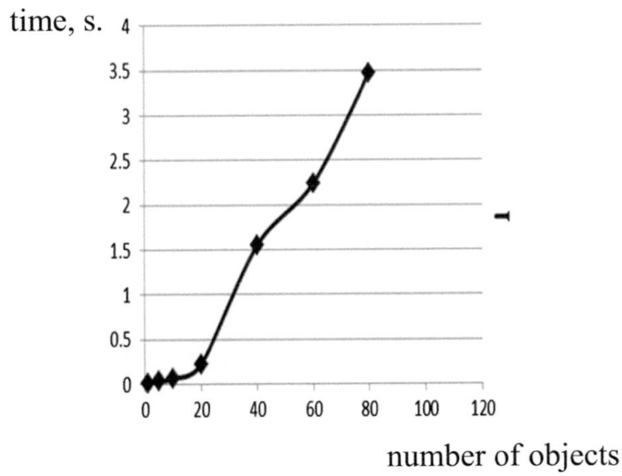

Fig. 25.14 Dependence of execution time on number of input objects

The results showing the scenario execution time as a function of the number of input objects (victims, hospitals, emergency teams, fire brigades) are presented in Fig. 25.14.

25.4 Conclusions

This chapter has discussed the problem of Information integration processes in a context-aware decision support system (CADSS) for emergency management. The CADSS comprises the following elements:

- An application ontology combining domain and problem-solving knowledge;
- An abstract representations of the emergency situation and its instantiation;
- A new configuration of the network of environmental sources with new capacities/capabilities;
- An extension of capabilities of decision executors;
- Refined information about the application domain and emergency situation;
- A set of emergency response plans;
- A set of evacuation plans.

The capabilities of the CADSS were illustrated through decision support in a simulated fire response scenario.

Acknowledgments The present research was partly supported by the projects funded through grants 12-07-00298, 13-07-00336, 13-07-12095, 13-07-13159, 14-07-00345, 14-07-00427 (the Russian Foundation for Basic Research), the Project 213 (the research program "Information, control, and intelligent technologies & systems" of the Russian Academy of Sciences (RAS)), the Project 2.2 (the Nano- and Information Technologies Branch of RAS), and grant 074-U01 (the Government of the Russian Federation).

References

Appriou A, Ayoun A, Benferhat S et al (2001) Fusion: general concepts and characteristics. Int J Intell Syst 16:1107–1134

Baumgartner N et al (2010) BeAware! – Situation awareness, the ontology-driven way. Data Knowl Eng 69:1181–1193

Blasch E, Bossé É, Lambert DA (eds) (2012) High-level information fusion management and systems design. Artech House, Boston

Dasarathy BV (2001) Information Fusion – what, where, why, when, and how? Information Fusion 2(2):75–76

Dey AK (2001) Understanding and using context. Pers Ubiquit Comput 5(1):4–7

Garcia J et al (2014) Context-based multi-level information fusion for harbor surveillance. Information Fusion 21(1):173–186

Gomez-Romero J, Patricio MA, Garcia J, Molina JM (2011) Ontology-based context representation and reasoning for object tracking and scene interpretation in video. Expert Syst Appl 38(6):7494–7510

Haghighat MBA, Aghagolzadeh A, Seyedarabi H (2011) Multi-focus image fusion for visual sensor networks in DCT domain. Comput Electr Eng 37(5):789–797

Holsapple CW, Whinston AB (1986) Building blocks for decision support systems. In: Ariav G, Clifford J (eds) New directions for database systems. Ablex Publishing Corp, Norwood, pp 66–86

Honkola J, Laine H, Brown R, Tyrkko O (2010) Smart-M3 information sharing platform. In: Proceedings of IEEE Symposium on Computers and Communications, IEEE Computer Society, pp 1041–1046, doi.ieeecomputersociety.org/10.1109/ISCC.2010.5546642

Kennewell JA, Ba-Ngu Vo (2013) An overview of space situational awareness. In: Proceedings of the 16th international conference on information fusion, Istanbul, Turkey, 9–12 July 2013, p 1029–1036

Kokar MM, Matheus CJ, Baclawski K (2009) Ontology-based situation awareness. Information Fusion 10(1):83–98. doi:10.1016/j.inffus.2007.0004

Landry BC, Mathis BA, Meara NM, Rush JE, Young CE (1970) Definition of some basic terms in computer and information science. J Am Soc Inf Sci 24(5):328–342

Masse T, O'Neil S, Rollins J (2008) Information and intelligence (including terrorism) fusion centers. Nova Science Publishers Inc., New York

Scherl R, Ulery DL (2004) Technologies for army knowledge fusion. Monmouth University, Computer Science Department, West Long Branch, Monmouth. Final report ARL-TR-3279

Smirnov A, Kashevnik A, Shilov N, Balandin S, Oliver I, Boldyrev S (2010) On-the-fly ontology matching in smart spaces: a multi-model approach. In: Smart Spaces and Next Generation Wired/Wireless Networking. Proceedings of the third conference on smart spaces, ruSMART 2010, and the 10th international conference NEW2AN 2010, St. Petersburg, Russia, 23–25 Aug 2010. Lecture notes in computer science, vol 6294, Springer, pp 72–83

Smirnov A, Levashova T, Shilov N (2015) Patterns for context-based knowledge fusion in decision support. Inform Fusion 21(1):114–129, http://dx.doi.org/10.1016/j.inffus.2013.10.010

Smirnov A, Pashkin M, Chilov N, Levashova T (2005) Constraint-driven methodology for context-based decision support. J Decis Syst 14(3):279–301

Smirnov A, Pashkin M, Chilov N, Levashova T, Haritatos F (2003) Knowledge source network configuration approach to knowledge logistics. Int J Gen Syst 32(3):251–269

Steinberg AN, Bowman CL (2013) Adaptive context discovery and exploitation. In: Proceedings of the 16th international conference on information fusion, Istanbul, Turkey, 9–12 July 2013, pp 2004–2011

Tsang E (1995) Foundations of constraint satisfaction. Academic, London

Waltz EL, Llinas J (1990) Multisensor data fusion. Artech House, Norwood, MA

Zins C (2007) Conceptual approaches for defining data, information, and knowledge. J Am Soc Inf Sci Technol 58(4):479–493

Index

G. Rogova, P. Scott (eds.), *Fusion Methodologies in Crisis Management*,
DOI 10.1007/978-3-319-22527-2

N

O

P

MIX
Papier aus verantwortungsvollen Quellen
Paper from responsible sources
FSC® C105338

If you have any concerns about our products,
you can contact us on
ProductSafety@springernature.com

In case Publisher is established outside the EU,
the EU authorized representative is:
Springer Nature Customer Service Center GmbH
Europaplatz 3, 69115 Heidelberg, Germany

Printed by Libri Plureos GmbH
in Hamburg, Germany